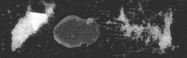

TAXONOMY OF CULTIVATED PLANTS
Third International Symposium

Symposium sponsored by

Royal Botanic Garden
Edinburgh

The Royal Botanic
Gardens, Kew

Royal Horticultural Society

INTERNATIONAL CODES OF NOMENCLATURE

International Code of Botanical Nomenclature (Tokyo Code). (1994). Adopted by the Fifteenth International Botanical Congress, Yokohama, August–September 1993. (*Regnum Vegetabile* vol. 131). 389 pp. Prepared and edited by W. Greuter *et al.* Koeltz Scientific Books, D-61453 Königstein, Germany. (ISBN 3-87429-367-X or 1-878762-66-4 or 80-901699-1-0).

International Code of Nomenclature for Cultivated Plants – 1995. (1995). Adopted by the International Commission for the Nomenclature of Cultivated Plants. (*Regnum Vegetabile* vol. 133). 175 pp. Prepared and edited by P. Trehane *et al.* Quarterjack Publishing, Wimborne, Dorset, UK. (ISBN 0-948117-01-X).

In addition to the rules concerning the formation and publication of cultivar and cultivar-group names, the Cultivated Plant Code includes a range of helpful lists and guides including Directories of International Registration Authorities and of Statutory Plant Registration Authorities, plus references to published checklists of ornamental cultivars.

An updated version of the Directory of International Registration Authorities is maintained on the ISHS (International Society for Horticultural Science) web site (http://www.ishs.org/sci/iradirec.html).

PROCEEDINGS OF PREVIOUS SYMPOSIA

Maesen, L.J.G. van der (ed.). (1986). First International Symposium on Taxonomy of Cultivated Plants, Wageningen, The Netherlands, 12–16 August 1985. *Acta Hort.* 182: 1–436.

Tukey, H.B. (ed.). (1995). Proceedings of the Second International Symposium on the Taxonomy of Cultivated Plants, Seattle, WA, USA, 10–14 August 1994. *Acta Hort.* 413: 1–201.

TAXONOMY OF CULTIVATED PLANTS

Third International Symposium

Proceedings of the Meeting held in
Edinburgh, Scotland
20-26 July 1998

Edited by
Susyn Andrews
Royal Botanic Gardens, Kew

Alan Leslie
Royal Horticultural Society, Wisley

Crinan Alexander
Royal Botanic Garden Edinburgh

Published by the Royal Botanic Gardens, Kew, 1999

First published 1999

Production Editor: S. Dickerson

Cover design by Johnny Willis Fleming, page make up by Media Resources, Information Services Department,
Royal Botanic Gardens, Kew

ISBN 1 900347 89 X

Printed in Great Britain
by
Whitstable Litho Printers Ltd., Whitstable, Kent

PREFACE

Given the encouraging renewed interest in the taxonomy of cultivated plants, it is good that the gap between the Second and Third International Symposia (1994–1998) is much shorter than between the first two, 1985 in Wageningen, The Netherlands and 1994 in Seattle, Washington, USA. It was also gratifying to see the much greater number of delegates at the 1998 symposium: 168 participants from 30 countries compared with well under 100 for the first two symposia. It seems that the establishment of the Horticultural Taxonomy Group (HORTAX) in 1988 has had the desired effect of bringing together in a more co-ordinated way the considerable number of people interested in the subject. The work of HORTAX together with the Dutch Nomenclature and Registration Working Group of the Vaste Keurings Commissie (VKC) also made a significant contribution to the 6th edition of the *International Code of Nomenclature for Cultivated Plants – 1995.* The informative *Hortax News,* which has appeared at irregular intervals since December 1993, now reaches 1300 people in 30 countries and has done much to stimulate work on the nomenclature of cultivated plants. It is good to see this healthy growth in an area which seemed to be on the wane until recently, yet which is so vital as the legal complications of names become more significant and horticulture becomes an ever more popular pursuit world-wide.

The symposium on which this book is based brought together a remarkable range of those working with cultivated plants, such as holders of national collections, international seed trade executives, nurserymen, gardeners, specialists in nomenclature, people involved with the registration of cultivars and database managers. The resulting exchange of views has undoubtedly made a considerable impact.

The increasing worldwide trade in cultivated plants together with stronger legal protection of new cultivars demands that names be precise, accurate and stable. At the Royal Botanic Gardens, Kew we are doing all we can to promote interest in horticultural taxonomy, because it is so important to take this often neglected field forward and to make use of new tools such as molecular techniques that are now available.

I commend the authors of papers and the editors of this volume for producing a milestone publication which will be of considerable use to all interested in the naming of cultivated plants.

Professor Sir Ghillean Prance, F.R.S. VMH
Director
Royal Botanic Gardens, Kew

ACKNOWLEDGEMENTS

This very successful symposium would never have taken place without generous financial support from our three main sponsors:– The Royal Botanic Garden Edinburgh, Royal Botanic Gardens, Kew and The Royal Horticultural Society. The funds provided allowed the Horticultural Taxonomy Group to bring important delegates from several different countries and also paid for advance publicity.

We should like to thank the then Regius Keeper, RBGE, Professor David Ingram for providing facilities for the supplementary sessions devoted to the International Code of Nomenclature for Cultivated Plants on 25–26 July, and for a spectacular evening reception and tours. The many staff, associates and students of RBGE who contributed to the success of the symposium include:– Ian Edgeler, Malcolm Newport, Lesley Taylor and Senga Andrews (Finance & Administration); Norma Gregory, Erica Schwarz, Kim Howell and Mark Parry (Print & Publications); David Wood, Elaine Carmichael, Clara Govier, Laura Condie and Charlotte Neilson (Development & Communications); Debbie White (Photography), and several members of RBGE messengerial and security staff. For advice and help with transport and general organisation we are indebted to Martin Gardner, Barry Unwin, Graham Stewart and Geoff Brookes (Tour Leaders); Neil Brummitt, David Cann, Rob Cubey, Rose Colangeli, Freya Cooper, Jill Harrison, Mark Hughes, Vlasta Jamnicky, Michelle McLaren, Suzanne Maxwell, David Rae, Phil Thomas and Kerry Walter.

Members of staff from the Scottish Agricultural Science Agency (SASA) were involved in providing, transporting and erecting display boards for the poster sessions.

Ann Hall and Alison Watson of the Conference Centre at the Pollock Halls of Edinburgh University deserve our thanks for their efficiency and patience in handling our fluctuating demands for accommodation, equipment, meals and refreshments. The audio-visual engineer from the University deserves special mention for his excellent and unobtrusive service while coping with our complex requirements.

At the RHS grateful thanks are due to Mr Gordon Rae, the former Director General and to Mrs Joyce Stewart, the Director of Horticulture, Science and Education. Patty Boardman and Diana Miller at RHS Wisley are to be thanked for all their work in producing the Abstracts volume.

Thanks also to the RHS and RBG Kew for providing the Reception in the Poster Hall.

Special thanks are due to the following at RBG Kew: the then Director, Professor Sir Ghillean Prance for support and encouragement; the former Director of Science and Horticulture, Professor Charles Stirton; the previous Keeper of the Herbarium, Professor Gren Lucas and the current Keeper, Professor Simon Owens. Additional mention should be made of David Gardner, Maureen Bradford, Nicholas Hind, Henk Beentje and Brian Schrire; to Sarah Smith for compiling the index; to Suzy Dickerson, the Production Editor and her team of Margaret Newman, Ann Lucas and Ann McNeil; to John Harris and Johnny Willis Fleming for Media Resources, RBG, Kew.

We would also like to thank all our reviewers for their helpful comments and advice regarding the manuscripts: C. Alexander, E. Braimbridge, W.A. Brandenburg, C.D. Brickell, P. Button, M. Chase, R.J. Cooke, P. Cribb, R. Cubey, A. Culham, J. Cullen, J.C. Davey, M. Flanagan, D.G. Frodin, D. Gardner, M.L. Grant, F.N. Green, S. Harris, Y.B. Harvey, A. Heitz, W.L.A. Hetterscheid, K. Hill, M.H.A. Hoffman, P. Hollingsworth, L. Jones, B. Jonsell, G. Kite, E. Knox, A.C. Leslie, M. Lock, W.A. Lord, B. Mathew, M. Maunder, J. McNeill, D.M. Miller, J. Parker, R.T. Pennington, B. Pickersgill, A.E. Plovanich-Jones, H.D.V. Prendergast, C.J. Quin, G. Rice, B. Schrire, A.R.J. Sier, L.

Skog, G. Staples, R. Staward, J. Stewart, C. Stirton, J.M. Strachan, N.P. Taylor, P. Thomas, S. Thomas, S.P. Thornton-Wood, A. Tramposch, M. Turner, K. van Ettekoven, L.W.D. van Raamsdonk, J. van Scheepen, M. van Slageren, J. Vaughan, A.C. Whiteley, J.H. Wiersema and P. Wysc-Jackson.

Finally the editorial committee would like to thank their fellow HORTAX members, without whom this symposium would never have taken place, namely Diana Miller, Adrian Whiteley, Piers Trehane, Allen Coombes, Niall Green, Sabina Knees, Tony Lord and Chris Brickell.

SYMPOSIUM PHOTOGRAPH

SYMPOSIUM DELEGATES, ROYAL BOTANIC GARDEN STAFF, AND OTHERS PRESENT AT THE GLASSHOUSE RECEPTION ON 20TH JULY 1998

1 Donglin Zhang
2 Rob Cubey
3 George Staples
4 Peter del Tredici
5 Helmut Knüpffer
6 Philipp Knüpffer
7 David Cann
8 Sean Butler
9 Charles Stirton
10 Suzanne Maxwell
11 Meg Alexander
12 Martin Gardner
13 Susyn Andrews
14 Kerry Walter
15 Morten Hulden
16 Tony Lord
17 Simon Thornton-Wood
18 Mike Sinnott
19 Barbara Becker
20 Klaus Dehmer
21 Dave Astley
22 Robert Treu
23 Valentinas Cirtautas
24 Colin Morgan
25 Andrew Sier
26 David Gardner
27 Wim Snoeijer
28 Niki Simpson
29 Elspeth Haston
30 Paul Fantz
31 Maria Bennett
32 Natalia Louneva
33 André Heitz
34 Tamara Smekalova
35 Hélène Bertrand
36 Mollie Perrin
37 Crinan Alexander
38 Vince Lea
39 Mike Grant
40 Richard Wilford
41 Bert Hennipman
42 Ros Ferguson

43 Dora Jakobsdottir
44 Adrian Whiteley
45 John Wiersema
46 Ann Cook
47 Melissa Simpson
48 Melinda McMillan
49 Alice Le Duc
50 Boris Kitek
51 Alenka Zupancic
52 Eva Thorvaldsdottir
53 Malcolm McGregor
54 Doreen Hunt
55 Peter Hunt
56 (not identified)
57 David Victor
58 Barry Greengrass
59 Graham Pattison
60 Heiko Parzies
61 Gabriel Saavedra
62 Pauline Churcher
63 Elizabeth Scott
64 Sally Kington
65 Campbell Davidson
66 Kees van Ettekoven
67 Julia Borys
68 Pentti Alanko
69 Tony Hender
70 Shahzad Chaudry
71 John Whiteman
72 Chris Brickell
73 John McNeill
74 Diego Rivera
75 Gwen Scoble
76 Conchita Obon
77 Peter Lewis
78 Anne Plovanich-Jones
79 Allen Coombes
80 Alison Lean
81 Judy Greengrass
82 Nynke Groendijk-Wilders
83 Willem Brandenburg
84 Christopher Tramposch

85 Chris Clennett
86 Margaret McKendrick
87 Ronald van den Berg
88 Marcel Bartels
89 Albert Tramposch
90 Wilbert Hetterscheid
91 Roger Spencer
92 Victoria Matthews
93 Piers Trehane
94 David Ingram
95 Zhou Zhekun
96 Bernard Baum
97 Barbara Pickersgill
98 Valéry Malecot
99 Tom Hazekamp
100 Teresa Vasconcelos
101 Maricela Rodríguez-Acosta
102 Yafa Paris
103 Markku Huttunen
104 Harry Paris
104a Barbara Davies
105 *Pitus withamii*
106 Sissel Tramposch
107 Adelaïde Stork
108 Sabina Knees
109 Binion Amerson
110 Morellen Thompson
111 Maria Teresa Schifino-Wittmann
112 May Sandved
113 Julian Shaw
114 Iain Dawson
115 Malcolm Manners
116 Jörg Ochsmann
117 Milton Nurse
118 Grace Price
119 Diana Miller
120 Alan Leslie
121 Maurice Fowler
122 Hugh McAllister
123 Elizabeth McClintock
124 Lynn Batdorf
125 Janice Strachan
126 Dick Munson

TABLE OF CONTENTS

Taxonomy of Cultivated Plants

D. CONSERVATION AND COLLECTIONS

E. PUBLICATIONS

INTRODUCTION

"Sad to say, we must look to books from Britain and Europe for correct names because there are so few authoritative American references."

(R. Kvaalen 1996)

Cultivated plant taxonomy is concerned with the identification, naming and classification of all cultivated plants. The recognition of cultivars and cultivar-groups is of special consequence in this respect as these categories are unique to cultivated plant taxonomy. However, a knowledge of the variation and distribution of the wild progenitors of any cultivated plant is often an essential pre-requisite for understanding their frequently complex development in cultivation. To these ends the use and development of all potential investigatory techniques and the storage, manipulation and informative reproduction of the data they produce, are all essential to the role of the cultivated plant taxonomist. This work needs to be made readily accessible to all users of cultivated plants and thus the development of user-friendly identification aids and clear explanations of why changes sometimes need to be made are of great importance. Due attention has also to be given to the interaction required with those bodies concerned with the legal regulation of the use of plants and their names, and to understanding the competing pressures and concerns of an international plant industry.

This volume brings together a series of papers and shorter papers delivered as posters presented at the **Third International Symposium on the Taxonomy of Cultivated Plants,** which was sponsored jointly by the Royal Botanic Garden Edinburgh, the Royal Botanic Gardens, Kew and The Royal Horticultural Society under the auspices of the International Society for Horticultural Science. The symposium was organised by HORTAX (Horticultural Taxonomy Group) and held at the Pollock Halls of the University of Edinburgh, from 20-26 July 1998.

In preparing the programme for the symposium, HORTAX members were concerned to cover the widest possible range of subjects and to bring together workers from the divergent areas involved with cultivated plants. This approach met with considerable success as the meeting was notable for its lively social gathering and a mixing of disciplines which sparked much enjoyable and productive debate. Clearly the exhaustive and detailed preparation before the event paid off!

Following the formal days of the symposium, two meetings were held relating to the *International Code of Nomenclature for Cultivated Plants.* The first was a remarkably well-attended open day for discussion and provoked much valuable debate. There was clearly considerable interest in the evolution of the Code, both in respect of the broader principles involved as well as in the practical detail. There are as yet no formal gatherings of cultivated plant taxonomists to discuss these issues and no clearly formulated procedures for flagging up potential changes to the Code for more open debate. This is surely an area to which the International Commission for the Nomenclature of Cultivated Plants may wish to give further consideration.

On the following day the Commission met in closed session to consider revisions to the Code, both in the light of the previous day's discussion and on the basis of experience gained since the publication of the 1995 edition (Trehane *et al.* 1995). A revised Code is expected to be published in 2000.

Also held at Pollock Halls, was the inaugural meeting of The International Association for Cultivated Plant Taxonomy (IACPT).

In the eleven years since HORTAX was founded, cultivated plant taxonomy has suddenly become respectable and of interest to many people around the world. HORTAX has played a major role in focusing this interest: locally, nationally and internationally. The 6th edition of the Cultivated Plant Code (Trehane *et al.* 1995) has also had a major influence and indeed institutions both in America and Europe have filled posts or expanded them into units for horticultural taxonomists, in contrast to the situation ten years ago when such positions were being closed down or filled by molecular biologists!

It is to be hoped that a Fourth International Symposium will continue this escalating momentum, possibly in North America early in the next century.

> "The horticultural industry's aim is to sell plants and lots of them, and subject to any legal considerations, names will be chosen to assist in that aim. Therefore those seeking to establish greater stability and clarity in this area must not make the rules too complicated, and must also ensure that user-friendly guidance is readily available. This symposium, involving so many people with different but overlapping areas of relevant expertise, both legal and technical should go a long way towards removing many of the existing misconceptions and achieving this aim."
>
> (Elizabeth Scott 1999)

References

Kvaalen, R. (1996). Help for the binomically challenged. *Amer. Gardener* 75(3): 48-49.

Scott, E.M.R. (1999). Plant Variety Rights trials for ornamentals: the international testing system and its interaction with the naming process for new cultivars. In S. Andrews, A.C. Leslie & C. Alexander (eds). Taxonomy of Cultivated Plants: Third International Symposium, pp. 89–94. Royal Botanic Gardens, Kew.

Trehane, P. *et al.* (1995). International Code of Nomenclature for Cultivated Plants – 1995. 175 pp. Quarterjack Publishing, Wimborne.

Susyn Andrews
Alan Leslie
Crinan Alexander

September 1999

PLANT AND GERMPLASM COLLECTIONS

1

McAllister, H.A. (1999). The importance of living collections for taxonomy. In: S. Andrews, A.C. Leslie and C. Alexander (Editors). Taxonomy of Cultivated Plants: Third International Symposium, pp. 3–10. Royal Botanic Gardens, Kew.

THE IMPORTANCE OF LIVING COLLECTIONS FOR TAXONOMY

HUGH A. MCALLISTER

Applied Ecology Research Group, School of Biological Sciences, The University, Liverpool L69 3BX, UK

Abstract

Although many species have been taken into cultivation, the plants grown represent only a tiny fraction of their genetic diversity. In many studies of the world's biodiversity cultivated material has usually proved to be of minor significance because so little of it is relevant. However, there are numerous taxonomic problems which can only be solved by the study of living plants, in which morphological characteristics become evident which are not detectable in dried specimens and which are required for most DNA and other molecular studies.

With rare species of very restricted distribution any cultivated plants are effectively of known wild origin. In contrast, undocumented individuals of commoner species and cultivars of garden origin are usually of very little scientific significance, though often of great importance to horticulture. The interests of scientists studying biodiversity and horticulturists beautifying the man-made environment are therefore rather divergent.

Introduction

Taxonomic work based on herbarium specimens is the essential first step in any systematic study, mainly because many more species, and provenances of each species, are represented in herbaria than in cultivation. However, the study of dried or preserved specimens can only take us so far. Living plants are required for chromosome counts, most molecular systematic work on DNA and isozymes, and they also reveal many taxonomically valuable characters which are either not evident or are difficult to detect in dead material. Many recent advances in systematics and taxonomy have depended on the use of living material and would not have been feasible without the collections held in botanic gardens (Morgan *et al.* 1994, Campbell *et al.* 1995).

Systematic research must be useful to its users, in this case growers of plants, or it will lose the respect of the biological research community (Scoble 1997). Far too much time is often spent on purely nomenclatural matters which has nothing to do with science and the advance of knowledge (Hawksworth 1992) but we must have a consistent naming system.

Why is Living Material often not Studied in Taxonomic Revisions?

When preparing taxonomic revisions there is often a tendency to go back to the same old original descriptions or herbarium specimens rather than look at living material.

3

This is often due to the lack of availability of, or effort required to trace and obtain, well documented living plants. When living material is available, it usually represents only a tiny fraction of the variation which exists in nature (Jeffrey 1982, De Jong 1996). Therefore, although living plants provide valuable information, herbarium specimens usually represent a much wider range of variation and type specimens provide an anchor for nomenclatural stability.

A. Botanists v. Gardeners

The reluctance of many taxonomists to work on living plants is, however, understandable given the regrettable divide between taxonomic botanists and horticulturists, even those working in botanic gardens. Taxonomists are often disliked by gardeners because they change plant names and they are perceived to have little horticultural knowledge and may not be very competent at growing plants. On the other hand, taxonomists often consider horticulturists to be interested only in growing 'pretty flowers' and to have little interest in less attractive plants or those whose potential is unknown — as with most new, known source introductions.

Another source of misunderstanding arises when horticulturists ask taxonomists for identifications; botanists are rarely interested in identifying precisely which cultivar a particular rose or camellia belongs to, and horticulturists often do not realise how time consuming — and of little scientific significance — such a task can be. Even plants grown under collectors' numbers in the most respected botanic gardens may not be trustworthy and should certainly not be accepted at face value without comparison with the herbarium specimen of the same collection.

B. Nature of Documented Material

A problem with, at least the older, introductions is that almost always only a single clone is retained for each collection number, and it is often not possible to know if two plants under the same number belong to the same clone or are derived from different seedlings. When we consider the large quantities of seed sent back, particularly by George Forrest, it is disappointing to find that only a single clone of each collection remains with source documentation. As, at least with small species, more than one plant is usually grown, it should be possible in future to ensure that each plant belongs to a different clone, retaining at least some of the genetic variation of the original population. This is particularly important with species which are, or are suspected to be, self-incompatible, a major consideration in the *ex situ* conservation of endangered species.

C. Problems with Undocumented Material – Common Species

The situation which often exists with named cultivated plants with no source data can be illustrated through several examples.

Undocumented birches, especially white-barked species, in botanic gardens are particularly problematic because hybridisation is possible both within and between ploidy levels (Brittain & Grant 1965, Dugle 1966) and any one tree might have European, east Asian and North American ancestry. Such a tree is not much use for the study of the variation in silver birches around the Northern Hemisphere. The birch distributed commercially as *Betula szechuanica* (C.K. Schneid.) Jansson bears very little resemblance to recent introductions from Sichuan and adjacent areas. It is possibly pure *B. pendula* Roth or perhaps a hybrid of *B. szechuanica* with *B. pendula* dating from the days when certain nurserymen raised such species from seed, ignoring the possibility of hybridisation. The so called *'Rhododendron ponticum'* naturalised in the British Isles is probably a hybrid swarm between *R. ponticum* L., *R. maximum* L. and *R. catawbiense* Michx. (Bean 1976, Robson 1991). Studying such cultivated or naturalised material is often compared to a Martian trying to study variation in *Homo sapiens* using one individual, or even a population sample, taken off the streets of London — he would not learn much about the distribution of the variants of *H. sapiens* over the Earth. Such situations indicate why taxonomic botanists rarely wish to work on undocumented plants in gardens and why such work may lead to invalid conclusions (Shambulingappa 1966). Thus, with widespread, variable species, unlocalised material in cultivation is of very little value for scientific study.

D. Undocumented Material — Very Rare Species
In contrast, with species of very restricted wild distribution or more widespread species known to have been introduced only once to cultivation, all plants are effectively of known wild origin. Species with restricted wild distributions include *Pinus radiata* D. Don, *Cupressus macrocarpa* Hartw. ex Gordon, *Metasequoia glyptostroboides* Hu & Cheng, *Rhododendron pemakoense* Kingdon-Ward, *Kolkwitzia amabilis* Graebn., *Campanula sartorii* Boiss. & Heldr. and many others; while *Rhododendron williamsianum* Rehder & E.H. Wilson, *Populus lasiocarpa* Oliv., *Salix magnifica* Hemsl., *Nepeta subsessilis* Maxim. and *Magnolia sprengeri* Pamp. are known to have been introduced only once to cultivation. In the extreme, but by no means unusual, cases of *Populus wilsonii* C.K. Schneid., *Cosmos atrosanguineus* (Hook.) Ortgies and *Santolina chamaecyparissus* L. (McAllister 1987) probably only a single clone is in cultivation.

Use of Living Material

A. Necessity for Herbarium Material of Living Collections
As the persistence of any living collection in cultivation cannot be guaranteed, it is essential that herbarium specimens be made so that others can check the identity of critical living taxa being worked on. In the late 1940s, Janaki Ammal *et al.* (1950) made chromosome counts of 350 *Rhododendron* L. collections, finding variation in section *Rhododendron* from 2n=26 (2x) to 2n=156 (12x). Unfortunately she failed to make voucher specimens, so the identifications of the plants from which the counts were made cannot now be verified and the work was therefore discounted in the recent revision (Cullen 1980). Thus, making herbarium specimens is essential even if studies are primarily on living plants, especially in difficult groups. However, her work indicated where further cytological studies may solve taxonomic problems in this horticulturally very important genus (Cubey in press).

B. Use of Living Material – Undocumented
Despite what has been said above, undocumented garden material can be of taxonomic value. Undocumented cultivated plants solved a taxonomic problem in *Lysimachia* L. in which two quite distinct taxa are commonly being grown under the name *L. punctata* L. (McAllister 1999). Only the *New Royal Horticultural Society Dictionary of Gardening* (Huxley *et al.* 1992a) and Leblebici (1978) make the distinction between *L. punctata* and *L. verticillaris* Spreng., largely on the basis of leaf shape and petiole length — easily detected characters in herbarium specimens but not too obvious in living plants. Ferguson (1972b) in *Flora Europaea* treats them as synonyms. Only the study of living material showed that the two species differ most conspicuously

in the presence of an orange spot at the petal base (the usual key character of the very different *L. ciliata* L.) and self-compatibility and therefore seed production in *L. verticillaris*. These two characters are not evident on herbarium specimens.

C. Information Obtainable only from Living Collections

In cultivation there are pairs of named 'species' which look very different in the living state and are therefore accepted as distinct by gardeners, often because of different flower colour, but which cannot be distinguished on the herbarium sheet. When two such closely related, often sympatric, 'species' are found to have the same ploidy level there is a suspicion that they should be treated as no more than forms. A good example of this situation exists with the candelabra primulas *Primula bulleyana* Forrest, *P. burmanica* Balf. f. & Kingdon-Ward and *P. beesiana* Forrest which hybridise readily with one another in cultivation (Shields 1979); (*P.* × *bulleesiana* Janson, Huxley *et al.* 1992b) and should be referred to a single species. It is interesting that gardeners in this case have concluded that various other species are involved in this hybrid complex (Huxley *et al.* 1992b) though experimental evidence shows that this is not possible (Shields 1979).

There has been much argument about the recognition of the genus *Micromeles* Decne. as distinct or otherwise from *Aria* (Pers.) Host (*Sorbus* L. subg. *Aria* Jacq.). The separation of these as genera has been largely based on *degree* of carpel fusion and the deciduousness or otherwise of the calyx (Kovanda & Challice 1981). However there are some anomalous species combining characteristics of both groups, and it is now generally agreed that the distinction cannot be maintained (Robertson *et al.* 1991). While handling trees in cultivation, it became evident that the natural relationships of the species followed quite a different pattern. The European and west Asian species had embryos which required a cold treatment to obtain germination, tear-drop shaped seeds, often fruits of a red colour never found in the more eastern species and the group had generated many polyploid apomicts. In contrast the Himalayan and east Asian species were all diploid, most had embryos which would germinate immediately on excision, flattened seeds and fruits which were never of the same red colour as found in most of the western species. This is a more natural subdivision of the whitebeams and follows a recognised phytogeographical pattern found by Phipps (1983) in *Crataegus* with relatively distantly related groups of species separated by the position of the ancient Turgai Straits (Tiffney 1985).

D. Importance of Chromosome Numbers

Where more than one base number is found within a family or genus this may give a clue as to where generic splits should perhaps be made. In the *Rosaceae* Juss., suggestions made on cytological grounds have been confirmed by DNA sequencing data (Morgan *et al.* 1994).

In the genus *Primula* L. most subgenera have a base number of x=11 with most species being diploid except in subgenus *Auriculastrum* Schott. Species of section *Cortusoides* Balf. f. with their very distinctive petiolate, more or less palmately veined leaves have x=12 (as has *P. forrestii* Balf. f., McAllister in mss.) as do species of *Cortusa* L. (Ferguson 1972a). This strongly suggests that these form a natural grouping of related species and *P.* sect. *Cortusoides* and *P. forrestii* should be united with *Cortusa*, or *Cortusa* combined with sect. *Cortusoides* of *Primula*. The birds' eye primroses (subg. *Aleuritia* (Duby) Wendelbo redefined to exclude the rather different species with n=11) have a base number of x=9 and species up to 14-ploid (2n=126), (Valentine & Kress 1972). Should we therefore accept *Aleuritia* as a separate genus?

In *Meconopsis* Vig., the problem of the fertile 'hybrid' blue poppies (Stevens 1998) seems to have been resolved with the finding that fertile *M.* × *sheldonii* G. Taylor has 2n=246. It is therefore a fertile allopolyploid derivative of *M. grandis* Prain (2n=164) and *M. betonicifolia* Franch. (2n=82) and can be described as a new species of garden origin similar to *Primula kewensis* W. Wats. and *Aesculus* × *carnea* Hayne.

I became interested in the pinnate leaved rowans (*Sorbus* subg. *Sorbus*) through the difficulties experienced in trying to identify a tree in cultivation (McAllister & Gillham 1980). I soon discovered that much naming was chaotic, with *S. aucuparia* L. being grown under a wide variety of names, probably due to the death of scions and the *S. aucuparia* stocks being grown on and getting into the trade under the name of the dead scion. The trade wisdom was that all *Sorbus* had to be grafted because hybridisation occurred so readily. Progeny raising showed that the red-fruited species, *S. scalaris* Koehne and *S. esserteauiana* Koehne, probably represented in cultivation by single clones, were almost totally self-incompatible so, with one exception, all seedlings raised were hybrids with neighbouring trees. This was probably the basis of the received wisdom.

No amount of herbarium work (Schneider 1906a & b, Koehne 1913, Yü & Kuan 1959–63, Gabrielian 1978) could have revealed the cause of the taxonomic difficulties in the pinnate leaved Chinese and Himalayan white-crimson fruited *Sorbus* (rowans). Chromosome counts and pollination studies on the unknown taxon at Ness showed it to be tetraploid and apomictic (all known triploids and tetraploids in the *Maloideae* Weber are apomictic). The discovery of the widespread occurrence of apomixis in the pinnate leaved *Sorbus* immediately led to an understanding of the taxonomic difficulties in the group — it consisted of a mixture of variable sexual species and apomictic microspecies (= clones) (Gillham and McAllister 1977, McAllister 1986, Phipps *et al.* 1990). It is amusing that one of these apomicts is represented by several so called selections; *S. hupehensis* 'Pink Form', 'November Pink', 'Rufus' and 'Pagoda Pink', all of which belong to the same clone. This situation is brought about by the lack of any requirement with most ornamentals to prove that a 'new' cultivar is distinct from existing named cultivars.

The taxonomy of ivies (*Hedera* L. species) has long been unsatisfactory and the account given by Webb (1968) is quite unworkable, partly because the trichome measurements in the key appear to have been switched. It was only the study of wild collected living material which resolved the problems (McAllister 1994), though there has been considerable resistance to recognition of the commonly cultivated *H. algeriensis* 'Gloire de Marengo' as distinct from the very different species native to the Canaries (McAllister 1988). In Spain and Portugal the most commonly cultivated ivy was recognised as a new species, *H. maroccana* McAllister, which had previously been 'lumped' with the very different *H. helix* L. (Rutherford *et al.* 1993). Where more than one ploidy level is found within a 'species', this indicates which entities to look for differences between. *Hedera helix* and *H. hibernica* (Kirchn.) Bean as now understood were differentiated in this way (McAllister & Rutherford 1990).

Why are there so Few Studies of Horticulturally Important, especially Woody, Genera?

An Iranian PhD student asked why all British taxonomic PhDs were on annuals. She was perhaps overstating the case but did have a point. Most projects require statistically significant samples from several populations of a species, and experimental breeding work. Only the various Royal Botanic Gardens and a very few other gardens have many wild source collections and in many cases, especially with collections made earlier this century, only a single clone of each collector's number. There is therefore no information on variation within populations and very few collections of any one species. As a result very little taxonomic work is done on horticulturally important perennial, and especially woody, genera of great commercial significance while PhDs are common on British native species (e.g. McAllister 1972). However, more recent accessions are being planted in larger numbers.

Unless the documented collections in botanic gardens are the subject of scientific work we risk losing them as questions will be asked as to their purpose. There are many classification problems in genera of considerable commercial, scientific, evolutionary and phytogeographical interest. However, it tends to require longer than the usual three year PhD to study such topics, and some agency needs to be found to provide the funding. The Dutch have similar problems but are more aware of how dependent their hardy nursery stock industry is on the diversity of material available (De Jong 1996). Major projects are currently underway to produce monographs of *Cotoneaster* Medik., *Betula* L. and *Alnus* Mill. which are likely to be of great interest to both the scientific and horticultural communities, but they are being undertaken by amateurs with very little assistance.

Conclusions

This demonstrates, I hope, that there is a limit to the kinds of systematic work which can be done with dried material and that many problems which seem intractable in the herbarium may be easily solved by work with living plants. Herbarium work is often seen as 'dry and dusty' by the general public and even disreputable (because of name changes) by gardeners. In contrast, people working with living plants can relate more readily to others who also grow plants — and co-operate through the exchange of living material and the introduction of new species to cultivation. Studies on living plants will therefore lead to greater scientific understanding and increase the usefulness of plants while at the same time be more understandable to the 'Man in the Street'. For such scientific work we need well documented wild source collections and sympathetic staff to manage them. For this to happen we must break down the barrier which exists between botanists and gardeners by training people with skills in both areas who understand the point of view of those with skills in only one area.

Acknowledgments

I am very grateful to the Horticultural Taxonomy Group (HORTAX) for inviting me to speak at this conference and air my views. The staff of Ness Botanic Gardens, University of Liverpool, have been most co-operative in providing space and care for large numbers of new accessions of wild source plants. I am grateful to Professor R.H. Marrs, the reviewers and the editor for constructive comments on the manuscript.

References

Bean, W.J. (1976). *Rhododendron ponticum.*Trees and shrubs hardy in the British Isles. Vol. 3. (Ed. 8). pp. 741–744. John Murray, London.

Brittain, W.H. and Grant, W.F. (1965). Observations on Canadian birch (*Betula*) collections at the Morgan Arboretum. I. *B. papyrifera* in eastern Canada. *Canad. Field-Naturalist* 79(3): 189–197.

Campbell, C.S., Donoghue, M.J., Baldwin, B.G. & Wojciechowski, M.F. (1995). Phylogenetic relationships in *Maloideae* (*Rosaceae*): evidence from sequences of the internal transcribed spacers of nuclear ribosomal DNA and its congruence with morphology. *Amer. J. Bot.* 82(7): 903–918.

Cubey, J. Cytotaxonomic revision of subsections *Saluenense* and *Maddenia* of *Rhododendron*. Unpublished PhD Thesis. University of Liverpool.

Cullen, J. (1980). Revision of *Rhododendron*. 1. Subgenus *Rhododendron* sections *Rhododendron* & *Pogonanthum*. *Notes Roy. Bot. Gard. Edinburgh* 39(1): 1–207.

De Jong, P.C. (1996). The genetic diversity of trees in cultivation. In D. Hunt (ed.). Temperate trees under threat. pp. 179–184. International Dendrological Society, Morpeth.

Dugle, J.R. (1966). A taxonomic study of western Canadian species in the genus *Betula*. *Canad. J. Bot.* 44(2): 929–1007.

Ferguson, L.F. (1972a). *Cortusa* L. In Tutin, T.G. *et al.* (eds). Flora Europaea. Vol. 3. p. 23. Cambridge University Press, Cambridge.

Ferguson, L.F. (1972b). *Lysimachia* L. In Tutin, T.G. *et al.* (eds). Flora Europaea. Vol. 3. pp. 26–27. Cambridge University Press, Cambridge.

Gabrielian, E. (1978). The genus *Sorbus* in Western (mistranslated as Eastern in the English summary) Asia and the Himalayas. 264 pp + 62 plates. Academy of Sciences of the Armenian SSR, Erevan, USSR. (In Russian with English summary).

Gillham, C.M. & McAllister H.A. (1977). Tree genera – 6. *Sorbus* Sect. *Aucuparia*. *Arbor. J.* 3(2): 85–95.

Hawksworth, D.L. (1992). The need for a more effective biological nomenclature for the 21st century. *Bot. J. Linn. Soc.* 109(4): 543–567.

Huxley, A. *et al.* (eds). (1992a). *Lysimachia* L. In *The New Royal Horticultural Society Dictionary of Gardening*. Vol. 3. pp. 143–145. Macmillan, London.

Huxley, A. *et al.* (eds). (1992b). *Primula* L. In *The New Royal Horticultural Society Dictionary of Gardening*. Vol. 3. pp. 708–724. Macmillan, London.

Janaki Ammal, E.K., Enoch, I.C., & Bridgwater, M. (1950). Chromosome numbers in species of *Rhododendron*. *Rhododendron Year Book 1950*. No. 5: 78–91.

Jeffrey, C. (1982). *Rhododendron* and classification – a comment. *Rhododendron 1982/83 Magnolias Camellias*: 48–51.

Koehne, E. (1913). *Sorbus* L. In C.S. Sargent. Plantae Wilsonianae Vol. 1(3). pp. 457–483. The University Press, Cambridge.

Kovanda, M. & Challice, J. (1981). The genus *Micromeles* revisited. *Folia Geobot. Phytotax.* 16(2): 181–193.

Leblebici, E. (1978). *Lysimachia* L. In P.H. Davis (ed.). Flora of Turkey. Vol. 6. pp. 135–138. University Press, Edinburgh.

McAllister, H.A. (1972). The experimental taxonomy of *Campanula rotundifolia* L. Unpublished PhD Thesis. Glasgow University.

McAllister, H.A. (1986). The rowan and its relatives (*Sorbus* spp.). 14 pp. University of Liverpool Botanic Gardens, Liverpool.

McAllister, H.A. (1987). Conservation and taxonomy of *Santolina chamaecyparissus* agg. *The National Council for the Conservation of Plants and Gardens Newsletter* 10: 7–10.

McAllister, H.A. (1988). Canary and Algerian Ivies. *Plantsman* 10(1): 27–29.

McAllister, H.A. (1994). *Hedera*. In The Common Ground of Wild and Cultivated Plants. *Bot. Soc. Brit. Isles Conference Report* 22: 145–150. Cardiff.

McAllister, H.A. (1999). *Lysimachia punctata* L. and *L. verticillaris* Sprengel (*Primulaceae*) naturalised in the British Isles. *Watsonia* 22: 279–281.

McAllister, H.A. & Gillham, C.M. (1980). *Sorbus forrestii. Bot. Mag.* 183 (1), Tab. 792.

McAllister, H.A. & Rutherford, A. (1990). *Hedera helix* L. and *H. hibernica* (Kirchn.) Bean (*Araliaceae*) in the British Isles. *Watsonia* 18(1): 7–15.

Morgan, D.R., Soltis, D.E. & Robertson, K.R. (1994). Systematic and evolutionary implications of *rbcL* sequence variation in *Rosaceae. Amer. J. Bot.* 81(7): 890–903.

Phipps, J.B. (1983). Biogeographic, taxonomic, and cladistic relationships between east Asiatic and North American *Crataegus. Ann. Missouri Bot. Gard.* 70(4): 667–700.

Phipps, J.B., Robertson, K.R., Smith, P.G. & Rohrer, J.R. (1990). A checklist of the subfamily *Maloideae* (*Rosaceae*). *Can. J. Bot.* 68(10): 2209–2269.

Robertson, K.R., Phipps, J.B., Rohrer, J.R. & Smith, P.G. (1991). A synopsis of genera in *Maloideae* (*Rosaceae*). *Syst. Bot.* 16(2): 376–394.

Robson, M.G. (1991). The *ponticum* problem. *Rhododendron 1991 Camellias Magnolias.* 43: 46–48.

Rutherford, A., Mc Allister H.A. & Mill, R.R. (1993). New ivies from the Mediterranean area and Macronesia. *Plantsman* 15(2): 115–128.

Schneider, C.K. (1906a). Handbuch der Laubholzkunde. Vol. I. pp. 667–700. Gustav Fischer, Jena.

Schneider, C.K. (1906b). Pomaceae sinico–japonicae novae – et adnotationes generales de Pomaceis. *Bull. Herb. Boissier ser. 2,* 6: 311–319.

Scoble, M.J. (1997). The transformation of systematics? *Trends Ecol. Evol.* 12(12): 465–466.

Shambulingappa, K.G. (1966). Cytomorphological studies of *Clematis hatherliensis* (*C. orientalis* × *C. tangutica*). *Caryologia* 19(4): 395–401.

Shields, C. (1979). The cytogenetics of *Primula* L. Unpublished PhD Thesis. University of Liverpool.

Stevens, E. (1998). *Meconopsis* × *sheldonii* 'Lingholm Strain'. *Rock Gard.* 25(4) no. 101: 399–402.

Tiffney, B.H. (1985). The Eocene North Atlantic landbridge: its importance in Tertiary and modern phytogeography at the Northern Hemisphere. *J. Arnold Arbor.* 66(2): 243–273.

Valentine, D.H. & Kress, A. (1972). *Primula* L. In Tutin, T.G. *et al.* (eds). Flora Europaea. Vol. 3. pp. 15–20. Cambridge University Press, Cambridge.

Webb, D.A. (1968). *Hedera*. In Tutin, T.G. *et al.* (eds). (1968). Flora Europaea. Vol. 2. p. 314. Cambridge University Press, Cambridge.

Yü, T.T. & Kuan, K.C. (1963). Taxa nova Rosacearum Sinicarum (I). 1. *Spiraea* L. *Acta Phytotax. Sin.* 8(3): 214–234.

Hazekamp, T, and Guarino, L. (1999). Taxonomy and the conservation and use of plant genetic resources. In: S. Andrews, A.C. Leslie and C. Alexander (Editors). Taxonomy of Cultivated Plants: Third International Symposium, pp. 11–17. Royal Botanic Gardens, Kew.

TAXONOMY AND THE CONSERVATION AND USE OF PLANT GENETIC RESOURCES

TOM HAZEKAMP[1] AND LUIGI GUARINO[2]

[1]Documentation, Information and Training Group, International Plant Genetic Resources Institute, Via delle Sette Chiese 142, 00145 Rome, Italy
[2]Regional Office for the America's, International Plant Genetic Resources Institute, c/o Centro Internacional de Agricultura Tropical, Apartado Aereo 6713, Cali, Colombia

Abstract

Increasing interest in the conservation and sustainable use of plant genetic diversity at a world-wide level is driving the need for more and better information on this important resource. Taxonomic data are an essential part of the documentation of the germplasm of crops and their wild relatives. They should provide managers and users of genetic resources with a rational, stable, widely-accepted framework within which to organise their activities. However, the taxonomy of many genepools is still unresolved, and alternative taxonomic schemes are in use even for important genepools. This can hinder the efficient and effective exchange and synthesis of information on collections held by different national, regional and international organizations. As international collaboration in plant genetic resources conservation and use increases, there will be a greater need to bring together information on the germplasm held in different collections. Thus, the case for uniform taxonomic approaches can only become stronger. In this paper the areas of plant genetic resources conservation and use where taxonomic issues have a strong impact are discussed. These include ecogeographic surveying, aspects of germplasm maintenance such as the implementation of appropriate procedures for germplasm rejuvenation and multiplication, and aspects of use such as the exchange of germplasm and the development of core collections. The crop network is highlighted as a concept that has helped the plant genetic resources community to address taxonomic issues.

Introduction

The conservation and use of plant genetic resources comprises a number of interlinked activities. A model has been developed showing how these elements — from the selection of the target taxon, through to the collecting, management and finally the use of germplasm — fit together and depend on one another (Maxted *et al.* 1997). Taxonomic information is crucial at various stages of this process. Some of its component activities require such information, others generate data with potential taxonomic importance. The taxonomy of many genepools is still unresolved, in particular for so-called neglected crops, many tropical forage and forestry species. Alternative taxonomic schemes are in use for genepools of staple crops such as sorghum. This can present considerable problems when information generated by the

11

conservation process needs to be exchanged or compared with information from other sources. In addition, the incomplete or erroneous identification of germplasm accessions further affects the use of this type of data. In this paper some of the aspects of plant genetic resources conservation and use where taxonomic issues have the greatest relevance are highlighted.

Germplasm Exploration and Collecting

To increase the probability of conducting a successful collecting mission, comprehensive planning is essential. In the course of such planning, different sources of information are consulted as part of a so-called "ecogeographic survey". In particular, the collector needs to be familiar with what is known about genetic variation within the target taxon (Hanelt & Hammer 1995). The collector will also gather information on geographic distribution patterns, ecological preferences, reproductive biology and ethnobotany (Maxted *et al.* 1995). These surveys enable collectors to target specific areas for action, because they are especially diverse, likely to harbour specific germplasm of interest or at risk of genetic erosion. Herbaria are very useful sources of ecogeographic information on wild species, but usually less so for cultivated plants (Engels *et al.* 1995). In addition, collectors consult experts, floras, monographs and the reports of previous germplasm collecting missions.

The ecogeographic survey requires a solid taxonomic foundation. Prospective collectors need to be aware of alternative taxonomic treatments of their target group in their target area and decide which to follow, to gain entry to, and exchange, the information available from experts, herbaria, the literature and databases.

Taxonomic custom provides well described, formal mechanisms for revision and international agreement. However, for both wild and cultivated species, frequent revisions and contesting taxonomic treatments, homonyms and synonyms, can complicate the compilation and matching of information on target taxa. From a genetic resources conservation point of view, it is important to know how formal taxonomic categories relate to each other within the crop genepool, which is the unit of conservation. The genepool concept as developed by Harlan & de Wet (1971), is based on the biological species concept and distinguishes three main categories defined by the extent to which gene transfer is possible with the selected target crop. The primary genepool includes the cultivated type, weedy forms and wild progenitors that are all completely interfertile. The secondary genepool includes taxa that will cross with the crop and produce a fertile progeny, although certain isolation barriers exist. The tertiary genepool includes taxa that can only be successfully crossed with the crop using special techniques such as embryo rescue. For many crops, the genepool concept has been successfully applied in crop improvement and conservation studies.

Even the well-prepared collector is likely to face a number of taxonomy-related problems in the field. For example, there are often problems in identifying the material encountered. Traditional taxonomic aids, such as keys, often rely on characteristics which are not available at the time of seed collection, in particular flowers. Taxonomists could make an important contribution to plant genetic resources conservation by producing keys and other identification aids specifically aimed at the germplasm collector, for example lateral keys based on seed or vegetative characters. Supplementary methods of identification will sometimes be necessary. For cultivar identification in perennial crops, e.g. *Hevea* Aubl., a portable field set for analysis of molecular markers has proved useful (Lanaud & Lebot 1997). Such an approach could be applied to other levels of the taxonomic hierarchy.

Germplasm Management

For the efficient management of germplasm collections, curators need to be familiar with the material in their care. This will include aspects such as vegetative development, reproductive biology, domestication status, associated pests, diseases and weeds, harvest techniques and post-harvest processing and seed storage behaviour. The specific characteristics of an accession determine the most appropriate management strategy. Passport data, and in particular taxonomic information, is the key that is used to collate relevant information from a variety of sources. For example, correct identification of a seed sample will allow the genebank curator to consult Hong *et al.* (1998) for information on the most appropriate method for long-term seed storage. Likewise, the characterization and evaluation that is carried out on an accession following standard genepool descriptor lists will depend on its correct initial identification. These data are especially important in that they provide the main entry point for potential users of the germplasm.

There are indications that the taxonomic data maintained in genebank documentation systems require careful consideration when used as a basis for germplasm management and evaluation practices. The taxonomic data are not always complete nor correct (Maxted & Crust 1995). This could be for a variety of reasons. In some cases the original collections were established with objectives in mind other than germplasm conservation, e.g. the material could have been collected as an input to a specific breeding programme. In these cases the emphasis might have been on the documentation of characterization and evaluation descriptors, while the accumulation of passport data, including the taxonomy, was of secondary importance. In other cases, exact identification was not possible at the time of collecting due to the absence of key structures. If this material was donated to a genebank, the taxonomic data might not have been verified by the genebank, often as a result of a lack of resources (money, time and/or taxonomic expertise) or the fact that passport data such as precise location of collection and ecographical data are simply missing and can no longer be retrieved from the original source. Although inprecise or incorrect taxonomic identification would not necessarily result in the loss of the sample as a consequence of incorrect management procedures, there is a danger that the sample would not receive an optimal treatment and its genetic integrity might be compromised.

Germplasm Use

FAO (Food and Agriculture Organisation) (1996) estimated that worldwide more than 6 million germplasm accessions are maintained *ex situ*. Lyman (1984) judged that at least half of the crop germplasm maintained in collections were duplicates. Considering the size of these collections and the potential duplication, methodologies are needed to enhance access to the wealth of biodiversity they represent. Core collections are under development for several genepools (e.g. barley, sugar beet, brassicas). Their aim is to identify a small subset that will represent a large proportion of the biodiversity present in a germplasm collection or genetic resources network. The idea is that the core collection would provide the means to rationalise management and use of crop genetic resources. Along with descriptors on geographical origin, ecological data, genetic markers and agronomic data, taxonomic information is used for allocation to groups from which samples are selected for the core collection (Brown 1995). Again, the role of the taxonomic descriptors depends on the degree to which the formal taxonomic ranks correspond with the underlying patterns of genetic diversity.

As noted in the section on germplasm management, the correct and complete taxonomic identification of germplasm in collections is not always achieved. From a use perspective, this can lead to substantial under-utilization of germplasm for several reasons. As users of genebank material, such as breeders, tend to be interested in intraspecific diversity, the species name is an important query field. Incorrect or incomplete classification would easily eliminate a sample from the result set. Also, incorrect classification could cause the sample to be documented against completely inappropriate characterization and evaluation traits. This would reduce the chances that its specific qualities could be put to good use.

Plant Genetic Resources Networks

Since 1980, IPGRI (International Plant Genetic Resources Institute) has been stimulating the establishment of crop or regional networks as platforms for international and regional collaboration in the area of plant genetic resources. Usually one of the first actions taken by a network is to establish a central crop database. Such a database contains selected data on the crop germplasm collections of the participating members and forms the basis on which collaborative activities are developed. Examples of such activities include joint collecting missions, germplasm multiplication and regeneration activities, analysis of germplasm duplication and the designation of core collections. Table 1 provides an overview of central crop databases established through the framework of the European Cooperative Programme for Crop Genetic Resources Networks (ECP/GR).

TABLE 1. Central crop databases established through the ECP/GR.

Cereals

Avena L.	Barley
Secale L.	Triticale
Wheat	Maize

Forages

Agropyron Gaertn.	*Agrostis* L.
Arrhenatherum elatius (L.) P. Beauv. ex J. Presl & C. Presl/ *Trisetum flavescens* (L.) P. Beauv.	*Bromus* L.
Dactylis L./*Festuca* L.	*Lolium* L.
Perennial *Medicago* L.	Other perennial legume forage species
Poa L.	*Phalaris* L.
Phleum L.	*Trifolium alexandrinum* L./*T. resupinatum* L.
Trifolium pratense L.	*Trifolium subterraneum* L./annual *Medicago* L.
Lathyrus latifolius L., *L. tuberosus* L., *L. heterophyllus* L. and *L. sylvestris* L.	Other *Vicieae* excepting the grain crops
Vicia spp.	

Fruit trees

Malus Mill.	*Prunus* L.

Grain legumes

Cicer L.	*Glycine* Willd.
Lens Mill.	*Lupinus* L.
Phaseolus L.	*Pisum* L.
Vicia faba L.	

Other crops

Allium L.	*Beta* L. (currently World *Beta* Network)
Brassica L.	

The value of the central crop database as a management tool is greatest if a consistent taxonomic treatment is applied to all accessions. In some cases, different approaches to the taxonomy of a crop exist, from highly dissected conventional hierarchies to pragmatic systems that abandon hierarchical botanical classification below the species level and use cultivar groups. There is a clear practical need to come to terms with these differences at the network level. The World *Beta* Network is a good example where the network has been very active in discussing the taxonomy of the crop (Frese *et al.* 1998). Although these discussions may not always lead to the application of a single taxonomic system by all members of the network, they do provide members with the insight to make an informed choice. Once the quality of data in a central crop database has been maximised in terms of completeness, consistency and accuracy, the database becomes a tool that can make a strong contribution to the work of the group and thus is a valuable source of information for a large number of potential users.

Conclusions

Any scientific discipline that is involved in the conservation and use of collections of biological material will be confronted with taxonomic issues. In the specific case of plant genetic resources, the focus is on genetic diversity in crop genepools, which involves infra- and intra-specific diversity of a crop and related wild species. How well the taxonomic ranks represent relevant genetic diversity determines, amongst other things, the usefulness of formal taxonomy as a framework to study this diversity. Usually, additional descriptors such as geographical origin, ecological data, agronomical traits, genetic markers and geneflow characteristics play an important role in determining the relative position and importance of specific germplasm in an integrated conservation and use strategy.

The conservation of plant genetic resources is a global concern. Its practical implementation is more and more taking the form of regional and global initiatives. These require clear and unambiguous information — including taxonomic information — to formulate scientific and policy decisions. However, frequent taxonomic revisions, the existence of contesting taxonomic approaches and in some cases a taxonomy that is still unresolved, have confused and frustrated many users of taxonomic data and complicated their practical use at the international level. A definite step forward was the realization that the nomenclature of wild and cultivated species needed to be considered in different contexts. This led to the development of the *International Code of Nomenclature for Cultivated Plants* in 1953 in addition to the *International Code of Botanical Nomenclature* (ICBN) which was first published in 1905.

In general, it should be realised that many users of taxonomic systems are not taxonomists. This means that the taxonomic systems need to be transparent and easy to apply if they are to realise their full potential by the global conservation community. In some cases this will mean that customised tools are needed to assist specific users in applying these systems. For example, taxonomic keys that rely on a combination of specimen characteristics that are expressed at distinct times are not practical for application in germplasm collecting.

There is also a clear need for a central list covering existing species names, such as the global plant checklist being developed by the International Organization for Plant Information (URL: iopi.csu.edu.au/iopi). Once completed the global plant checklist could serve as a taxonomic backbone on some 300,000 vascular plant species to which users can append their more specialised information. This checklist will form part of the Species 2000 project.

Change with time is inevitable. Taxonomic insights will emerge as new concepts are developed and new technologies become available. It is important that these changes are made transparent to users of taxonomic information, including plant genetic resources conservationists and users, so that they may make informed decisions on how these developments can help them to improve their work. In the genetic resources field, crop networks can provide a platform for this. In addition, Internet-based information systems and discussion groups could also play a role in enhancing the understanding and use of taxonomy for large groups of users.

Acknowledgement

The authors would like to thank Dr Toby Hodgin (IPGRI) for his valuable comments on earlier drafts of this manuscript.

References

Brown, A.H.D. (1995). The core collection at the crossroads. In T. Hodgkin, A.H.D. Brown, Th.J.L. van Hintum & E.A.V. Morales (eds). Core Collections of Plant Genetic Resources, pp. 3–19. Wiley and Sons, Chichester.

Engels, J.M.M., Arora, R.K. & Guarino, L. (1995). An introduction to plant germplasm exploration and collecting: planning, methods and procedures, follow-up. In L. Guarino, V.R. Rao & R. Reid (eds). Collecting Plant Genetic Diversity, pp. 31–63. CAB International.

FAO (1996). Report on the State of the World's Plant Genetic Resources. International Technical Conference on Plant Genetic Resources, Leipzig, Germany, 17-23 June 1996. FAO Report ITCPGR/96/3. FAO, Rome, Italy.

Frese, L., Panella, L., Srivastava, H.M. & Lange, W. (eds). (1998). International *Beta* Genetic Resources Network. A report on the 4th International *Beta* Genetic Resources Workshop and World *Beta* Network Conference held at the Aegean Agricultural Research Institute, Izmir, Turkey, 28 February–3 March 1996. International Crop Network Series 12. International Plant Genetic Resources Institute, Rome, Italy.

Hanelt, P. & Hammer, K. (1995). Classification of intraspecific variation in crop plants. In L. Guarino, V.R. Rao & R. Reid (eds). Collecting Plant Genetic Diversity, pp. 113–120. CAB International.

Harlan, J.R. & de Wet, J.M.J. (1971). Towards a rational classification of cultivated plants. *Taxon* 20: 509–517.

Hong, T.D., Linington, S. & Ellis, R.H. (1998). Compendium of information on seed storage behaviour. Vol. 1. pp. 1–400, Vol. 2. pp 401–901. Royal Botanic Gardens, Kew.

Lanaud, C. & Lebot, V. (1997). Molecular techniques for increased use of genetic resources. In A.W. Ayad, T. Hodgkin, A. Jaradat & V.R. Rao (eds). Molecular genetic techniques for plant genetic resources. Report of an IPGRI workshop, 9-11 October 1995, Rome, Italy, pp. 92–97. International Plant Genetic Resources Institute, Rome, Italy.

Lyman, J.M. (1984). Progress and planning for germplasm conservation of major food crops. *FAO/IBPGR Pl. Genet. Resources Newslett.* 60: 3–21.

Maxted, N., Hawkes, J.G., Guarino, L. & Sawkins, M. (1997). Towards the selection of taxa for plant genetic conservation. *Genet. Resources Crop Evol.* 44: 337–348.

Maxted, N., & Crust, R. (1995). Aids to taxonomic identification. In L. Guarino, V.R. Rao, & R. Reid (eds). Collecting Plant Genetic Diversity, pp.181–194. CAB International.

Maxted, N., Slageren, M.W. van & Rihan, J.R. (1995). Ecogeographical surveys. In L. Guarino, V.R. Rao & R. Reid (eds). Collecting Plant Genetic Diversity, pp. 255–286. CAB International.

Gardner, M. (1999). Managing *ex situ* conifer conservation collections. In: S. Andrews, A.C. Leslie and C. Alexander (Editors). Taxonomy of Cultivated Plants: Third International Symposium, pp. 19–23. Royal Botanic Gardens, Kew.

MANAGING *EX SITU* CONIFER CONSERVATION COLLECTIONS

MARTIN GARDNER

International Conifer Conservation Programme, The Royal Botanic Garden Edinburgh, 20A Inverleith Row, Edinburgh EH3 5LR, UK

Abstract

Today an important remit of many botanic gardens is the management of their living collections for the conservation of genetic resources. Although the protocols for this have been developed few botanic gardens have fully implemented them. The International Conifer Conservation Programme (ICCP), which is based at the Royal Botanic Garden Edinburgh, has spent the last seven years testing and modifying some of these protocols for the *ex situ* conservation of threatened conifers, and in some cases their associated broad-leaved species. The ICCP has successfully networked with over 100 dedicated sites throughout Britain and Ireland in order to accommodate the large numbers of individuals often necessary to create a broad genetic base. Many of these plants and the historical introductions of conifers are the focus of research by the ICCP, which includes monitoring growth rates, assessing genetic diversity and early cone induction.

Introduction

When I returned to RBG Edinburgh in 1991 to co-ordinate the International Conifer Conservation Programme (ICCP) (after a break of some 17 years) I was struck by two aspects of the living collections. Firstly, how diverse and well documented they were. For example, in 1995 an assessment of the collections revealed that they comprised 21,578 taxa from 336 families and 3,020 genera with almost 57% of the 39,578 accessions being of known wild origin (Walter *et al.* 1995). Over 7.5% (1600) of the taxa were listed as threatened by IUCN (International Union for Conservation of Nature and Natural Resources) (Walter *et al.* 1995). Secondly, I was encouraged by how well utilised some of the key collections were for taxonomic study, although little active research had been carried out on the conservation of important species. It soon became apparent though that Edinburgh was perhaps one of the exceptions in having a progressive attitude to researching their living plant collections.

Recent research suggests that 50% of the world's botanic gardens use less than 10% of their living plant collections for taxonomic related studies; it is interesting to note that 59% of the gardens approached carry out taxonomic research (Rae 1995). Relatively recently botanic gardens have realised their potential role as guardians of threatened plant germplasm and this has resulted in 87% of them specifically cultivating plants for conservation purposes (Rae 1995). Some gardens have a focused approach such as those which form part of the Center for Plant Conservation network in the US, whereby each participating garden is responsible for growing specific threatened North American native plants. The majority of gardens, however, have inadvertently amassed most of their accessions of threatened plants and these merely play a passive conservation role.

The ICCP was established in 1991 as an externally funded initiative based at the Royal Botanic Garden Edinburgh (RBGE) where it has been able to extend and research the already comprehensive collections of living conifers. However, even with Edinburgh's 450 acres, which are shared between four different widely dispersed sites (Benmore, Dawyck, Inverleith and Logan), this still does not allow sufficient space to accommodate *ex situ* conservation collections of large growing trees such as most conifer species. The remit of the ICCP has been, therefore, to develop a network of dedicated sites outwith Edinburgh. The ICCP has made every effort to conform within the framework of the Convention on Biological Diversity (CBD). Its collaborative research in Chile, using funding from The Darwin Initiative for the Survival of Species, has enabled it to develop a meaningful collaboration which has included "the fair and equitable sharing of the benefits arising out of the utilization of genetic resources".

Networking to Provide 'Safe Havens'

Like so many groups of valuable timber trees, conifer species are threatened in the wild. Of the world's 805 recognized taxa, 356 (44%) are threatened under the new IUCN criteria (Farjon 1998). For over 130 years many of these species have traditionally formed an integral part of our public parks, gardens and private estates. Commonly cultivated examples which are listed as being threatened in the wild include, *Araucaria araucana* (Molina) K.Koch, *Cedrus libani* A.Rich., *Chamaecyparis lawsoniana* (A. Murray bis) Parl., *Picea omorika* Pančič and *Sequoia sempervirens* (D.Don) Endl.

The ICCP is building on the great tradition in Britain and Ireland of cultivating conifers and this widely spread network is providing a valuable sanctuary for threatened conifers. Managing a disparate collection of trees scattered from south-east England to northern Scotland and west across to Ireland, is no easy task and it is dependent on a lot of good will on the part of the collaborating landowners. It is important that these sites have a commitment to conservation and a long-term future (the latter being something which is often very difficult to predict) and that the staff have the relevant horticultural skills to maintain the collections. It is the firm belief of the ICCP that the management of conservation resources should not necessarily be confined to botanic gardens. Networking with dedicated sites has many advantages, one of which is to heighten the awareness of conservation among both professional gardeners and the wider general public. By applying and developing the necessary protocols for the management of threatened plants, will help to set standards for the collaborating sites to adopt for their own collections. Careful attention is given to accurate record keeping and a strict procedure is used when collecting in the wild. For instance, where possible, seed or cuttings are taken from a single mother-tree and assigned a unique field collecting number. Each propagule type is assigned a unique accession when it arrives at RBGE and the progeny, (whether this is seedlings or rooted cuttings) are allocated a qualifier in the form of a letter. This enables individual plants from a single accession to be monitored thereon. Similar standards have to be adopted by more collections holders if their collections are going to have any relevance for conservation. To date the network comprises 110 sites which provide an opportunity to represent many threatened conifers with a large number of genotypes from a range of known wild provenances of the same taxon. Over 10,000 plants have been planted in these sites and these comprise 2,124 different accessions. Of these, 633 accessions are represented by between 5 and 10 genotypes and 400 accessions are represented by more than 10 genotypes. Over 50% of the world's conifer taxa are managed by the ICCP, 86% of the conifers are growing in its network of sites outwith RBGE.

The way in which these collections are made in the wild is extremely important. The protocols for sampling wild populations have largely been adopted from the recommendations of Falk & Holsinger (1991). In reality the sampling is frequently influenced by seed availability, which in many species is very scarce. In the case of the South American conifer *Fitzroya cupressoides* (Molina) Johnst., for example, seed production is extremely erratic or access to cones difficult due to the height of the trees, therefore vegetative material is collected. Sampling from native forests is not necessarily undertaken as a 'one off' seed collecting expedition. For example, Chilean conifers have been sampled during five separate expeditions. The reasons for this include the time it takes to negotiate difficult terrain, for example species such as *Pilgerodendron uviferum* (Don) Florin, has a 1300 km linear north-south distribution much of which includes Chile's southern archipelago. The advantages of sampling over several visits also means that samples have the potential of being much broader in their scope, e.g. some genotypes cone erratically, therefore a single visit may not coincide with good seed production. Furthermore, it is not always possible to plant a large number of genotypes from a single taxon in a 'safe site' in one operation due to the lack of available space. The alternative of storing seed until space does become available is not always an option especially in the case of recalcitrant seeds such as taxa from the *Araucariaceae* Henkel & W. Hochst. and *Podocarpaceae* Endl.

Samples from each mother-tree are kept separate and assigned a unique field collecting number. When field sampling precise location details are recorded by using a Global Positioning System. Each collection is allocated an eight digit accession number at RBGE and once this material has become established, i.e. seed germinated or vegetative material rooted, each genotype is assigned a qualifier in the form of a letter. The assignment of a qualifier will enable each genotype to be monitored from thereon. Monitoring involves annual measurements of growth and observations on biological changes such as coning and transition from juvenile to adult foliage, etc. Traditionally few collection holders undertake phenological recording and yet this is one way in which much can be learnt about cultivated plants. There seems to be a good knowledge concerning the heights of our tallest trees yet little data exists on how these trees performed during their formative years and beyond. The ICCP's network acts as a focus for accumulating such information and it is assembling an invaluable data set on the relative growth patterns and biological changes of individual species in a wide range of climates and soil types.

Integrating *in situ* with *ex situ* Conservation

If *ex situ* conservation is to play any sort of meaningful role then it has to be linked with *in situ* conservation and initiating research which integrates the two has been an important aim of the ICCP. Investigating the genetic diversity of native forests and comparing this with historical introductions to cultivation of the same species has been undertaken. The first conifer to be chosen for this research was *Fitzroya cupressoides*, which is native to a restricted area of southern Chile and a small part of adjacent Argentina. It was first introduced to cultivation in 1849 by William Lobb and since this date it has been cultivated in specialist arboreta in Britain and Ireland. Samples from 23 landowners who were known to grow plants from what was thought to be Lobb's introduction were requested. The molecular research carried out on these samples was part of a broader investigation to discover the genetic diversity of the wild populations which was funded by the Darwin Initiative for the Survival of Species. This research involved collaboration with the University of Edinburgh, and the universities of Chile and Valdivia. The results of the cultivated material clearly showed that all the historical trees were of a single clone, (Alnutt 1998), therefore it is presumed that Lobb's introduction was possibly as a single living plant. This is certainly feasible, since the odds are probably against his visit coinciding with a good seed production year. Resampling *Fitzroya* from across its natural range has now been undertaken by the ICCP. Again following joint research between Edinburgh based research institutes and those in Chile on the genetic diversity of the wild *Fitzroya* populations, (Newton *et al.* in press) the sampling strategy of the ICCP can now be guided by these results. A second area of research has been developed which has a practical application to assist the reintrodution of plants from *ex situ* collections in order to enforce depleted wild populations. The research, between the ICCP and Kim Tripp, the Vice-President for Horticulture, New York Botanical Garden, USA and the Arnold Arboretum of Harvard University involves manipulating cone formation and seed development by using gibberelin. Selected conifer populations have been chosen at the Younger Botanic Garden, Benmore and groups under glass at RBGE Inverleith and the Arnold Arboretum. To date positive results have been observed in several populations including, *Abies densa* Griff., *Calocedrus decurrens* (Torr.) Florin, *Fitzroya cupressoides* and *Podocarpus nakaii* Hayata. The longterm aim is to establish the necessary protocols for all threatened conifer taxa so that these can be applied to assist *in situ* conservation activities.

The great debate concerning the value of *ex situ* conservation will no doubt continue for a long time with many *in situ* conservationists decrying it, among other things, as expensive and a distraction from the efforts of *in situ* conservation. Heywood's statement (1992) that 'If we are to face the challenges of preserving a reasonable sample of plant diversity for future generations, all available methods must need to be employed' is certainly one which the ICCP has taken on board. In Britain and Ireland and further afield we have the ability to grow plants to a high level of expertise; with a slight modification to how we manage these collections, we can help to assist the needs of *in situ* conservation. Merely holding a stamp collection of threatened plant species of known wild origin does not necessarily mean that this is playing a meaningful role in conservation. The cultivated *Fitzroya* example discussed earlier, whereby all historically cultivated plants were of the same clone, is likely to be repeated in numerous other species. The ICCP recognises that if *ex situ* conservation is going to have any substantial impact on the conservation of wild genetic resources then there has to be a greater focused effort; the lead of which in botanic gardens has been taken by the Botanic Gardens Conservation International (BGCI).

Acknowledgments

Generous financial support was given by the Darwin Initiative for the Survival of Species which is administered by the Department of the Environment, for the research on *Fitzroya cupressoides*. I would also like to acknowledge the invaluable help from many colleagues of the Universidad de Chile, Santiago, the Universidad Austral de Chile, Valdivia and the University of Edinburgh for joint research projects in Chile. The early cone induction research is supported by the Stanley Smith Horticultural Trust to whom I am extremely grateful.

References

Allnutt, T.R., Thomas, P., Newton, A.C. & Gardner, M.F. (1998). Genetic variation in *Fitzroya cupressoides* cultivated in the British Isles assessed using RAPDS. *Edin. J. Bot.* 55(3): 329–341.

Falk, D.A. & Holsinger, K.E. (1991). Genetic conservation of rare plants. 283 pp. Oxford University Press.

Farjon, A. (1998). World checklist and bibliography of conifers. 298 pp. Royal Botanic Gardens, Kew.

Heywood, V.H. (1992). Conservation of germplasm of wild plant species. In Sandlund *et al.* (eds). Conservation of Biodiversity for Sustainable Development. pp. 189–203. Scandinavian University Press.

Newton, A.C., Allnutt, T.R., Lara, A., Armesto, J.J., Premoli, A., Vergara, A. & Gardner, M.F. (in press). Genetic variation in *Fitzroya cupressoides* (alerce), a threatened South American conifer. *Conservation Biol.*

Rae, D. (1995). Botanic gardens and their live plant collections: present and future roles. Unpublished PhD Thesis, University of Edinburgh.

Thomas, P. & Tripp, K. (1998). *Ex situ* conservation of conifers: a collaborative model for biodiversity preservation. *Public Garden* 13(3): 5–8.

Walter, K.S. *et al.* (eds). (1995). Catalogue of plants growing at the Royal Botanic Garden Edinburgh 1995. 477 pp. Royal Botanic Garden Edinburgh.

Oakeley, H.F. (1999). The National Collection of *Lycaste* Lindl. and *Anguloa* Ruíz & Pav. — its value to taxonomy. In: S. Andrews, A.C. Leslie and C. Alexander (Editors). Taxonomy of Cultivated Plants: Third International Symposium, pp. 25–28. Royal Botanic Gardens, Kew.

THE NATIONAL COLLECTION OF *LYCASTE* LINDL. AND *ANGULOA* RUÍZ & PAV. — ITS VALUE TO TAXONOMY

HENRY FRANCIS OAKELEY

77 Copers Cope Road, Beckenham, Kent BR3 1NR, UK

Abstract

Lycaste and *Anguloa* are two related genera of orchids from Latin America. This paper outlines how the recording within a National Collection of not only the structure but also the function, pollination, habitats, distribution, commercialism, cultivation, flowering patterns, fragrance, DNA, chromosome counts, evolution, discovery, history, their art and literature, along with breeding and conservation programmes gives the botanist a wealth of data for use in taxonomy, and makes a National Collection more than just a number of plants at the bottom of the garden.

Introduction

This amateur collection was begun in 1956 and became a National Collection under the auspices of the National Council for the Conservation of Plants and Gardens (NCCPG) in 1990. It is grown in a winter-heated and summer-cooled greenhouse at the bottom of the garden, reproducing the humid conditions of their habitats that one encounters in the cloud forest and other zones where they grow, from southern Mexico to Brazil.

There are three main groups of lycastes. The first is the yellow-flowered deciduous, principally meso-American Section *Xanthanthanae* Fowlie as exemplified by *Lycaste aromatica* (Graham ex Hook.) Lindl. The second, moving south, is the geographically wide-spread and closely related Section *Macrophyllae* Fowlie of which the most well known is *L. skinneri* (Bateman ex Lindl.) Lindl., now a rare lithophyte and epiphyte in its native habitat of Coban in Guatemala. The third group are the very distinct purely South American, Section *Fimbriatae* Fowlie exemplified by *L. lanipes* Lindl.

Anguloa Ruíz & Pav. is a closely related genus widely distributed in Colombia, Peru, Ecuador and Venezuela. The species have the same leaf and annually incremented pseudobulb structure as *Lycaste* Lindl. but the flowers are cup-shaped, ranging from the yellow *A. clowesii* Lindl. and the red *A. brevilabris* Rolfe, to the white *A. virginalis* Linden. They grow terrestrially in moist, lightly-shaded woodland.

The National Collection

The Collection now includes all except for 5 of the 68 *Lycaste* species and natural hybrids, including 20 new species, 8 new natural hybrids, 9 out of the possible 11 *Anguloa* species, 51 varieties and 135 of the approximately 230 registered hybrids within and between the two genera. This alone makes it a potentially important taxonomic database. In total about 800 plants are grown with a further 2000 seedlings and plants with a friend near Stoke-on-Trent, and several very large hybrid collections in Germany, Australia and Malaysia that originate from the breeding and conservation programme. Horticulturally it has been used to establish the growing conditions and identify many of the pests, including viruses that affect the leaves. Electron microscopy has separated out those leaf markings which are, and which are not, virus-induced.

Research within the Collection

Habitats of the majority of the species within Latin America have been visited, and the national herbarium collections studied and recorded both in notebooks and photographically. These notebooks form an important part of the Collection and every plant of a species that flowers in the collection, or is seen in the wild, is recorded in them. They comprise approximately 400 pages of field notes and drawings and 300 detailed drawings. Additionally there are some 60 drawings and paintings by the botanical artists, Ann Swan, Gillian Barlow, Cherry-Ann Lavrih and Camilla Speight. The Collection has been used to identify several new species including *Lycaste jarae* D.E. Benn. & Christenson (Bennett & Christenson 1996), *L. maxibractea* D.E. Benn. & Oakeley, *L. diastasia* D.E. Benn. & Oakeley (Bennett & Oakley 1994), *L. fragrans* Oakeley (Oakley 1993), *L. michellii* Oakeley (Oakeley 1991), *L. grandis* Oakeley (Oakeley 1993), *L. nana* Oakeley (Oakeley 1994) and *L. dunstervillei* G. Bergold (Bergold 1996), as well as another dozen taxa awaiting publication. It also helped to identify the 90 year mystery of *L. lata* Rolfe and to put into synonymy many long-established names (Oakeley in press).

An example of this clarification is instanced by *L. trifoliata* F. Lehm. ex Mast., a species described by M. Masters in 1895 (Masters 1895) but lost to science for much of this century until a Peruvian collected plant was seen by the author in a Colombian nursery in 1998. The identity was established from comparison of plants in the Collection along with his records of early paintings, and herbarium sheets.

Two paintings of this plant exist in the Kew Herbarium: one by John Day in 1883, is labelled as *L. cobbiana sine auct.* and the other by F.C. Lehmann, the German Consul to Colombia, in 1906 is labelled *L. trifoliata.* Janet Ross, painted a similar, but different, plant from her husband's collection in Italy as *L. cobbiana sine auct.* in 1885. Further work on these taxa has shown (Oakeley in press) that *L. cobbiana sensu* John Day was published as *L. cobbiana* B.S. Williams in 1885 (Williams 1885), thus putting *L. trifoliata* F. Lehm. ex Mast. into synonymy. While there is insufficient data to clearly identify *L. cobbiana sensu* Janet Ross with a plant currently in commerce, that had been erroneously given the name *L. trifoliata* Hort., the latter plant is now identified as a previously undescribed species, *L. jamesii* Oakeley (in press). All these taxa are present in the Collection and once the identity of *L. cobbiana* B.S. Williams was established from living material, further work has revealed that *L. mesochlaena sensu* Dodson & P.M. Dodson *non* Rchb.f. and *L. mesochlaena sensu* Fowlie *non* Rchb.f. are all now in synonymy (Oakeley in press).

The Value of Photography and an Herbarium

All the plants in the Collection are photographed as well as drawn and show the plant with its bulbs; the flower from the front; a cutaway to show the small lateral lobes of the labellum and how the nectar is trapped by the footplate in the case of *L. cobbiana*, and the tepals at the base of the mentum; the tangled, lacinate margin to the labellum, with the fimbriations in all planes, and the callus on it with three pronounced and two minor longitudinal keels in cross-section and from above. Close-up photography shows the shape of the stigmatic cavity at the tip of the under surface of the rostellum; the hinged operculum exposing the pollinia; and the shape of the viscidium, stipe and pollen masses after removal. A quick comparison of *L. cobbiana* with the much smaller *L. jamesii* shows it to have shorter pseudobulbs, an untangled lacinate margin to the labellum, an almost smooth callus with only a vestigial keel and no lateral lobes to the labellum. These among other differences are sufficient to give it separate specific status. A similar range of photographs (all in triplicate) record the colour and structure of every species and natural hybrid, making identification of new species very easy. Artificial hybrids are also recorded, but usually in less detail, although the knowledge of dominant features in primary hybrids is extremely useful in recognising natural hybrids.

The photographs, cross-indexed with the drawings and the living plants are central to the collection's taxonomic value. They record everything from guttation patterns on the sepals of lycastes to the movement of the pollinia after removal from the rostellum. The photographs record the explanation as to why the viscidium points upwards in *L. aromatica* and its allies — it allows it to be removed by the cloacal orifice of the visiting euglossine bee and avoids being stuck to the head of other insects. The collection now has over 25,000 slides of *Lycaste* and *Anguloa* species and hybrids and their habitats, which are stored in filing cabinets along with five hundred 20 × 30 cm Cibachromes.

An accompanying herbarium specimen, either dissected and mounted or half dissected or unmounted, which are stored in conventional herbarium boxes both alphabetically and by country, has been prepared for nearly every species. There are approximately 200 herbarium sheets and 300 bottles of pickled flowers in Copenhagen mixture all cross-referenced to the working drawings and the plants. Nearly all the species in both genera have been illustrated by Camilla Speight and Ann Swan, and are part of a collection of 70 *Lycaste* and *Anguloa* drawings and paintings as well as 30 plates from dismantled books and numerous photographic records of early illustrations.

Finally, Angela Ryan has been studying the collection for her PhD, and the scent of the majority of the species and many hybrids have been assessed by gas chromatography and their relationship reviewed using DNA cladograms. Cytological and molecular analyses indicate that the atypical, pendulous *L. dyeriana* Sander ex Rolfe belongs to the *Fimbriatae* Group. They also suggest that this southern group is more closely related to *Anguloa* than to the other groups of *Lycaste*. The very different structural features, such as the rostellum and labellum of *Anguloa* indicate that the southern *Fimbriatae* Group of *Lycaste* cannot be merged with *Anguloa* but should be segregated into a separate genus, *Ida ined.*(Oakeley & Ryan in press).

Publishing and Education based on the Collection

Trying to find time to write and publish gets harder and harder but 40 articles and a small book (Oakeley 1993) with 80 colour photographs and brief details of all the *Lycaste* species have been produced in the past decade.

In all of this there is the fun of exhibiting, (e.g. Chelsea exhibits 1990, 1991, 1992) and trying to educate the public (e.g. educational display of *Anguloa*, *Lycaste* and primary *Lycaste* hybrids at the Royal Horticultural Society Show, Westminster, June 1998), the breeding and flasking of up to 30 seed pods per year to raise seedlings of rare species (such as *L. jarae*, only one plant of which was known in 1997) and new hybrids. Primary hybrids were studied to help understand the role of natural hybridisation in the process of speciation. In addition, the histories of the plant hunters like Consul Lehmann and their introductions, and the lives of the botanists like H.G. Reichenbach add interest for this National Collection holder. Lecturing, publications, exhibiting, and letter writing disseminate the information collected.

Reichenbach (1879) wrote that studying lycastes was as much fun as stroking hedgehogs, but I believe he would have found it easier if he had had all the plants, habitat details, illustrations, photographs, herbarium specimens, scent analyses, DNA studies, chromosome counts, and other records, with much of the world literature for the past two centuries in one place. This is why a National Collection is so valuable.

Acknowledgements

The National Collection of *Lycaste* and *Anguloa* is designated a Scientific Collection by the National Council for the Conservation of Plants and Gardens (NCCPG), RHS Gardens, Wisley, Surrey GU23 6QB. The London Group of the NCCPG has generously supported this Collection with funds for photographic materials.

References

Bennett, D.E. & Christianson E.A. (1996). A new *Lycaste* species from Peru. *Orchid Digest* 60(1): 14–17.

Bennett, D.E. & Oakeley, H.F. (1994). *Lycaste* Lindl. In D.E. Bennett, Jr. & E.A. Christenson. New species of Peruvian *Orchidaceae* II. *Brittonia* 46(3): 243–249, figs. 10–12.

Bergold, G. (1966). *Lycaste dunstervillei* Bergold *spec. nov.*, el nacimiento de una neuva especie. *Orquideophilo* 4(1): 37–38.

Masters, M. (1895). Orchid Committee. *Gard. Chron.* 17: 529.

Oakeley, H.F. (in press). *Lycaste* Lindl., *Ida* A. Ryan & Oakeley, *Anguloa* Ruíz & Pav., a bibliographic checklist. *Lindleyana*.

Oakeley, H.F. (1994). Green, brown and hairy lycastes. *Orchid Digest* 58(1): 19–27.

Oakeley, H.F. (1993). *Lycaste* species, the essential guide. 36 pp. Vigo Press Ltd., Penge.

Oakeley, H.F. (1991). A survey of lycastes – 1: *Lycaste aromatica*, its allies and their hybrids. *Amer. Orchid Soc. Bull.* 60(3): 222–231.

Oakeley, H.F. & Ryan, A. (in press). *Ida* A. Ryan & Oakeley, *gen nov. Lindleyana*.

Reichenbach, H.F. (1879). New Garden plants. *Lycaste locusta. Gard. Chron.* 11: 524.

Williams, H. (ed.). (1894). *Lycaste*. In B.S. Williams. The Orchid-growers manual. (Ed. 7. revised & enlarged). 471–479 pp. Victoria and Paradise Nurseries, London.

Davies, B.J. (1999). The National Collections of *Nymphaea* – U.K. and France. In: S. Andrews, A.C. Leslie and C. Alexander (Editors). Taxonomy of Cultivated Plants: Third International Symposium, pp. 29–30. Royal Botanic Gardens, Kew.

THE NATIONAL COLLECTIONS OF *NYMPHAEA* – UK AND FRANCE

BARBARA J. DAVIES

PO Box 62438, 8046 Paphos, Cyprus and Etablissements Botaniques Latour-Marliac S.A., 'Bateau',rue des Nenuphars, 47110 Le Temple-sur-Lot, France

Abstract

The National Collections of *Nymphaea* in the United Kingdom and France are discussed. Some historical background is given along with recent work carried out by the current curators of these collections.

Certified by the National Council for the Conservation of Plants and Gardens (NCCPG) in the U.K. since 1989 and the Conservatoire Français des Collections Vegetales Specialisées (CCVS) in France since 1995, these National Collections are curated by Stapeley Water Gardens and Ets. Botaniques Latour-Marliac, respectively. The Collections each contain over 350 different *Nymphaea* L. species and cultivars, both hardy and tropical.

Three sites currently house the collections. Stapeley Water Gardens displays hardy *Nymphaea* in their display garden pools, tropicals in The Palms Tropical Oasis complex at Stapeley, and makes them available to purchase in their retail area. Since 1976, Stapeley also curates a display of 100 hardy *Nymphaea*, at the charitable trust Burnby Hall Gardens, near York. The NCCPG's work and Stapeley's *Nymphaea* collections are made further accessible to the general public at Royal Horticultural Society exhibitions, most recently at the 1998 RHS Chelsea Flower Show and RHS Hampton Court Flower Show in 1997. They also mounted a joint exhibition at Les Journées des Plantes de Courson in May 1997.

Archives dating from the late 18th century are maintained by Ray and Barbara Davies, the Collection curators. Serving as back-up reference for the Collections, the archives contain early and contemporary botanical illustrations, monographs and other books referring to the genus *Nymphaea* in German, Italian, French and English, early and contemporary aquatic plant nursery catalogues from various countries, magazines, periodicals, newspapers and personal correspondence. Of particular interest are the extensive correspondence copies of Joseph Bory Latour-Marliac. Some correspondence has great popular appeal because it documents the supply of *Nymphaea* to the Impressionist painter, Claude Monet.

In 1875, on the site of his bamboo nursery Latour-Marliac founded the first aquatic plant nursery, situated mid-way between Bordeaux and Toulouse. There, he gathered all the known hardy and tropical species of *Nymphaea*, creating and introducing the first hardy *Nymphaea* hybrid, *N.* 'Marliacea Chromatella' (1877), a cross between *N. alba* L. and *N. mexicana* Zucc. Additionally, Latour Marliac was able to harness the mutable red gene in *N. alba* var. *rubra* Lönnr. to hybridise the first consistently red-blooming hardy

Nymphaea, beginning with *N.* 'Marliacea Rubra Punctata' in 1889. Latour-Marliac's nursery and archives were acquired from his descendants by Stapeley Water Gardens in 1991. That facility and the archives were an invaluable aid in continuing the research begun by Ray Davies in 1989 on hardy *Nymphaea* nomenclature. Davies' research culminated in his 1993 publication on the identification of hardy *Nymphaea*.

All the archival references were translated where necessary, then logged on disk by genus or species, in date order. In this way it was possible to review the many hundreds of entries to trace when descriptions went 'adrift' through inaccurate translation, unscrupulous trade practices or journalism which did not go back to original sources for information. Two botanists were hired and given office space, alongside two botanical illustrators, who jointly gathered morphological details and accurately documented 180 *Nymphaea* species and cultivars. The resulting monograph serves not just as a reference tool, but also as the registration documentation for most of the cultivars it describes.

Representative of the taxonomic confusion existing before this research was begun, was the interchanging in commerce of *N.* 'Mrs Richmond', introduced by Latour-Marliac in 1910, and *N.* 'Fabiola', introduced by his nursery in 1913. Their identities have been clarified, but they were also confused with *N.* 'Attraction', also introduced by Latour-Marliac in 1910, and distinct to an observant grower.

Follow-on work was undertaken by both Davies in June 1996 in Sweden, Finland and Northeastern America, with research on *N. alba* var. *rubra*, *N. candida* J. Presl. var. *rubra sensu auct.*, *N. tetragona* Georgi var. *rubra sensu auct.*, *N. leiberghii* Morong and variants of *N. odorata* Ait. These field studies, undertaken in co-operation with the Environment Office of Uppsala, Sweden, the Natural History Museum of Kuopio, Finland and with Dr. C. Barre Hellquist, North Adams State College, Massachusetts, produced additional water-colour botanical illustrations, chromosome DNA analysis and pollen photomicroscopy of the Swedish and Finnish *Nymphaea*, morphological data recordings and, in some cases, permit-agreed acquisition of rhizomes and seed for cultivation.

Reports on these field studies are to be published in a forthcoming issue of the *Water Garden Journal*, the quarterly publication of the International Water Lily and Water Gardening Society (IWGS) which is the registration authority for the genera *Nymphaea* and *Nelumbo* Adans.

Preliminary work has now been begun by both the Davies with Professor Mark Chase of the Royal Botanic Gardens, Kew, to do DNA analysis of these and other hardy species of *Nymphaea*, particularly *N. tetragona*, *N. leiberghii* and variants of *N. odorata*. It is hoped that the additional information to be discovered will aid in the conservation efforts desperately needed to preserve the last few remaining plants of *N. alba* var. *rubra*, *N. candida* var. *rubra* and *N. tetragona* var. *rubra* in Finland and Sweden.

References

Davies, R. (1993). International Water Lily Society — Identification of Hardy *Nymphaea*. 380 pp. Stapeley Water Gardens Ltd. for the IWGS.

Twibell, J.D. (1999). Chemotaxonomy of plants by vapour profiling. In: S. Andrews, A.C. Leslie and C. Alexander (Editors). Taxonomy of Cultivated Plants: Third International Symposium, pp. 31–34. Royal Botanic Gardens, Kew.

CHEMOTAXONOMY OF PLANTS BY VAPOUR PROFILING

J.D. TWIBELL

31 Smith Street, Elsworth, Cambridge, CB3 8HY, UK

Abstract

Scented plants produce volatile oils which consist of numerous individual organic components. The particular blend of components gives rise to the characteristic odour of the species or variant. Amongst many species within the genus *Artemisia* there are a number of regional variants which can be readily segregated by odour differences. Clearly odour sensation is highly subjective, varying between individuals and over time periods, and thus forms no basis for scientific records. Analysis of the vapours from such plants does, however, enable the subtle differences to be exploited and recorded and the characteristic vapour profiles form a basis for chemotaxonomy.

A representative sample of the plant material is placed in a nylon bag and heated for several minutes to vaporise the volatiles which are then collected by drawing a sample through a tube containing an absorbent packing. The tubes are then analysed on an automated thermal desorber which introduces the volatiles onto a gas chromatography column. Individual vapour components emerging from the column can be analysed in a mass spectrometric detector which can give further information, allowing the individual components to be identified.

Whilst some species can be separated and identified into sub-groups or regional variants by this technique, other species appear more stable and show little variation across their geographical distribution. Although the overall vapour profiles for a given variant remain similar throughout the growing season, changes do appear to occur in the relative ratios of some components. To avoid this and other potential false discrimination problems, plants should ideally be grown in the same location and should be sampled at the same time of year.

The technique will be illustrated by representative chromatograms or vapour profiles from a number of variants within the *Artemisia arborescens* complex (Twibell 1991).

Some individual components of plant vapours are now known to influence the growth or predator response of other plants even of different genera and the headspace ATD-MS method may be useful as a means of screening plants for potential phytoactive volatiles. The method of sampling is discussed.

General Introduction

The National Council for the Conservation of Plants and Gardens (NCCPG) was set up almost 20 years ago to safeguard cultivars of garden plants, many of which had hitherto been lost when they fell from gardening fashion. The National Collection scheme aims to conserve cultivated plants for posterity by maintaining living material of all available cultivars. Many collection holders are also interested in conserving wild species, several of which are under threat in their native habitat. The recent CITES (Convention on International Trade in Endangered Species of Wild Fauna and Flora) legislation may however be counter-productive to conservation, as it is now being used by botanic gardens in some countries as a reason not to supply material. If a plant is under threat, it can make no sense to restrict its availability, particularly to collections which have a direct interest in safeguarding and propagating the material.

It is vital to save as much genetic diversity as possible as many plant species have known medicinal uses and others may be sources of as yet undiscovered drug compounds. Within the genus *Artemisia* L. the most obvious example is that of *A. annua* L. which has been used for centuries by Chinese herbalists to treat malaria and fevers. Recent research on this plant has isolated chemical compounds which are indeed active against the malaria parasite and new drugs are now in use. There is now a strong impetus in synthesising similar compounds with the same active grouping to try to make even more effective drugs.

Whilst a major aim of the collection is to supply plant material for medicinal research this does create its own problems. Drug companies are understandably very secretive about their research and are reluctant to allow access to potential competitors. Most are keen to obtain material from the National Collection but are not willing to donate specimens of strains or cultivars, which have been found to be particularly rich in the active compounds, back to the Collection, as they realise that it may be passed on to a competitor. Collections need to adapt within commercial realities and constraints.

Research within a Collection

One of the advantages of having a large collection of plant material at a single site is that close comparisons can be made between various species, subspecies or cultivars at the same stages of development and exposed to similar soil and climatic conditions. Field botanists are usually not able to see or gather specimens of plants at different stages of their life cycle or growing under comparable conditions. Clearly much research can be done to discriminate or identify plants in collections from their habit, leaf and flower forms throughout the yearly cycle. In this case, the potential of vapour profiling as a means of producing a characteristic fingerprint of the vapours produced by different artemisias in the collection was explored.

Artemisias are often very variable in leaf form and habit and yet many species are very similar with the described differences being at what could be considered to be a minor level. It is often difficult therefore to exactly identify newly collected material. Vapour profiling adds a new parameter for potentially equating or discriminating between similar material.

Vapour Profiling

Over a long period the author has been interested in the wide range of scents present within the genus *Artemisia*. Most species and many of the variants within a species have characteristic odours. Such odours are subjective and difficult to describe and so it is necessary to devise a scientific method of recording and comparing the data. Several years ago, as a forensic scientist with a background in devising methods for detecting low levels of fire accelerants such as petrol, paraffin, etc. in fire debris samples, he decided to see if he could analyse artemisias in the same way. The analytical method was restricted in that he had to use the same method as was used in accelerants analysis forensic casework and hence he could not optimise the method for analysing plant material. Basically the plant material was placed in a nylon bag and heated to vaporise the volatiles and the headspace was then drawn through a tube packed with Tenax adsorbent to collect and concentrate the vapours. The tubes were then desorbed on the Perkin Elmer ATD50, following which the vapours were separated by gas chromatography followed by mass spectrometric detection (Barberio & Twibell 1991). Other workers are not limited by such constraints and have developed more direct analytical systems such as placing small pieces of plant material directly within empty ATD tubes which cuts out the first stage of sample collection. Although the extra stage introduced by the nylon bag method may have some limitations, it does allow a larger sample, such as a fresh shoot, to be examined. This may give a more representative sample for the plant as it eliminates any differences caused by sampling leaves at different stages of their development, when they may show differences in the volatiles produced.

The initial work was devoted to the *A. arborescens* complex and attempts to find the origin of the garden cultivar 'Powis Castle' which has been suggested to be a hybrid between *A. arborescens* L. and *A. absinthium* L. The former occurs in numerous variants throughout the Mediterranean and whilst some native populations may remain on high ground, the plant has been widely redistributed by man in coastal areas. To the west, the arborescent form appears to have evolved into the island endemics *A. thuscula* Cav. of the Canary Islands, *A. argentea* L'Hér. of Madeira and *A. gorgonium* Webb of the Cape Verde Islands. To the east, plants show characters tending towards *A. absinthium*, and if further material is collected it may ultimately be shown that the two intergrade. The recent history of the cultivar 'Powis Castle' was that it was first found in a garden in Yorkshire from whence it was taken to Powis Castle in Wales. It is more hardy and long-lived in England than typical *A. arborescens* forms and is extremely unlikely to be a hybrid occurring in Yorkshire. Thus, it could be a plant which had been wild collected abroad and brought back to England at some time but for which the field notes are now lost.

Chromatograms of several *A. arborescens* taxa and other species were shown in the talk. Thujones are the major peaks in various *A. absinthium* and 'Powis Castle' but there are other differences between these. Cultivars 'Faith Raven' (Twibell 1994) and 'Brass Band' are almost identical to 'Powis Castle' in their habit, appearance and vapour profiling and thus are either the same plant or are from similar wild collections. There appear to be several chemotypes of *A. arborescens*, only some of which show prominent thujone peaks.

This work has been extended to other species and unknown taxa within *Artemisia* and is proving useful in linking unidentified plants. Interest has reawakened in plant volatiles following recent reports that plants can communicate with each other or even from genus to genus by releasing volatiles such as methyl jasmonate (Farmer & Ryan 1990, Staswick 1992).

Acknowledgements

I wish to thank NCCPG for a grant towards buying analytical time, and to the Forensic Science Service and colleagues for their time and for use of ATD equipment. I also thank Dr Alan Braithwaite, Patricia Gregory and others at Trent University for their more recent analyses currently under consideration.

References

Barberio, J. & Twibell, J.D. (1991). Chemotaxonomy of plant species using headspace sampling thermal desorption and capillary GC. *J. High Resolution Chromatog.* 14: 637–639.

Farmer, E.E. & Ryan, C.A. (1990). Interplant communication: Airborne methyl jasmonate induces synthesis of proteinase inhibitors in plant leaves. *Proc. Natl. Acad. Sci. U.S.A.* 87: 7713–7716.

Staswick, P.E. (1992). Jasmonate, genes and fragrant signals. *Pl. Physiol.* 99: 804–807.

Twibell, J.D. (1992). Plant identification from vapour analysis. *Plantsman* 14(3): 184–190.

Twibell, J.D. (1994). *Artemisia* 'Faith Raven'. *Plantsman* 15(4): 255–258.

Lewis, P.E. (1999). The establishment of a National Collection of *Campanula*. In: S. Andrews, A.C. Leslie and C. Alexander (Editors). Taxonomy of Cultivated Plants: Third International Symposium, pp. 35–38. Royal Botanic Gardens, Kew.

THE ESTABLISHMENT OF A NATIONAL COLLECTION OF *CAMPANULA*

PETER E. LEWIS

Padlock Croft, Padlock Road, West Wratting, Cambridge CB1 5LS, UK

Abstract

A summary of the genus *Campanula*, its distribution and uses of cultivated species and the origins of its cultivars is given. The background of the joint holders, the establishment and mode of operation of this privately-held collection, the virtues, disadvantages and safety measures are detailed. The scope of the *Campanula* collection, and the approach to its cultivation in the garden has proved culturally and visually acceptable. Sources of information for identification and of acquisition of living material are given. The hybridisation of closely related species in cultivation, and the colours of *Campanula* cultivars present identification problems. A computer database showing accessions and synonyms, as well as a documentary and slide library, and a herbarium and seed bank have been assembled. Material and information have been made available to inquirers, a nursery has been established, a display garden opened to visitors and an authoritative book published. Finally, the value and uses of a National Collection are discussed.

Introduction

Campanula L., the type genus and the best known of all the cultivated genera in the family *Campanulaceae* Juss., is comprised of some four hundred species of herbaceous annuals, biennials and perennials native to the Northern Hemisphere. An arc stretching from the European Alps through the Balkans to Greece, Turkey, the Caucasus and Russia is particularly rich in taxa. Within the genus the differences in habitat and morphology is extremely variable. Height, for instance, can vary between 2 cm and 2 m. *Campanula* appears to have an alpine affinity, and many species are well adapted to growth in poorer soils with rapid drainage. The genus has long been grown for other purposes; some have been used for food, including the root of *C. rapunculus* L., the Brothers Grimms' Rapunzel, and folklore offers *C. rapunculoides* L. as a cure for rabies, but experimentation is not advised! The secretion of latex is a generic character and trials have been carried out to grow crops for rubber production, but this is nowhere near an economic proposition. Thus there remains the decorative factor. As campanulas or bellflowers have so long been known and grown in gardens, opportunities for variation have been many, hence the proliferation of cultivars arising from genetic aberrations and mutations which result in colour variants and/or flower-doubling.

The National Collections of *Campanula*

The joint holders of this NCCPG (National Council for the Conservation of Plants and Gardens) National Collection, do not have the advantages, or disadvantages, of formal horticultural training or qualifications. Medically-related backgrounds, and practical experience of agriculture and horticulture in Britain, Europe and New Zealand gave something of a scientific and systematic but unprejudiced outlook. The collection is grown in a 0.5 hectare garden near Cambridge. Being keen gardeners, some 50 taxa of *Campanula* had been amassed when the newly-formed NCCPG were approached. National Collection status was approved in 1982 and was one of the first in private hands. Since then the holders have cultivated nearly 300 taxa. One advantage of private ownership, as opposed to collections in public or corporate hands, is continuity, the collection is not subject to promotion or change of staff or management. At the same time, they have been able to promote the establishment of two other approved NCCPG National Collections of *Campanula*, with whom they maintain very good relations. These are in Driffield, E. Yorkshire, and Bucknell, Shropshire, near the Welsh border, giving a range of geographic, climatic and edaphic conditions.

All species, subspecies, forms and cultivars are considered worthy of attention, provided they are of current or potential horticultural interest; in general little notice has been given to annuals, few of which are garden worthy. However, numerous biennials and monocarpics are grown. The starting point for each species has been to create in the garden a habitat as near as possible to that found in the wild, subsequently making adaptations as experience dictates. On the whole this method has worked well and also provided satisfying visual impact in the garden.

By the terms of reference to the parent organisation, the NCCPG, the concern is garden plants. This will, of course, include all the taxa grown in the garden, some of which are of threatened status both in cultivation and in the wild. *The Plant Finder* has been used as a general guide since its inception in 1987, but there are many more plants in cultivation than are offered commercially in this or any other publication. Much information has been acquired from other growers, especially the more serious plantsman. Membership of many British, European and American societies, and the subscribing to a range of periodicals and other publications has also been extremely useful.

Verifying the Collection

One of the primary sources of information for the identification and distribution of species, though now somewhat outdated, has been Crook's 1951 monograph. Other major sources have been Fedorov & Kovanda (1976), Fedorov (1972), Greuter *et al.* (1994), Damboldt & Phitos (1978), and Rechinger & Schimann-Czeika (1965). Individual floras from Europe, Eastern Mediterranean, North Africa, Japan and North America have combined to give the bulk of the database, together with their descriptions supporting the identification of individual species. Full access to the Cory Library at the University Botanic Garden Cambridge has provided an immense amount of information.

Cultivars present greater problems. Many of the older publications, the serious as well as the superficial, have extolled the virtues of their plants without giving many clues as to identity. To be informed in flowing and even poetic language that a plant is about eighteen inches tall with hanging bells of a beautiful mid-blue does not in fact give much useful information. Thousands of plants have been observed in an attempt to identify with certainty some of the more popular cultivars, but several names must remain at best doubtful. This predicament would less easily occur if the RHS Colour Chart (Anon. 1986) was used, and a more systematic approach to cultivar description employed.

Cultivation Notes

Cultivation in the relatively confined conditions of the garden present open-pollination problems, but these are not as common as has been supposed by some writers and the products of such hybridisation are usually easily identified. The case of closely related species, such as the Greek, Aegean and some Mediterranean monocarpics, or the *C. tridentata* Schreb. aggregates from Turkey, Armenia and the Caucasus is more difficult. The ease with which these taxa cross may well have so diluted their gene pool that the individual species are no longer to be found in cultivation, and the names thus given are of little value. This is carried to extremes in the case of the aggregate of the circumpolar harebell, *C. rotundifolia* L. Here, the names in commerce are virtually valueless, as all hybridise haphazardly in cultivation. It is worth remembering that this is due to the lack of discrimination of pollinating bees rather than the promiscuity of *Campanula*.

Living material is acquired from every possible source, both amateur and professional. Species are obtained from botanic garden seed lists and from seed-hunting expeditions, and both species and cultivars from commercial sources and amateur exchange. Full accession and cultivation records have been kept on computer.

The Value and Uses of a National Collection

A useful function of a recognised National Collection is as a bridge between the amateur or professional grower and the taxonomist. The conscientious systematic collector needs to understand both points of view and be in the position to mediate, e.g. to explain unpopular name changes. These are many mis-identifications, mis-namings, misunderstandings, and just plain mistakes, which come down to a question of taxonomy. Collection holders are consulted far and wide by inquirers, both plantsmen and botanists, but less often from the commercial sector. As an example, the alpine *C. cochlearifolia* Lam. has produced many cultivars, among them 'Tubby', a neat and very aptly named epithet. However, when some commercial interest produces a *C. cochlearifolia* 'Tubby Alba' (*sic*) and has already printed thousands of labels ready for distribution, unfortunately the protests of a National Collection holder do not carry far.

The Collection holders have accumulated a documentary library relevant to the genus in the wild and the garden. There are some 800 slides which are frequently used in lectures on the genus and its cultivation. There is a database for *Campanula* and other *Campanulaceae* which is often consulted to solve problems of synonymy. The beginnings of an herbarium have been formed, though this solves few if any of the colour queries and problems. A seed bank is maintained with seeds refrigerated under silica gel, from which one can expect long-term viability. Material from this, as well as other live material has been made available to gardeners and others by the establishment of a small nursery which, together with the display garden, is open to the public. Visitors have been received from over 25 different countries. Contributions to

37

several of the more serious amateur publications have been made (Lewis 1994, 1995, 1996a, 1996b, 1998) and a book has been published (Lewis & Lynch 1989) which went to a second edition in 1998.

Conclusion

A recognised, well organised and operated National Collection has a number of invaluable services to offer all sectors. It acts as a germplasm bank, with the potential conservation value that this implies; as a data bank, and as an information source. It is listed, accessible to botanists, gardeners, commercial organisations such as nurseries and seed companies, and to conservation bodies, including, for example, the National Trust. Contact has been made with each of these, the information and material going both ways.

References

Anon. (1986). R.H.S. Colour Chart in association with the Flower Council of Holland. The Royal Horticultural Society, London and the Flower Council, Leiden.

Crook, H. Clifford. (1951). Campanulas, their cultivation and classification. 256 pp. Country Life Ltd., London.

Damboldt, J. & Phitos, D. (1978). *Campanula*. In P.H. Davis (ed.). Flora of Turkey and the East Aegean Islands. Vol. 6. pp. 2–64. Edinburgh University Press, Edinburgh.

Fedorov, A.A. (1972). *Campanula*. In B.K. Shishkin (Chief ed.). Flora of the U.S.S.R. Vol. 24. pp. 97–238. Israel Program for Scientific Translations, Jerusalem.

Fedorov, A.A. & Kovanda, M. (1976). *Campanula*. In Tutin, T.G. *et al.*, (eds). Flora Europaea, Vol. 4. pp. 74–93, Cambridge University Press, Cambridge.

Greuter, W., Burdet, H.M. & Long, G. (eds). (1984). *Campanula* in Med – Checklist, Vol. 1. pp. 123–145. Conservatoire et Jardin botaniques, Ville de Genève.

Lewis, P.E. (1994). Cottage garden bellflowers. *The Cottage Gardener* 48: 11–14.

Lewis, P.E. (1995). National Collection of *Campanula*. The National Plant Collections Directory 1995: 14–16.

Lewis, P.E. (1996a). Cottage garden campanulas, further tintinnabulations. *The Cottage Gardener* 56: 7–10.

Lewis, P.E. (1996b). Plant impropagation. *The Garden* 121(3): 160–161.

Lewis, P.E. (1998). Belles of the border. *The Garden* 123(6): 436–439.

Lewis, P.E. & Lynch, M. (1989). Campanulas. 149 pp. Christopher Helm, London & Timber Press, Oregon in association with the Hardy Plant Society.

Lewis, P.E. & Lynch, M. (1998). Campanulas: a gardener's guide. (Ed. 2). 176 pp. Batsford Ltd.

Rechinger, K.H. & Schimann-Czeika, H. (1965). *Campanula*. In K.H. Rechinger (ed.). Flora Iranica 13: 7–38.

NOMENCLATURE IN THE
ORNAMENTAL SEED-TRADE

2

Hender, A.B. (1999). An introduction to the flower seed industry. In: S. Andrews, A.C. Leslie and C. Alexander (Editors). Taxonomy of Cultivated Plants: Third International Symposium, pp. 41–44. Royal Botanic Gardens, Kew.

AN INTRODUCTION TO THE FLOWER SEED INDUSTRY

A.B. Hender

Floranova Ltd., Norwich Road, Foxley, Dereham, Norfolk NR20 4SS, UK

Abstract

The flower seed industry is truly international, operating in most countries of the world. The chain starts with the breeders who create new cultivars. This process can take up to 10 years from the first concept to actual launch in the market place. The cultivars may be F1 hybrids or open-pollinated, depending on the type of plant concerned. Once the breeders have completed their work the cultivar undergoes productivity trials and comparison trials to determine if it is an improvement over existing cultivars. The seed multiplication stage is then undertaken; this may be done in various countries in the world depending upon the conditions best suited to the cultivars involved. Once sufficient seed has been produced, the breeder/producer sells the seed to the distribution companies for worldwide sale to professional growers or home gardeners.

Main Markets

The main markets for flower seeds are the USA, Europe and Japan but sales are made in most countries of the world. The total value of the market for flower seeds is estimated at US$250–350 million at the wholesale level. About 80% of this is for bedding plants, 15% pot plants and 5% perennials and cut flowers.

Major Flower Seed Taxa

The major flower seed genera, by value, are *Viola* L. and *Impatiens* L., (see list below). Like most of the other genera their cultivars are all derived from F1 hybrids. The cultivars of a few important species, such as *Salvia splendens* Sellow ex Roem. & Schult., are derived from open pollination and although F1 hybrids of these could be produced, it would not be viable to do so.

VIOLA	–	*Viola* × *wittrockiana* Gams
IMPATIENS	–	*Impatiens walleriana* Hook. f.
PRIMULA	–	*Primula vulgaris* Huds.
PELARGONIUM	–	*Pelargonium* × *hortorum* L.H. Bailey
BEGONIA	–	*Begonia semperflorens* Hort.
PETUNIA	–	*Petunia* × *hybrida* Hort. Vilm.-Andr.
SALVIA	–	*Salvia splendens* Sellow ex Roem. & Schult.
MARIGOLD	–	*Tagetes patula* L./ *Tagetes erecta* L.
CYCLAMEN	–	*Cyclamen persicum* Mill.

Major Flower Seed Breeders

There are a number of breeders worldwide, the major ones are listed below, each specialising in a number of genera:

Goldsmith (USA)	*Petunia* Juss., *Impatiens* L., *Viola* L. (pansy), *Pelargonium* L'Hér. ex Aiton
Sakata (Japan)	*Viola* L. (pansy), *Primula* L.
Pan American Seeds (USA)	*Impatiens* L., *Petunia* L.
Daehnfeldt (Denmark)	*Begonia* L.
Benary (Germany)	*Begonia* L. and herbaceous perennials
Floranova (UK)	*Salvia* L., *Pelargonium* L'Hér. ex Aiton, *Petunia* Juss.
Bodger (USA)	*Tagetes* L., *Impatiens* L.
Kieft (The Netherlands)	Herbaceous perennials and cut flower plants

Production

Production is carried out worldwide (see Fig. 1). The range of plants involved means that the ideal climate to achieve the quality of seed required will vary from one genus to another. The economics of the production process will also be a factor in determining the location used.

Types of Production

- **F1 Hybrids** – This involves the crossing of two pure inbred lines. Most mother plants are hand emasculated by a highly skilled work force under protected cropping conditions. Some use is made of male-sterility to eliminate the need for emasculation in certain cases, e.g. *Pelargonium* × *hortorum*. Each inbred line has to be maintained true to type.
- **Open Pollinated** – In these cases the crops are mostly grown in the open field and pollinated by bees. The production fields need to be isolated by about 1000 metres from the same taxon to avoid cross-pollination. The seed may be harvested by machine or hand.

For both types of production the seed undergoes a cleaning process and germination tests to make sure it meets a set standard.

Flower Seed Trials

There are also a number of trialing organisations worldwide, e.g. Fleuroselect in Europe, and All American Selections, in the USA, who assess new cultivars against control cultivars and make independent judgements as to whether the novelties are an improvement. In the United Kingdom the Royal Horticultural Society trials at Wisley assess a number of cultivars each year to evaluate garden performance. All these trials provide a valuable service and are open to the public to view. In addition to these trials there are also official seed testing trials in many countries worldwide.

Many of the other companies involved in the flower seed industry also have their own trials. These are primarily for checking the quality of the lots grown or purchased and in some cases they may also be used for evaluation or publicity purposes.

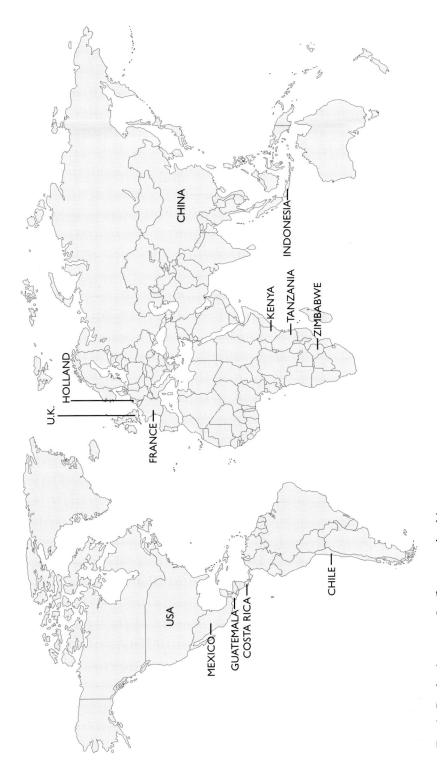

FIG. 1. Production areas for flower seed cultivars

Sales and Marketing

This is carried out at various levels in the market place. The flower seed breeders, wholesalers and producers are the start of the chain and they in turn sell the seed, or in some cases young plants, to professional growers or the home gardeners.

Nomenclature and its Importance

New cultivars take many years to develop and a wide range of taxa are involved. As there is a long chain from breeder to consumer it is vital that at all levels there is a clear and consistent point of reference, ideally the cultivar name. This name has to be acceptable to all parties involved — the seedsmen, the trialing authorities, the phytosanitary authorities and both the marketing sector and the consumer.

Chisholm, D. (1999). Breeding and maintenance of seed-raised decorative cultivars with observations on commercial naming practice. In: S. Andrews, A.C. Leslie and C. Alexander (Editors). Taxonomy of Cultivated Plants: Third International Symposium, pp. 45–48. Royal Botanic Gardens, Kew.

BREEDING AND MAINTENANCE OF SEED-RAISED DECORATIVE CULTIVARS WITH OBSERVATIONS ON COMMERCIAL NAMING PRACTICE

DAVID CHISHOLM

Floranova Ltd., Norwich Road, Foxley, Dereham, Norfolk NR20 4SS, UK

Abstract

Seed-raised decorative cultivars are produced as F1 hybrids, F2 hybrids or open-pollinated populations. Most development work today is carried out on F1 hybrids in a restricted range of species. Much effort is put into developing a wide range of colours well-matched for flowering time and plant habit, in order to allow their being grown as mixtures. Competition is intense.

Cultivars are sold as part of a matched series and are named using a series denomination combined with a colour description. Cultivars may be sold as single colours or a part of a mixture. A high proportion of sales are as formula mixtures with several cultivars blended together, in contrast to the normal agricultural practice of growing a monocrop.

Parental stock may be maintained either vegetatively or by seed. Every effort is made to maintain high uniformity and quality in the parental stock.

Introduction

The breeding of seed-raised decoratives is a worldwide business. Most commercial breeders are based in Europe, USA and Japan. Seed production likewise is global albeit mostly in the tropics.

The ornamental seed industry has developed many unique features, related to the lack of legal plant variety protection (until recently), the requirement for colour mixtures to be grown, and the breeding systems of particular species. This account gives a brief overview of the main features.

The Concept of Cultivar

The accepted meaning of a cultivar is a genetically and phenotypically uniform entity which can pass the DUS criteria of Distinctness, Uniformity and Stability. These cultivars — at least in the agricultural and vegetable fields — are normally grown as monocultures. Indeed, if cultivars are mixed together they are generally prohibited from being sold at all.

This concept can also be readily applied to vegetatively propagated material. It tends to break down, however, for seed-raised decoratives. It is true that many of the products on sale would pass the DUS test. Others, such as F2 hybrids, by definition would not. Others still — the majority of open-pollinated cultivars — are genetically variable and therefore marginal. In many cases, primula being a good example, the breeding system does not allow uniformity without a commercially unacceptable loss of vigour.

Further, it is normal commercial practice to not only grow together mixtures of cultivars of the same species, but to grow blends of species together also. This practice has led to some unique features within the bedding plant industry.

The Concept of Series

The concept of cultivar series has matured over the past 20–30 years. A cultivar series is a grouping of individual cultivars which are matched for all aspects of performance except colour. They can thus be grown together as mixtures and give a uniform performance, both in the pack for garden centre sales and later when planted out. Normal naming practice is to have a series name (which has to be international) and a colour description — e.g. 'Horizon Deep Scarlet', 'Horizon Salmon'. An "international" name is one that can be used in any country without giving offence.

Ideally, cultivars within a series should be isogenic except for colour. In theory this can be achieved by backcrossing, but in practice this ideal is rarely reached, with colours within a series being bred in parallel for similar performance. Usually when a series is first released the "match" is not perfect and breeding work to improve it is ongoing, with continual incremental improvements. Over a 5 year period the series and colour names may remain the same, but the cultivars underneath them will have changed. The term employed is "superisation" or "silent replacement".

A great deal of marketing capital is placed in developing an awareness of the series within the market place, although this awareness tends to diminish the nearer the final customer. A series name is a "promise" that under most growing conditions a cultivar mixture can be grown which will give near identical performance between colour components. There are many examples of this, from all commercial breeders and in all major species.

Importance of Series in the Marketplace

It is difficult to generalise, but a significant proportion of seed is sold as formula mixtures: cultivars produced as separate colours and then blended to a given proportion of colours. In the United Kingdom, this is more than 50% of the market. Different countries have different colour preferences — the USA prefers bold colours, Japan pastels, for example. Where colours are sold separately, they are often used in designs with other colours from the same series. A relatively small proportion of seed is sold for use as an isolated planting. If a series is not well matched, it is simply not commercially viable.

Types of Mixtures

Mixtures may be grown or a formula. A formula mix is a blend of separately produced single colours which are combined after harvesting to give a standard and repeatable mixture. A fixed proportion for each component defines the formula used. Normally F1 hybrid and hand-produced seed are used.

A grown mix results from harvesting seed from plants of many colours grown together and is usually open-pollinated. The proportion of colours in the product is known approximately, but is not exactly repeatable between productions.

Breeding Techniques and Objectives

Cultivars may be F1 hybrids, F2 hybrids or open-pollinated.

F1 hybrids: most breeding work is focussed on this type of product. The benefits of F1s for uniformity and vigour are well established. Not all species are capable of generating F1s economically. In some cases (geranium, marigold) male-sterility exists and facilitates the production of seed. However in most cases hand emasculation and hand pollination are required. Most production is under cover with insects excluded.

F2 hybrids: there is a market demand for more economical seed from field production, usually in produced mixes. An F1 hybrid is grown on a large scale for open-pollination. Properties of the resulting harvest can be predicted with reasonable precision.

Open-pollinated: until the mid-1960s all flower seeds were of this type. A uniform stock is selected, bulked up to a large scale, and seed harvested from insect pollination. Considerable effort is expended to make the nuclear stock as uniform as possible.

Species which allow self-pollination are more easy to work than ones that are self-incompatible. In some cases, such as salvia, the morphology of the flower makes F1 hybrids impractical, hence all cultivars on the market are open-pollinated. Where possible, however, F1 hybrids are targeted.

Breeding Systems

A high degree of uniformity — and hence predictability — is essential to commercial growers. Some species are able to comply with this requirement, others are not. This depends critically upon the breeding system employed by the plant. Self-incompatible plants are slow to inbreed (since inbreeding can be only via sib-pollination) and tend to lose vigour quickly. The prime example of this is primrose, where the degree of uniformity is only now becoming acceptable after many years of effort. Self-compatible species are much more amenable, and suffer far less from inbreeding depression.

Parental Maintenance

Parental stock may be maintained vegetatively or as seed. Vegetative maintenance of parents may be either through cuttings or micropropagation. Seed maintenance of parents requires high quality nuclear stocks, which must be reselected frequently for uniformity and quality. Seed yield and germination vigour are critically important. F1 hybrids require that two parents be maintained, and crossed together to give rise to the product. Seed production may be carried out anywhere in the world that has a suitable climate.

Seed Production

Seed production is carried out throughout the world. Increasingly the focus for production of many plants is based in the tropics to allow all year round production. Some plants, where either daylength requirements or cool temperature requirements are the over-riding consideration, are produced only in northern Europe. Examples of plants produced in the tropics are impatiens and petunia. Examples of plants produced only in higher latitudes are primroses and wallflowers.

F1 hybrid cultivars are produced almost exclusively by hand pollination under insect-proofed greenhouses. Some male-sterility systems do exist but more typically hand emasculation is performed. Seed is collected by hand when ripe.

Open-pollinated and F2 hybrid cultivars are grown either in field conditions or under protection, but insect pollination is used. Seed may be collected by hand or by a small combine harvester.

Seed Technologies

After harvest, seed usually requires processing in order to achieve commercially acceptable standards of germination. Processing techniques range from simple cleaning and grading to pre-conditioning of seed through various priming processes. Smaller seeded species are now more normally sold as pellets, for ease of sowing by machine.

The Sale of the End Product

Seed is traded internationally. There are relatively few breeding companies worldwide — between 10 and 20 involved in seed-raised decoratives — but there are many more distributor companies. Breeding companies tend to be international in scope, distributor companies tend to restrict their activities to a single country. Distributors generally fall into two main groups — those who sell primarily to the home gardener, and those who sell primarily to the professional grower.

Bartels, M.J. (1999). Distinctness, uniformity and stability in Fleuroselect trials. In: S. Andrews, A.C. Leslie and C. Alexander (Editors). Taxonomy of Cultivated Plants: Third International Symposium, pp. 49–51. Royal Botanic Gardens, Kew.

DISTINCTNESS, UNIFORMITY AND STABILITY IN FLEUROSELECT TRIALS

MARCEL J. BARTELS

Executive Director, Fleuroselect, Parallel Boulevard 214d, NL 2202 NT Noordwijk, The Netherlands

Abstract

The comparative trialing of new seed-raised ornamental cultivars by Fleuroselect is described. New cultivars must pass tests of distinctness, uniformity and stability and be provided with a suitable name.

Fleuroselect Concept

In the association known as Fleuroselect, flower seed breeders and distributors from all over the world come together to encourage the breeding and introduction of new flower cultivars from seed. The best encouragement is money or profit, although we like to call it 'return on investment'. For flower breeders profit starts with the breeding of unique and beautiful cultivars. But there are more considerations than this in producing a profit, namely production, protection and promotion. The members of Fleuroselect have mutual agreements on the exclusive production and mutual protection and marketing of their best new seed cultivars.

Fleuroselect Standards

Fleuroselect members test, compare and judge each other's new flowers on ornamental and technical qualities. Distinctness, Uniformity and Stability (DUS) are essential. The worst that can happen to a cultivar (and to its breeder) is that it cannot be recognised. Therefore it must be at least as distinct, uniform and stable as comparable existing cultivars. Otherwise, introduction of the cultivar will be blocked. Moreover, only cultivars that are better than similar cultivars against which they are compared in one or more ornamental or cultural aspect are regarded as valuable novelties to be protected or even awarded.

Procedures for the Introduction of a New Cultivar

Trials

The number, spread and anonymity of the trials guarantee objectivity and expertise. Each entry should be judged by at least 10 trial holders throughout Europe. Entries and comparison cultivars must be treated equally. The breeder is responsible for adequate information on the cultivation and breeding history of the entry. He indicates the novelty aspect, which is a major criteria for judgement. For optimal testing he can choose from five different trial categories.

Distinctiveness

The entry must be clearly different from the wild species from which it is derived, from commercial cultivars and from cultivars tested by Fleuroselect before. Differences identified by observation or cultivation must be an improvement at market level. F1 hybrids can be compared to open-pollinated cultivars, since this characteristic is not a visual difference. In principle if a veto on the introduction of a cultivar is proposed because it is too similar to another cultivar, then at least one, or up to 25% of the judges present should be obliged to prove this. If there is a potential conflict of interest (e.g. an existing patent) vegetative cultivars can be used for comparison as well.

Uniformity

A new cultivar must be at least as uniform as the best existing seed-raised cultivar. The best existing cultivar sets the standard. For example: *Lobelia erinus* L. generally shows 10–15% off coloured flowers. If an entry in our trials shows 5% off colours, this is a valuable improvement and breeding accomplishment. For this species and colour the new standard for later introductions will be a maximum of 5% deviation.

Stability

Cultivars are stable as long as the parent lines remain unchanged. Professional selection of parent lines of hybrids as well as open-pollinated cultivars makes them more uniform. In principle selection may not change the original cultivar, but should improve the original characteristics (colour, habit, etc.). An improved cultivar will not lose its Fleuroselect protection if it still fits the original description and adds to the quality and reliability. If conflicts occur, the "violating cultivar" is compared to current samples of the protected cultivar. Fleuroselect keeps samples of the original entry for potential comparison in case of complaints about the quality or selection. Mixtures of a new, uniform and stable composition of new colours of similar habit can be recognised and protected as well (e.g. in *Nemesia* Vent.). Mixtures that are composed of individually produced cultivars cannot be protected.

Introduction

Cultivars that are new (distinct), uniform, stable and improved are registered by Fleuroselect. All registered cultivars are "protected", meaning that production remains the exclusive right (and duty) of the original breeder. Members may not reproduce them nor buy these cultivars from other than the breeder. The best new cultivars are awarded and promoted by the organisation and distributed by all its members. The cultivar name must be considered acceptable by the Fleuroselect Board. This name will be promoted world-wide by the organisation and therefore, may not be changed, either when used for the individual product or when it is incorporated as a part of a series or mixture.

Identification

Flowers must be attributed to a genus and species before we can even think of a nice, short cultivar epithet. Then we add a colour indication if it is part of a series. This full name (or denomination) must be in conformity with the International Codes of Nomenclature and with different national instructions for cultivar denomination. Since botanical names are in Latin, which even in catholic circles is hardly spoken anymore, we give it a common name instead. Finally, we check that it does not conflict with existing cultivar names, trademark registrations, etc. We may still need additional indications for habit, etc. to clearly identify them. Take for example petunias: at grower level the distinction between Grandiflora, Multiflora and Pendula Groups is as relevant as adding that a plant is an F1 hybrid. They are quite different, e.g.

Petunia (Pendula Group) 'Revolution'	known as Surfinia Purple Mini
Petunia (*Calibrachoa*) 'Sunbelkubu'	known as Million Bells Blue
Petunia (Multiflora Group) 'Fantasy'	known as Junior Petunia

Sometimes the genus or species to which a plant belongs has changed and we must indicate this:

Alyssum maritimum (L.) Lam. is now *Lobularia maritima* (L.) Desv.
Chrysanthemum parthenium (L.) Bernh. (*Matricaria* L.) is now *Tanacetum parthenium* (L.) Sch.Bip.
Lisianthius Hort. *non* P. Browne is now *Eustoma* Salisb.

Summary of Objectives

- Workable DUS standards for flowers grown from seed.
- Workable names for cultivars. These should be correct, distinct, unique, stable and nice.
- Communication between organisations responsible for describing, testing, protecting, registering, naming and marketing of new ornamental cultivars.
- Combined use of facilities and knowledge. Fleuroselect does not seek to take over official, legal or governmental tasks.
- Concentrate administrative processes to save costs and energy.

Waltrip, J. (1999) Nomenclature in the North American seed-trade. In: S. Andrews, A.C. Leslie and C. Alexander (Editors). Taxonomy of Cultivated Plants: Third International Symposium, pp. 53–56. Royal Botanic Gardens, Kew.

NOMENCLATURE IN THE NORTH AMERICAN SEED-TRADE

J. WALTRIP

10392 Boulder Court, Ventura, California 93004, USA
President, All-America Selections and Director of Seminis Garden
(Seminis Vegetable Seeds)

Abstract

In North America there is no formal requirement to register the names of new cultivars of seed-raised vegetables or flowers. This paper outlines the importance of good marketing names and recommends a procedure for establishing the acceptability of new names in co-operation with the American Seed Trade Association and the United States Department of Agriculture.

Introduction

Good marketing names are very important in stimulating sales of a new cultivar to the hobby or home gardening trade. But misnaming products can also have disastrous effects. The following are some examples of product names or slogans that turned out to be less than helpful in marketing them:

- Clairol introduced a curling iron called "Mist-Stick" in Germany only to find out that "mist" is slang for manure.
- Scandinavian vacuum cleaner manufacture, Electrolux, began a sales campaign in the US with the slogan "Nothing sucks like an Electrolux".
- US chicken marketer Frank Purdue's slogan, "It takes a strong man to make a tender chicken", was translated into Spanish as, "It takes an aroused man to make a chicken affectionate".
- Colgate introduced a toothpaste in France called "Cue", the name of a notorious pornographic magazine.

Names can certainly affect sales and marketing and the cultivar names we use in the United States may not be the best names for marketing the same product in Jordan or Thailand, Russia or Japan.

Cultivar Name Registration in North America

There is really no name registration process for flowers in North America and the one used for vegetables is voluntary, and loose, to say the least.

When new cultivars of ornamental flowers are named, there are usually two or three people at a seed company sitting in a room who come up with a name that sounds good to them. Then they try to remember if there is any cultivar of that species on the market with that name. If not, bingo! it is named. No clearance process, no fees, no registration, no regulations. The only thing that keeps a company from using a name that is already in use is a gentlemen's agreement between flower seed companies.

Furthermore if a customer chooses to rename a cultivar for marketing purposes, there are no laws against it. The originator may take action by refusing to supply the seed, but that is a different story.

In vegetables, as with grains and soybeans, we are governed by the United States Federal Seed Act, which says that a cultivar must be sold under only one name and a cultivar name must not be used for more than one cultivar within a species or kind. A name must not be misleading or should not be so similar to an existing name as to cause confusion. This is all. There is no compulsory registration process for vegetable cultivar names.

The American Seed Trade Association (ASTA) has some guidelines on naming, but they are only recommendations. They are as follows:

- Select simple, distinctive names.
- Numerals or symbols alone or a combination with firm or trade names make very unsatisfactory names.
- Names should not begin with abbreviations, initials, numbers or symbols.
- Avoid names that are hard to write, spell or pronounce.
- Avoid the use of superlatives or words that exaggerate.
- Do not use the term "Hybrid" or "F1" as part of the name.
- Foreign names are normally retained, but it is permissible to use a translation to avoid linguistic difficulties.

Priority of vegetable cultivar names is based on the date of valid publication in catalogues, bulletins and technical publications. In other words, if you're first on the market, that name is yours.

To protect yourself, a cultivar registration card should be mailed to ASTA, and they will add the name to a computer listing of names they maintain. If your card is dated prior to someone else's, you will have priority unless the other cultivar is already on the market and referenced in print. Registration cards, in addition to a name, include such headings as experimental designation, originator/breeder, first retail sale date, category (open-pollinated or hybrid), distinguishing characteristics, disease resistance and reference to a published description. ASTA is a clearing house only and exercises no authority, which is good, since they have none. Authority rests with the government agriculture departments. In the United States this is the United States Department of Agriculture (USDA), in Canada the Canadian Department of Agriculture.

Names may be re-used within a species if the old cultivar has been out of commerce for a minimum of 20 years according to ASTA, but USDA disagrees. The USDA enforces the Federal Seed Act with respect to naming so following USDA rules is the wiser part of valour. Company names may be used as part of the name, as long as it's part of the originally assigned legal name. An examples is 'Burpee's Big Boy' hybrid tomato. If you advertise this tomato without using the word 'Burpee's', you are in violation. You cannot label the seed packets simply with 'Big Boy'. They must state 'Burpee's Big Boy'.

Imported seed cannot be renamed, according to USDA, if the original name of the seed is in the Roman alphabet. This is not in agreement with American Seed Trade guidelines which suggest you rename the cultivar in order to give the name the same meaning in English as in the original language.

Vegetable seed mixtures containing more than one cultivar must be labelled with the name and percentage of each cultivar present, but the mixture may be given a name. An example would be a leaf lettuce blend we have named 'Bon Vivant'.

Everyone does not follow the same procedure for naming new vegetable cultivars, but they should, even though not required by law. Here is what I recommend for agricultural and vegetable seed.

1. Fax or e-mail ASTA saying you want to use the name 'Gang Green', for instance, for a new cucumber. They will likely reply that they have no record of that name being used for cucumber.
2. When cleared by ASTA, send a fax or letter to USDA telling them the same thing, and ask if they know of any conflict. It takes about 3 weeks for USDA to respond.
3. USDA will check their own cultivar name listing log, and also check to see if the name is a registered trademark in use by a firm in the vegetable produce business or a related area. This search takes a little time, but this is a free service.
4. Assuming that 'Gang Green' is not a registered trademark and has never been used as a cultivar name for cucumber, USDA will tell you that in their opinion it is okay to use the name for your new cultivar. If it is a registered trademark, they will recommend you not to use it, but you do not have to follow their advice. You can use 'Gang Green' and take a chance on getting sued.
5. You should now feel free to assign the name 'Gang Green' to your new cucumber, and while not required, it would be sensible to mail in a registration card to ASTA. You do not need to send anything more to USDA since they would have already added the name to their log or system.
6. That's it. No fee. You're finished.

The Importance of a Good Name

Now I just want to briefly cover the importance of a good name in marketing vegetable seeds to the home garden hobby market. Here are some names we have used very successfully:

- 'Red Express' cabbage - the name is descriptive because it says it's early and red.
- 'Small Miracle' broccoli has dwarf, space saving plants, yet produces what doctors often describe as a miracle food in terms of nutrition.
- 'Brocoverde' cauliflower has green heads and looks like broccoli.
- 'Mucho Nacho' is one of the largest jalapeño peppers.
- 'Sweetness' carrot is our best selling cultivar for home gardening. This is because of the name, and not necessarily because of cultivar performance.
- When I needed a name for a carrot that was 50% higher in beta carotene, I came up with the name 'Healthmaster'. It helped to make that cultivar a big seller.
- 'Beefmaster', a giant beefsteak type tomato, was not always called 'Beefmaster'. It was originally called 'Beefeater'. The makers of Beefeater gin threatened to sue us, so we changed the name to 'Beefmaster'. Our attorney said it would cost us $100,000 to find out if we could continue with the name 'Beefeater', and much more if we lost.
- Calloway Golf came after us when they found out that we were calling a bell pepper 'Big Bertha'. But they backed off when they discovered that our pepper cultivar is much older than their golf club.

- 'Spacemiser' is a good name for a zucchini that takes one third less garden space.
- 'Pasta' seemed to be the perfect name for a spaghetti squash.
- USDA, once told me I shouldn't name a new Roma-type pear-shaped tomato 'Viva Italia' because there was already a cultivar called 'Viva'. 'Viva' was actually a processing tomato that was unavailable to home gardeners, so I ignored USDA and named our new cultivar 'Viva Italia'. It was the right choice, as this is now a big seller.

Finally, I want to tell you about the name coup of the century. Back in the late 1960s, my company was trying to name a new tomato. It had large, perfectly smooth fruits, indeterminate plants, great taste and a nice package of disease resistance. At that time, 'Burpee's Big Boy' was the number one seller, and one of the people in our meeting said "Let's call ours 'Better Boy' as a joke. But it soon became clear that this was the perfect name to compete with 'Burpee's Big Boy'. Today, about 30 years later, 'Better Boy' is still America's best selling home garden hybrid tomato, even though 'Big Beef' and 'Celebrity' should be. This shows you the power of a name in America, combined with a good product. In conclusion may I offer a piece of philosophical advice:

"Before you criticise a man, walk a mile in his shoes. That way, when you do criticise him, you'll be a mile away and have his shoes."

INTELLECTUAL PROPERTY RIGHTS AND PLANTS

3

Heitz, A. (1999). Plant variety protection and cultivar names under the UPOV Convention. In: S. Andrews, A.C. Leslie and C. Alexander (Editors). Taxonomy of Cultivated Plants: Third International Symposium, pp. 59–65. Royal Botanic Gardens, Kew.

PLANT VARIETY PROTECTION AND CULTIVAR NAMES UNDER THE UPOV CONVENTION

ANDRÉ HEITZ

Director-Counsellor, International Union for the Protection of New Varieties of Plants, 34, chemin des Colombettes, 1211 Geneva 20, Switzerland

Abstract

The development of the International Convention for the Protection of New Varieties of Plant is described, together with the nature and scope of the protection offered. The recommendations concerning the provision of cultivar names for protected plants are discussed and the relationship between cultivar names and trademarks is noted.

Introduction

"I have been for years in correspondence with leading breeders, nurserymen, and Federal officials and I despair of anything being done at present to secure to the plant breeder any adequate returns for his enormous outlays of energy and money. A man can patent a mousetrap or copyright a nasty song, but if he gives to the world a new fruit that will add millions to the value of the earth's annual harvests he will be fortunate if he is rewarded by so much as having his name connected with the result. .... I would hesitate to advise a young man, no matter how gifted or devoted, to adopt plant breeding as a life work until [the United States Congress] takes some action to protect his unquestioned rights to some benefit from his achievements."

This statement was written at the beginning of the 20th century by the well known plant breeder Luther Burbank.

The efforts to secure rights to plant breeders over the commercial exploitation of their new cultivars took a long time to come to fruition. It was not until 1930 that the United States Congress adopted an amendment to the Patent Act, known as the "Plant Patent Act". This was limited to vegetatively propagated plants (except potato and Jerusalem artichoke) and thus benefited only, in the main, breeders of fruit and ornamental plants. In Europe, the efforts led to the exploration of various avenues during the 1930s and after World War II: straight recourse to the patent law (mostly through cleverly drafted applications), adoption of an amendment similar to the United States Plant Patent Act, organization of the seed and plant trade through catalogues of cultivars authorized for sale and seed certification, use of a seal or certificate to distinguish authentic plant material originating from the breeder from other material, and finally introduction of a special protection system for cultivars.

The results achieved remained, however, broadly unsatisfactory. By way of illustration, the patent offices of Belgium, France, Germany and Italy granted patents for new cultivars, essentially for roses and carnations (or for processes concerning such cultivars), whereas others refused this as a matter of principle; as a result, the patent protection granted in the mentioned countries was all but secure.

This situation led ASSINSEL, the Association of Plant Breeders for the Protection of Plant Varieties (founded in 1938), to request, at its annual congress held in Semmering (Austria) in 1956, the organization of an international conference to sort out the matter. The idea was taken up by the French authorities, more specifically the Ministry for Agriculture, and led to the negotiation, between May 1958 and December 1961, of the International Convention for the Protection of New Varieties of Plants (the "UPOV Convention") and its adoption on December 2, 1961.

The UPOV Convention sets out the basis for a system of intellectual property protection that includes essential elements of agricultural policy. In particular, it sets out basic principles for the naming of plant cultivars which greatly contribute to market regulation.

The UPOV Convention and UPOV

Basic Features
The UPOV Convention is an international treaty concluded by and adhered to by States (and since 1991 certain intergovernmental organizations). Its aim is described as follows in the Preamble to the original text:

"THE CONTRACTING STATES,

Convinced of the importance attaching to the protection of new varieties of plants not only for the development of agriculture in their territory but also for safeguarding the interests of breeders,

Conscious of the special problems arising from the recognition and protection of the right of the creator in this field and particularly of the limitations that the requirements of the public interest may impose on the free exercise of such a right,

Deeming it highly desirable that these problems to which very many States rightly attach importance should be resolved by each of them in accordance with uniform and clearly defined principles ..."

The UPOV Convention was revised on November 10, 1972, in respect of some administrative provisions, on October 23, 1978, in respect of all provisions but without substantial changes and on March 19, 1991, in respect of all provisions but with substantial changes designed to clarify certain rules in the light of the experience of the UPOV member States in operating the Convention since 1961, to strengthen the protection offered to the breeder in specific ways and to reflect technological changes.

Membership
UPOV currently (June 8, 1999) has 43 member States:

- On the basis of the 1961/1972 Act: *Belgium, Spain.*
- On the basis of the 1978 Act (the States marked with an asterisk have a law that conforms entirely or substantially with the 1991 Act): *Argentina, Australia*,

Austria, Bolivia, Brazil, Canada, China, Czech Republic, Chile, Colombia*, Ecuador*, Finland*, France, Hungary, Ireland*, Italy*, Kenya, Mexico, New Zealand, Norway, Panama, Paraguay, Poland*, Portugal, Slovakia*, South Africa*, Switzerland, Trinidad and Tobago, Ukraine, Uruguay.

- On the basis of the 1991 Act: *Bulgaria, Denmark, Germany, Israel, Japan, Netherlands, Republic of Moldova, Russian Federation, Sweden, United Kingdom, United States of America.*

The following States — and the European Community — have initiated the procedure for becoming members by tabling before the Council of UPOV a request for advice on the conformity of their laws with the provisions of the Act which they wished to adhere to: *Belarus, Costa Rica, Croatia, Estonia, Georgia, Kyrgystan, Morocco, Nicaragua, Romania, Slovenia, Venezuela, Zimbabwe.*

There is considerable interest in plant variety protection and in membership in UPOV from a great number of States. Some 60 further States have laws or draft laws based upon the UPOV Convention. That interest is essentially generated by the increased awareness of the benefits that are to be drawn from a plant variety protection system — fostered in many developing countries by the current trend towards privatization of the plant variety and seed sector in an effort to make it more effective.

In addition, membership in the World Trade Organization (WTO) entails, under Article 27.3(b) of the Agreement on Trade-Related Aspects of Intellectual Property Rights — the TRIPS Agreement — an undertaking to:

"provide for the protection of plant varieties either by patents or by an effective *sui generis* system or by any combination thereof."

There is no doubt that the vast majority of those WTO members which have not yet fulfilled that obligation will introduce a specially designed protection system for plant varieties based upon the UPOV Convention. Indeed a great many countries exclude plant varieties *per se* from patent protection, either through specific provisions enshrined in their patent laws or through administrative practice, sometimes sanctioned by judicial decisions (however, most of them accept patents on inventions consisting of products such as genetic constructs or of processes such as genetic engineering processes, which patents may deploy their effects at the level of plant varieties); the United States of America is the only country in which patents for plant varieties *per se* have acquired some significance.

The Institutional Setting
The circumstances surrounding the establishment of the Convention and its subsequent evolution also led to the creation of a separate intergovernmental organization, the International Union for the Protection of New Varieties of Plants (UPOV), which maintains close cooperation with the World Intellectual Property Organization (WIPO). In particular, the WIPO Director-General, Dr Kamil Idris (Sudan), is at the same time the Secretary-General of UPOV. UPOV's offices are located in the WIPO building in Geneva, Switzerland, and UPOV draws upon the support services of WIPO. The Office is thus quite small; its regular staff includes four professionals and five secretaries and clerical staff.

Some Basic Features of the Protection under the UPOV Convention

The Scope of Protection

All Acts of the UPOV Convention describe a minimum scope of protection that provides a right over the commercial exploitation of the protected cultivar and certain other cultivars.

Under the 1978 Act, the minimum right implies that the authorization of the right holder is required, in essence, for the production for purposes of commercial marketing, the offering for sale and the marketing of propagating material of the protected cultivar. The right also extends to cultivars whose commercial production requires the repeated use of the protected cultivar, typically to hybrid cultivars. States may provide a more extensive right covering, in particular, the marketed product.

This definition has two important consequences:

- The reference to "purposes of commercial marketing" implies that a farmer may produce his own seeds or plants for use, on his holding, in the subsequent growing season. This is usually referred to as the "farmers' privilege." The privilege does not exist, under the Convention, for ornamental plants. Nor does it exist, either generally or in respect of certain other crops, in certain States which have granted more extensive rights to breeders.
- The protected cultivar may be freely used as an initial source of variation for the purpose of creating other cultivars. This is usually referred to as the "breeders' exemption" and represents the cornerstone of the protection system based upon the UPOV Convention: having been able to freely use the cultivars of his predecessors in his own breeding programme, the successful breeder has to accept — and he happily does — that his protected cultivar be freely available to his successors. The principle of the exemption is expressly set out in the Convention.

The provisions of the 1991 Act pertaining to the (minimum) scope of protection are much more detailed, and the following might be highlighted:

- The description of the acts of exploitation that require the right holder's authorization is more detailed, but does not entail a real extension of the right.
- The description no longer includes the reference to "purposes of commercial marketing," so that any production of propagating material would require the holder's authorization; the "farmers' privilege" has not been abolished, however, being treated as an exception that any State may introduce under national legislation.
- The minimum right granted to the breeder now extends to the harvested material, but only where the breeder has not had reasonable opportunity to exercise his right in relation to the propagating material from which it has been produced.
- The right granted in respect of a cultivar also extends to essentially derived cultivars, for instance, to a mutant or a genetically engineered cultivar.

Whatever the formulation of the scope of protection, the objective of the right granted to the breeder is to enable him, during the term of protection:

- To collect a remuneration in respect of each cycle of exploitation of his cultivar.
- To organize, at least to some extent, an optimal exploitation of his cultivar.

The Conditions for Protection

Protection is granted upon an application and is subject to the payment of fees. The persons entitled to protection are the breeder or his successor in title. "Breeder" includes the person who has discovered and developed a cultivar, e.g. someone who spotted a mutation in a bed of plants and ensured the isolation and propagation of the mutation as a new cultivar. Given the importance of such cultivars, it is essential for the protection system to encompass both the cultivars that are "bred" in the restricted sense of the word and those that are "discovered."

To be eligible for protection, a cultivar must be:

- Clearly distinguishable from any other cultivar whose existence is a matter of common knowledge at the time of filing of the application.
- Sufficiently uniform (homogeneous) in its relevant characteristics, having regard to the particular features of its propagation.
- Stable in its relevant characteristics, that is, remain unchanged after repeated propagation or, for instance in the case of a hybrid, at the end of each particular cycle of propagation.
- Commercially new, subject to some "grace periods".

The standards established by UPOV for "DUS" (distinctness, uniformity and stability) have gained wide acceptance for matters such as cultivar listing and seed certification. In effect they have become standards for the cultivar concept.

The Cultivar Name

The UPOV Convention also requires that a cultivar that is the subject of an application for protection be given a cultivar name ("denomination"). The cultivar name is proposed by the applicant and registered by the relevant protection office to which the application is made.

The cultivar name is the "generic designation" of the corresponding cultivar. "Generic" is to be understood as a concept of trademark law (not taxonomy). Most of the consequences of "genericity" are spelled out in the Convention and further explained in the recommendations on cultivar names adopted by the Council of UPOV. They are as follows:

- The designation that is to be registered as a cultivar name may not contain any element that would hamper the free use of the name in connection with the cultivar. In particular, it may not be identical with or similar to a trademark or trade name belonging to the breeder or the applicant, or a third party, and applying to products that are similar to those of the cultivar within the meaning of trademark law. It may not be identical with a family name in certain countries or be in conflict with the name or acronym of a State or organization. Conversely, the registration of a designation as a cultivar name will prevent the registration of the same or a similar designation as a trademark. These principles are not affected by the expiration of the right. Where an unsuitable cultivar name slips through the net and there is a conflict with a prior right, a new name must be proposed and registered.
- The cultivar name must enable the cultivar to be identified. It may not consist solely of figures, except where this is an established practice for designating cultivars.

- The cultivar name must not be liable to mislead or to cause confusion concerning the characteristics, value or identity of the cultivar or the identity of the breeder. For instance, comparative and superlative names are not permitted; nor are names which suggest that the cultivar concerned derives from another when that is not the case.
- The cultivar name must be different from every cultivar name which designates an existing cultivar of the same plant species or of a closely related species. The basic rule applied within UPOV is that all species of a given genus are considered to be "closely related" for naming purposes. Certain genera are pooled together (for instance *Avena* L., *Hordeum* L., *Secale* L., × *Triticale* Wittm. and *Triticum* L., or *Sorghum* Moench and *Zea* L.); a few others are split (for instance the agricultural and vegetable *Beta* L.). The rule is relaxed in several countries to enable the re-use of a cultivar name where the older cultivar is no longer in cultivation and its name has not acquired any particular significance.
- The cultivar name must not be (too) difficult to recognize or reproduce in speech and writing, or contrary to public policy. It may not include elements used in trade to designate matters such as quantity, quality or type of plant or seed and may not be able to create confusion in that respect.

The cultivar name must be unique within UPOV. A cultivar name proposed in a given member State must therefore be proposed in all others. A synonym will only be permitted if the first proposed cultivar name proves unsuitable in another country. To guarantee the smooth functioning of the system, an international checking system has been put in place within UPOV. A cultivar database is also published at regular intervals on a CD-ROM.

Any person who commercializes propagating material of a protected cultivar must use the cultivar name. This obligation remains in force after the expiration of the right.

However, the obligation to use the cultivar name in the course of trade with seeds or plants does not prejudice the right to use a trademark, trade name or a similar indication in association with the cultivar name, provided that the latter remains easily recognizable. Two strategies are used in this respect:

- Certain breeders, particularly breeders of agricultural crops and vegetables, are happy to use the cultivar name and will endeavor to find a "fancy" name that does the job in the various countries concerned.
- Other breeders, particularly breeders of ornamental plants, are very keen to use trademarks, in conjunction with a coded, universally usable cultivar name; a common practice is to take the first three letters of the breeder's name and to attach to them one or several syllables chosen to a large extent at random. Since the trademark has to be attractive, it may differ from one national or regional market to the other. Moreover the trademark is usually given prominence over the cultivar name in view of its role in the marketing strategy.

Conclusion

Plant variety protection provides a legal and economic framework within which plant breeders and their partners in the plant and seed sector can deploy their activities for the benefit of the farming community and society at large. The UPOV Convention provides a framework for effective national and regional plant variety protection legislation that respond both to the need for international harmonisation and the need for adaptation to local circumstances. It has also led to approximation and unification in other sectors; its most prominent achievement is perhaps the clarification of the cultivar concept.

With respect to the naming of cultivars, the UPOV Convention has created the basis for uniformity, accuracy and fixity, and also the legal framework to ensure that cultivar names are effectively used, without prejudice to the right of the various economic parties to use proprietary signs such as trademarks or other indications.

However, when breeders use trademarks in the manner described above, it is not surprising that the cultivar tends to be known and referred to under the trademark rather than the cultivar name; some people may even, in that process, forget that there is a cultivar name. There is thus a great need for promoting a better understanding of the rules of the Convention and of their rationale, and also for promoting compliance with those rules by the seed and nursery industry, by scientists, journalists, etc., and even by breeders and national authorities.

Editors' note: In these Proceedings the editors have, where relevant, replaced authors' use of the terms "variety" and "denomination" with "cultivar" and "cultivar name" respectively, in the belief that they can be regarded as equivalents but acknowledging that the former terms would consistently be employed by all those dealing with Plant Breeders' Rights.

Strachan, J.M. (1999). Plant variety protection in the USA. In: S. Andrews, A.C. Leslie and C. Alexander (Editors). Taxonomy of Cultivated Plants: Third International Symposium, pp. 67–72. Royal Botanic Gardens, Kew.

PLANT VARIETY PROTECTION IN THE USA

JANICE M STRACHAN

Examiner, Plant Variety Protection Office, NAL Building, Room 500, 10301 Baltimore Avenue, Beltsville, Maryland 20705, USA

Abstract

Development of a new cultivar, either by traditional breeding methods or by modern molecular modification, requires a large input of time and effort. To recover the costs of this research and development, the breeder may seek to obtain exclusive marketing rights for the new cultivar. Various ways of doing this exist, such as keeping trade secrets, marketing only F1 hybrids, or obtaining plant patents, utility patents, or Plant Breeders' Rights (plant variety protection or PVP). In the United States, plant variety protection has been available for most seed-reproduced plants since 1970. Protection for six other crops was added in 1980, and protection for first generation hybrids and tuber-reproduced crops was added in 1994. The rationale for this programme existing separately from the US Patent Office is explained and the procedure for obtaining a PVP certificate described. A summary of the US Federal Seed Act, which governs the naming of cultivars in some crops, is presented.

Introduction

Development of a new cultivar, either by traditional breeding methods or by modern molecular modification, requires a large input of time and effort. Some companies estimate that it takes 10–15 years to develop a new cultivar. In order to speed up this process, companies use winter breeding sites and genetic manipulation. Although these practices may produce new cultivars faster, they also increase costs. To recover the costs of research and development, the breeder may seek to obtain exclusive marketing rights for the new cultivar. Various ways of doing this exist, such as keeping trade secrets, marketing only F1 hybrids, or obtaining plant patents, utility patents, or Plant Breeders' Rights (plant variety protection or PVP).

History

In the United States, the 1930 Plant Patent Act first allowed for patenting of asexually reproduced cultivars (except tubers). Tubers and seed-reproduced crops were excluded because they are the major source of food, and they were believed to be non-uniform and unstable. By the 1960s, some European countries had enacted plant breeders' rights laws. It was demonstrated that sexually reproduced cultivars were uniform and stable enough to be included in these laws. During the 1960s several attempts were made to enact similar protection in the United States, including a proposal to revise the Plant Patent Act to include sexually reproduced plants. These early attempts were unsuccessful.

The Plant Variety Protection (PVP) Act was enacted on December 24, 1970. Its purpose is to encourage the development of novel cultivars of sexually (i.e. seed) reproduced plants, other than fungi and bacteria, by providing their owners with exclusive marketing rights to them in the United States. Protection for okra, celery, peppers, tomatoes, carrots, and cucumbers was added in 1980. In 1994, the United States PVP Act was amended to comply with the 1991 UPOV Convention (UPOV is the International Union for the Protection of New Varieties of Plants). At that time, tuber-reproduced plants were specifically added to the scope of eligibility and an exclusion against F1 hybrids was removed.

The PVP Office is responsible for administration of the PVP Act. It is organised within the Agricultural Marketing Service of the United States Department of Agriculture. The Commissioner heads the PVP Office staff, which includes five plant variety examiners and five associate examiners. Prior to 1994, the PVP Office received 300 new applications per year. Since the amendments were enacted, the number of applications filed each year has increased to 400. Tubers and F1 hybrids do not account for all of this growth. Since 1971, over 4,020 Certificates of Protection have been issued in over 170 crops. Some certificates are no longer in effect due to being abandoned, cancelled, expired, or withdrawn.

Eligibility

The requirements of protection are that the cultivar be new, uniform, stable, and distinct from all other cultivars. All information used to determine whether a cultivar meets these criteria is gathered and reported by the applicant to the PVP Office. Site visits are not required. Examiners base their decisions on the descriptive information and trial data supplied by the applicant. Clarifying or supplementary data may be requested from the applicant during the course of the examination. This may require that the applicant conduct additional trials.

In order to be considered new, a cultivar may not have been sold or otherwise disposed of for more than one year in the United States or for more than four years (six years in the case of trees and vines) in a foreign country. This means that propagating or harvested material of the cultivar existed, was publicly known or a matter of common knowledge, that it was on an official register of cultivars or had been filed for plant breeders' rights. Sale or disposal of hybrid seed is considered to be sale or use of the parental lines. This must be considered when dealing with certain crops, such as maize. Exceptions for these definitions exist for cultivars while they are undergoing testing or experimentation.

How to Apply for Protection

To request protection for a new cultivar, the applicant completes an application form and an application packet. The complete packet must contain the following items: the fees, a seed sample, and Exhibits A, B, C and E.

The application form needs (a) to have an original signature, (b) to be accompanied by the fees for filing the application (US$300) and for conducting the search (US$2,150), (c) to have the required documentation attached, and (d) to be accompanied by 2,500 viable (i.e. 85% germination or better), untreated seeds to serve as the voucher specimen. The sample is stored at the National Seed Storage Laboratory in Ft. Collins, Colorado. The applicant may be asked to replenish this sample if the germination rate or sample size fall below adequate levels during the protection period. Another item that should accompany the application packet is an affidavit or letter which states that the cultivar name is acceptable to naming authorities in the United States.

Seed sample requirements for tuber crops or first generation hybrids are slightly different than for other crop kinds. The voucher 'seed sample' of a tuber will be requested when the certificate is ready to be issued. At that time, the deposit procedures and associated costs will be sent to the applicant. With the application, the applicant should send a document that states that a tissue sample will be deposited when the certificate is issued. Seed samples for first generation hybrids need to include seed of the hybrid *and* seed of all parental lines needed to propagate the cultivar.

Exhibit A needs to show (a) the complete breeding history back to commercially or publicly available lines, cultivars, populations, etc. (b) the steps taken during the breeding history including what traits were used as selection criteria at each stage of selection, what breeding methods were used, how many generations of inbreeding and roguing were performed, etc., (c) whether there are genetic variants that are to be expected during normal maintenance of the cultivar (not off-types which derive from external contamination), the description of the variants, and their frequency, and (d) a statement concerning the uniformity (all plants in the population look the same) and stability (all traits of the cultivar are maintained over generations) of the cultivar.

The basis for distinctness is presented in Exhibit B. Since the United States PVP Office does not perform grow-out trials, applicants need to gather and report all information that is required to complete the application. The simplest way to present distinctness is to name one cultivar which is **most similar** to the application cultivar in genetic background and morphology. Other methods are acceptable but more difficult to carry out, such as either naming the group to which the new cultivar belongs, naming all cultivars in the group and describing one-at-a-time how the new cultivar differs from each cultivar in the group, or naming all cultivars in the crop and one-at-a-time describing how the new cultivar differs from each. The applicant also needs to provide a complete description of the most similar comparison cultivar, if one is not already on file in the PVP Office.

After this comparison statement is made, the applicant needs to **contrast** the application cultivar with the cultivar to which it is most similar, citing specific character states to support the contrasts. Differences in quantitative characters (such as plant size, leaf size, and flower size) between the new cultivar and its most similar comparison cultivars must be given as numerical data obtained from paired comparisons with statistical tests showing degree of significance. Colour differences should be referenced with a standard such as the Munsell Book of Color or Royal Horticultural Society Colour Chart. Complex traits, such as yield, are not useful in establishing distinctness. However, traits which contribute to yield (fruit weight, fruit size, number of fruits per plant, etc.) may be used to establish a clear difference. Colour photographs (prints) showing the contrasting traits of both the new cultivar and its comparison cultivar should also be included. At least one year of trials should be conducted in the United States in a region where the cultivar will be grown.

Exhibit C is the Objective Description of the cultivar using forms created by the PVP Office. Breeders and other knowledgeable persons are consulted before a draft form is finalised. These forms are used to standardise a complete botanical description of the cultivar, making it easier to determine differences between cultivars. The form should be completely filled-in unless a particular character has no relevance to your crop. Newer forms may have room for descriptions of the application cultivar and a comparison cultivar, plus simple descriptive statistics and colour chart values.

Exhibit D is optional and can include anything not included elsewhere in the application packet. Information in this section may include test-cross results, trial data, isozyme or other molecular test results, photographs, possible uses for the cultivar or its products, specific descriptive information not disclosed elsewhere in the application, or anything the applicant feels may be useful. This section may be omitted if the data is placed in another Exhibit.

The Exhibit E should describe how the applicant obtained ownership of the cultivar and whether anyone else can claim any rights regarding this cultivar. This document helps us determine that the original breeder and the applicant, if they are different, are eligible to apply for protection in the United States.

How Applications are Processed

Once the applicant has submitted a complete application to the PVP Office, the application is assigned to an examiner. The examiner conducts a literature search of the crop and gathers descriptive information on cultivars from grow-out trials, release notices, seed catalogues, PVP applications, and other published sources. The examiner maintains the cultivar descriptions in computerised crop databases. Over 50,000 different cultivars in more than 170 crops are currently in the system. The examiner then uses the appropriate database to determine the novelty of the application cultivar. The description of the application cultivar (from the Exhibit C form) is compared to the database to help determine distinctness. These searches are not 'point' searches, but rather look for values in a reasonable range surrounding the value reported. This helps to account for environmental influences. For example, we cannot distinguish 'intermediate' from either 'low' or 'high', but we can distinguish 'low' from 'high'.

If questions arise during the examination of the application, the examiner corresponds with the applicant. Questions may be raised concerning data conflicts, uniformity and stability, inadequate statistical support for distinctness, or the existence of cultivars which are indistinguishable from the application cultivar (based on the search of the crop database). The applicant is asked to provide additional information. This information (with supporting evidence) may come from prior knowledge or from additional grow-out trials.

When the examination is complete, the examiner summarises the process and its results. If the cultivar is found to be eligible for protection, then the examiner recommends this to the Commissioner. The Commissioner verifies the findings of the examiner and writes to the applicant, requesting that the certificate issuance fee be paid. The certificate holder may specify that the cultivar be sold by cultivar name only as a class of certified seed, as defined in the Federal Seed Act. Once so specified, the designation cannot be reversed. The final certificate is signed by the Commissioner and the Secretary of Agriculture, is bound with a green ribbon, and is embossed with a golden seal of the PVP Office.

Applications are processed in the order in which they are received. Historically, the Office has been able to process 70% of the applications (from receipt to issuance of a certificate) in less than 24 months. The most difficult case took ten years to process. The total time needed to process an application depends on the amount of research needed to create or maintain the crop database, the completeness of the application packet, and the ease of distinguishing the application cultivar from all other cultivars of the crop. Eligibility and procedural requirements may also cause delays, thus lengthening the examination period.

Rights Granted

Once granted, the applicant may exclude others from selling or marketing the cultivar, conditioning or stocking the cultivar, offering it for sale or reproducing it, importing or exporting it, or using it to produce (as distinguished from develop) a hybrid or different cultivar, or any other transfer of title or possession of the cultivar. Certificates of Protection are effective for 20 years (25 years for vines and trees) from the issuance of the certificate.

These rights extend to essentially derived cultivars, indistinct cultivars, harvested materials and cultivars which require repeated use of the protected cultivar. There is an exemption which allows farmers to save seed of the protected cultivar for use on their own farm, but this exception does not allow the farmer to transfer ownership of the saved seed to another person for reproductive purposes. Researchers also have an exemption so that the protected cultivar can be used in plant breeding or other research.

A certificate holder has certain responsibilities. The seed sample must be replenished when requested. The PVP Office must be informed of changes to the address of the certificate holder, or the person who is authorised to act on the certificate holder's behalf. The cultivar name must be used even after the certificate expires. The version of the PVP Act under which the certificate was issued must be included on all labels. The certificate holder also must notify the public that the cultivar is protected, using appropriate language.

Cultivar Naming Regulations

Cultivar names or denominations are proposed on the application form. The applicant may choose to use a temporary name until the certificate is issued and an appropriate permanent name has been cleared with the proper naming authorities. For vegetable and agricultural crops, cultivar names are regulated by the Federal Seed Act. This Act is administered by, and thus the naming authority in the United States is, the Seed Regulatory and Testing Branch (SRTB) in Beltsville, Maryland. Applicants may submit proposed names to this office. They check the name against three criteria: (a) the name should not be prohibited by the Federal Seed Act, (b) the name must not be a registered trademark and (c) the name must never have been used in commerce.

The Federal Seed Act states that the first name used in commerce is the name of the cultivar. A cultivar may not be renamed unless the name is illegal. Some cultivars on the market prior to 1956 may be known by allowable synonyms. A cultivar name must be unique from other cultivar names of the same crop kind. The same name may be used if the crops are not closely related. Cultivar names cannot be re-used if they were ever used in the past. If a company name is part of the cultivar name, everyone who markets the cultivar must use that company name. To avoid confusion on this issue, the company name and cultivar name should be separated if the company name is not part of the cultivar name. Although discouraged, descriptive terms may be used as long as the terms are not misleading. Cultivar names must be clearly different in spelling and sound. Hybrid designations may be names or numbers, however, the same rules apply to hybrid and non-hybrid cultivars. Seed imported unto the United States cannot be renamed if the original name uses the Roman alphabet. A brand or trademark must never take the place of a cultivar name.

Although the proposed name must meet these criteria, the SRTB has a difficult time ensuring that a name is cleared. The United States does not have a crop registration list, and companies may market plants without clearing the name prior to offering the cultivar for sale. The SRTB routinely checks seed catalogues and advertisements in industry journals for new cultivar names. These are entered into a names database when they are found, but SRTB is not certain that they receive all catalogues and journals. Therefore when a proposed name is checked against this database, the SRTB cannot be certain that the name has not be used before.

The Federal Seed Act does not cover ornamental crops. For ornamental crops, breeders are encouraged to follow the same guidelines as agricultural and vegetable crops. However, the SRTB does not maintain a names database for ornamentals. Since these crops are outside of their jurisdiction, they will not check the trademarks database. Currently their recommendation is to check the proposed name with a crop group, for example the Marigold Society of America for cultivars of *Tagetes* L. Other sources that may be useful include the UPOV CD-ROM and the United States Trademark Office.

Of the 400 new applications received in the PVP Office each year, less than 3% are for ornamental crops. Agricultural crops make up 76% of applications and vegetable crops account for 21% of applications. It is doubtful that the laws will be changed unless ornamental breeders take an interest in this situation and lobby for a change in the law.

Plant breeding is a dynamic industry, mingling the old methods with new technology to its best advantage. The PVP Office also mingles the old and new methods when determining the novelty of a cultivar. Their effectiveness shows in the growth of the breeding industry since 1970.

Tramposch, A. (1999). Introduction to trademarks: loss of trademark rights for generic terms. In: S. Andrews, A.C. Leslie and C. Alexander (Editors). Taxonomy of Cultivated Plants: Third International Symposium, pp. 73–79. Royal Botanic Gardens, Kew.

INTRODUCTION TO TRADEMARKS: LOSS OF TRADEMARK RIGHTS FOR GENERIC TERMS

ALBERT TRAMPOSCH

Director, Industrial Property Law Division, World Intellectual Property Organization (WIPO), 34 chemin de Colombettes, CH-1211, Geneva 20, Switzerland

Abstract

The definition of a trademark, its function and the elements of which it may be comprised are discussed. The requirement for use of a registered mark are underlined and the consequences of non-use or improper use are noted. For a trademark owner it is essential to ensure that the trademark does not become a generic term.

Introduction

Trademarks have become a key factor in the modern world of international trade and market-oriented economies. Industrialization and the growth of the system of the market-oriented economy allow competing manufacturers and traders to offer consumers a variety of goods in the same category. Clearly consumers need to be given the guidance that will allow them to consider the alternatives and make their choice between the competing goods. Consequently, the goods must be distinctively marked. The medium for marking goods on the market is the trademark.

By enabling consumers to make their choice between the various goods available on the market, trademarks encourage their owners to maintain and improve the quality of the products sold under the trademark, in order to meet consumer expectations. Thus trademarks reward the manufacturer who constantly produces high-quality goods, and as a result they stimulate economic progress.

Definition of Trademark

"A trademark is any sign that individualises the goods of a given enterprise and distinguishes them from the goods of its competitors." This definition comprises two aspects, which are sometimes referred to as the different functions of the trademark, but which are, however, interdependent and for all practical purposes should always be looked at together.

In order to individualise a product for the consumer, the trademark must indicate its source. This does not mean that it must inform the consumer of the actual person who has manufactured the product or even the one who is trading in it. It is sufficient that the consumer can trust in a given enterprise, not necessarily known to him, being responsible for the product sold under the trademark.

The function of indicating the source as described above presupposes that the trademark distinguishes the goods of a given enterprise from those of other enterprises; only if it allows the consumer to distinguish a product sold under it from the goods of other enterprises offered on the market can the trademark fulfill this function. This shows that the distinguishing function and the function of indicating the source cannot really be separated. For practical purposes one can even simply rely on the distinguishing function of the trademark, and define it as "A sign which serves to distinguish the goods of one enterprise from those of other enterprises."

Signs which may serve as Trademarks

It follows from the purpose of the trademark that virtually any sign that can serve to distinguish goods from other goods is capable of constituting a trademark.

The following types and categories of signs can be imagined:

- Words: this category includes company names, surnames, forenames, geographical names and any other words or sets of words, whether invented or not, and slogans.
- Letters and Numerals: examples are one or more letters, one or more numerals or any combination thereof.
- Devices: this category includes fancy devices, drawings and symbols and also two-dimensional representations of goods or containers.
- Combinations of any of those listed above, including Logotypes and Labels
- Coloured Marks: this category includes words, devices and any combinations thereof in colour, as well as colour combinations and colour as such.
- Three-Dimensional Signs: a typical category of three-dimensional signs is the shape of the goods or their packaging. However, other three-dimensional signs such as the three-pointed Mercedes star can serve as a trademark.
- Audible Signs (Sounds Marks): two typical categories of sound marks can be distinguished, namely those that can be transcribed in musical notes or other symbols and others (e.g. the cry of an animal).
- Olfactory Marks (Smell Marks): imagine that a company sells its goods (e.g. writing paper) with a certain fragrance and the consumer becomes accustomed to recognizing the goods by their smell.
- Other (Invisible) Signs: examples of these are signs recognised by touch.

Criteria of Protectability

The requirements which a sign must fulfill in order to serve as a trademark are reasonably standard throughout the world. Generally speaking, two different kinds of requirement are to be distinguished. The first kind of requirement relates to the basic function of a trademark, namely, its function to distinguish the products or services of one enterprise from the products or services of other enterprises. From that function it follows that a trademark must be distinctive or distinguishable among different products. The second kind of requirement relates to the possible harmful effects of a trademark if it has a misleading character or if it violates public order or morality. These two kinds of requirement exist in practically all national trademark laws.

Requirement of Distinctiveness

A trademark, in order to function, must be distinctive. A sign that is not distinctive cannot help the consumer to identify the goods of his choice. The word "apple" or an apple device cannot be registered for apples, but it is highly distinctive for computers. This shows that distinctive character must be evaluated in relation to the goods to which the trademark is applied. The test of whether a trademark is distinctive is bound to depend on the understanding of the consumers, or at least the persons to whom the sign is addressed. A sign is distinctive for the goods to which it is to be applied when it is recognised by those to whom it is addressed as identifying goods from a particular trade source, or is capable of being so recognised.

The distinctiveness of a sign is not an absolute and unchangeable factor. Depending on the steps taken by the user of the sign or third parties, it can be acquired or increased or even lost. Circumstances such as (possibly long and intensive) previous use of the sign have to be taken into account when the registrar is assessing whether the sign lacks the necessary distinctiveness, that is, if it is regarded as being not inherently distinctive.

There are, of course, different degrees of distinctiveness, and the question is how distinctive a sign must be in order to be registrable. In that connection a distinction is generally made between certain typical categories of marks — fanciful or coined trademarks which are meaningless and the others. These trademarks may not be the favourites of the marketing people, since they require heavy advertising investment to become known to consumers. However, they inherently enjoy very strong legal protection.

Common words from everyday language can also be highly distinctive if they communicate a meaning that is arbitrary in relation to the products on which they are used. The same is true of the corresponding devices. Examples are the famous CAMEL trademark for cigarettes (and the equally famous device mark) and the previously-mentioned APPLE mark (both the word and the device) for computers.

Marketing people are generally fond of brand names that generate a positive association with the product in the mind of the consumer. They tend therefore to choose more or less descriptive terms. If the sign is exclusively descriptive, it lacks distinctiveness and cannot be registered as such as a trademark. However, not all signs that are neither meaningless nor arbitrarily used necessarily lack distinctiveness: there is an intermediate category of signs that are suggestive, by association, of the goods for which they are to be used, and of the nature, quality, origin or any other characteristic, of those goods, without being actually descriptive. Those signs are registrable. The crucial question in practice is whether a trademark is suggestive or descriptive of the goods applied for. This question has to be judged according to the local law and jurisprudence of the country and all the circumstances of the specific case. As a general rule, it can be said that a descriptive term is distinctive for the goods concerned if it has acquired a secondary meaning, that is, if those to whom it is addressed have come to recognise it as indicating that the goods for which it is used are from a particular trade source.

In case of doubt as to whether a term is descriptive or suggestive, the very fact that the mark has been used in the course of trade for a certain period of time may be sufficient for accepting it for registration. However, the more descriptive the term is, the more difficult it will be to prove secondary meaning, and a higher percentage of consumer awareness will be necessary.

Lack of Distinctiveness

If a sign is not distinctive, it cannot function as a trademark and its registration should be refused. The applicant normally need not prove distinctiveness. It is up to the registrar to prove lack of distinctiveness, and in the case of doubt the trademark should be registered. Some trademark laws, such as the British Trade Marks Act 1938 (and laws in countries which have followed the British approach) put the onus on the applicant to show that his mark ought to be registered. This practice may be considered strict, however, and sometimes prevents the registration of marks that are demonstrably capable of distinguishing their proprietor's goods. And yet the modern trend is clearly to treat lack of distinctiveness as a ground for refusing an application for registration of a trademark.

The following criteria govern the refusal of registration for lack of distinctiveness:

Generic Terms

A sign is generic when it defines a category or type to which the goods belong. It is essential to the trade and also to consumers that nobody should be allowed to monopolise such a generic term. Examples of generic terms are "furniture" (for furniture in general, and also for tables, chairs, etc.) and "chair" (for chairs). Other examples would be "drinks," "coffee" and "instant coffee," which shows that there are larger and narrower categories and groups of goods, all having in common that the broad term consistently used to describe them is generic.

These signs are totally lacking in distinctiveness, and some jurisdictions hold that, even if they are used intensively and may have acquired a secondary meaning, they cannot be registered since, in view of the absolute need of the trade to be able to use them, they must not be monopolised.

Descriptive Signs

Descriptive signs are those that serve in trade to designate the kind, quality, intended purpose, value, place of origin, time of production or any other characteristic of the goods for which the sign is intended to be used or is being used. In line with the definition of the distinctive sign given earlier, the test to be applied must establish whether consumers are likely to regard a sign as a reference to the origin of the product (distinctive sign) or whether they will rather look on it as a reference to the characteristics of the goods or their geographical origin (descriptive sign). The term "consumer" is used here as an abbreviation denoting the relevant circles to be considered in a specific case, namely those to whom the sign is addressed (and in certain cases also those who are otherwise reached by the sign).

The fact of other traders having a legitimate interest in the fair use of a term can therefore be used as a kind of additional ground when making the decisive test of whether consumers are likely to regard the sign as a reference to origin or as a reference to characteristics of the goods. It should not, however, be used on its own as a basis for a decision to refuse the registration of a term when it is not clear that consumers are also likely to regard the term as descriptive.

Exclusions from Registration on other Grounds—Public Interest

Deceptiveness

Trademarks that are likely to deceive the public as to the nature, quality or any other characteristics of the goods or their geographical origin do not, in the interest of the public, qualify for registration. The test here is for intrinsic deception, inherent in the trademark itself when associated with the goods for which it is proposed. This test should be clearly distinguished from the test for the risk of confusing customers by the use of identical or similar trademarks for identical or similar goods.

It is true that fanciful trademarks or marks with an arbitrary meaning for the goods proposed cannot be deceptive. And yet trademarks that have a descriptive meaning, even if they are only evocative or suggestive and therefore distinctive, may still be deceptive. Such trademarks have therefore to be examined from two angles: first they must be distinctive, and secondly they must not be deceptive. As a rule, it can be said that the more descriptive a trademark is, the more easily it will deceive if it is not used for the goods with the characteristics described.

Partial Deceptiveness
We have seen that the question whether or not a trademark is inherently deceptive must be examined in relation to the goods in respect of which the application is made. Depending on the list of goods, therefore, an application may be distinctive for some, descriptive for others and/or deceptive for still others. In such cases the examiner has to require a limitation of the list of goods. Should the applicant not agree to such limitation, the examiner refuses the whole application in some countries. In others, he accepts the application only for the goods for which, in his opinion, the mark is not deceptive and refuses it for the others.

Signs Contrary to Morality or Public Policy
Trademark laws generally deny registration to signs that are contrary to morality or public policy.

Protection of Trademark Rights

A trademark can be protected on the basis of either use or registration. Both approaches have developed historically, but today trademark protection systems generally combine both elements. The Paris Convention places contracting countries under the obligation to provide for a trademark register. Over one hundred and fifty States have adhered to the Paris Convention. Nearly all countries today provide for a trademark register, and full trademark protection is properly secured only by registration.

Use does still play an important role, however: first of all, in countries that have traditionally based trademark protection on use, the registration of a trademark merely confirms the trademark right that has been acquired by use. Consequently, the first user has priority in a trademark dispute, not the one who first registered the trademark.

Use Requirements

Need for an Obligation to Use
Trademark protection is not an end in itself. Even though trademark laws generally do not require use as a condition for the application for trademark registration, or even the actual registration, the ultimate reason for trademark protection is the function of distinguishing the goods on which the trademark is used from others. It makes no economic sense, therefore, to protect trademarks by registration without imposing the obligation to use them. Unused trademarks are an artificial barrier to the registration of new marks. There is an absolute need to provide for a use obligation in trademark law.

At the same time trademark owners need a grace period after registration before the use obligation comes into effect. This is especially true of the many companies that are active in international trade. In order to avoid loopholes in the protection of their new trademarks of which competitors could take advantage, they must from the very beginning apply for the registration of their new trademarks in all countries of potential future use. Even in their own countries companies often need several years before they can properly launch a newly-developed product on the market. This is especially true of pharmaceutical companies, which have to make clinical tests and have to apply for approval of their product by the health authorities. The grace period granted in trademark laws that provide for a use obligation is sometimes three years, but more often five years.

Consequences of Non-Use

The principal consequence of unjustified non-use is that the registration is open to cancellation at the request of a person with a legitimate interest. There is moreover a tendency to require of the registered owner that he prove use, since it is very difficult for the interested third party to prove non-use. In the interest of removing "deadwood" from the register, such reversal of the burden of proof is justified. The burden of proof should be on the trademark owner not only in cancellation proceedings but also in any other proceedings where the owner is alleged to have taken advantage of his unused trademark right (opposition procedure, infringement action).

No evidence of use should be required for the renewal of a trademark registration, however. This is an administrative complication which is unnecessary in view of the fact that an interested person can at any time at all take appropriate action against an unused trademark registration.

Non-use does not always lead to invalidation of the trademark right. Non-use can be justified in the case of *force majeure*, and any other circumstance that is not due to fault or negligence on the part of the proprietor of the mark, such as import restrictions or special legal requirements within the country.

Proper Use of Trademarks: how a Trademark becomes a Generic Term

Non-use can lead to the loss of trademark rights. Improper use can have the same result. A mark may become liable for removal from the Register if the registered owner has provoked or tolerated its transformation into a generic name for one or more of the goods or services in respect of which the mark is registered, so that, in trade circles and in the eyes of the appropriate consumers and of the public in general, its significance as a mark has been lost.

Basically, two things can cause genericness: namely, improper use by the owner, provoking transformation of the mark into a generic term, and improper use by third parties that is tolerated by the owner. In order to avoid improper use, everyone in the company owning the trademark who is involved in advertising or publicising the brand must follow some rules.

The basic rule is that the trademark should not be used as, or instead of, the product designation. By systematically using a product designation in addition to the trademark, the proprietor clearly informs the public that his mark identifies a specific product as one in a certain category. This is especially important if the trademark proprietor has invented a totally new product which at the outset is the only one in the category.

A second important rule is that trademarks should always be used as true adjectives and never as nouns, in other words the trademark should not be used with an article, and the possessive "s" and the plural form should be avoided.

Furthermore, it is advisable always to highlight the trademark, that is, to make it stand out from its surroundings.

Finally, a trademark should be identified as such by a trademark notice. Only a few laws provide for such notices, and making their use on goods compulsory is prohibited by Article 5D of the Paris Convention. Trademark law in the United States of America allows the use of a long statement (such as "Registered with the United States Patent and Trademark Office") to be replaced by a short symbol, namely, the circled R(®). Over the years this symbol has spread throughout the world and become a widely recognised symbol for a registered trademark. Its use is recommended for registered trademarks as a warning to competitors not to engage in any act that would infringe the mark.

However, it is not enough just to follow these rules: the trademark owner must also ensure that third parties and the public do not misuse his mark. It is specifically important that the trademark should not be used as or instead of the product description in dictionaries, official publications, journals, etc. If the registered owner has provoked or tolerated the transformation of a mark into a generic name for one or more of the goods or services in respect of which the mark is registered, the mark becomes liable for removal from the register. The cancellation of a trademark registration is a serious matter for its owner, as it leads to a loss of his rights under the registration.

Gioia, V.G. (1999). Trademark rights — a sometimes overlooked tool for plant variety (marketing) protection. In: S. Andrews, A.C. Leslie and C. Alexander (Editors). Taxonomy of Cultivated Plants: Third International Symposium, pp. 81–87. Royal Botanic Gardens, Kew.

TRADEMARK RIGHTS — A SOMETIMES OVERLOOKED TOOL FOR PLANT VARIETY (MARKETING) PROTECTION

VINCENT G. GIOIA

Christie, Parker & Hale, L.L.P., 5 Park Plaza, Suite 1440, Irvine, California 92614, USA

Abstract

Plant variety protection enables breeders and introducers of new cultivars to protect their investment of time, effort and funds in developing new plants. Most people in the business of developing and introducing new cultivars would like to enjoy exclusive marketing opportunities or income through licensing. Although breeders' rights and plant patents, or the like, come to mind as the first "line of defence", trademark rights may provide the best "real world" opportunity to protect this investment at lowest cost. Furthermore, trademark rights can provide a useful mechanism to discourage importation of new cultivars into home markets.

Introduction

Let's say you're a nurseryman and have developed a new cultivar. It's a superior plant and you believe there is a great market for the cultivar in your country. You have a significant investment in the new cultivar because it is only one of several plants with commercial potential out of many that you developed and tested over a period of time. The development programme itself incurred out-of-pocket costs, nursery space and a lot of your time and the time of employees. Of course, you have already paid these costs but now here is an opportunity to recover some of your investment.

But you have to be careful. If you just release the cultivar and begin selling plants your competitors can jump on the bandwagon. They can get some plants, propagate and grow as many as they want and there goes your market — and return on investment.

So how do you deal with this situation? Of course, you would like to protect your investment and income potential, if you can. You are aware that new plants can be protected with plant patents (or breeders' rights) and you contact a patent attorney, hopefully one that knows something about plants. In short order you apply for and obtain a plant patent for this wonderful new cultivar. Now, you think, I'm all right — the plant patent will enable me to exclude others from propagating, growing and selling plants of my patented cultivar. And, legally speaking, you are absolutely correct. The only problem is if an unscrupulous competitor decides to infringe your patent and sells the patented plant, your recourse is to become involved in an expensive game of Monopoly with real money. Patents, including plant patents, are enforced by suing the culprit in an appropriate United States federal district court. Let the games begin.

Quickly you learn that skilled litigation lawyers are very expensive, as is the entire process. Unless you're willing to spend several hundred thousand dollars, with some risk of losing the case, you may feel discretion is indeed the better part of valour. Now, this is not to say that there isn't the possibility that circumstances are such that you have no choice but to do what you can to enforce your patent to protect your market and your business. In fact, we have represented patent owners in situations where the market opportunity, and consequent damage due to loss of business, was considered great enough to justify the cost of patent litigation. But the reality is that in most situations a patent owner either doesn't have pockets deep enough to finance a law suit or the lost relevant market isn't large enough to justify the cost.

Fortunately, most businessmen are ethical and will not encroach on your legitimate rights. Protecting the new cultivar enables you to profit from your investment directly or indirectly through licensing since without a plant patent or other breeders' rights there is no possibility of winning the new cultivar development game.

In the best case scenario, you have a good cultivar that you protected and are enjoying the exclusivity to which you are entitled because of your patent rights. Things are going along fine and then you learn that growers in another country have discovered your good thing and start offering for sale, and selling, your patented cultivar into your country. To make matters worse, your foreign competitor may be using the benefits of our cyber age and offers your plants in a web site on the Internet. Since the Internet is available throughout your country, your patented plants are available to everyone, and without the need for the infringer to publish and distribute a catalogue.

This is indeed a serious problem. Your plant patent is being infringed and you have the right to sue, but litigation is expensive and you (and your lawyer) still have to figure out how, or if, you can get jurisdiction in a United States court.

Trademarks may be the Answer

Well, take heart, all is not lost. Consider trademarks, often overlooked as a tool for plant variety (marketing) protection.

Unlike plant patents, which protect a patent owner from unauthorised reproduction and sale of a new cultivar, trademarks protect the marketing interests of a plant introducer. By using trademarks on or in connection with plants you denote the source of the plant, and acquire a valuable right that can be used to prevent others from using the mark without your approval. And, importantly, the cost of enforcing trademark rights is comparatively low and other enforcement mechanisms besides litigation are available.

Therefore, a good strategy is to select and use a trademark as a commercial designation for plants of the new cultivar. However, the trademark cannot be the cultivar name. Unlike cultivar names, which are generic plant identifiers, trademarks are used to indicate the source or origin of plants sold in the marketplace. Cultivar names may be freely used worldwide to identify a particular plant, but trademarks may not be used in commerce without authorisation of the trademark owner. Trademark rights are established very easily; simply by using the mark correctly. Once you have established trademark rights they can continue indefinitely as long as the trademark is used. Although not necessary to establishing trademark rights, where possible trademarks should be registered with the United States Patent and Trademark Office, or other national trademark office, because trademark rights are significantly enhanced by registration of the mark.

Protection of Trademarks under United States Law

In the United States, Federal and State trademark law protects a mark whether or not it is registered. That is to say, even in the absence of federal registration, the owner of a trademark will have rights to the sole use of the mark. Such rights are referred to as "common law" trademark rights in the United States but most countries do not recognise common law rights by use. To identify your mark as proprietary it is useful to use the ™ symbol (or ˢᴹ symbol of services) following the word, name or symbol constituting the mark, in print and labels. Even so, it is better, without exception, to obtain governmental registration of the mark when possible, rather than rely on common law rights. Among the advantages of registration are a presumption of ownership and validity of the mark, and the exclusive right to use the mark in commerce. Since January 1, 1996, it has been possible to apply for a Community Trademark Registration covering 15 countries in the European Community and the same benefits are obtained by community registration.

To show the mark has been registered the ® symbol should be used with the mark. (This symbol can only be used with registered marks). A federal trademark registration protects the owner and the public against a likelihood of confusion, mistake or deception about the source, affiliation or sponsorship for approval of goods or services, by commercial activities of another. In addition to trademark infringement, unauthorised use of your registered trademark is considered a misrepresentation under the law.

As a general rule, it is an infringement of trademark rights for someone else to use the same or a confusingly similar term on the same or closely related goods in the same geographical area, or in some cases, within a natural area of expansion. Moreover, since trademark infringement is a form of unfair competition, the infringing activity is subject to unfair competition laws.

Trademark Rights Useful to Protect Domestic Markets

As you can see, if you have trademark rights in the designation used to commercially designate plants of the new cultivar, your marketing protection is significantly enhanced. Your foreign competitor not only infringes your plant patent when he sells plants, your trademark rights are also infringed if the infringing plants are identified by your trademark. Invariably, there is little or no possibility that infringing plants can be sold without your trademark because you made the market for the plant and the foreign exporter hasn't.

The plant industry in the United States, and perhaps other countries as well, has been losing its profitability and competitiveness due to unfair trade practices, particularly due to the infringement of copyrights, trademarks, patents, and other intellectual property rights. Infringement is not just a United States problem. The United States International Trade Commission (ITC) estimated in 1986 that American companies in all industries lose between US$43 and US$63 billion per year because of the absence of, or inadequate foreign protection of, intellectual property rights. Much of this infringing merchandise makes its way to, and through, United States borders.

The United States Government can be on your Side

When you have trademark rights you get a very important ally in your war against the infringer — the United States government. The Customs Service in the United States is in a unique position to prevent these illegal importations. In order to protect American industry from unfair trade practices and create a fair trading environment worldwide, the Customs Service has developed a mechanism to stem the importation of infringing merchandise. Current Customs regulations and internal directives, as well as established procedures, can be used in this effort. United States Customs will help you protect trademarks and trade names that have been recorded with the Customs Service Intellectual Property Rights Branch (IPR Branch) under various Customs Regulations by the simple process of making an appropriate application and paying a recordation fee. Once subject to Customs Service protection, infringing products can be stopped at the border.

Both registered trademarks and unregistered trade names can enjoy Customs protection. To be eligible for recordation with Customs, trademarks to be recorded are either those registered on the Principal Register in the United States Patent and Trademark Office, Department of Commerce, or an unregistered mark (referred to in the regulations as a "trade name"). Customs recordation of registered trademarks usually takes about one week. Recordation of trade names, which are not registered, or registrable, at the Patent and Trademark Office, can also be recorded but recordation takes a little longer because notice of tentative recordation must be published in the United States Federal Register and the Customs Bulletin. Interested parties are then given the opportunity to oppose the recordation. Notice of final approval or disapproval is published after consideration of all claims, rebuttals, and relevant evidence bearing on the recordation.

Once a trademark or trade name has been recorded, Customs has the authority to deny entry to, or seize, goods that infringe upon the recorded right. The Customs Service inspects shipments and can seize merchandise considered to infringe trademark rights. Seized goods bearing an infringing mark may be sent back to the place of origin.

Although the Customs Service does not provide application forms, anyone interested in this mechanism, may contact me for further information. As a general rule, an application to record trademarks with Customs must include certain information to satisfy the requirements of the Customs Regulations and to provide adequate and effective border enforcement of United States trademark rights. While a separate application for each trademark is not required, information provided to Customs must be accurate for all trademarks in the application, and fees must be paid for each trademark and for each class of goods.

The application, whether for a registered mark or a trade name, should include the applicant's business address and citizenship. For registered marks, a certified copy of the registration from the United States Patent and Trademark Office and evidence showing that the applicant owns the mark should be provided. If the trademark owner resides or operates in different states or countries, information concerning these locations is also required. This requirement applies to all entities having more than 50% ownership interest. Also required are the names of each parent or subsidiary company, or other related foreign company which uses the trademark abroad. It is also useful to provide information regarding any authorised use of the registered mark so Customs can distinguish from unauthorised users. Copies of contracts, licenses or corporate affiliation documents are also usefully included with the application. It is additionally important to indicate the principal business address(es) where the goods bearing the trademark are legally manufactured with the consent of the applicant, for example, places where other parties are authorised to use the mark, as well as information about where the trademark infringing goods are being manufactured without approval of the trademark owner.

Recording trade names which are not registered trademarks with the United States Custom Service will also earn assistance of Customs to prevent importation of goods bearing the trade names into the United States. In applications to record trade names similar information must be provided as required in applications to record registered trademarks. This information includes the trade name to be recorded and the name address and citizenship of the trade name owner. It is also required to indicate whether the trade name has been used for at least six months. The goods with which the trade name is associated should be described and, as with the recording of registered marks, a fee must be submitted with the application.

As a result of provisions in the General Agreement on Tariffs and Trade (GATT) and North American Free Trade Agreement (NAFTA) treaties, the United States. Customs Service will notify trademark owners when goods are seized bearing marks that have been recorded with Customs for enforcement. Trademark owners can receive a sample of the product and useful information such as the quantity involved and the name and address of the "manufacturer". Upon receipt of a seizure notice it is worthwhile to ask Customs for additional information.

Well, you now have a useful and realistic process to protect your mark, and your investment. Your trademark rights enable you to get the Customs Service to help you without spending a lot of money. Very often putting your trademarks to proper use will be a sufficient deterrent to unfair foreign competition, but if not, and you want to go further, there are some other important things to know that can help you.

Guarding against Deceptive Trade Practices

The United States law provides protection against many deceptive commercial practices, including false designation of origin or source. This is known as "product infringement" and false description or representation. All these come within the category of false advertising. The court in the case described below held that a failure to designate a product as foreign made is a violation of the statute and is an act of unfair competition.

The federal law in the United States that provides for registration of marks also offers other rights to the domestic industries. For example, in a decision by a Federal Court in New York, applicable to the plant industry, foreign manufacturers who sell their products in the United States through importers may be liable for unfair competition if they fail to mark the goods as foreign made. The court held that foreign manufacturers having knowledge and intention that their products will be resold to consumers in this country must comply with provisions of the same law that provides rights to trademark owners, but under sections of the law dealing with unfair trade practices. The failure to designate the country of origin of the product is a violation of both the United States Tariff Act and the Unfair Competition Section of the trademark statute. The basis for this decision is that consumers encountering goods with no markings as to country of origin will assume that they are made in the United States, thereby creating a likelihood of confusion as to which goods are, in fact, made in the United States. A foreign manufacturer that sells goods made abroad without an indication of origin to importers with knowledge that an importer will resell them in the United States, may be liable. Although the decision pertained to other goods, this decision also clearly applies to the plant industry, including the importation into the United States of parts of plants such as cut flowers and fruit, as well as plants.

This case is not the only one to deal with this subject, just one of the more recent. A prior decision by the United States Supreme Court relied upon in the recently decided case extended liability of trademark infringement beyond those who actually mislabel goods with the mark of another. Therefore, even if a manufacturer does not directly control others in the train of distribution, it can be held responsible for their infringing activities. In such a situation, the manufacturer or distributor is contributorily responsible for any harm done as a result of the deceit.

Although false advertising clearly encompasses fraudulent representations that goods marketed have "ingredients or qualities" that they, in fact, do not have, it also applies to advertising that gives the impression that all of the advertised goods, including other foreign made goods referred to in the advertising, are actually produced in the United States. Where a foreign manufacturer is aware of or otherwise engages in such advertising, the manufacturer as well as the distributor or seller may be also guilty of false advertising. In other words, the manufacturer and others in the chain of marketing may be enjoined from stating explicitly or creating the impression that foreign made products are manufactured in the United States.

The Internet Pirate

What about that foreign cyber age pirate, the one that sells your cultivars over the Internet through a web site? Well, if the web site lists your patented cultivars and/or your trademarks, and plants are offered for sale, you can go after the scoundrel at your neighbourhood Federal District court. There's a good chance the foreign supplier will not want to spend the money, especially the higher cost of proceeding with litigation near your home and far from theirs. Very likely you will be able to get some sort of court enforceable settlement agreement, or judgment including an injunction, to prevent further damage to your business.

This is not a mere speculative remedy. On behalf of our client Weeks Wholesale Rose Growers, Inc., we recently filed a suit in Los Angeles, California, against a company in Canada, for patent and trademark infringement, as well as unfair competition.

The Canadian company has a very lovely web site. It lists many rose cultivars available from the company, page after page after page. The list includes many cultivars introduced by many United States companies which are the result of expensive breeding programmes of the introducers or others from whom they license the cultivars and to whom they pay royalties. To my knowledge this company has no breeding programme, it seems to merely identify cultivars with market potential from whatever source, grows the plants and sells them as they can.

Unfortunately for this Canadian grower the cultivars offered for sale on their web site included some patented in the United States by Weeks as well as others identified by Weeks' trademarks. The grower has no license from Weeks and is not likely to get one.

The law in the United States regarding Internet issues is still evolving but it is clear that courts recognise the Internet can be used as a vehicle for patent and trademark infringement. Although this defendant is a Canadian enterprise, by their actions via the Internet they subject themselves to the jurisdiction of the courts, in this case the United States Federal District Court for the Central District of California in Los

Angeles. There is the likelihood that the Internet will also confer jurisdiction on infringers, including this Canadian company, in other countries as well.

(Note added in press: since the presentation and submission of this paper the court has passed judgement in favour of Weeks Wholesale Rose Grower Inc. This judgement enjoins the Canadian company from sale or offer for sale of Weeks' patented cultivars and use of Weeks' trademarks).

Conclusion

If you struggled through this tome thus far, the conclusion should be obvious. To me the contribution to the trade and the public by those responsible for new cultivars is significant and substantial. Their efforts are entitled to be rewarded and they should be able to support their work, and profit from it, in a free enterprise system. Society recognises this by enabling breeders to protect new cultivars in different ways. Patents and breeders' rights provide protection against unauthorised propagation and sale for a limited period of time and trademark rights, often overlooked as a tool, are equally useful as a mechanism for protecting investment in promotion and marketing.

It's worth considering.

Scott, E.M.R. (1999). Plant Breeders' Rights trials for ornamentals: the international testing system and its interaction with the naming process for new cultivars. In: S. Andrews, A.C. Leslie and C. Alexander (Editors). Taxonomy of Cultivated Plants: Third International Symposium, pp. 89–94. Royal Botanic Gardens, Kew.

PLANT BREEDERS' RIGHTS TRIALS FOR ORNAMENTALS: THE INTERNATIONAL TESTING SYSTEM AND ITS INTERACTION WITH THE NAMING PROCESS FOR NEW CULTIVARS

ELIZABETH M.R. SCOTT

Head of Ornamental Plants Section, National Institute of Agricultural Botany (NIAB), Huntingdon Road, Cambridge CB3 OLE, UK

Abstract

Plant Breeders' Rights have existed in the United Kingdom for nearly 35 years, during which time it has been possible to protect cultivars of an ever widening range of ornamental plant species; as of May 1998 such protection is now available to all genera. To ensure new plants meet internationally agreed criteria for granting of rights — novelty, distinctness, uniformity and stability, or 'DUS' — a growing trial is carried out, either at Cambridge, or by agreement in an equivalent test centre. In general, under a network of international bilateral agreements established over the last 30 years, most important ornamental species are now tested in a single designated site: for example the United Kingdom has responsibility for *Chrysanthemum* and many herbaceous perennials, whereas the Bundessortenamt in Germany tests *Pelargonium* among others. An overview of this international trials procedure and the way the testing system interacts with the naming process for new cultivars is presented.

Introduction

The task of the Ornamental Plants Unit at the National Institute of Agricultural Botany (NIAB) is to organise growing trials for new cultivars of cut flowers, pot plants, garden plants and ornamental trees which are the subject of Plant Breeders' Rights (PBR) applications, to check whether they meet the internationally agreed criteria for the grant of rights. It is also involved in the associated evaluation of cultivar names.

As will be explained, the work of a trials centre involves contact with bodies involved in all aspects of plant nomenclature, both legal and technical, and therefore gives a unique overall view of some of the practical problems currently associated with the naming of protected cultivars.

Plant Breeders' Rights for Ornamentals

Plant Breeders' Rights are sometimes regarded by the general public and even by some sectors of the horticultural world as a bit of a mystery. Perhaps this is because in general, people only become aware of them when they buy a protected garden plant and read the label, or see some story in the press of instant riches "found on the compost heap". In reality, protected cultivars are everywhere in the form of cut flowers, pot plants and basket plants, in other words, the products of the international plant breeding industry — but in buying the finished plant or bouquet one is often unaware of this.

A PBR is a form of intellectual property which enables the creator of a new plant cultivar to gain a financial reward for his or her skills by charging a royalty on sales of the plant. It covers a defined area (e.g. the United Kingdom or the Netherlands) and lasts for a limited period of time (25 or 30 years, depending on species and subject to the payment of renewal fees, although most are terminated much earlier).

To be eligible for a grant of rights, a plant cultivar must meet certain criteria defined by the UPOV Convention. UPOV, the International Union for the Protection of New Varieties of Plants, is an intergovernmental organisation with headquarters in Geneva which currently has some 37 countries as members, all of whom agree to provide protection for new cultivars of plants under the same conditions.

Each UPOV member state has a national office responsible for plant breeders' rights matters, including the legislation, administration and the grant of rights; in the United Kingdom it is the Plant Variety Rights Office (PVRO) in Cambridge. European Union (EU) plant breeders' rights are the responsibility of the Community Plant Variety Office(CPVO), and are administered from its central headquarters in Angers, France. All such offices produce regular Gazettes giving information about applications made and granted or refused, proposed and approved denominations and changes in the legislation and other administrative announcements. Information about applications is also now available from UPOV in the form of an extremely useful bi-monthly CD-ROM containing gazette information from most UPOV countries and the EU.

For a grant of PBR, the UPOV Convention lays down that a cultivar must be new, distinct from all other cultivars in common knowledge at the time of the application, uniform and stable (normally abbreviated to 'DUS'). It must also have an acceptable denomination (cultivar name). These are the only criteria: there is no assessment of merit or value, or whether the plant is a 'good' or 'garden-worthy' cultivar.

A breeder can choose to apply for protection in one country or in several, depending on the likely market for the plant. Since 1995 however there has been a big change, in that it is also now possible to apply for rights covering the whole European Union in one go. Breeders of ornamentals have already made extensive use of this new development, with some 60% of new EU applications being in this category.

Another important change for breeders of ornamentals has taken place in the United Kingdom: national plant breeders' rights have existed here for over thirty five years, but in May 1998, the 1997 Plant Varieties Act came into force: this Act swept away the former slightly cumbersome lists of 'protectable species' and extended protection to all genera and species, also making other changes in the scope and strength of the right in line with the 1991 UPOV Convention. Similar changes are taking place in many other countries, while the EU offered protection to all genera and species from the start.

However, whether a breeder wants to make one or more individual national applications or an EU application, the process is broadly the same. An application form is filled in, covering administrative details, and also a technical questionnaire which, with a photo, describes the plant; these are sent to the relevant Plant Variety Rights (PVR) office together with the appropriate fees. Provided there are no problems with the administrative part of the application, the next step is a growing test to establish whether the plant is DUS.

The Trials Network

In general, most PVR offices have a technical unit responsible for trials, or have appointed separate approved technical institutes to carry out the work. Such is the case with NIAB. Applications for United Kingdom rights are made to the PVRO; in the case of ornamentals the technical details are then passed to NIAB for a growing trial to be organised. NIAB in turn are responsible for making a technical recommendation to the PVRO at the end of the trial. If the recommendation is positive and all other factors are in order, United Kingdom rights will be granted.

This is the essence of the system but because the horticultural trade is international so are the trials. In the 30 or more years since the coming into force of the first UPOV convention, the member states have gradually developed a co-operative network whereby under a system of bilateral agreements, trials on most important horticultural taxa take place at one test centre only, for better use of resources and so that all candidate cultivars can be directly compared side by side. The result is a network of test centres in UPOV states, each centres of expertise in certain genera and species, maintaining large databases and reference collections.

Under this system the United Kingdom has assumed international responsibility for many species, most importantly in *Chrysanthemum* L., whereas, for example, *Pelargonium* L'Hér. ex Aiton is tested in Germany, *Gerbera* L. in the Netherlands and *Aster* L. in Israel.

Applicants apply to the PVR office where they are seeking protection, but the technical details of the application are then passed to the test centre via the PVR offices. After the trial, the test centre will make a recommendation to the PVR office, but it is the decision of the PVR office whether to grant rights or not. In this way the cultivar can be the subject of several national applications or just one, for the EU, but there will only be one growing test.

Those undertaking the work are generally, like myself, delegates to the relevant UPOV Technical Working Party which meets at least once a year to discuss and agree Technical Guidelines and test methods. Thus wherever they take place, all DUS trials are conducted using the same UPOV methods and standards.

Growing Test

The growing trial has to establish whether the cultivar is DUS and sufficiently novel taking into account the application date, while also describing the cultivar according to UPOV methods. There is no assessment of merit or relative value.

The test is conducted according to UPOV methods and principles and broadly follows the same path for all species. The starting point is the technical information provided by the breeder. After evaluation of this, the cultivar is allocated to the first suitable trial alongside selected existing similar cultivars.

During each flowering period or growing season the plants are evaluated for:

- Uniformity and Stability. Regular (daily, weekly, etc. depending on species) assessment is made against the agreed UPOV standards for the species, taking into account the propagation method.
- Distinctness. A detailed comparison is made of the cultivar with the existing similar cultivars, and consultation and searches take place to see if other perhaps closer cultivars exist which may need to be obtained for further trials (the original opinion may change after viewing the living plant, as opposed to the photograph

and description). This part of the trial takes into account all known existing cultivars, not just protected ones as is sometimes thought.

- Description. The cultivar characteristics are measured and recorded according to UPOV guidelines and methods. If the cultivar meets the criteria for a grant of rights, a full UPOV description is prepared.
- Novelty. A check is made to see if the cultivar is sufficiently novel, taking account of the application date.

If any problems are found during the course of the trial, the breeder is invited to visit to see the plants so that he has full information for any discussions with the PVR office.

The end result of the trial is a recommendation to the relevant PVR office(s) for either grant of PBR, or a further test for distinctness, uniformity or stability, or refusal. If all other aspects of the application are in order, the PVR office will use a positive recommendation as the basis for a grant of rights.

After the trial the test centre generally maintains a reference sample of the plants.

Cultivar Names

Evaluation of the denomination (cultivar name) proposed by the breeder takes place as a separate but interlinking process. It is generally carried out by the PVR office with technical advice from the test centre, or in some cases by the test centre direct. The proposed cultivar name is evaluated according to UPOV recommendations. If it appears suitable, it will be published in the Gazette of the relevant PVR office, for all other PVR offices and interested parties to check. Objections must generally be lodged with the publishing office within 3 months.

If there is a valid objection, another name must be proposed. Since a cultivar must have an approved name for a grant of rights, any delay (for instance if several proposals are rejected) can lead to a delay in the grant or in extreme cases, possibly the refusal of the application.

Any further applications for the same cultivar in other UPOV states must use the same name.

Areas of Confusion in the Naming of Protected Cultivars

There is much confusion over the correct naming of plants in horticulture, but I would like to conclude with an overview of those areas which cause particular practical problems in the trialling of cultivars for a grant of PBR.

Lack of Stability in Scientific Names

This presents particular problems in the areas of legislation, administration, database management and the checking of cultivar names, particularly if any part of the work is done by staff without formal training in taxonomy. For example, if one is using the UPOV CD-ROM to search for information about florists' chrysanthemums, one needs to be aware that the records appear under some 14, possibly 16 (2 are ambiguous!) different categories, as they have been provided from the different national databases. This is perhaps an extreme example but it can easily lead to information being missed or wrongly indexed.

What is 'Registration'?

Confusion often arises between the two entirely separate processes of application for legal PBR protection for a plant cultivar, and an application to an International Registrar to register a cultivar name. It is preferable to avoid the use of the term 'registered cultivar' which may be ambiguous and misleading.

Lack of Awareness

UPOV member states and the EU provide PBR applicants with guidance notes on the choice of an appropriate cultivar name, but in general there seems to be very little awareness among breeding companies of *The International Code of Nomenclature for Cultivated Plants — 1995* (Trehane *et al.* 1995) and the advice and information it contains about the appropriate naming of cultivars.

Selling Names

Used in its loosest sense to cover all forms of trade designation whether trademarked or not, this is the area that causes the greatest problems.

Under the UPOV system a cultivar must have an approved name for a grant of rights; UPOV recognises that some breeders may like to use a trade designation in addition to this approved name but such designations are not officially recognised by UPOV or recorded by most PVR offices. Trials centres however, must maintain extensive databases with cultivar names and trade designations cross-referenced in order to be fully informed about cultivars in the market.

This system is at best slightly cumbersome, but perhaps understandable in view of the background. In naming the new cultivar, the applicant has to satisfy not only the legal requirements of the PBR system but also the demands of the marketing department which will be selling the plants; with names for ornamentals being such a part of the 'image' of the plant and perhaps also a selling point, it is not surprising that it is sometimes impossible to reconcile the two.

The consequent use by many applicants of a 'nonsense' epithet for use as the cultivar name later followed by a trade designation causes much confusion among certain sectors of the horticultural world. It is perhaps not appreciated that while some applicants may release only one new cultivar every few years (a variegated sport found by a grower for instance) and thus have plenty of time to give to matters such as the evaluation of the cultivar, choice of suitable name, registration and so on, big breeding companies are making perhaps hundreds of PBR applications per year. Such is the pressure in this highly competitive world, particularly in species where a high proportion of new cultivars arise from mutation, that applications are usually made before the cultivar is fully evaluated, 'just in case'. About 20% of applications are withdrawn before completion of test, and many more are never launched onto the market. Against this background it is hardly surprising that the applicants either wait until the last minute before proposing a cultivar name, or use a meaningless but acceptable one, adding a trade designation if the cultivar turns out to be a good one. This way they avoid the waste of a 'good' name.

It should also not be forgotten that the primary aim of the horticultural industry is to promote and sell plants, and get a good return on the sales. A name which engenders huge turnover in one country may have quite the opposite effect in another: history is littered with the stories of marketing men who chose names for new products without checking any connotations they might have in the launch country. Such a name would obviously cause problems, so the solution of two trade designations linked by a common cultivar name seems not unreasonable.

So while use of this system goes somewhat against the basic idea that a cultivar should have only one name world-wide (and applicants should be encouraged to propose a cultivar name which will also be the name under which the plant is sold if at all possible) it appears to be workable so long as the cultivar name and the trade designation always appear together, particularly in catalogues and on labels. How else is the plant trade to know that the cultivar is protected by PBR?

Many breeders have adopted this system without difficulty, but there is still some way to go. With so many people involved in the supply chain from breeder to public (a recent application for PBR involved seven different companies across three continents) information tends to get lost.

The situation has been exacerbated by PBR laws conforming to the 1991 UPOV Convention, which allow a cultivar to be sold on the market for up to one year before making a PBR application. This means in practice that the cultivar may have become well known under its trade designation before an application for rights is made. Thus it is possible that there might be a recommendation to compare the candidate (under its cultivar name) against an apparently similar existing cultivar, which turns out to be the same plant under its trade designation. This is quite possible if information about the plant is not freely available.

In conclusion: the horticultural industry's aim is to sell plants and lots of them, and subject to any legal considerations, names will be chosen to assist in that aim. Therefore those seeking to establish greater stability and clarity in this area must not make the rules too complicated, and must also ensure that user-friendly guidance is readily available. This symposium, involving so many people with different but overlapping areas of relevant expertise, both legal and technical, should go a long way towards removing many of the existing misconceptions and achieving this aim.

Reference

Trehane, P. *et al.* (eds). (1995). International Code of Nomenclature for Cultivated Plants — 1995. 175 pp. Quarterjack Publishing, Wimborne.

REGISTRATION OF PLANT NAMES

4

Austin, J.E. (1999). An overview of Plant Variety Rights and National Listing in the United Kingdom in relation to the nomenclature of cultivated plants. In: S. Andrews, A.C. Leslie and C. Alexander (Editors). Taxonomy of Cultivated Plants: Third International Symposium, pp. 97–100. Royal Botanic Gardens, Kew.

AN OVERVIEW OF PLANT VARIETY RIGHTS AND NATIONAL LISTING IN THE UNITED KINGDOM IN RELATION TO THE NOMENCLATURE OF CULTIVATED PLANTS

JOHN EDWARD AUSTIN

Ministry of Agriculture, Fisheries and Food, Plant Variety Rights Office and Seeds Division, White House Lane, Huntingdon Road, Cambridge CB3 OLF, UK

Abstract

The United Kingdom system for plant variety rights is based on the International Convention for the Protection of New Varieties (the UPOV Convention), whilst the requirements for National Listing are laid down in European Community (EC) Council Directives on the Common Catalogues. Plant Variety Rights were first established by the 1964 Plant Varieties and Seeds Act but National Listing was not a requirement until the United Kingdom entered the European Community in 1972.

Plant Breeders' Rights are now available for all genera and species whereas National Listing covers only the most important agricultural and vegetable crops.

The United Kingdom, along with most other European Union (EU) members, adopts the UPOV guidelines on naming. There are no clearly definable rules for naming in the EC Common Catalogue Directives which lay down the procedures for National Listing. Member States vary in their interpretation of naming rules and whilst the potential for difficulties are high the number of problems are, in practice, few. In 1997 voluntary detailed rules on naming were developed by an EU Working Group and the Community Plant Variety Office.

Introduction

As a member of the European Union (EU) the United Kingdom has to conform with the legislation concerning plants and seeds. The main requirements for National Listing are contained in Council Directives 70/457/EEC (Common Catalogue of Varieties of Agricultural Plant Species) and 70/458/EEC (Marketing of Vegetable Seed). The provisions made in the Directives set out the results to be achieved within a stated period but leave implementation to national Governments. In the United Kingdom these two Directives are implemented through the provisions of the 1964 Plant Varieties and Seeds Act and the Regulations made under the 1964 Act. They came into force in 1972 when the United Kingdom entered the European Community. Changes to and revisions of the Directives occur from time to time and these are achieved through discussion in a series of Committees in which Member States and the Commission are represented.

In a similar way, the Council Regulation 2100/94 establishing a Community plant protection system is applicable in the United Kingdom and is based on the provisions of the UPOV 1991 Convention. The United Kingdom national legislation has recently been updated to bring our national system into line with UPOV 1991. The Plant Varieties Act 1997 came into force on 8 May 1998.

The Plant Variety Rights Office (PVRO) is responsible for the administration and grant of Plant Breeders' Rights in the United Kingdom. The Plant Variety and Seeds Division (PVS) of the Ministry of Agriculture is responsible for the administration of the National Listing (NL) systems. The technical work is contracted to the National Institute of Agricultural Botany in England and Wales, the Scottish Agricultural Science Agency and the Department of Agriculture for Northern Ireland.

The Common Catalogue Directives lay down the conditions for acceptance in a common catalogue of cultivars of the most important agricultural and vegetable species. There are two Common Catalogues, one for cultivars of agricultural species and one for cultivars of vegetable species. The Common Catalogues are formed by combining the National Lists of cultivars of each Member State. Each Member State is required to produce National Lists of cultivars officially accepted for certification and marketing. The conditions for acceptance for a National List are that a cultivar must be Distinct, Uniform and Stable (DUS) and, in the case of agricultural cultivars only, have a Value for Cultivation and Use (VCU). There is no equivalent VCU requirement for vegetable species. In the United Kingdom all new cultivars that have been entered on the National List or granted Plant Breeders' Rights are indicated in the *Plant Varieties and Seeds Gazette* which is published monthly. An annual publication called the 'Special Edition' summarises all valid grants and other listings. It is illegal to market seed of any agricultural or vegetable cultivar in the European Union unless the cultivar is on a Member State's National List or the EC Common Catalogue.

Legislative Provision for Naming Cultivars

Council Directives 70/457/EEC and 70/458/EEC do not lay down clear guidance or detailed rules on naming. Neither do they indicate that naming should be in conformity with the revised UPOV Convention (1991) or the Council Regulation 2100/94 on Community Variety Rights. However, there is currently a proposal to amend these Directives to require that the suitability of cultivar names should be determined in accordance with Article 63 of Council Regulation 2100/94 on Community Plant Variety Rights. This amendment will also enable detailed implementing rules to be adopted by the Standing Committee on Seeds. As this proposal is not contentious it is expected that it will be adopted under the Austrian Presidency in the latter half of 1998. When adopted these provisions should join together the general principles on naming for National List purposes with those agreed for Community Plant Variety Rights.

Rules for Cultivar Names

Article 63 of Council Regulation (EC) No. 2100/94
Articles 63.3 and 63.4 indicate that a cultivar name is suitable provided:

- Its use in the territory of the Community is not precluded by the right of a third party.
- It does not cause users difficulties as regards recognition or reproduction.

- It is not identical or confused with a cultivar name under which another cultivar of the same or closely related species is entered in an official register or

  "has been marketed in a Member state or in a Member State of the International Union for the Protection of New Varieties of Plants, unless the other variety no longer remains in existence and its denomination has acquired no special significance"

- It is not confused with designations for marketing of goods which are commonly used or have to be kept free under other legislation.
- It does not give offence in one of the Member States or is contrary to public policy.
- It is not misleading nor causes confusion concerning the characteristics, the value or the identity of the cultivar, breeder or other parties.
- The cultivar name being already entered in a register in one of the Member States or a member of UPOV or in another State which has been granted equivalence under the Common Catalogue Directives and has been marketed for commercial purposes.

The importance of Article 63 is the relationship between naming rules developed in the EU, UPOV and Third Countries which have been granted equivalence of maintenance. It is intended that the name for a plant awarded EU plant variety rights does not conflict with any other name registered in the same cultivar class.

Working Rules

Early in 1997 the Commission established a small Group of Rapporteurs to develop a harmonised system of naming, both for National Listing and Community Variety Rights. The Group met a number of times, examined the various difficulties and presented a report to the Standing Committee on Seeds in June 1998. This report made a number of recommendations as well as proposing working rules for naming. It recommended that the Administrative Council for Community Plant Variety Rights adopt the rules and procedures of the Group of Rapporteurs and that a legal basis for rules on cultivar names be created in Council Directives 70/457/EEC and 70/458/EEC. This would enable working rules and procedures to be adopted by the Standing Committee on Seeds.

As indicated earlier in this paper, a legal basis to amend Directives 70/457/EEC and 70/458/EEC require that the suitability of cultivar names should be determined in accordance with Article 63 of Council Regulation 2100/94 on Community Variety Rights is expected to be agreed under the Austrian Presidency later in 1998.

In addition the Community Plant Variety Office has produced Guidelines on Variety Denominations which have been presented to the Administrative Council for adoption. These guidelines closely follow the working rules agreed in the Group of Rapporteurs. They are based on the following criteria and utilise examples of names that are not permissible. The criteria are as follows:

- The need to be recognisable as cultivar names.
- The need to be capable of reproduction.
- The exclusion of prohibited cultivar names.
- The exclusion of cultivar names liable to mislead or confuse as to either the characteristics or value of the cultivar or the identity of the breeder or other party.

- The refusal of identical or similar cultivar names.
- The need to prevent the re-use of cultivar names.
- The recognition of UPOV cultivar classes and complementary classes.

When the Common Catalogue Directives are amended the working rules established by the Standing Committee on Seeds, care will have to be taken to ensure that these are in accordance with the criteria now developed by the Community Plant Variety Rights Office.

Working Procedures in the United Kingdom

In the United Kingdom we use the UPOV (1991) rules on cultivar names. The working procedures for naming are explained in the guides to National Listing and PBR available from the PVRO. These refer to Article 20 of the revised UPOV Convention (1991). Applicants must supply a name for their cultivar either at the time of application or later during the testing period. If a name is not submitted within the time period specified then there is no obligation to take any further steps with the application.

Names are checked in our computer naming system whose database is updated regularly. After passing our internal audit, acceptable names are published in the monthly *Plant Varieties and Seeds Gazette* and a period of three months is allowed for any interested parties to make representations. Our gazettes go to all UPOV member countries and their equivalent publications are received under reciprocal arrangements. Any person may object to the approval of a name proposed for a cultivar if they feel that the name may lead to confusion with that of an existing cultivar or to the ownership or characteristics of the cultivar. A fee for making an objection is payable on delivery of the written representations.

From the date on which rights are granted, reproductive material of the cultivar must only be sold or offered for sale using the name registered in the United Kingdom. Whilst there is the potential for many difficulties surprisingly few arise in practice and these can usually be resolved.

Sadie, J. (1999). Cultivar registration for statutory and non-statutory purposes in South Africa. In: S. Andrews, A.C. Leslie and C. Alexander (Editors). Taxonomy of Cultivated Plants: Third International Symposium, pp. 101–106. Royal Botanic Gardens, Kew.

CULTIVAR REGISTRATION FOR STATUTORY AND NON-STATUTORY PURPOSES IN SOUTH AFRICA

JOAN SADIE

National Department of Agriculture, Directorate Genetic Resources, Private Bag X5044, Stellenbosch 7599, South Africa

Abstract

In South Africa three different registration systems are used to provide for cultivar names, protection of material and plant improvement. A cultivar can be registered by using one or two of the systems, depending on the crop and purpose of registration. Two of these systems are statutory, based upon the Plant Breeders' Rights Act, 1976 (no. 15 of 1976), which provides the option for protection and the Plant Improvement Act, 1976 (no. 53 of 1976), which requires obligatory registration. The third system is non-statutory and provides for international registration of *Protea* cultivar names according to the rules of the *International Code of Nomenclature for Cultivated Plants*. All three systems are managed by the Directorate Genetic Resources.

Background to the Registration Systems

Plant Breeders' Rights

Plant Breeders' Rights are granted under the Plant Breeders' Right Act, 1976 (Act no. 15 of 1976, as amended). After amendments in 1996, the Act is based on the 1991 UPOV Convention (UPOV is the French acronym for International Union for the Protection of New Varieties of Plants). Although South Africa was the tenth member of UPOV after signing the 1978 Convention in October 1978, the first Plant Breeders' Right was already granted in 1970 for a garden bean cultivar. Currently Plant Breeders' Rights are available for cultivars of 204 genera of which 122 genera are ornamental.

To qualify for a Plant Breeders' Right, a cultivar must pass the DUS test (Distinctness, Uniformity, and Stability). It must be novel, i.e. not being commercially available for more than one year in the country where the application is filed and more than four years, or six years for trees and vines, elsewhere. The cultivar must also have an acceptable name. The costs for registering a Plant Breeders' Right consist of an application and examination fee resulting in a total amount of R 1,464 or R 1,854 for trees and vines.

The purpose of Plant Breeders' Rights is to provide for protection of the plant breeders' propagating material.

National Listing

The national variety list is maintained in terms of the Plant Improvement Act, 1976 (Act no. 53 of 1976, as amended). Part of the Plant Improvement Act is based on the rules of the International Seed Testing Association (ISTA). National listing is obligatory, since the Plant Improvement Act states that propagating material of a cultivar may only be sold if the name is included in the national list. This of course is only applicable to genera declared in terms of the Act. The most important crops for South Africa are declared in terms of the Act, 56 genera in total, of which the majority are agricultural crops.

Cultivars must comply with the DUS test and have an acceptable name in order to be included in the national list. In order to qualify for selling, the propagating material must comply with certain physical and phytosanitary requirements prescribed by the Act. The cost of registration in the national list consist of an application and examination fee resulting in a total amount of R 1,000 or R 1,400 for trees and vines.

Certification schemes for seed crops, deciduous fruit, table and wine grapes and potatoes are in force and have been published under the Plant Improvement Act. Unofficial certification schemes are also available for strawberries and citrus and should be published in the near future, following negotiations. Seed of certain cultivars may only be sold if the seeds are certified. Cultivars with this limitation are listed in a table additional to the national list. The management and application of the certification schemes have been privatised. The South African National Seed Organisation (SANSOR) is responsible for the seed certification scheme while the Deciduous Fruit Plant Improvement Association and Vine Improvement Association are responsible for the deciduous fruit and wine grapes respectively.

The purpose of the Plant Improvement Act is to ensure that only propagating material of high quality is sold in South Africa. National listing ensures control over cultivars due to the requirement of evaluating cultivars before inclusion in the list. Therefore we have the opportunity to minimise the possibility that material of cultivars illegally obtained, will be sold. The list is also of help when compiling evaluation trials for Plant Breeders' Rights. All cultivars available in South Africa are listed; therefore similar cultivars can be easily selected. We only have difficulty where no list exists for a crop, e.g. *Medicago polymorpha* L. The cultivars in circulation in the country have to be identified by searching the importation lists and by liaison with the seed companies. All of these cultivars have to be planted to ensure that a plant breeders' right is not granted erroneously to an already existing cultivar.

International *Protea* Registration

Protea registration had its origin in 1973 when the Horticultural Research Institute and the then Division of Plant and Seed Control jointly initiated a cultivar registration program. In 1978 Dr G. J. Brits negotiated with the Commission for Nomenclature and Registration at the conference of the International Society of Horticultural Science in Sydney, Australia to appoint an International Registration Authority (IRA) for all cultivars derived from South African genera of *Proteaceae* Juss. The Division Variety Control, which already dealt with Plant Breeders' Rights, was subsequently appointed as IRA in 1980.

The first five registrations were done in 1974, including Plant Breeders' Rights. Up to 1984 another fourteen cultivars were added to the International *Protea* Register (IPR), all of them from South Africa. No official publication appeared, except for the *South African Plant Variety Journal* no. 30 (Anon. 1986), which included a list of the names registered. In 1992 a new attempt was made to activate international *Protea* registration. A huge input was made to find all cultivar names that had been published. The applications for the International *Protea* Register received until then were processed and a combined register and international checklist was produced as the preliminary first edition of the IPR (Anon. 1993). This was circulated informally to delegates at the seventh conference of the International *Protea* Association in Zimbabwe, 1993. The fifth edition was published in 1998 (Sadie 1998).

Registration in the IPR is free of charge and mainly based on the acceptability of the name and distinctness of the cultivar, based on photographs and a short description. *The International Protea Register* which incorporates an *International Checklist* is published annually and distributed worldwide to approximately 120 people in fifteen countries.

International *Protea* registration is voluntary, which places an extra burden on the IRA to convince breeders to register their cultivars. In South Africa this problem has been overcome to a large extent by including the application forms for IPR with the forms for Plant Breeders' Rights. The South African breeders have responded very positively to the call for registration of cultivar names and submit their applications for the IPR even when no Plant Breeders' Rights are involved. In other countries we have no jurisdiction and have to rely on the co-operation of the breeders, responding to calls for registration. Good responses have been received from Zimbabwe and Australia.

Comparison of the Registration Systems

Similarities
The most obvious similarity between the systems is the requirements for an acceptable name. The requirements for national listing are based on that for Plant Breeders' Rights. The requirements for names in Plant Breeders' Rights are based on that of the UPOV Convention. The rules of the UPOV Convention are similar to that of the *International Code of Nomenclature for Cultivated Plants* or Cultivated Plant Code (Trehane *et al.* 1995), which makes one think that the rules of the former are based on the latter. UPOV was established in 1961, while the first edition of the Code was published in 1953.

The three systems also have a common goal, namely to promote order and stability in both the international and national industries. This is to a lesser extent valid for Plant Breeders' Rights, because the main aim there is to provide for protection of plant breeders' propagating material.

A similarity between the IPR and national listing is that once a cultivar name has been registered, there is no limitation to the period of registration, as there is with Plant Breeders' Rights. Also the novelty requirement is not applicable for either of the two systems. Cultivar names can still be registered after they have been published and commercialised, provided that no duplication will result from this.

Evaluations for Plant Breeders' Rights and national listing are exactly the same. The UPOV guidelines form the basis of the descriptions and the DUS test is performed for both. The protection afforded by these two systems is only valid in South Africa.

Differences
When working with the IPR, we refer to cultivars, while for Plant Breeders' Rights and national listing, reference is made to varieties.

Plant Breeders' Rights are optional, but when granted the protection has statutory force. The IPR on the other hand has no statutory powers and has to rely on the positive co-operation and attitude of the participants.

Plant Breeders' Rights and national listing are only valid in South Africa where the legislation was drafted, while international *Protea* registration is not limited to a country. Evaluations for Plant Breeders' Rights and national listing are done by performing a full morphological description according to the UPOV guidelines. This is necessary due to the effect climatic conditions have on the expression of characteristics. Evaluation of an application for the IPR is based on the short description and photographs provided.

One of the prerequisites for Plant Breeders' Rights is that the applicant must have a residential address in South Africa. For the IPR this is not necessary.

A qualification distinguishing Plant Breeders' Rights from national listing and international *Protea* registration, is the novelty requirement. As already mentioned, a cultivar may not have been sold for longer than one year in South Africa or longer than four years, or six years in the case of trees and vines, in other countries.

Protection obtained by Plant Breeders' Rights is only valid for a period up to 25 years, while there is no time limit for national listing and international protea registration. Although the latter two systems have no time limit connected to the registered name, in the case of national listing a cultivar name is deleted from the list if the cultivar is no longer in cultivation. In the case of the IPR, a name is only used once, unless the IRA decides no confusion will result from the re-use of the name.

Discussion

Plant Breeders' Rights

A problem experienced with both Plant Breeders' Rights and national listing is the increasing difficulty in distinguishing between cultivars, due to the development of new breeding techniques and a decreasing gene pool. We also experience increased pressure by the industry to use new techniques such as DNA methods to prove distinctness. We do not have the expertise, equipment or funds to perform these tests. Also UPOV does not accept these new methods and will not accept it until a sufficiently stable method has been developed. The same recommendation was made by ISTA at the 1998 Congress in South Africa, that the new techniques cannot be accepted and implemented before stable methods have been developed.

Testing the first cultivar in a species or genus as well as testing cultivars of genera from other countries are also problematic in Plant Breeders' Rights evaluations. It is very difficult to determine the novelty of such cultivars. Usually the right is granted and one must rely upon the people in the industry to identify cultivars for which Plant Breeders' Rights were granted erroneously.

Another problem is that old cultivars as well as their descriptions disappear from the scene with time. Should they still be considered for comparison with new cultivars in Plant Breeders' Rights trials or not? If they must be considered, how will the comparison be conducted, because it is difficult to distinguish on the basis of a description only. It has already happened that a Plant Breeders' Right was granted to a pear, which was, after about a year, pointed out to be an old variety dating back to the 1950s. Needless to say, the right was cancelled immediately.

National Listing

The advantage of national listing is that we know which cultivars are in the country and this makes Plant Breeders' Rights evaluations much easier. The main disadvantage, according to our fruit industry, is that selling of propagating material of the new cultivars is prohibited when the denomination is not in the list. The industry is thus hampered. From our point of view, the national listing serves to prevent illegal propagation and selling of cultivars that would otherwise qualify for Plant Breeders' Rights. In most cases plant improvement organisations evaluate material under local climatic conditions before applying for PBR. Selling and distribution of propagating material could in these cases counteract applications when it is done for a period longer than one year before the application is filed.

Regarding the selling of propagating material, the Plant Improvement Act seems to be contradictory to the Plant Breeders' Rights Act. In terms of the Plant Improvement Act, no material may be sold unless the name of the cultivar is in the national list, while no such prohibition exists in the Plant Breeders' Rights Act. This is causing great frustration in our industry. We have the problem however, that if selling is permitted whilst an application for PBR is being assessed, what will the consequences be when the right is not granted or it is discovered that the cultivar is indistinguishable from another cultivar already protected by Plant Breeders' Rights.

International *Protea* Registration

The biggest difficulty experienced with the IPR, is that the industry we serve is spread from Hawaii and California (USA) in the west to Israel and Spain in the north and Australia and New Zealand in the south/east. It is extremely difficult to reach the appropriate people across these distances and get co-operation from them. The distances cause extra effort regarding correspondence, resulting in turn in elevating costs.

Due to the fact that we deal with Plant Breeders' Rights, there is a tendency to also perform the DUS test on IPR applications. According to the Cultivated Plant Code, it is not the duty of the IRA to prove distinctness, but how can order in the industry be achieved if there is a chance that indistinguishable cultivars are allowed to exist in different countries under different names? It is easy to accept cultivars with Plant Breeders' Rights, because the DUS test has already been performed, but can the problem be solved with the other cultivars not involved in Plant Breeders' Rights?

Another desirable objective of the IRA is to collect herbarium specimens of the registered cultivars. Again the trouble is to find the right person in the relevant countries to prepare the specimens and send them to the Registrar. I suppose the answer is to keep the correspondence up and find the right persons!

Conclusions

It is important to note that the three registration systems have to be dealt with separately. Although there are some similarities, they have different aims and objectives. Consultation with other IRAs and statutory registration authorities could provide answers to the questions posed. Regarding international co-operation for the IPR, persistent promotion of cultivar registration must be undertaken although this is time-consuming and expensive. A possible solution to the huge distances between the countries involved in protea registration, is the Internet homepage which is planned for the Division Variety Control. This should be to the advantage of all three systems by making information and application forms more accessible to applicants, including those who are not involved in proteas.

References

Anon. (1986). Additional Information. *S. African Pl. Var. J.* 30: 10.

Anon. (1993). International *Protea* Register. Preliminary 1st edition. Directorate Plant & Quality Control, Stellenbosch.

Sadie, J. (1998). The International *Protea* Register, Ed. 5. 58 pp. Directorate of Plant and Quality Control, Stellenbosch.

Trehane, P. *et al.* (eds). (1995). International Code of Nomenclature for Cultivated Plants — 1995. 175 pp. Quarterjack Publishing, Wimborne.

Dawson, I. (1999). Cultivar registration in Australia. In: S. Andrews, A.C. Leslie and C. Alexander (Editors). Taxonomy of Cultivated Plants: Third International Symposium, pp. 107–111. Royal Botanic Gardens, Kew.

CULTIVAR REGISTRATION IN AUSTRALIA

IAIN DAWSON

Australian Cultivar Registration Authority, Australian National Botanic Gardens, GPO Box 1777, Canberra, ACT 2614, Australia

Abstract

The Australian Cultivar Registration Authority is the International Registration Authority for genera that are endemic, or predominantly endemic, to Australia. To date 401 cultivars have been registered from 64 genera. Most new cultivars are stabilised forms selected from wild populations. A major problem is the diversity of the Australian flora, and our imperfect knowledge of it. Current projects include the compilation of a colour code database, the publication of all cultivar descriptions on the internet, and promotion of the benefits of stable nomenclature with commercial plant growers.

Introduction

This year marks the 40th anniversary of the conception of the Australian Cultivar Registration Authority (ACRA). It is a happy coincidence that the very first letter in our files, dated 4th December 1958 and which dealt with the procedure for obtaining international recognition of our organisation, was written by the secretary of the International Commission for the Nomenclature of Cultivated Plants, H.R. Fletcher, whose address was the Royal Botanic Garden Edinburgh. For the record the gestation period was five years as we commenced operations in 1963.

At the Second International Symposium on the Taxonomy of Cultivated Plants in Seattle in 1994, the then registrar, Geoff Butler, presented a paper (Butler 1995) on the workings of the Australian Cultivar Registration Authority. His paper dealt with how the Authority was established, how we register new cultivars, the achievements of ACRA and the challenges it faced at that time. My purpose is not to repeat that paper but rather to give you an update and to identify the challenges that we face and how our role is changing.

ACRA

ACRA is a committee formed by representatives of each of the major regional (State) botanic gardens, the Society for Growing Australian Plants, and the Nursery Industry Association of Australia. In addition we can appoint other members who we feel represent special interests or who have expertise that is of value to us. The office is maintained at the Australian National Botanic Gardens (ANBG) in the national capital, Canberra. ANBG grows only our native flora. To date the living collection includes about 7000 taxa.

ANBG provides office space and services, the facilities of the Australian National Herbarium, and 10% of my time. As well as being responsible for ACRA I am also in charge of horticultural research and the seedbank at ANBG. My research interest for the past 16 years has been the economic development of the Australian flora so I have experience and close contacts with the various plant based industries whose needs ACRA serves. Most of the work is done by volunteers. This includes the clerical, record keeping side of cultivar registration and mounting herbarium specimens. The main task at the moment is checking and revising all our records prior to internet publication.

ACRA is the International Registration Authority (IRA) for Australian plant genera excluding those covered by other authorities. This includes:

- All Australian endemic genera.
- All predominantly Australian genera.

In addition to our IRA responsibilities we also register:

- All new varieties/cultivars derived from indigenous species which have been accepted by the Australian Plant Breeders' Rights Office.

There are also some species that belong to genera that are not predominantly Australian for which we have accepted registrations. *Helichrysum* Mill., *Syzygium* Gaertn. and *Microlaena* R.Br. are some examples.

To date we have registered 401 cultivars from 64 genera (Table 1). Nearly a third are *Proteaceae* Juss., and most of those are *Grevillea* R.Br. ex Knight. Other significant *Proteaceae* are *Banksia* L. and *Telopea* R.Br. The next largest group are the *Myrtaceae* Juss., with nearly a quarter of all registrations. These include *Agonis* (DC.) Sweet, *Baeckea* L.,

TABLE 1. ACRA Registrations (including PBR) at 28 June 1998.

Grevillea R.Br. ex Knight	112	*Eriostemon* Sm.	2
Anigozanthos Labill.	35	*Lechenaultia* R.Br.	2
Callistemon R.Br.	32	*Lophostemon* Schott	2
Chamelaucium Desf.	30	*Macadamia* F. Muell.	2
Boronia Sm.	12	*Ozothamnus* R.Br.	2
Correa Andrews	12	*Scaevola* L.	2
Crowea Sm.	12	*Acmena* DC.	1
Leptospermum J.R. Forst. & G. Forst.	10	*Actinodium* Schauer	1
Bracteantha Anderb.	9	*Astartea* DC.	1
Telopea R.Br.	9	*Bauera* Banks ex Andr.	1
Acacia Mill.	8	*Callitris* Vent.	1
Banksia L.	8	*Chrysocephalum* Walp.	1
Eucalyptus L'Hér.	7	*Clematis* L.	1
Brachyscome Cass.	6	*Diplarrhena* Labill.	1
Hardenbergia Benth.	6	*Epacris* Cav.	1
Melaleuca L.	5	*Hakea* Schrad.	1
Pimelea Banks & Sol.	5	*Hypocalymma* (Endl.) Endl.	1
Agonis (DC.) Sweet	4	*Kennedia* Vent.	1
Verticordia DC.	4	*Kunzea* Rchb.	1
Baeckea L.	3	*Leucaena* Benth.	1
Eucryphia Cav.	3	*Lomandra* Labill.	1
Pandorea (Endl.) Spach	3	*Macropidia* J.L. Drumm. ex Harv.	1
Prostanthera Labill.	3	*Melia* L.	1
Pultenaea Sm.	3	*Myoporum* Banks & Sol. ex G. Forst.	1
Westringia Sm.	3	*Olearia* Moench	1
Blechnum L.	2	*Plectranthus* L'Hér.	1
Brachychiton Schott & Endl.	2	*Spyridium* Fenzl	1
Ceratopetalum Sm.	2	*Tetratheca* Sm.	1
Cyathea Sm.	2	*Themeda* Forssk.	1
Danthonia DC.	2	*Thryptomene* Endl.	1
Eremophila R.Br.	2	*Xanthostemon* F. Muell.	1
		TOTAL	401

Callistemon R.Br., *Chamelaucium* Desf., *Eucalyptus* L'Hér, *Leptospermum* J.R. Forst. & G. Forst., *Melaleuca* L. and *Verticordia* DC. The *Rutaceae* Juss. are the next biggest group and important genera are *Boronia* Sm., *Crowea* Sm. and *Eriostemon* Sm. Another family of major significance to ACRA is the *Haemodoraceae* R.Br., with 9% of our registrations being *Anigozanthos* Labill. or kangaroo paws.

Variation in Native Species

A major difficulty for us is the diversity of the Australian flora and our inadequate knowledge of it. We have perhaps as many as 25,000 species of vascular plants (George 1981) with about 18,000 of these adequately described. I say adequately because since the last comprehensive attempt at a national flora, Bentham's *Flora Australiensis* of 1863–1878 (which described 8,125 species) the description of our native flora has been piecemeal. The new *Flora of Australia* is very underresourced and in nearly 20 years only 19 volumes out of an anticipated 51 or more have been published. Even some of the published versions are unsatisfactory, at least for the practical purposes of this user. There is an enormous amount of work still to be done even to make a comprehensive inventory of our flora. The discovery in 1994 of the Wollemi pine (*Wollemia nobilis* W.G. Jones, K.D. Hill & J.M. Allen in the *Araucariaceae* Henkel & W. Hochst.) within 150 km of the centre of Sydney, a city of nearly four million people, is a reminder that we are not necessarily only likely to find small, relatively insignificant new species. Remember that Australian forests and woodlands are typically dominated by angiosperm species so you would think that a group of 35 metre tall pine trees would be rather obvious! Who knows what is waiting to be discovered in the extensive rainforests of the tropical north-east of the continent?

Of the 17,590 species listed in the most recent *Census of Australian Vascular Plants* (Hnatiuk 1990) only about 9% are naturalised. There is also a high degree of endemism, perhaps as high as 90% of the native species. At the last count there were 557 endemic genera, comprising 4038 species. A further 104 genera are mainly endemic to Australia, containing 3,487 species out of 3862. Of the 16,000 or so Australian native species about 2000 species are commonly cultivated (ANBG cultivates 7000 taxa). We have detailed research on the horticulture of only a few of these species. About 350 species are picked from wild populations for the cut flower industry, mainly because we do not know how to cultivate them.

The consequences for cultivar registration, and for plant property rights such as Plant Breeders' Rights (PBR), are that if we do not know what occurs naturally how do we determine what is a novel form? We often do not even know what the 'normal' flower colour range is for a particular species, yet so many applications are based on this character. We have started to compile an inventory of this by describing the natural colours that we have growing in our botanic gardens but so far we have only got data for about 1,000 taxa and we certainly do not have the range described for any of them.

To date we have asked applicants to demonstrate that their cultivars differ from the known (published) range of variation for the species. The Australian Plant Breeders' Rights Office has also accepted applications for stabilised forms selected from wild populations, although this may change in the near future due to a legal challenge. Another problem is that so many species of interest to horticulturists are woody and may have very long juvenile phases. This slows down the registration process particularly for PBR where comparative trials are required. Another characteristic of our registrations is that for the most part they are selections rather than deliberately bred. The main exceptions are *Anigozanthos*, *Leptospermum* and *Chamelaucium* where there have been a number of successful breeding programs.

Use of Plant Resources

I have so far dealt with the botanical resource and I would now like to talk about the people who want to exploit this resource in broadly horticultural ways. They include the nursery industry, the cut flower industry, the forest industries, and the 'bush foods' industry. In addition we have the pastoral industry which for various ecological reasons (such as soil acidification) wants native pasture species. We have a thriving conservation industry, and because of the environmental movement there are political imperatives for mining and public utility companies to use native plants for revegetation. Often it is not possible to use unimproved forms of native plants — for example, large quantities of germinable seed from common Australian grasses such as species of *Themeda* Forssk. and *Danthonia* DC. may be difficult to collect and use because of lack of uniformity in ripening so there is a need for selection against this characteristic. My point is that there are a diversity of economic and legal needs that must be met and for which the establishment and authorisation of cultivar names are part of the process.

Nomenclature

In the preface to the *International Code of Botanical Nomenclature* (Greuter *et al.* 1994) we are reminded that plant nomenclature "entirely depends on user consensus for its universal application and implementation". Unfortunately, the users of cultivated plant names are generally not the public spirited, democratic, rational people that scientists and even some plant taxonomists are. Many of them in fact do not care what the name is as long as they can make a profit from it. If our aim is to promote stability in nomenclature we obviously need to spend a lot of time educating our trade users. Unfortunately, if they only judge benefits in terms of economic rationalism and there is no enforceable legislation for them to operate under, this task is difficult. PBR and other property rights have this regulatory framework. ACRA only has it in so far as we can influence those that grant enforceable property rights.

Take the example of *Chamelaucium* cultivars. Over the past 20 years this genus has risen to be one of the biggest flower crops in the world in terms of both value and volume of production. We have 31 registered cultivars, and of these 17 are protected by PBR. I have photographs of 44 distinct forms and my guess is that there are well over 100 forms in Australia that are potentially registerable, as well as an unknown number in Israel. I have also observed at least four forms that are being grown as *Chamelaucium* 'Purple Pride'. There is obvious scope for chaos.

We can only assess this problem by close liaison with other plant name regulators (such as the PBR Office, patents and trademarks, label manufacturers and publishers) and through education. We also need closer international cooperation.

Fortunately ACRA has been fully consulted at least in the establishment of PBR in Australia, and we have an ongoing relationship with the PBR Office, which means that we provide them with an advisory service for Australian native plant applications as well as herbarium facilities for storage of their plant specimens. We also have excellent relations with the Society for Growing Australian Plants who, apart for the PBR Office, provide most of our applications. We also have reasonably good communications with the major plant label manufacturers and many of the major nurseries. Our main concern is whatever promotional name is used on a label the registered name should also appear.

Education and Resources

Promoting the benefits of stable nomenclature with our users is an ongoing task. We have to persuade them that property rights are *not* compromised by standard nomenclature but are in fact strengthened. It has become apparent in Australia that many in the nursery industry are concerned about pirating of cultivars. We are able to point out that whilst cultivar registration does not prevent others from growing a cultivar, it prevents them from taking out a Plant Breeders' Right on it and thereby stopping the originator from growing it.

Many Australian growers are concerned about overseas growers using Australian native plants without any benefit being gained by Australia. I remind you that under Article 15 of the International Convention on Biological Diversity, access to genetic resources must be subject to prior informed consent and on mutually agreed terms. Again, the proper documentation of the resource helps in the enforcement of those provisions.

Another aspect of our education role is to make our databases accessible. We are in the process of constructing a web site (accessible via the ANBG web site) that will make all our cultivar descriptions, colour code and plant name databases easily available to the public. In addition there are links to other plant name regulators.

Our major problem is that we are under-resourced. We rely on voluntary work, the contributions of staff time by various institutions, and cash donations from the organisations that our committee represents. Like all organisations to survive we have to keep re-inventing ourselves to ensure that we are relevant to contemporary needs, and then make sure that we promote the benefits of what we do so that we are valued by the community.

References

Bentham, G. (1863–78). Flora Australiensis: a description of plants of the Australian territory. 7 vols. L. Reeve & Co., London.

Butler, B. (1995). The Australian Cultivar Registration Authority. *Acta Hort.* 413: 125–129.

George, A.S. (1981). The background to the *Flora of Australia*. In Flora of Australia. Vol. 1. pp. 3–24. Australian Government Publishing Service, Canberra.

Greuter, W. *et al.* (eds). (1994). International Code of Botanical Nomenclature (Tokyo Code). 389 pp. Koeltz Scientific Books, Königstein.

Hnatiuk, R.J. (1990). Census of Australian Vascular Plants. Australian Flora and Fauna Series. No. 11. 650 pp. Australian Government Publishing Service, Canberra.

Singh, B., Panwar, R.S. & Voleti, S.R. (1999). The origin, registration and identification of *Bougainvillea* cultivars. In: S. Andrews, A.C. Leslie and C. Alexander (Editors). Taxonomy of Cultivated Plants: Third International Symposium, pp. 113–116

THE ORIGIN, REGISTRATION AND IDENTIFICATION OF *BOUGAINVILLEA* CULTIVARS

BRIJENDRA SINGH, R.S. PANWAR AND S.R. VOLETI

Division of Floriculture and Landscaping, Indian Agricultural Research Institute, New Delhi 110012, India.

Abstract

Bougainvillea is an important ornamental plant in tropical and sub-tropical cultivation. The origin, registration and identification of the cultivars in this genus are discussed.

Introduction

Bougainvillea Comm. ex Juss. (*Nyctaginaceae* Juss.) is a versatile and important ornamental genus of *c.* 14 species from Central and South America. Plants can be grown outside in tropical and sub-tropical regions and are widespread in Indian horticulture. There are two seasons of flowering in India, February/March and October/November, but a few cultivars (e.g. 'Mary Palmer', 'Thimma', 'Shubhra') give flowers almost all year round. Bougainvilleas find a wide variety of uses in landscape gardening, particularly as roadside plants or on roofs, walls or climbing up other woody plants. They are also used as hedges and even as a subject of bonsai cultivation. Four taxa have contributed the majority of cultivars: *B. glabra* Choisy, *B. peruviana* Humb. & Bonpl., *B. spectabilis* Willd. and *B.* × *buttiana* Holtt. & Standl. (*B. glabra* × *B. peruviana*).

International Registration Authority

The Division of Floriculture and Landscaping was designated the International Registration Authority (IRA) for *Bougainvillea* at the 17th International Horticultural Congress in Maryland, USA, in 1966. As the IRA, the Division is responsible for listing all uses of cultivar names in the genus and for establishing priority of use. It attempts to prevent the duplication of names for new cultivars and registers all new names together with standard descriptions.

More than 100 cultivars had been collected at the Division and these were described in detail. Scientists from the Division also travelled all over India to study other *Bougainvillea* cultivars and to prepare further descriptions based upon a detailed form. Descriptions have also been culled from catalogues and other literature as well as by direct correspondence with growers. In all, 313 cultivars have been recorded and *The International Bougainvillea Check List* was published in 1981 (Choudhury & Singh 1981). A second edition has now been published (Singh *et al.* 1999).

Cultivar Characteristics

Following the initial research, further work has been carried out on the characters most useful in distinguishing between cultivars and determining parental origin. The size and morphology of pollen grains proved to be a valuable guide to the four principal taxa listed above (Swarup & Singh 1964). A study of leaf indumentum also revealed that variation in hair density, especially on the midrib of the leaves, was also helpful in this respect. The most recent research on biochemical characteristics has shown further significant markers to distinguish species and cultivars and characterisation of registered cultivars using these markers is now underway (Singh *et al.* 1999).

Origin of *Bougainvillea* Cultivars

New cultivars of *Bougainvillea* have arisen in a number of different ways. These included natural and controlled hybridisation, bud sports and the development of polyploids by artificial means (Gupta & Shukla 1974, Datta 1985, Deng-Hong 1990).

Bud Sports

Bud sporting is the most common origin of new *Bougainvillea* cultivars. Such sports can be propagated and, when considered worthy, are given new cultivar names. Table 1 shows the extent of variation in the occurrence of named bud sports in the four principal botanical taxa. Some cultivars and their derivatives are particularly prone to sporting and 'Mrs H.C. Buck' has directly or indirectly given rise to more than most other cultivars. Sports of 'Mrs H.C. Buck' itself have been named 'Mary Palmer', 'Soundarys' and 'Vishakha', whilst 'Mary Palmer' has gone on to produce at least seven further named sports ('Mary Palmer Enchantment', 'Munivenkatappa', 'Mataji Agnihotri', 'Odisee', 'Shubhra', 'Smoky' and 'Thimma'). Several of these have variegated foliage.

TABLE. 1. Bud sports of *Bougainvillea*, showing the number of leaf and bract variants for the principal taxa.

Species	Leaf Variant	Bract Variant
B. × *buttiana* Holtt. & Standl.	20	28
B. *peruviana* Humb. & Bonpl.	10	5
B. *spectabilis* Willd.	5	5
B. *glabra* Choisy	2	4
unknown	1	1

Most of these sports have occurred naturally but induced mutations have also given rise to cultivars such as 'Jaya' (a mutation from 'Jajalakshmi') and 'Arjuna' (a mutation from 'Partha').

Hybridisation

Both natural hybridisation and controlled crosses have resulted in the development of new cultivars (see Table 2). In the case of natural hybridisation, one, both or neither parent may be known.

TABLE 2. Examples of *Bougainvillea* cultivars derived from planned (*) or natural hybridisation.

Parentage	Named Hybrid Cultivar
B. peruviana × *B. glabra*	'Begum Sikander'* 'Mrs Butt'
B. peruviana × *B. spectabilis*	'Wajid Alishah'* 'Purple Gem'
B. × *buttiana* × *B. peruviana*	'Chitra'*
B. × *buttiana* × *B. glabra*	'Barbara Karst'
B. spectabilis × *B.* × *buttiana*	'Dr R.R. Pal'
(*B. spectabilis* × *B. glabra*) × *B. spectabilis*	'Daniel Bacon'* 'Margaret Bacon'* 'Lady Watts'*

Polyploids

In *Bougainvillea*, the natural and induced occurrence of polyploids has also given rise to new cultivars (Ohri & Zadoo 1986). The basic chromosome number is n = 17 and both triploid (2n = 3x = 51) and tetraploid (2n = 4x = 68) cultivars are recorded, together with some aneuploids.

References

Choudhury, B. & Singh, B. (1981). The International *Bougainvillea* Check List. 106 pp. Indian Agricultural Research Institute, New Delhi.

Datta, S.K. (1985). The study of *Bougainvillea* cv. Partha on its gamma ray induced mutant "Arjuna". *J. Nuclear Agric. Biol.* 14(4): 159–163.

Dong-Hong, S.S. (1990). Study of radiation induced mutation for *Bougainvillea spectabilis*. *J. China Agric. Univ. (China)* 11(1): 89–93.

Gupta, M.N. & Shukla, R. (1974). Mutation breeding in *Bougainvillea*. *Indian J. Genetics Pl. Breed.* 34: 1295–1299.

Ohri, D. & Zadoo, S.N. (1986). Cytogenetics of cultivated *Bougainvillea*. IX. Precocious centromere division and origin of polyploid taxa. *Pl. Breed. (New York)*. 97(3): 227–231.

Singh, B., Panwar, R.S., Voleti, S.R., Sharma, V.K. & Thakur, S. (1999). The New International *Bougainvillea* Check List. 79 pp. Division of Floriculture and Landscaping, Indian Agricultural Research Institute, New Delhi.

Singh, B., Panwar, R.S., Voleti, S.R., Sharma, V.K., Thakur, S. & Singh, N. (1999). Esterase isozymes in *Bougainvillea* are useful tools for identification of varieties. Presented at National Symposium on "Emerging Scenario in Ornamental Horticulture in 2000 AD and beyond. 21–22 July, 1999, New Delhi. Abstract: 56.

Swarup, V. & Singh, B. (1964). Pollen morphology and leaf hairs in classification of *Bougainvillea*. *Indian J. Hort.* 21: 155–164.

Manners, M.M. (1999). Rose registration: cultivar names, code names and selling names. In: S. Andrews, A.C. Leslie and C. Alexander (Editors). Taxonomy of Cultivated Plants: Third International Symposium, pp. 117–124. Royal Botanic Gardens, Kew.

ROSE REGISTRATION: CULTIVAR NAMES, CODE NAMES, AND SELLING NAMES

MALCOLM M. MANNERS

Dept. of Citrus and Environmental Horticulture, Florida Southern College, 111 Lake Hollingsworth Drive, Lakeland, Florida 33801-5698, USA

Abstract

In recent years, there has been much confusion, dissatisfaction, and argument, over the nature of rose nomenclature and the process of cultivar registration through the International Registration Authority for roses. Commercial growers, hobbyist rose growers and exhibitors, and taxonomists have different goals and philosophies of how and why roses should be named. Much misinformation has been disseminated, and many large producers of roses have stopped registering new cultivars altogether. This paper discusses the history of those concerns, and how trademarks, official designations for exhibition and judging, and registered cultivar names are involved in the situation.

Introduction

The nomenclature of cultivated roses (*Rosa* L. hybrids) is confused and confusing. There are at least four groups of people involved, who have different goals and ideas about the problem, and it is to a great extent their disagreements which have brought about the current situation.

The first group is made up of people who have an interest in the giving of a single, unambiguous name to each rose cultivar, and in the maintenance of an archival record of those names. The second group includes the breeders and marketers of roses, whose major interests at present are the protection of property rights, including trademarks. Names are important to this group only in that they provide another potential tool for making a profit. This group is represented by two organizations: the Communauté Internationale des Obtenteurs de Plantes Ornementales et Fruitières de Reproduction Asexuée (CIOPORA) worldwide, and the All America Rose Selections (AARS) in the United States. The third group are amateur gardeners who exhibit their roses in shows judged by the American Rose Society's rules. This group is politically powerful within that Society. The fourth group is the American Rose Society (ARS, not to be confused with AARS) in its capacity as the International Registration Authority for Roses (IRAR). Each group has legitimate interests and concerns, and to date, no way has been found to make all four groups completely happy. This has resulted in power struggles, with each group trying to force its ideas upon the other groups.

It is the purpose of this paper to describe the history of the situation, the viewpoints of the groups involved, and current developments in rose nomenclature.

The American Rose Society as IRAR

The American Rose Society began recording rose names in 1912, and by 1916, had published a set of rules for the naming of roses. The ARS has been the IRAR since 1955. At present, there are three publications in which the ARS publishes cultivar registrations. The *American Rose* magazine, published monthly except December, and the *American Rose Annual*, a much larger issue of the magazine, published every December, list new registrations. The third publication, where all registrations eventually get published, is *Modern Roses*, a hardbound book, which is revised and updated periodically. The most recent edition is *Modern Roses 10* (Cairns 1993), which lists more than 16,500 rose cultivars.

The political structure of the ARS is the source of part of the current problem with rose registrations. The Society elects a new vice president every three years. The outgoing vice president then automatically moves into the position of president, for the next three years. The president appoints chairmen for the various committees, including the Committee on Rose Registrations, and that chairman selects the other members of the committee, subject to the president's approval. While the executive director of the society (a permanent, non-elected position) is officially the Registrar for the IRAR, it is the Committee on Rose Registrations which actually processes the applications for registration. Because of this system, the registration process is very much controlled by the current president and his political leanings. The Committee on Rose Registrations, its philosophy of registration, and its role as IRAR can completely change, as often as every three years. Much of the commercial growers' concern about the ARS as IRAR is likely due to the ephemeral nature of registration policies.

The typography used in the *Modern Roses* series and other ARS publications has changed from one edition to the next. *Modern Roses 8* (Meikle 1980) and *Modern Roses 9* (Haring 1986) list the registered cultivar name in bold type, without quotation marks or other marks. They list trade designations in plain type and in parentheses, following the registered name. Both editions list code names as trade designations, with only the first letter capitalised. Code names also appear as separate entries in both editions, in this case with the first three letters capitalised in *Modern Roses 8* and with only the first letter capitalised in *Modern Roses 9*, cross-referenced to the registered cultivar names. *Modern Roses 10* (Cairns 1993) continues to list registered cultivars in bold type and without quotation marks. It lists code names with their first three letters capitalised, and it is the first edition to label trademarks as "TM" or "®," where known. An example would be **Scarlet Gem**® (MEIdo; Scarlet Pimpernel). In this case, Scarlet Gem is both the registered cultivar name and a registered trademark. MEIdo is the breeder's code name for the cultivar, and Scarlet Pimpernel is a trade designation, not known to be a trademark.

After intense discussions in 1995–1996 between AARS and the ARS, concerning protection of property rights, the ARS publication typography was changed to list the registered cultivar name first, in single quotation marks and underlined. Trade designations and trademarks are then listed in parentheses, with the "TM" or "®" symbol attached if appropriate. The ARS Approved Exhibition Name (discussed later in this paper), which may be the cultivar name, a trade designation, or a trademark, is shown in bold type. An example from the *American Rose* magazine is 'WEKcryland' (Cadillac DeVille; **Moonstone**™), in which 'WEKcryland', a code name, is also the registered cultivar name. Cadillac DeVille is a trade designation, and Moonstone is an unregistered trademark, also serving as the ARS Approved Exhibition Name for the cultivar.

For *Modern Roses 11*, now being edited for publication in spring 1999, the ARS executive committee has, at the request of the editor, approved listing registered cultivar names in small capitals and without single quotation marks (e.g., WEKCRYLAND). This further movement away from the system of the *International Code of Nomenclature for Cultivated Plants* (Trehane 1995) (hereinafter abbreviated "the Code") is being protested and may not actually occur.

Code Names

There has been a trend in recent years, to register roses under a code name, usually consisting of the first three letters of the name of the person or company registering the rose, written in capital letters, followed by some word or other string of letters. Research by Robert B. Martin, Jr. and Marily Young (database co-ordinator for the *Modern Roses 11* project) found that the oldest code name appearing in the American Rose Society literature was GAUdino, registered as 'Renaissance'®, in 1945, by Gaujard. Kordes registered 'Iceberg' and included the code name KORbin in 1958. Meilland of France adopted the practice in 1960 with 'Clair Martin'® (MEImont). MacGredy followed in 1965 with 'Handel'® (MACha), and Jackson & Perkins introduced the practice in the USA, in about 1969, with 'White Masterpiece' (JACmas). The usage spread, and today the code name is used by most of the world's major rose breeders. It should be noted that in each of the above cases, it was the fancy name, and not the code name, that was actually registered by the American Rose Society as the cultivar epithet.

Until 1994, the IRAR understood the *International Code of Nomenclature for Cultivated Plants* (Brickell 1980), Article 27, to mean that code names were not acceptable as cultivar epithets, since they were not fancy names and not really words. Therefore, when a cultivar registration was received with a code name as well as one or more fancy names, it was the first fancy name listed that actually got registered (Carol Spiers *pers. comm.*). At that point, the use of trademarks for roses was new and poorly understood by ARS, so they did not realise that registration of such marks as cultivar epithets could endanger the trademark's legal status. Commercial organisations, especially CIOPORA and AARS, complained that the IRAR was interfering with their choice of names and trying to control or approve their trademarks. In 1981, Vincent Gioia, an attorney dealing with plant patents and trademarks and then vice-president of ARS, wrote to the ARS Board of Directors, explaining the problem and asking that they accept code names as legitimate registered cultivar denominations. However, since the meaning of Article 27 of the 1980 Code was still not clearly understood, some roses continued to be registered under fancy names, even if submitted with a code name, and even if that fancy name was a trademark. As far as this author has been able to determine, none of the cultivars registered in this way has ever had its registration modified later, to use the code name; in every case, it is the original fancy name, whether or not a registered trademark, which remains the registered cultivar name.

Registration Boycott

In late 1987, the German firms of Kordes and Tantau stopped registering their new cultivars with the IRAR, partly in protest of the fee then being levied on the process, but also because they claimed that they were being prevented from registering roses with code names. This is in spite of the fact that both companies had been successfully registering code-named roses for several years. Kordes registered 26 code names in 1985, 5 in 1986, and 18 in 1987. Tantau registered 20 code names in 1985. IRAR policy concerning registration of code names at that time seems to have been to accept them if no fancy name was also submitted (Carol Spiers *pers. comm.*).

Then in early 1993, most CIOPORA members in Europe stopped or severely curtailed registering new cultivars, claiming that their trademarks were being endangered by ARS publication policies. CIOPORA officially promoted that boycott. At the autumn 1994 meeting of the ARS Board of Directors, a petition was presented, signed by representatives of most of the larger United States growers, asking that future publications be done in such a way as not to hinder the acquisition of a trademark. The petition also demanded that ARS form an *ad hoc* committee, with AARS representation, to work out the details of trademark issues for ARS publications. A new ARS president was beginning his term of office, and he asked the author to be a member of the *ad hoc* committee, as well as to chair the Committee on Rose Registrations for the next three years. It was quickly agreed by the *ad hoc* committee that trademarks would no longer be registered as cultivar names, and that code names would be fully accepted as legitimate cultivar names. Most AARS members had joined the registration boycott by that time, and many did not resume registering until 1998.

The author selected members for the new Committee on Rose Registrations, who were knowledgeable about rose nomenclature and history, and who had no pecuniary interest in the issue. Soon after the committee membership was announced, AARS declared that the committee was unacceptable; AARS *must* be given at least 50% membership on the registration committee and they, not ARS, would choose those members. Otherwise, AARS-affiliated growers would never register a rose with ARS again, and AARS would actively seek to have ARS removed as IRAR and to have themselves named the new IRAR. While the ARS leadership felt that a marketing organization had no right to make such demands, they accepted those conditions, in the hope of getting the registration process restarted.

During the years 1995–1997, the non-AARS members of the committee made sure that the Code was scrupulously followed, at first in its proposed form, and then in the final form when it was published, in regard to the acceptance of cultivar names. Code names were freely accepted as cultivar names. Trademarks were never used as cultivar names. Some applications continued to be submitted, requesting trademarks as cultivars, but in every such case, the IRAR returned those forms to the applicant, asking for clarification and pointing out that a trademark could not be so used. Some of those improper applications were likely due to problems with reading the English-language form; however, several such applications were submitted from the United States and the United Kingdom, from experienced commercial growers, who apparently still did not understand the nature of their own trademarks.

From 1995 through 1997, CIOPORA continued its assault on the IRAR, claiming that trademarks were still being endangered, and that ARS continued to try to control or approve trademarks. The claims were false, but CIOPORA members were told otherwise by their management, as recently as the minutes of their Annual General Meeting for 1997 (CIOPORA 1997).

CIOPORA has also continually tried to evade the use of the Code. In a letter to the IRAR (25 August 1997), The Secretary General of CIOPORA stated, 'Registration rules must be simple; breeders already have to abide by numerous and rather constraining rules when they want to protect their varieties by patents/breeders' rights and they do not wish to be submitted to more limitations under non legislative regulations like e.g. the *International Code of Nomenclature*.' Later in the same letter, the Secretary General referred to the rule about the length of a cultivar name — '...should also delete the second requirement: '*must consist of no more etc...*' because this goes beyond the basic compulsory requirements of the UPOV Convention and rose breeders do not wish to be imposed supplementary constraints. It is again requested to delete '*Consult the International Code of Nomenclature...*' because the International Code may contain restrictions that go beyond what is required under the UPOV Convention and breeders need not have supplementary constraints. And such a mention is not necessary for the purpose of the registration of 'varieties'.

The minutes of the 1997 meeting (CIOPORA 1997) also state, prophetically, that 'Contacts were still going on with the American Rose Society and because of an expected change in the Management of the ARS it was strongly hoped that a more positive cooperation would result between ARS and CIOPORA.' Of course, cooperation in this case likely meant departure from the Code since, for the previous three years, the IRAR was absolutely as cooperative as the Code allowed. In October 1997, when the new ARS president took office, the members of the Committee on Rose Registrations who had upheld the Code were replaced with AARS/CIOPORA representatives or supporters. The committee is now chaired by an AARS member. The boycott on registrations has been lifted and registrations are again being submitted by CIOPORA and AARS members. Only time will tell whether the Code will continue to play any significant role in the registration process.

Misuse of Trademarks as Cultivar Names

At the Second International Symposium on the Taxonomy of Cultivated Plants, attorney Vincent Gioia explained the correct use of trademarks, and how such use relates to cultivar names (Gioia 1995): 'A trademark is a term adopted and used by a manufacturer or merchant to identify goods and distinguish them from those manufactured or sold by others.... The main function of a trademark is to indicate the origin of goods with which the trademark is associated.... When a trademark becomes "generic" and is used to denote the goods themselves and not the origin, then the exclusive right to use the mark may be lost.... It is incumbent on those wishing to employ a protectable trademark to use trademarks for plants properly.'

It appears to this author that trademarks are being incorrectly used as names for roses. Contrary to what Mr Gioia said should be the case, trademarked rose names are almost always used uniquely to identify an individual rose cultivar, giving no information at all that would associate the trademark with its owner. Indeed, the industry tends to use the trademark as the actual name of the cultivar, instead of the registered cultivar name. Some companies list the trademark with a "TM" or "®", followed by the registered cultivar name, on the nursery tag or in a catalogue, but many do not. Certainly the vast majority of the rose-buying and growing public are completely unaware that the trademark is anything other than the accepted name of the cultivar. In the database for *Modern Roses 11*, there are only three trademarks listed which refer to more than one cultivar. Two of those, Carefree Beauty™ and Vision® are claimed by different companies in different countries, for their different cultivars. So only one trademark (Renaissance®), referring to two of Gaujard's roses, appears to have been used to refer to more than one cultivar, by one company. Admittedly, the *Modern Roses 11* database is incomplete, due to the boycott on registrations, but it appears to be standard practice to assign a code name to a cultivar, whether that name is registered or not, and then to trademark the only name under which the rose will actually be marketed and known to the public.

121

Approved Exhibition Names

The advent of code names as registered cultivar names, along with the common use of trademarks as if they were accepted cultivar names, caused another major debate in the ARS. Until 1996, ARS-sanctioned rose shows required that only roses that had appeared in an ARS publication (*Modern Roses*, the *American Rose* magazine, or the *American Rose Annual*) were eligible to be shown, and that those roses must be exhibited under their registered cultivar names. The rule worked well, as long as roses were given fancy names, but having to show a rose under a code name was annoying and confusing. Exhibitors, show judges, and the public who viewed the shows were unfamiliar with the meaningless codes under which the roses had to be shown. So in 1995, the Board of Directors of the ARS instituted the use of an "Approved Exhibition Name," for shows held after 1 January 1996. This name is usually the term under which the rose is marketed in the United States. In some cases, it is the registered cultivar name. In most cases, it is a trade designation and often a trademark. While it seems to the author that such use could endanger a trademark far more than any previous ARS actions, the commercial industry appears to be happy with the situation. When growers apply to register new cultivars, they usually specify their preferred Approved Exhibition Name as well, and they often list a trademark for that use.

Any rose can be given an Approved Exhibition Name; formal registration of a cultivar name is not a prerequisite. Whether registered as a cultivar or not, the Approved Exhibition Name is then published in one of the official ARS publications, and the rose may then be exhibited for judging.

Since Approved Exhibition Names are not subject to the Code, more confusion among names is certain to result. Recent examples include The McCartney Rose, the Approved Exhibition Name for a modern Hybrid Tea. But the term "Macartney Rose" has been in use as the common name for *Rosa bracteata* Wendl., for generations. Also, Temptress is being used as the Approved Exhibition Name for one rose, although it is the registered cultivar name (and presumably also the Approved Exhibition Name) of a different rose.

Ongoing Problems

There are several parts of the Code, with which the IRAR still has problems:

Article 17.7 'Cultivar status is to be indicated by enclosing the cultivar epithet within demarcating single quotation marks....' Unless the recent decision is rescinded, beginning with *Modern Roses 11*, the ARS/IRAR will publish cultivar epithets in small capitals, without quotation marks.

Article 17.9 '...a new cultivar epithet published on or after 1 January 1959 must be a word or words in a modern language....' The code names, as currently used, are not a part of any modern language. A recent change in policy permits code names that are not pronounceable as words.

Article 17.10 '...new cultivar epithets must consist of no more than 10 syllables and no more than 30 letters or characters overall....' While this requirement has not yet been struck from the IRAR registration form, it is one of the areas to which CIOPORA has expressed strong opposition. Now that the president of ARS, the vice president (and editor of *Modern Roses 11*), and the entire registration committee are in strong support of AARS/CIOPORA, the requirement appears to be in imminent danger.

Over many years, one of the most flagrant violations of the Code has been the practice of declaring cultivars extinct in order to reuse their names (Article 26.1–26.4).

Beginning in 1995, the practice was completely stopped, since it would be virtually impossible to prove that a cultivar was truly extinct, and in fact, several so-called extinct roses have reappeared in cultivation and even in the commercial trade. Yet it is the expressed goal of the commercial industry to reuse names, and in many cases, extinct seems to be conveniently defined as 'not appearing to be in commerce this year.' Again, it will be interesting to see if the IRAR resumes declaring extinctions, in the next few years.

The database for the *Modern Roses 11* project shows one incidence of a single name having been used for five different cultivars. There are six occurrences of a single name having been used for four cultivars. And there are 32 instances of a name having been used for three cultivars. As recently as 1992, a cultivar was registered using a name for the third time.

Summary and Conclusions

The IRAR has gone through more than a decade of extreme pressure from groups with opposing viewpoints. Despite meticulous adherence to the Code in its approach to the proper registration of cultivar names since 1994, and despite there having not been an instance, since late 1994, of a trademark having been registered as a cultivar epithet nor of a legitimate code name having been turned down for registration, CIOPORA and AARS have continued to make false accusations, resulting in continuation of the boycott of the registration process, and ultimately in large changes in the political structure of the Committee on Rose Registrations. There is a strong move among commercial interests to abandon the use of the Code completely. And it is those commercial interests who now control the IRAR.

There appears to be no immediate solution to the problem of naming roses — growers who bother to register their cultivars at all will submit code names, which will be used nowhere else. Roses will continue to be marketed under trademarks, as if those trademarks were accepted names for the plants. ARS-sanctioned rose shows will use Approved Exhibition Names for roses to be exhibited. And it would appear that commercial marketing organizations will be in control of the IRAR for at least the next six years.

Acknowledgements

I would like to thank the following people for their valuable help with research for this paper. Ms Marily Young, database coordinator for the ARS *Modern Roses 11* project, cheerfully provided accurate data from the rose registration records. Mr Robert B. Martin, Jr, attorney and rosarian; Dr Charles A. Walker, Jr, horticulturist, rose historian, and president of the Heritage Rose Foundation; Ms Carol Spiers, registration clerk for IRAR and Assistant to the Executive Director at ARS; Ms Darlene Kamperman, Administrative Assistant at ARS; and Mr Pete Haring, past president of the ARS; also provided valuable information on the history of rose registration.

References

Brickell, C.D. (ed.). (1980). International Code of Nomenclature for Cultivated Plants — 1980. Regnum Vegetabile, 104, 32 pp. Utrecht.

Cairns, T. (ed.). (1993). Modern Roses 10. 740 pp. The American Rose Society, Shreveport, Louisiana.

CIOPORA. (1997). Minutes of the annual general meeting. Communauté Internationale des Obtenteurs de Plantes Ornementales et Fruitières de Reproduction Asexuée, Geneva.

Gioia, V.G. (1995). Using and registering plant names as trademarks. *Acta Hort.* 413: 19–25.

Haring, P.A. (ed.). (1986). Modern Roses 9. 402 pp. The American Rose Society, Shreveport, Louisiana.

Meikle, C.E. (ed.). (1980). Modern Roses 8. 580 pp. The McFarland Company, Harrisburg, Pennsylvania.

Trehane, P. *et al.* (eds). (1995). International Code of Nomenclature for Cultivated Plants — 1995. 175 pp. Quarterjack Publishing, Wimborne.

PRINCIPLES OF CULTIVATED
PLANT CLASSIFICATION

5

Hetterscheid, W.L.A. (1999). Stability through the culton concept. In: S. Andrews, A.C. Leslie and C. Alexander (Editors). Taxonomy of Cultivated Plants: Third International Symposium, pp. 127–133. Royal Botanic Gardens, Kew.

STABILITY THROUGH THE CULTON CONCEPT

W.L.A. HETTERSCHEID

VKC, Linnaeuslaan 2a, 1431 JV Aalsmeer, The Netherlands

Abstract

Switching emphasis from a plant-centered to a human-centered taxonomy of cultivated plants may be the basis for an improved stability of nomenclature. The application of the culton concept (cultonomy) provides the necessary reasons for and implementation of this switch. An important tool in this respect is the current *International Code of Nomenclature for Cultivated Plants* (Cultivated Plant Code). It is argued that a safe distance should be kept from an excessive influence of the *International Code of Botanical Nomenclature* (Botanical Code), which introduces unnecessary instability into the system of cultivated plant taxonomy.

Introduction

During the ages of plant domestication by man, from basic selection from the wild to the 'high-tech' breeding of today, the aim has always been to preserve advantageous characters or character combinations. These characters cover a wide spectrum, including medicinal, food and ornamental properties. The ultimate reward of domestication is the stabilisation of such characters in groups of plants (usually cultivars). This stabilisation means that the characters remain present in the plants upon propagation, either vegetatively or by seed. Elaborate systems of breeding have been devised to enhance this stability of character representation in the domesticated plants. The final 'product', the cultivar, is nothing more nor less than an 'industrial' product, irrespective of whether it is a simple selection or the result of a highly complex process of breeding. One could say that the cultivar is the sum-total of selected (including unconsciously selected) characters materialised in a selected group of plants through artificial breeding, selection, and maintenance.

One would expect that such industrial products, be they landraces or highly stable modern cultivars, would be labelled with simple and stable names or some other device to identify them quickly. But nothing is further from the truth. Domesticated (cultivated) plant nomenclature is still an amalgamation of past and recent systems of naming plants in the wild and naming plants in cultivation. The most problematic systems of cultivated plant nomenclature are those arising from efforts to synthesize classification philosophies of wild and domesticated plants (Hetterscheid *et al.* 1996). The complexity of nomenclature arising from these efforts is obvious and the acceptance of the underlying taxonomies by a larger audience has not usually been achieved (Hetterscheid *et al.* 1996). None of them has been universally established with

Editorial note: A number of substantial points raised during pre-publication review were rejected by the author of this paper. The editors decided to include it in the form insisted on by the author rather than omit it from the proceedings.

one exception: the Linnaean binomial system. But it has been suggested that that system is inadequate as a general option for a stable domesticated plant nomenclature (Hetterscheid & Brandenburg 1995a & b). This conclusion may be drawn from merely listening to the many complaints one hears incessantly about changing names of domesticated plants.

Cultonomy: *en route* to a Solution

With the introduction of the culton concept (Hetterscheid & Brandenburg 1995) an effort was made to construct a different way for taxonomists to look at the context in which man domesticates plants. The prevailing view of taxonomists has always been plant-centered. Brandenburg and the present author advocated a more human-centered view as a basis for domesticated plant systematics. The plant-centered view of the vast majority of taxonomists is symptomatic. Plant taxonomy these days and during much of its history has been and is all about plants in their natural settings. Plants in this view are the constrained results of an array of evolutionary processes. According to a majority view, their classification and resultant names have to mirror this evolutionary history and many decades of discussions about the best way to reach this goal have followed. The discussion still goes on. The classification units are named taxa and they are classified in a highly specialised hierachical system, which, according to those who use it, is a perfect tool to represent attempts to reconstruct parts of the 'Tree of Life'. The names to label these taxa are formed in accordance with an equally specialised tool, the *International Code of Botanical Nomenclature*, or Botanical Code (ICBN, Greuter *et al.* 1994). This tool is specifically devised to label taxa in hierarchical systems, and cannot cope with essentially non-hierarchical systems (*teste* the discussions around varietas and forma).

Taxonomists adhering to the plant-centered view, start their classification philosophy (if they have one) for domesticated plants from the idea that these plants are primarily the result of their evolutionary history and that human intervention is of secondary importance. Their procedure in classifying such plants logically follows the same mind steps: first there is the 'evolutionary' basis for a classification and then there is that part reflecting the domestication. In their view, the names of classes of domesticated plants, should primarily reflect this evolutionary historical component, to which name-parts are added that reflect their domesticated history.

The resulting classifications are crammed into the hierarchical system used to represent evolutionary history (thought to be a system of processes that creates a hierachical pattern of descent), whereas the actual classes (units) in the system carry all the characters of the domestication history (which has usually involved much 'messing up' of the evolutionary make-up of the plants) and hence are not evolutionary units at all. Therefore the choice of the 'taxonomic' hierarchy as a backbone for classifying the results of domestication is in fact an error of logic. It will also be shown to be a choice for instability.

Cultonomy, introduced as a nickname set against 'taxonomy' (Hetterscheid & van den Berg 1996), focuses primarily on man's influence on the existing botanical evolutionary diversity and starts by examining the plants as representatives of domestication and NOT evolution. The main emphasis in this system of thought lies with man's INTENTIONAL actions of breaking the natural sequence of evolutionary events and moulding it so as to reach a certain goal. Unconscious selection resulting in non-intended change of characters is a by-product of domestication not relevant to the content of this paper. Domestication as an historical process directed by man, leads to a new kind of 'bio'diversity, with selected groups of individual organisms (plants in this case) representing this diversity. These plants are different in essence from their evolutionary counterparts and need human attention in order to exist at all, or remain in existence. If human maintenance of this diversity subsides, the system falls or finds a new 'equilibrium' (e.g. as in many crop-weed complexes) in which the balance between evolution and man's influence on it has shifted away from the latter.

Cultonomy therefore needs a system of classification that does justice to man's all-important influence. It is not evolution or its results that need cultonomic classification but the results of man's intentional actions. Rather than classes of domesticated plants based on a plant-centered (evolutionary) view, cultonomic classifications focus primarily on creating classes that serve man's need. This is a strictly logical approach, since domestication is a process that has been created by man himself to fulfil certain needs. The fact that he uses and manipulates products of evolution is relatively unimportant. It is the results of the manipulations that count and need classifying. It will be shown below that the cultonomic approach, in addition to being logical, also creates the foundations for much-needed stability in domesticated plant nomenclature.

Stability is Twofold

If we regard plant breeding and selection as an industrial activity, we should not be surprised to find that the products comply with a number of pre-set standards. The resulting plants are grouped into discrete classes we have termed cultivars. As such, cultivars are not different from cars or other industrial products, in that man has set out certain quality demands for them. For cultivars a very important quality demand is stability of characters. And here we hit upon an important aspect of stability. A cultivar has to be 'immutable'. Everytime we order a cultivar or refer to it by a name, we must be confident which cultivar we have in mind, or want to get. This is a typical user-driven quality. A breeder will have to take care that his cultivars are stable, and a grower must be certain that this stability remains when he propagates the cultivar. This wish for stable cultivars is best filled in countries where plant breeding has reached a 'high-tech' level. In other countries, the stability of cultivars may be less well-guaranteed but even in less developed agricultural systems there is always the need for a cultivar to retain at least those characters for which it is maintained in the first place. Without this minimal level of stability, agriculture, horticulture and forestry would not have developed in the way they have.

In addition to this stability of content of products (cultivars), we need stability of reference to the products, i.e. their nomenclature. The majority of users need a simple, one-to-one reference system in which they can identify cultivars and groups of cultivars and use their names to communicate about them in the widest sense of the word. The basic element of the present-day system of nomenclature for domesticated plants is the cultivar-epithet. *The International Code of Nomenclature for Cultivated Plants*, or Cultivated Plant Code (Trehane *et al.* 1995) deals with this. In order for this Code to be workable, the '*cultivar-as-a-class*' is defined therein as well. For discussion on this see Hetterscheid & Brandenburg (1995a & b). Stability of cultivar epithets is thus entirely in the hands of those responsible for the rules set out in the Cultivated Plant Code.

The Advantage of Distancing the Botanical Code

The nomenclature system of the Cultivated Plant Code is as much based on binomials as that of the Botanical Code but with a very fundamental difference. The binomial basis to the Cultivated Plant Code is a combination of genus name (better: denomination class) and cultivar epithet. In principle, every name-part added to that combination is secondary. Cultonomic philosophy supports this system of nomenclature because it creates a fundamental and for reasons of stability, much-needed distance between it and the Botanical Code. The distance is much greater than some of us would like to believe. The important fact that ONLY the (notho-)genus name is the one fundamental tie between the two Codes, is not always appreciated by taxonomists. To those embracing the culton concept the Botanical Code is additional to using the Cultivated Plant Code and NOT the reverse. In their view the Botanical Code is the source of classificatory and nomenclatural instability in cultivated plants. This is inherent to the Botanical Code and the subjects of its rules, namely taxa. Whereas the basic taxon of the Botanical Code and its practitioners is the species, the basic (AND basal!) culton of the Cultivated Plant Code is the cultivar. Hetterscheid & Brandenburg (1995a & b) have tried to show, hopefully convincingly, that species and cultivars are entities radically different from each other, each requiring its own system of classification and nomenclature. The Botanical Code works only for taxa, which by definition are classified in an obligatory hierachical system. Taxa as historical evolutionary entities are variable by definition, being the results of evolution. The Cultivated Plant Code however supports an open system (not fundamentally hierachical) of entities with fixed definitions (classes). Their descriptions (definitions) are fixed because human society requires them to be.

An important entity figuring prominently in both systems is the genus. But even so in each system the genus has a different meaning. In the Botanical Code it is just one of the many hierarchically arranged classification categories, whereas in the Cultivated Plant Code it is the anchor to which a cultivar SHOULD be tied in order to create the necessary binomial. The genus thus serves as an important, so called 'denomination class'. The latter being the nomenclatural unit in which a cultivar-epithet is allowed to be established only once. Of course we also use the genus name as a tool to get a first impression of what a cultivar may look like when we see its full binomial for the first time. Learning from the name that a cultivar belongs to *Begonia* L. and not, e.g. *Triticum* L., is a seismic difference.

Instability caused by different taxonomic opinions about a plant genus is relatively rare, although some unfortunate examples are known, e.g. *Rhododendron* L. (vs. *Azalea* L.), or *Gaultheria* Kalm ex L. (with or without *Pernettya* Gaudich.), or more recently *Dendranthema* (DC.) Des Moul. vs. *Chrysanthemum* L. More complications and much instability are introduced into domesticated plant nomenclature by the use of the species epithet as part of a full cultivar's name. As has been stated above, its use is not a necessity to fulfill the primary requirements of the Cultivated Plant Code in establishing a reference system of unique cultivar names. However to many, the use of the species epithet has its advantages, and in some cases this is indeed true. In genera where cultivar epithets were previously linked to full Latin binomials, many now prove to be contrary to the 'denomination class + cultivar-epithet must be unique' requirement of the Cultivated Plant Code. In other genera most cultivars are proven to be established through direct selection from known species and hybridisation between species was and is a rare tool. In such genera, most users are happy to maintain the full species names in order to distinguish different types of cultivars in a particular genus.

The differences between the groups of cultivars in such genera are a direct result of the existing differences in characteristics of the species to which they belong.

However in the plant-centred view of domesticated plant classification, the species epithet has become far too important and may lead to an unstable system of names. For instance, when selling perennials, it has been customary in some circles merely to suggest a species epithet in the full cultivar name when the exact origin of the cultivar is unknown. The name of the best 'look-alike' species was then taken up and used as part of the full cultivar name. Important organisations like UPOV demand a species-based nomenclature for domesticated plants. In wider circles it has been, and still is, customary to demand a Latin binomial when a cultivar is known to be of hybrid origin. In this way a multitude of hybrid species names has been created over the years; they are still being created to this day.

This is all very mystifying. Whereas everybody begs for stability of nomenclature for domesticated plants, many still reach for a system that is inherently unstable. The creation of hybrid species epithets links a cultivar to the Botanical Code at an additional level and introduces the full cultivar name to the mechanics of the Botanical Code at the species level as well as to the results of taxonomic investigations at the species level. Such investigations often enough lead to name changes. The Botanical Code also has its own technical rules that affect the stability of the hybrid species epithet. And yet we do NOT need the hybrid species as a classification category for cultivated plants!

To fill the need for a category in which similar cultivars can be assembled within a genus, without having to resort to using the species or to creating phony (hybrid) species, the Cultivated Plant Code has undergone an important improvement in its latest edition (Trehane *et al.* 1995). The old catch-all category of 'group' to encompass a multitude of domesticated plants that could not be assigned to species, has been replaced by a much more strictly defined category, the cultivar-group. The strict definition is a logical consequence of the 'cultonomic' effort to create and use well-defined categories of domesticated plants, which is the only basis upon which a stable nomenclature can be built. With this improvement, the need for a Botanical Code category in which to group cultivars, has disappeared for all those who appreciate the usefulness of the cultivar-group in its now more precise definition (Cultivated Plant Code 1995, Art. 4.1, excluding Note 1), the latter being a severe and detrimental undermining of the stability of the category.

In order to illustrate much of the above, I have chosen an example of the stabilising effect (ex Bean 1976) of domesticated plant classification and nomenclature.

Philadelphus in Cultivation: a Case in Point

In a recent paper, Hoffman (1996), presented a suggestion for improving stability to classification problems of cultivars in *Philadelphus* L., brought about by adherence to hybrid species names of olden days. In this problem-area, Hoffman has chosen to solve issues by using and/or proposing cultivar-groups and discarding the old and established hybrid species epithets as grouping devices. One example may suffice to prove the point. *Philadelphus* × *virginalis* Rehder is a collective name used to cover cultivars of the cross between *P.* × *nivalis* 'Plenus' and *P.* × *lemoinei* (Lemoine) Rehder, the parents themselves being respectively crosses of *P. coronarius* L. with *P. pubescens* Loisel., and *P. coronarius* with *P. microphyllus* A. Gray. The binomial *P.* × *virginalis* was established and a description was given. Of major importance were the semi-double or double flowers. However, recently single-flowered mutations and seedlings have been found, e.g. cultivars 'Limestone', 'Kasia' and 'Burfordensis'.

131

Using the Botanical Code approach, these cultivars would be given the full name *P.* × *virginalis* '.......'. Customers used to the double or semi-double flowers of *P.* × *virginalis* cultivars, would be misguided when ordering one of the single-flowered cultivars upon seeing their names in a catalogue or other literature. Looking upon *P.* × *virginalis* as a proper taxon with a Botanical Code name, it is no problem at all to add to the morphological characters of this hybrid 'species' the new character of single flowers. In a short while there would be confusion as to the 'proper' use of *P.* × *virginalis* in cultivation. Three flower types could be covered by it, and the unambiguous reference of the name to qualities of certain cultivars would be lost.

The option, chosen by Hoffman, provides for the necessary stability. He chose to use the cultivar-group option introduced for *Philadelphus* by Bean (1976). The cultivar-group in this example being the *P.* Virginalis Group. The name of this group preserves the reference to what was widely known as *P.* × *virginalis* but stabilises the description so as to include ONLY cultivars with semi-double or double flowers. In this way, the reference quality of the name *P.* Virginalis Group is stabilised. The single-flowered cultivars, selected from within cultivars of the *P.* Virginalis Group are re-allocated to cultivar-groups containing single flowers. Hoffman established a *P.* Burfordensis Group for a selection of single-flowered cultivars with additional characters in common that did not fit any of the other groups. The net result of Hoffman's work is a perfectly clear cultivar-group classification of *Philadelphus*, into which newly raised cultivars can be fitted with ease and in which any new cultivars are given names that automatically include the names of stabilised cultivar-groups. Those names are informative for users and prevent any further confusion from the use of hybrid species names. At the same time, any destabilisation inherent to the use of taxon names is eliminated. Whatever taxonomists may do to the hybrid names in *Philadelphus*, based on new taxonomic insight or due to the technicalities of the Botanical Code, it is irrelevant to the cultonomic classification of *Philadelphus* and nobody will be bothered by it.

Conclusions

In conclusion I believe that the culton concept, in conjunction with most of the 1995 edition of the Cultivated Plant Code is a good foundation for a full-scale swing from the plant-centered view of domesticated plant taxonomy of past days to a modern and human-centered approach, which van den Berg and I teasingly baptised cultonomy. The most important result is a safe distance from the Botanical Code and a full-scale use of the Cultivated Plant Code, with its better appreciation of and answer to the need for stability in domesticated plant nomenclature for breeding, trading, legislation, etc. An important step towards creating the necessary distance from the Botanical Code, would be to eliminate the possibility of creating hybrid binomials for artificially hybridised plants. This requires a redrafting of the Hybrid Appendix. Proposals to this effect will be submitted.

Whether or not Martians will understand the difference between cultonomic classifications and taxonomic ones (Waters 1998) is beside the point. WE are the users and creators of such classifications, not alien life-forms. And if they are able to decode our script, I hope they read this paper.

References

Bean, W.J. (1976). *Philadelphus. Philadelphaceae.* Trees & shrubs hardy in the British Isles. Vol. 4. (Ed. 8). pp. 127–146. John Murray, London.

Greuter, W. *et al.* (eds). (1994). International Code of Botanical Nomenclature (Tokyo Code). 389 pp. Koeltz Scientific Books, Königstein, Germany.

Hetterscheid, W.L.A. & Brandenburg, W.A. (1995a). Culton versus taxon: conceptual issues in cultivated plant systematics. *Taxon* 44(2): 161–175.

Hetterscheid, W.L.A. & Brandenburg, W.A. (1995b). The culton concept: setting the stage for an unambiguous taxonomy of cultivated plants. *Acta Hort.* 413: 29–34.

Hetterscheid, W.L.A. & Berg, R.G. van den (1996). Cultonomy of Aster. *Acta Bot. Neerl.* 45(2): 173–181.

Hetterscheid, W.L.A., Berg, R.G. van den & Brandenburg, W.A. (1996). An annotated history of the principles of cultivated plant classification. *Acta Bot. Neerl.* 45(2): 123–134.

Hoffman, M.H.A. (1996). Cultivar classification of *Philadelphus* L. (*Hydrangeaceae*). *Acta Bot. Neerl.* 45(2): 199–209.

Trehane, P. *et al.* (eds). (1995). International Code of Nomenclature for Cultivated Plants —1995. 175 pp. Quarterjack Publishing, Wimborne.

Waters, T. (1998). A comment on the culton concept. *Hortax News* 1(4): 13–15.

Berg, R.G. van den (1999). Cultivar-group classification. In: S. Andrews, A.C. Leslie and C. Alexander (Editors). Taxonomy of Cultivated Plants: Third International Symposium, pp. 135–143. Royal Botanic Gardens, Kew.

CULTIVAR-GROUP CLASSIFICATION

RONALD G. VAN DEN BERG

Plant Taxonomy, Wageningen Agricultural University, P.O. Box 8010, 6700 ED Wageningen, The Netherlands

Abstract

The cultivar-group, being a 'culton' rather than a taxon, is considered to serve the purpose of classifying cultivated plants better then e.g. infraspecific or hybrid taxa. The two classification systems (taxonomic and cultonomic) should be seen as separate but linked at some point. It is argued that this link should be at the level of 'crop', and not at the level of cultivar-group or cultivar. Although different crop situations might necessitate special provisions, the advantage of having one systematic approach for all situations (analogous to the classification of wild plants) should outweigh possible drawbacks. A number of practical considerations are discussed.

Introduction

The cultivar-group is a category that has provoked much discussion as can be seen in the newsletter of the Horticultural Taxonomy Group (*Hortax News*). In volume 1, no. 2 (February 1996) we find a number of contributions:

– Cultivar-groups – a sensible approach, by Allen Coombes
– Cultivar-groups – a systematic approach, by Wilbert Hetterscheid
– Cultivar-groups – a concept in need of clarification, by Piers Trehane

The main issue these authors concerned themselves with was whether or not a cultivar-group should contain only named cultivars or might also accommodate other plant material without cultivar status, unnamed populations with horticultural value, or even botanical taxa in cultivation. In a later contribution to *Hortax News*, Trehane (1997) pointed out that the term cultivar-group is being used in more than one sense and that this issue should be resolved before the next edition of the *International Code of Nomenclature for Cultivated Plants* (Cultivated Plant Code). It seems that the present edition of the Cultivated Plant Code, in article 4.1 already gives a clear definition (assemblages of NAMED cultivars) and Hetterscheid (1996) has argued convincingly for this restricted definition in order to arrive at precisely defined categories.

Cultivar-groups are Culta, not Taxa

At the risk of antagonising the anti-culton lobby I will take for granted the suitability of cultivar-groups (and their superiority over infraspecific or hybrid taxa) to accommodate groups of cultivated plant material. This means that I will also adhere to the distinction between taxa and culta. The cultonomic point of view is that the classification of cultivated plant material is better off with culta than with taxa for reasons that have been exhaustively explained by Hetterscheid & Brandenburg (1995) and Hetterscheid *et al.* (1996). Botanical categories are not suitable to accommodate groups of cultivated material. Attempts to do this have resulted in extensive hierarchical systems containing many levels. Examples include the classification of *Beta* L. (Helm 1957, compare with Lange *et al.* 1999, see Table 1) and *Brassica* L. (Oost & Toxopeus 1986, see Table 2). Many varieties and forms in those systems are in fact cultivars or cultivar-groups and should not be treated as taxa in a closed classification. In some cases however, if a grouping is already there, one can conveniently translate the classification into a cultivar-group classification, even using the Latin variety names as a basis for cultivar-group epithets, since these Latin names never really referred to botanical varieties in the first place. For the rest, it seems best to keep our distance from the *International Code of Botanical Nomenclature* (Botanical Code) and its epithets with their inherent threat of nomenclatural instability. Only when one is sure that the taxon name refers to the same entity that one wishes to treat as a culton, is adopting the Latin name for the cultivar(-group) epithet warranted.

TABLE 1. Classification of Cultivated *Beta*.

Helm 1957	**Lange *et al.* 1999**
Beta vulgaris L.	*Beta vulgaris* L.
subsp *vulgaris*	subsp. *vulgaris*
convar. *vulgaris*	Leaf Beet Group
provar. *vulgaris*	Garden Beet Group
f. *virescens* Beck.-Dill.	Fodder Beet Group
f. *vulgaris*	Sugar Beet Group
provar. *flavescens* Lam. & DC.	
f. *leucopleura* (Alef.) Voss	
f. *flavescens*	
f. *rhodopleura* (Alef.) Voss	
f. *variocicla* (Alef.) Helm	
convar. *crassa* Alef. *s.l.*	**UPOV**
provar. *crassa*	denomination class 1
f. *crassa*	Fodder Beet Group
f. *lutea* Hegi	Sugar Beet Group
f. *rosea* (Moq.) Beck.-Dill.	denomination class 2
f. *incarnata* Meissn.	Spinach Beet Group
provar. *altissima* Döll	Beetroot Group
provar. *lutea* Lam. & DC.	Mangold Group
provar. *conditiva* Alef. *s.l.*	denomination class 3
f. *conditiva*	other
f. *zonata* (Spenn.) Beck	
f. *metallica* Voss	
f. *dracaenaefolia* Voss	

TABLE 2. Classification of Cultivated *Brassica*.

Mansfeld 1986

Brassica oleracea L.
  subsp. *oleracea*
    convar. *fruticosa* (Metzg.) Alef.
      var. *ramosa* DC.
      var. *gemmifera* DC.
    convar. *acephala* (DC.) Alef.
      var. *viridis* L.
      var. *sabellica* L.
      var. *medullosa* Thell.
      var. *costata* DC.
    convar. *caulorapa* (DC.) Alef.
      var. *gongylodes* L.
    convar. *capitata* (L.) Alef.
      var. *sabauda* L.
      var. *capitata* L.
    convar. *botrytis* (L.) Alef.
      var. *italica* Plenck
      var. *botrytis* L.

Brassica oleracea L.
  Acephala Group
  Laciniata Group
  Medullosa Group
  Gemmifera Group
  Gongylodes Group
  Capitata Group
  Pyramidalis Group
  Sabauda Group
  Botrytis Group
  Italica Group

Oost & Toxopeus 1986

Brassica rapa L.
  Vegetable Turnip
  Fodder Turnip
  Winter Turnip Rape
  Spring Turnip Rape
  Yellow Sarson
  Pe Tsai
  Pak Choi
  Mizuna

Toxopeus, Yamagishi & Oost 1987

Brassica rapa L.
  Vegetable Turnip
  Fodder Turnip
  Winter Turnip Rape
  Spring Turnip Rape
  Chinese Cabbage
  Pak Choi
  Mizuna
  Taatsai
  Leaf Turnip
  Saishin
  Brocoletto

The Cultonomic and Taxonomic Classification Systems must be Linked at some Point

It cannot be over-emphasised that cultonomic (cultivar-group) classification is not part of taxonomic (botanical) classification, but exists beside it, separate, and to a large extent autonomous. However, the two systems must be linked at some point.

Starting from the Botanical Code hierarchy (genus, species, subspecies, variety, forma) one can deduce to which level cultivated material can be assigned. Hetterscheid & Brandenburg (1995) argue that the only relevant level is that of the (notho)genus, disregarding the extra information that could be derived from the use of lower categories. Mabberley (1997), on the other hand, in his recent classification of edible *Citrus* L. (see Table 3), considers the Linnaean system to be more informative, when there is enough information about the history of the cultivated material. In many cases, however, there is a severe lack of knowledge about the true nature of the taxa associated with a crop, as exemplified by the taxonomy of shallots, where names such as *Allium cepa* L., *A. ascalonicum* auct. non Strand, *A. cepa* var. *ascalonicum* Backer and *A. cepa* var. *aggregatum* G. Don are used (Rabinowitch & Brewster 1990, see Table 4).

TABLE 3. Classification of Cultivated *Citrus*.

Mabberley 1997	
1. *Citrus medica* L.	citron
a. *Citrus* × *limon* (L.) Osbeck	lemon
b. *Citrus* × *jambhiri* Lush.	rough lemon (bush lemon)
2. *Citrus maxima* (Burm.) Merr.	pomelo (pummelo)
a. *Citrus* × *aurantiifolia* (Christm.) Swingle	lime
b. *Citrus* × *aurantium* L.	orange, grapefruit
3. *Citrus reticulata* Blanco	tangerine, mandarin, clementine

These approaches start with the categories of the Botanical Code and ask at which taxonomic level the link should be made. It is more practical first to determine which cultonomic category should be linked to the botanical classification. On consideration this is surely NOT the cultivar, frequently confused with forms and botanical varieties, and also not the cultivar-group, which is no more than a group of cultivars and thus can never be equated to a taxon.

There thus arises a need for a category to contain cultivar-groups and un-grouped cultivars. It is at the level of such a category that the classification of cultivated plants can be linked to the botanical system. The general term 'crop' could very well fill this need; also the 'denomination class', by definition the **taxon** in which cultivar epithets may not be duplicated, could be used. Unfortunately, the list of denomination classes in the present Cultivated Plant Code includes classes which are not taxa (notably within *Beta vulgaris* L. where the denomination classes are sets of cultivar-groups, see Table 2). Furthermore, there is no alignment with the classes designated by UPOV (the International Union for the Protection of New Varieties of Plants); this would be highly desirable to say the least.

Special attention should be given to cultigens. These paradoxical entities (such as *Triticum aestivum* L. and *Solanum tuberosum* L.) are defined as taxa comprising cultivated material only, lacking wild representatives. They are conveniently treated as taxa although they do not represent true species, and crops can be assigned to them for the sake of convenience.

TABLE 4. Classification of Cultivated *Allium* (Edible).

Allium ascalonicum L.	shallot
Allium cepa L.	onion
Allium fistulosum L.	Welsh onion, Japanese bunching onion
Allium porrum L.	leek
Allium sativum L.	garlic
Rabinowitch & Brewster 1990	
Allium cepa L. var. *cepa*	*Allium* Common Onion Group
Allium cepa L. var. *ascalonicum* Backer	*Allium* Aggregatum Group

One Classification System should be Applicable to all Situations

The philosophy of both codes of plant nomenclature implies that one classification system should be applicable to all situations. This position might well lead to problems in the application of the same set of categories to different crop situations found in agriculture, horticulture, and silviculture. An example of such a problematic group is plantain (*Musa* L.), where groups based on bunchtype and the use of a system of phrase names have recently been proposed (Rossel 1998, see Table 5). However, the advantage of having one systematic approach should outweigh any possible drawbacks.

This principle does no more than demand the same respect for the categories in the Cultivated Plant Code as is shown for those of the Botanical Code. After all, although it is accepted that taxa of wild plants can differ enormously in their nature, especially with regard to population structure and breeding system, they are nevertheless classified in the same categories (genus, species, subspecies, variety and form). It is not always easy to fit the perceived patterns of variation into these categories but in order to ensure effective communication we do the best we can. The same should apply to cultivated plants where we have a very simple and limited set of categories (crop, cultivar-group and cultivar). In time we hope to see the acceptance of cultivar-group to

TABLE 5. Classification of Cultivated *Musa*.

Rossel 1998 (ammended)				
species	group	subgroup	bunchtype	cultivars
Musa				
acuminata Colla	AA			
balbisiana Colla	AAA			
	AB			
	ABB			
	AABB			
	AAB	Plantain	French	
			French Horn	
			False Horn	
			Horn	
		(several other subgroups)		

classification based on:

bunch type
pseudo-stem size
pseudo-stem colour
bunch orientation

system of phrase names   French Medium Green Pendulous
False Horn Medium Green Subhorizontal
Horn Medium Green Subhorizontal

replace grex (see, however, the complex situation described by Rice (1996) for annuals and bedding plants), and the use of these categories for cultivated material in other major taxa (fungi, strains of yeast, etc.).

Some Practical Considerations

Cultivar-groups should only be used IF NECESSARY, e.g. in large assemblages of cultivars, where some sort of ordering system is needed. Article 4.1 of the Cultivated Plant Code ("Assemblages of two or more similar, named cultivars ... may be designated as cultivar-groups") indicates a different view on this matter, but it is surely not the intention to promote a multitude of cultivar-groups, each consisting of only two cultivars without any need for such groups.

It is an essential part of this 'open' classification, as opposed to the closed classification of the Botanical Code, that there is no obligation to classify cultivars in groups. Subspecies and botanical varieties, by their nature, belong to higher taxa. Cultivar systematics is essentially non-hierarchical and only when helpful should cultivar-groups be established. Also, open classification implies that it is not necessary to group ALL extant cultivars. When several cultivar-groups are established in a crop, it may well be that some cultivars are not accommodated in any of the groups. These do not then automatically form a group themselves!

Cultivar-groups are only useful if they are used. The establishment of cultivar-groups is not an academic exercise engaged in by those who happen to study the group and think they have arrived at the 'best' solution. Rather, a cultivar-group classification should be tested in practice and be subsequently adapted if there is a need for change, as was done by Hetterscheid & van den Berg (1996) in *Aster* L.

The circumscription of cultivar-groups should be based on a few, user-orientated characters rather than on a multivariate analysis of many characters. If a clear-cut classification using a few, easily detectable characters is not possible, the question arises whether a classification in groups is justified at all. This is not to say that a multivariate approach cannot be useful as an initial approach to see which groups appear after cluster analysis or principal component analysis. If such groups are not supported however, by clear characters or combinations of characters, they are probably not suitable as cultivar-groups.

Hierarchical structure, typical of the closed classifications of taxonomic systems, is contrary to the whole philosophy of cultivar-group classifications, which aim at simple subdivision for practical purposes, and need not reflect complex interrelationships among (groups of) cultivars. A single-level classification should be sufficient in the majority of cases, and multi-level cultivar-group classification (such as proposed in *Tulipa* L., van Raamsdonk & de Vries 1996, see Table 6) can easily be translated into a single-level system, avoiding the use of terms like 'supergroup'. The establishment of formal culta on a level higher than that of cultivar-group, or between cultivar and cultivar-group would lead to an unwanted emphasis on hierarchical structure. Furthermore, the use of only a few characters to distinguish cultivar-groups should avoid the establishment of extensive hierarchies.

Interesting issues arise following developments in the grouping of a crop when cultivar-groups need to be revised, combined or split-up. To regulate the use of cultivar-group epithets, the standard specimen might play the same role as the type specimen in botanical nomenclature, i.e. to decide which group retains the original cultivar-group epithet when a group is split. Another issue involves the coexistence of alternative classifications in cultivar-groups within the same crop, each with its own user-orientated purpose, and the inherent overlap between cultivar-groups this would involve. Both these issues await application in actual practice when problems (and their possible solutions) will become apparent.

Finally, a remark on the formation of cultivar-group epithets. Article 19 of the 1995 Cultivated Plant Code sets out in (possibly too much) detail how to create acceptable names. Trehane (1996) has suggested that cultivar-group epithets can best be based on:

- a well-known, representative cultivar
- an obvious descriptive term or,
- the validly published epithet of a botanical taxon

This last option seems to me ill-advised as it again establishes dependence of Cultivated Plant Code categories on those of the Botanical Code. The same position is taken in article 16.2 of the Cultivated Plant Code:

> "When two or more groups of plants are considered to contain the same individuals and no others, they are said to be coextensive. If a cultivar or cultivar-group is considered to be coextensive with a taxon named under the provisions of the ICBN (Botanical Code), it is to bear the same epithet. Ex. 2. (...) Adoption of the cultivar name *Erica tetralix* 'Alba' or the cultivar-group name *Erica tetralix* Alba Group implies coextension with the botanical taxon".

TABLE 6. Classification of Cultivated *Tulipa*.

Classified List and International Register of Tulip Names

Early flowering
      Single Early Tulips
      Double Early Tulips
Mid-season flowering
      Triumph Tulips
      Darwin Hybrid Tulips
Late-flowering
      Single Late Tulips
      Lily-flowered Tulips
      Fringed Tulips
      Viridiflora Tulips
      Rembrandt Tulips
      Parrot Tulips
      Double Late Tulips
Species
      T. kaufmanniana Regel
      T. fosteriana Irving
      T. greigii Regel
      others

van Raamsdonk & de Vries 1996

Tulipa gesneriana L.
      supergroup Early
      supergroup Mid Season
      supergroup Late-flowering

In my view, this is a prime example of desperately hanging on to the Botanical Code, thus linking cultivar classification with botanical classification at levels lower than the crop category, and smearing the taxon/culton distinction in the process. Apart from the fact that coextension is next to impossible to prove or disprove, this is not a relevant issue. We should not burden cultonomy with the same problems that have weighed down taxonomy (cf. the enlightening discussion on the problem of tautonyms in the context of the Biocode). The title of Hetterscheids' paper (1999) points to the very heart of the problem: in order to ensure stability we should distance ourselves from the Botanical Code. Article 16.2 (and, for that matter, the whole of article 16) does just the opposite, trying to hold on to old crutches which we really do not need. The 1995 Cultivated Plant Code is a huge step forward on the way to an independent and adequate classification system for cultivated plants. But we're not there yet. Let's move forward and not look back!

References

Coombes, A.J. (1996). Cultivar-groups – a sensible approach. *Hortax News* 1(2): 5–7.

Helm, J. (1957). Versuch einer morphologisch-systematischen Gliederung der Art *Beta vulgaris* L. *Der Züchter* 27(5): 203–222.

Hetterscheid, W.L.A. (1996). Cultivar-groups – a systematic approach. *Hortax News* 1(2): 7–9.

Hetterscheid, W.L.A. (1999). Stability through the culton concept. In S. Andrews, A.C. Leslie & C. Alexander (eds). Taxonomy of Cultivated Plants: Third International Symposium, pp. 127–133. Royal Botanic Gardens, Kew.

Hetterscheid, W.L.A & Brandenburg, W.A. (1995). Culton versus taxon: conceptual issues in cultivated plant systematics. *Taxon* 44: 161–175.

Hetterscheid, W.L.A. & Berg, R.G. van den (1996). Cultonomy of *Aster* L. *Acta Bot. Neerl.* 45(2): 173–181.

Hetterscheid, W.L.A, Berg, R.G. van den & Brandenburg, W.A. (1996). An annotated history of the principles of cultivated plant classification. *Acta Bot. Neerl.* 45(2): 123–134.

Lange, W., Brandenburg, W.A. & De Bock, Th. S.M. (1999). Taxonomy and cultonomy of beet (*Beta vulgaris* L.) *Bot. J. Linn. Soc.* 130: 81–96.

Mabberley, D.J. (1997). A classification for edible *Citrus* (*Rutaceae*). *Telopea* 7(2): 167–172.

Mansfeld, R. (1986). Verzeichnis landwirtschaftlicher und gärtnerischer Kulturpflanze. Vol. 1. 577 pp. Springer Verlag, Berlin.

Oost, E.H. & Toxopeus, H. (1986). Scope and problems of cultivar group formation as exemplified in *Brassica rapa* L. *Acta Hort.* 182: 117–123.

Raamsdonk, L.W.D. van & Vries, T. de (1996). Cultivar classification in *Tulipa*. *Acta Bot. Neerl.* 45(2): 183–198.

Rabinowitch, H.D. & Brewster, J.L. (1990). Onions and allied crops. 277 pp. CRC Press, Florida.

Rice, G. (1996). Regularising the names for annuals and bedding plants. *Hortax News* 1(2):12–19.

Rossel, G. (1998). Taxonomic-linguistic study of plantain in Africa. PhD Thesis. Wageningen Agricultural University.

Trehane, P. (1996). Cultivar-groups – a concept in need of clarification. *Hortax News* 1(2): 10–11.

Trehane, P. (1997). The Code – contentious issues to be resolved. *Hortax News* 1(3): 10–16.

Brandenburg, W.A. (1999). Crop-weed complexes and the culton concept. In: S. Andrews, A.C. Leslie and C. Alexander (Editors). Taxonomy of Cultivated Plants: Third International Symposium, pp. 145–157. Royal Botanic Gardens, Kew.

CROP-WEED COMPLEXES AND THE CULTON CONCEPT

WILLEM A. BRANDENBURG

CPRO, PO Box 16, 6700AA Wageningen, The Netherlands

Abstract

Elucidating relationships within crop-weed complexes is essential for understanding the domestication process. Although crop-weed complexes are recognised by many authors as a most important starting point for further breeding programmes, molecular or traditional, their analysis suffers from lack of understanding of the true nature of their components. The cultivated plant that arises from domestication rather than evolution is in this respect misunderstood.

Application of the newly introduced term 'culton' to the categories of cultivated plants may aid the understanding of crop-weed complexes and contribute towards better classifications.

Introduction

Crop-weed complexes have been subjected to extensive systematic study, as can be learned from publications (de Wet & Harlan 1975; Hanelt 1986; Harlan 1965, 1975; Pickersgill 1981, 1986; van Raamsdonk & van der Maesen 1996). The interactions between domesticated plants, weeds and their wild relatives are complex and yet due to their economic importance urgently need to be understood. On the one hand, well-documented and classified crop-weed complexes are an important source of new cultivars; on the other hand their often complicated classification hampers their effective use due to misinterpretation and endless discussion on their (infraspecific) classification (Styles 1986). Despite the fact that the new and powerful tools of molecular systematics are now available, the classification of crop-weed complexes is still liable to controversy through lack of consensus on species concepts and the processes of evolution and domestication. In order to create an unambiguous starting point to their classification it is necessary to distinguish evolution (adaptation to natural circumstances) from domestication (adaptation to man-made conditions making plants more or less dependent on maintenance by man) and to examine the consequences for plant classification.

Schwanitz (1967) defines cultivated plants as follows:

'Die Kulturpflanzen sind das Ergebnis von Evolutionsvorgängen, die sich in vorgeschichtlicher und in geschichtlicher Zeit bis in unsere Tage hinein, teils unter dem unmittelbaren, teils unter dem mittelbaren einfluß des Menschen vollzogen haben und heute noch vollziehen.'

(Cultivated plants are the result of evolutionary processes, proceeding directly and indirectly in prehistoric, historic and even modern times, that continue to this day).

However, Schwanitz's definition shows ambiguities. On the one hand, classification of cultivated plants starts with evolutionary processes; on the other, it includes human influence, albeit in part indirect. It stresses the very beginning of the domestication process, whereas the current process of domestication (formulating the specification of a new plant, breeding and reproducing it) is under-emphasized. A more modern definition of cultivated plants according to Hetterscheid & Brandenburg (1995a), adapted from Trehane (*pers. comm.*): 'A cultivated plant is one, whose origin or selection is due to the activities of mankind. Such plants may arise either by deliberate or chance (garden!) hybridisation or by further selection from existing cultivated stock, or may be selected from a wild population and maintained as an entity by continuous cultivation.' Whereas domestication refers to changes in adaptation giving fitness in habitats specially prepared by man for his cultigens' (de Wet 1981), evolution leads to clades, speciation being the response to occasional events in earth history or to slight modifications in ecological conditions. In view of the above statement, speciation under domestication is a contradiction in terms. Such statements easily lead to confusion. The fact that speciation does not occur under domestication does not mean that one may decide to create 'species of convenience' for certain culta, such as *Triticum aestivum* L., described as a cultigen. This problem was addressed by Bailey (1918, 1923a). After Bailey, a range of taxonomists (Grebenschikov 1949; Jirasek 1961; Mansfeld 1953, 1954 and Pangalo 1948) tried to fit the complex, sometimes reticulate, relationships between cultivated plants and their weedy and wild relatives into a closed, botanical, and therefore hierarchical classification system. Even in the last two decades, in which the possibility of designing and creating novel cultivated plants has dramatically increased through molecular plant breeding, many authors (Hanelt 1986; van Raamsdonk 1993; van Raamsdonk & van der Maesen 1996 and Zohary 1984) argue that there is in effect no significant difference between evolution and domestication. Interestingly, they implicitly work with a biological species concept, that focusses on hybridisation boundaries rather than on evolutionary lineages.

Crop-weed Complexes

In studying crop-weed complexes three simple scenarios can be distinguished:

1. wild → weedy → cultivated populations

2. wild → cultivated → weedy populations

3. wild ↗ weedy → cultivated populations
    ↘ cultivated → weedy populations

(after Pickersgill 1981, 1986)

These simple scenarios can be combined into various complicated ones (Small 1984), outlining the complex interactions between wild, weedy and domesticated populations. Classification in genera such as *Brassica* L. (Bailey 1923b, 1930; Branca & Iapichino 1997; Darmency 1994; Dias *et al.* 1992; Gladis & Hammer 1992; Gupta *et al.* 1995; Gustafsson 1995; Helm 1963; Lanner 1997; Metzger 1833; Mithen *et al.* 1987; Oost 1986; Oost & Toxopeus 1986; Oost *et al.* 1987, 1989; Perrino *et al.* 1992; Prakash & Hinata 1980; Schulz 1919; Snogerup *et al.* 1990; Toxopeus *et al.* 1986; Warwick 1997 and Warwick & Black 1991); *Capsicum* L. (Eshbaugh 1974; Gonzalez & Bosland 1991; Loaiza Figueroa *et al.* 1989; McLeod *et al.* 1983; Pickersgill 1977, 1986, 1997; Pickersgill & Heiser

1977 and Pickersgill *et al.* 1979); *Cucumis* L. (Helm & Hemleben 1997; Kuriachan & Beevy 1992) and *Cucurbita* L. (Jeffrey 1980; Nabhan 1985 and Nee 1990) is confused by the simultaneous use of the evolutionary species concept and 'species of convenience'.

In other cases the use of pairs of species names for cultivated as opposed to wild or weedy populations has prevented real understanding of crop-weed complexes:

- *Brassica rapa* L. (cultivated) vs. *B. campestris* L. (weedy). In their respective protologues the name *B. rapa* was reserved for the crop plants, whereas *B. campestris* described the weedy populations next to the fields (Linnaeus 1753). Metzger (1833) decided to unite both species, but the confusion continued until the work of Oost and co-workers resulted in a cultivar-group classification of *B. rapa* (Oost 1986; Oost & Toxopeus 1986; Oost *et al.* 1987; Toxopeus *et al.* 1986).
- *Lactuca sativa* L. (cultivated) vs. *L. serriola* L. (wild and weedy). The history of lettuce in domestication is relatively well known (de Vries 1997, de Vries & Jarvis 1987; Oost 1980), although the origin of the heading lettuces is not yet revealed. *Lactuca serriola* is known from the Middle East and Egypt and is currently invasive throughout Europe. Careful examination of wild and cultivated lettuce leads to the conclusion that only one species is involved. For nomenclatural reasons this is known as *L. sativa* (Frietema de Vries *et al.* 1994; Frietema de Vries 1996). By concentrating on wild and weedy variation in the species, and excluding the variation in cultivation, we can re-assess its infraspecific classification. In the Middle East this complex initially followed scenario 1 of Fig. 2. In Europe, however, the true nature of the invasive populations is not yet clear and is the subject of study (Prince & Carter 1997; van der Ham 1981).
- *Ribes uva-crispa* L. (cultivated) vs. *R. grossularia* L. (wild). Economically not a very important complex; interestingly, weedy populations of this complex can be found, e.g. in the coastal dunes along the Dutch North Sea coast. Again, examination of the variation suggests that there is just one species involved, called *R. uva-crispa*. Comparison of natural populations and weedy populations should reveal their similarity and consequently their classification.

These pairs of species all date back to Linnaean times.

A third category of problems may arise through classifying the cultivated plants in a separate subspecies. This has been advocated by several authors (de Wet 1981; Pickersgill 1986):

- *Beta vulgaris* L. subsp. *vulgaris* (cultivated) vs. *B. vulgaris* subsp. *maritima* (L.) Arcang. (wild). In a recently submitted paper, elaborating on the work of Evans & Weir (1981) and Letschert (1993), Lange *et al.* (1999) point out that separating the cultivated beets from the wild and weedy ones results in a transparent classification, acceptable for different users. Crop-weed complexes in beet started from scenario 2 of Fig. 2.
- *Daucus carota* L. subsp. *sativus* (Hoffm.) Schübl. & G. Martens (cultivated) vs. *D. carota* L. subsp. *carota* (wild). By looking at domestication processes in carrot (Banga 1957a, 1957b, 1963; Brandenburg 1981; Small 1978; Wijnheijmer *et al.* 1989 and Zeven & Brandenburg 1986), there is an obvious advantage in separating domesticates from the subspecies described, resulting in a subspecies classification of wild and weedy populations in the biological sense (Fuchs 1958; Meikle 1957). The map in Fig. 1 illustrates a good example of scenario 3 of Fig. 2.

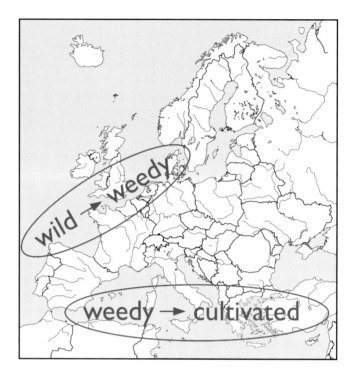

FIG. 1. Interactions in the *Daucus carota* wild-weedy-domesticate complex.

Early attempts to resolve the nature of crop-weed complexes started with comparative morphological and ecological studies, subsequently combined with hybridisation and introgression studies (Anderson 1949; de Wet & Harlan 1975; Frietema de Vries 1996; Harlan 1965, 1975; Nee 1990; Small 1984 and Zohary 1984). Today, the rapid developments in molecular biology have made it possible to reveal population structures at isozyme, DNA and chromosome level. Multidisciplinary approaches are especially powerful in unravelling the often complex crop-weed complexes; good examples are seen in:– *Beta* L. (Letschert 1993; Boudry *et al.* 1993, 1994); *Brassica* L. (Darmency 1994; Prakash & Hinata 1980; Warwick 1997 and Warwick & Black 1991); *Capsicum* L. (Davis & Gilmartin 1985; Loaiza-Figueroa *et al.* 1989; Moscone *et al.* 1993, 1996 and Pickersgill *et al.* 1979); *Lactuca* L. (Koopman & de Jong 1996 and Koopman *et al.* 1993, 1998) and *Lolium* L. (Hayward *et al.* 1995; Loos 1993, 1994 and Stammers *et al.* 1995). By revealing population structures and combining these with morphological data and ecological preferences, it should be possible to achieve baseline studies of the cultivated, weedy and wild plants of one complex.

Crop-weed Complexes and the Culton Concept

Due to the different nature of cultivated plants (domesticates) on the one hand, and weedy and wild populations on the other, the classification of crop-weed complexes inevitably leads to very complicated schemes or obvious misinterpretation.

However, the mechanism of botanical classification cannot cope with the dynamics of the development of variation in cultivated plants, irrespective of whether this development is merely the selection of desired genotypes in the initial phase of domestication or more advanced plant breeding (hybridisation, selection, cell and molecular biology) as is predominant in most current domestication processes.

In order to create an unambiguous classification system for cultivated plants, the principle of open classification in which entities are not only described but also circumscribed has been suggested (Brandenburg 1984, 1986a, 1986b; Brandenburg *et al.* 1982). The cultivar-group was proposed as an important tool for classifying cultivars irrespective of the species involved (Brandenburg & Schneider 1988). During preparations for a proposal to thoroughly revise the Cultivated Plant Code (*International Code of Nomenclature for Cultivated Plants – 1995* (Trehane *et al.* 1995)), the Vaste Keurings Commissie (VKC) working group on Nomenclature and Registration felt it necessary to distinguish between the underlying concept of botanical classification, the taxon, (botanical, closed, hierarchical classification, types – basal rank species) and that for cultivated plants, the culton (open, non-hierarchical classification, standards – basal term cultivar). The culton concept is extensively introduced and discussed by (Hetterscheid & Brandenburg 1995a, 1995b; Hetterscheid *et al.* 1996). Meanwhile, cultivar-group classification has often been used (see van den Berg 1999). Looking again at the simple scenarios for crop-weed complexes, by applying the culton concept, it is possible to reveal the cross-overs between classification concepts as shown below (see also Fig. 2):

1. wild [taxon] → weedy [taxon] → cultivated populations [culton]

2. wild [taxon] → cultivated [culton] → weedy populations [taxon]

3. wild [taxon] ⟋ weedy [taxon] → cultivated populations [culton]
   ⟍ cultivated [culton] weedy populations [taxon]

Adopting the culton concept for the classification of cultivated plants implies that this classification is at least interlinked with botanical classification at the rank of genus, or in certain cases at the rank of species, as is the case for *Beta vulgaris*. The cultivar-groups will be classified under *B. vulgaris*, whereas the recognised subspecies reflect spontaneous variation in the species (Lange *et al.* 1999):

- subsp. *maritima*: spontaneous variation along west European coast
- subsp. *vulgaris*: subspontaneous, weedy variation throughout Europe, mostly close to arable land with cultivated beets
- subsp. *adanensis* (Pamuk.) Ford-Lloyd & J.T. Williams: spontaneous, geographically and morphologically distinct variation

Separating the classification of the cultivated plants has led to a study of the botanical variation in *Beta* especially *B. vulgaris*, thus revealing the spontaneous and subspontaneous variation. The resulting classification scheme has the advantage of being accepted by the users and of being a fact-based classification of botanical variation within the species (Letschert 1993; Lange *et al.* 1999). The spontaneous and subspontantaeous variation, being in separate subspecies (taxa), suggest that weedy plants amid crop plants represent a subspecies (taxon). At the same time, escapes in coastal areas are subject to natural selection and can consequently be regarded as a taxon as well. In this case, both are classified within *B. vulgaris* subspecies *vulgaris*.

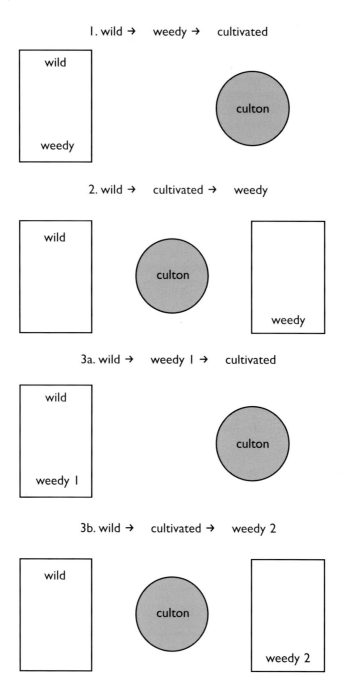

FIG. 2. Survey of simple scenarios in crop-weed complexes and their consequences for classification.
    Scenario 1: wild + weedy: conspecific
    Scenario 2: wild + weedy: different species or subspecies
    Scenario 3: wild + weedy 1: conspecific
            wild + weedy 2: different species or different subspecies

Classifications, such as proposed by Ford-Lloyd & Williams (1975) and later revised by Ford-Lloyd (1983, 1986), demonstrate clearly all the ambiguities in classifying crop-weed complexes in which different classification concepts are forced into the framework of botanical classification. A hypothetical phylogenetic reconstruction has been combined with the circumscription of current cultivated beets. The same holds for classifications, such as that proposed by Helm for *Lactuca sativa* (1954), *Beta vulgaris* (1957), *Brassica oleracea* L. (1963). Despite the extensive descriptions of the variation then known in the crop plants concerned, the strictly hierarchical nature of the resulting classifications into convarieties and cultivars is not sufficiently flexible to follow the dynamic changes in selections due to plant breeding – either by conventional or by molecular means. At the same time, these classifications deal with subspontaneous variation in different ways:

- No separate entities for spontaneous and subspontaneous variation, although there are distinct characters (e.g. *Brassica* spp. and *Lactuca sativa* vs. *L. serriola* merging into one species *L. sativa*, though the proposed infraspecific classifications does not fit).
- Separate entities (of convenience) for spontaneous and subspontaneous variation, where there is no firm distinction (*Citrus* L. cf. Swingle 1946; Tanaka 1954; *Colocasia* Schott cf. Greenwell 1947; Pancho 1959; Plucknett *et al.* 1970; Wang 1983).
- A mixture of both above approaches (*Cucurbita* cf. Andres 1995, Paris 1989; *Cucumis* subg. *Melo* cf. Naudin 1859).
- Separate entities for spontaneous and subspontaneous variation, distinguished by different character states (*Beta vulgaris*).

At first sight, it may seem paradoxical to aim at a uniform approach for the classification of crop-weed complexes by accepting two starting points of classification:

- Botanical classification via the taxon concept, in which the rank of species is basal.
- Classification of cultivated plants via the culton concept, in which the cultivar is the primary entity, thus allowing open classification of cultivar-groups.

Separating off the classification of cultivated plants in this way provides a better classification of spontaneous (wild) and subspontaneous (weedy) variation (Fig. 2). A consequence of this approach is that populations naturalised (escaped) from cultivation are again subject to botanical classification and thus become taxa. A comparison with the relevant spontaneous and subspontaneous populations has to combine sufficiently distinct character states and distribution areas to allow classification at one of the following ranks:

- *At species level* if due to complex hybridisation many species have been involved in the formation of the cultivated plant; as a consequence, the naturalised population cannot be assigned to any existing species because of its combination of character states.
- *At subspecies level* if there is one principal species involved in the formation of the cultivated plant, and the naturalised population has its own combination of character states and distribution area within the variation range of the species.
- *At the rank of botanical variety* if the naturalised population has a unique combination of character states but no distinct distribution area; this cannot be generally recommended because the lack of a distinct distribution area may lead to insufficient distinction after repeated sexual recombination.

One may wonder, however, whether there is a need for the introduction of a special concept (culton) for classifying cultivated plants (Alexander 1997; McNeill 1998 and Waters 1998). The culton stands for an open classification scheme involving cultivars and cultivar-groups (Trehane *et al.* 1995). Consequently, the conscious use of the term 'culton' makes the classification philosophy behind it explicit(cf. Hetterscheid & Brandenburg 1995a, 1995b; Hetterscheid *et al.* 1996). The application of open classification procedures in grouping cultivars is then called cultonomy (Hetterscheid & van den Berg 1996 and Lange *et al.* 1999). Most important, however, is the application of open classification, to establish cultivars and cultivar-groups and to use them in conjunction with the botanical classification of wild and weedy relatives to reveal the structure of crop-weed complexes.

Conclusions

Application of the newly introduced term culton to the classification of cultivated plants facilitates a better understanding of crop-weed complexes and, consequently, an improved classification.

Separating the open classification of cultivated plants from the closed botanical classification provides a better classification of spontaneous (wild) and subspontaneous (weedy) variation.

Naturalised populations are again subject to botanical classification with their ranking according to the established criteria of botanical classification.

References

Alexander, C. (1997). Do we need the term 'culton'? *Hortax News* 1(3): 6–9.

Anderson, E. (1949). Introgressive hybridization. 109 pp. John Wiley, New York.

Andres, T.C. (1995). Complexities in the infraspecific nomenclature of the *Cucurbita pepo* complex. *Acta Hort.* 413: 65–71.

Bailey, L.H. (1918). The indigen and the cultigen. *Science* 47: 306–309.

Bailey, L.H. (1923a). Various cultigens, and transfers in nomenclature. *Gentes Herb.* 1(3): 113–115.

Bailey, L.H. (1923b). The cultivated Brassicas I. *Gentes Herb.* 1(2): 53–108.

Bailey, L.H. (1930). The cultivated Brassicas II. *Gentes Herb.* 2(5): 211–267.

Banga, O. (1957a). Origin of the European cultivated carrot. *Euphytica* 6: 54–63.

Banga, O. (1957b). The development of the original European carrot material. *Euphytica* 6: 64–76.

Banga, O. (1963). Main types of the western carotene carrot. and their origin. 153pp. Tjeenk Willink, Zwolle.

Berg, R.G. van den (1999). Cultivar-group classification. In S. Andrews, A.C. Leslie & C. Alexander (eds). Taxonomy of Cultivated Plants: Third International Symposium. pp. 135–143. Royal Botanic Gardens, Kew.

Boudry, P., Mörchen, M., Saumitou-Laprade, P., Vernet, Ph. & Dijk, H. van (1993). The origin and evolution of weed beets: consequences for the breeding and release of herbicide-resistant transgenic sugar beets. *Theor. Appl. Genet.* 87(4): 471–478.

Boudry, P., Wieber, R., Saumitou-Laprade, P., Pillen, K., Dijk, H. van & Jung, C. (1994). Identification of RFLP markers closely linked to the bolting gene B and their significance for the study of the annual habit in beets (*Beta vulgaris* L.). *Theor. Appl. Genet.* 88(6–7): 852–858.

Branca, F. & Iapichino, G. (1997). Some wild and cultivated *Brassicaceae* exploited in Sicily as vegetables. *Pl. Genet. Resources Newslett.* 110: 22–28.

Brandenburg, W.A. (1981). Possible relationships between wild and cultivated carrots (*Daucus carota* L.) in the Netherlands. *Kulturpflanze* 29: 369–375.

Brandenburg, W.A. (1984). Biosystematics and hybridization of horticultural plants. In W.F. Grant (ed.). Plant Biosystematics, pp. 617–632. Academic Press, Don Mills, Canada.

Brandenburg, W.A. (1986a). Objectives in classification of cultivated plants. In B.T. Styles (ed.). Infraspecific classification of wild and cultivated plants, pp. 87–98. Clarendon Press. Oxford.

Brandenburg, W.A. (1986b). Classification of cultivated plants. *Acta Hort.* 182: 109–115.

Brandenburg, W.A., Oost, E.H. & Vooren, J.G. van de (1982). Taxonomic aspects of the germplasm conservation of cross-pollinated cultivated plants. In E. Porceddu & G. Jenkins (eds). Seed regeneration in cross-pollinated species, pp. 33–41. Balkema, Rotterdam.

Brandenburg, W.A. & Schneider, F. (1988). Cultivar grouping in relation to the International Code of Nomenclature for Cultivated Plants. *Taxon* 37(1): 141–147.

Darmency, H. (1994). The impact of hybrids between genetically modified crop plants and their related species: introgression and weediness. *Molec. Ecol.* 3(1): 37–40.

Davis, J.I. & Gilmartin, A.J. (1985). Morphological variation and speciation. *Syst. Bot.* 10(4): 417–425.

Dias, J.S., Lima, M.B., Song, K.M., Monteiro, A.A., Williams, P.H. & Osborn, T.C. (1992). Molecular taxonomy of Portuguese tronchuda cabbage and kale landraces using nuclear RFLPs. *Euphytica* 58(3): 221–229.

Evans, A. & Weir, J. (1981). The evolution of weed beet in sugar beet crops. *Kulturpflanze* 29: 301–310.

Eshbaugh, W.H. (1974). Variation and evolution in *Capsicum pubescens* Ruiz. & Pav. *Am. J. Bot.* 61 (5, Suppl.): 42.

Fuchs, H.P. (1958). Historische Bemerkungen zum Begriff der Subspezies. *Taxon* 7: 44–52.

Ford-Lloyd, B.V. (1983). Progress in beet germplasm utilization. *Genetika (Belgrade)* 15(2): 269–272.

Ford-Lloyd, B.V. (1986). Infraspecific variation in wild and cultivated beets and its effects upon infraspecific classification. In B.T. Styles (ed.). Infraspecific classification of wild and cultivated plants, pp. 331–344. Clarendon Press. Oxford.

Ford-Lloyd, B.V. & Williams, J.T. (1975). A revision of *Beta* section *Vulgares* (*Chenopodiaceae*) with new light on the origin of cultivated beets. *Bot. J. Linn. Soc.* 71(2): 89–102.

Frietema de Vries, F.T. (1996). Cultivated Plants and the Wild Flora – effect analysis by dispersal codes. 222 pp. Thesis, Rijksherbarium/Hortus Botanicus, Leiden.

Frietema de Vries, F.T., Meijden, R. van der & Brandenburg, W.A. (1994). Botanical files on Lettuce (*Lactuca sativa*) — on the chance for gene flow between wild and cultivated Lettuce (*Lactuca sativa* L., including *L. serriola* L., *Compositae*) and the generalized implications for risk-assessments on genetically modified plants. *Gorteria* 20(1): 1–44.

Gladis, Th. & Hammer, K. (1992). The Gaterslebener *Brassica*-Kollektion – *Brassica juncea, B. napus, B. nigra* und *B. rapa. Feddes Repert.* 103(7–8): 467–507.

Gonzalez, M.M. & Bosland, P.W. (1991). Strategies for stemming genetic erosion of *Capsicum* germplasm in the Americas. *Diversity* 7: 1–2.

Grebenscikov, I. (1949). Zur morphologische-systematische Einteilung von *Zea mays* L. unter besonderer Berücksichtigung der südbalkanischen Formen. *Züchter* 19(10): 302–311.

Greenwell, A.B.H. (1947). Taro — with special reference to its culture and uses in Hawaii. *Econ. Bot.* 1(3): 276–289.

Gupta, S.K., Sharma, T.R., & Chib, H.S. (1995). Evaluation of wild allies of *Brassica* under natural conditions. *Cruciferae Newslett.* 17: 10–11.

Gustafsson, M. (1982). Germplasm conservation of wild (n=9) Mediteranean *Brassica* species. *Sveriges Utsadesforen. Tidskr.* 92: 133–142.

Ham, R.W.J.M. van der (1981). Gifsla (*Lactuca virosa* L.) en kompassla (*Lactuca serriola* L.) in Nederland. *Gorteria* 10: 179–184.

Hanelt, P. (1986). Formal and informal classifications of the infraspecific variability of cultivated plants – advantages and limitations. In B.T. Styles (ed.). Infraspecific classification of wild and cultivated plants, pp. 139–156. Clarendon Press. Oxford.

Harlan, J.R. (1965). The possible role of weed races in the evolution of cultivated plants. *Euphytica* 14: 173–176.

Harlan, J.R. (1975). Crops and man. 295 pp. Madison, Wisconsin.

Hayward, M.D., Degenaars, G.H., Balfourier, F. & Eickmeyer, F. (1995). Isozyme procedures for the characterization of germplasm, exemplified by the collection of *Lolium perenne* L. *Genet. Resources Crop Evol.* 42: 327–337.

Helm, J. (1954). *Lactuca sativa* L. in morphologisch-systematischer Sicht. *Kulturpflanze* 2: 72–129.

Helm, J. (1957). Versuch einer morphologisch-systematischen Gliederung der Art *Beta vulgaris* L. *Züchter* 27(5): 203–222.

Helm, J. (1963). Morphologisch-taxonomischer Gliederung der Kultursippen von *Brassica oleracea* L. *Kulturpflanze* 11: 92–210.

Helm, M.A. & Hemleben, V. (1997). Characterization of a new prominent satellite DNA of *Cucumis metuliferus* and differential distribution of satellite DNA in cultivated and wild species of *Cucumis* and in related genera of *Cucurbitaceae*. *Euphytica* 94: 219–226.

Hetterscheid, W.L.A. & Brandenburg, W.A. (1995a). Culton versus taxon: conceptual issues in cultivated plant systematics. *Taxon* 44(2): 161–175.

Hetterscheid, W.L.A. & Brandenburg, W.A. (1995b). The culton concept: setting the stage for an unambiguous taxonomy of cultivated plants. *Acta Hort.* 413: 29–34.

Hetterscheid, W.L.A., Berg, R.G. van den & Brandenburg, W.A. (1996). An annotated history of the principles of cultivated plant classification. *Acta Bot. Neerl.* 45(2): 123–134.

Jeffrey, C. (1980). A review of the *Cucurbitaceae*. *Bot. J. Linn. Soc.* 81(3): 233–247.

Jirasek, V. (1961). Evolution of the proposals of taxonomic categories for the classification of cultivated plants. *Taxon* 10: 34–45.

Koopman, W.P.M. & Jong, J.H. de (1996). A numerical analysis of karyotypes and DNA amounts in lettuce cultivars and species (*Lactuca* subsect. *Lactuca*, *Compositae*). *Acta Bot. Neerl.* 45(2): 211–222.

Koopman, W.P.M., Jong, J.H. de & Vries, I.M. de (1993). Chromosome banding patterns in lettuce species (*Lactuca* subsect. *Lactuca*, *Compositae*). *Plant Syst. Evol.* 185(3–4): 249–257.

Koopman, W.P.M., Guetta, E., Wiel, C.C.M. van de, Vosman, B. & Berg, R.G. van den (1998). Phylogenetic relationships among *Lactuca* (*Asteraceae*) species and related genera based on ITS-1 DNA sequences. *Amer. J. Bot.* 85(11): 1517–1530.

Kuriachan P. & Beevy, S.S. (1992). Occurrence and chromosome number of *Cucumis sativus* var. *hardwickii* (Royle) Alef. in South India and its bearing on the origin of cultivated cucumber. *Euphytica* 61(2): 131–133.

Lange, W., Brandenburg, W.A. & Bock, T.S.M. de (1999). Taxonomy and cultonomy of Beet (*Beta vulgaris* L.). *Bot. J. Linn. Soc.* 130(1): 81–96.

Lanner, C. (1997). Genetic relationships within the *Brassica oleracea* cytodeme. comparison of molecular marker systems. *Acta Univ. Agric. Sueciae Agraria* 39, 57 + 65pp.

Letschert, J.P.W. (1993). *Beta* sect. *Beta*: biogeographical patterns of variation, and taxonomy. *Agric. Univ. Wageningen Pap.* 93–1. 155pp.

Linnaeus, C. (1753). Species plantarum. 1099 pp. Holmiae, Stockholm.

Loaiza-Figueroa, F., Ritland, K., Laborde Cancino, J.A. & Tanksley, S.D. (1989). Patterns of genetic variation of the genus *Capsicum* (*Solanaceae*) in Mexico. *Plant Syst. Evol.* 165(3–4): 159–188.

Loos, B.P. (1993). Allozyme variation within and between populations in *Lolium* (*Poaceae*). *Plant Syst. Evol.* 188: 101–113.

Loos, B.P. (1994). Allozyme differentiation of European populations and cultivars of *Lolium perenne* L., and the relation with ecogeographical factors. *Euphytica* 80: 49–57.

Mansfeld, R. (1953). Zur allgemeinen Systematik der Kulturpflanzen I. *Kulturpflanze* 1: 138–155.

Mansfeld, R. (1954). Zur allgemeinen Systematik der Kulturpflanzen II. *Kulturpflanze* 2: 130–142.

McLeod, M.J., Gutman, S.I., Eshbaugh, W.H. & Rayle, R.E. (1983). An electrophoretic study of evolution in *Capsicum* (*Solanaceae*). *Evolution* 37(3): 562–574.

McNeill, J. (1998). Culton: a useful term, questionably argued. *Hortax News* 1(4): 15–22.

Meikle, R.D. (1957). "What is the subspecies?" *Taxon* 6: 102–105.

Metzger, J. (1833). Systematische Beschreibung der kultivirten Kohlarten mit ihre Spielarten, ihrer Kultur und oekonomischen Benutzung. 68pp. Heidelberg.

Moscone, E.A., Lambrou, M. & Ehrendorfer, F. (1996). Fluorescent chromosome banding in the cultivated species of *Capsicum* (*Solanaceae*). *Plant Syst. Evol.* 202(1–2): 37–63.

Moscone, E.A., Lambrou, M., Hunziker, A.T. & Ehrendorfer, F. (1993). Giemsa C-banded karyotypes in *Capsicum* (*Solanaceae*). *Plant Syst. Evol.* 186(3–4): 213–229.

Mithen, R.F., Lewis, B.G., Heaney, R.K. & Fenwick, G.R. (1987). Glusosinolates of wild and cultivated *Brassica* species. *Phytochemistry* 26(7): 1969–1973.

Nabhan, G.P. (1985). Native crop diversity in Aridoamerica: conservation of regional gene pools. *Econ. Bot.* 39(4): 387–399.

Naudin, Ch. (1859). Essais d'une monographie des espèces et des variétés du genre *Cucumis*. *Ann. Sci. Nat., Bot.* Ser. 4, 11: 5–87.

Nee, N. (1990). The domestication of *Cucurbita* (*Cucurbitaceae*). *Econ. Bot.* 44 (3 suppl.): 56–68.

Oost, E.H. (1980). Domesticatie en verdere ontwikkeling van sla, witlof en andijvie. 37 pp Landbouwhogeschool, vg. Plantenveredeling. Wageningen.

Oost, E.H. (1986). A proposal for an infraspecific classification of *Brassica rapa* L. In B.T. Styles (ed.). Infraspecific classification of wild and cultivated plants, pp. 309–315. Clarendon Press. Oxford.

Oost, E.H., Brandenburg, W.A. & Jarvis, C.E. (1989). Typification of *Brassica oleracea* L. (*Cruciferae*) and its Linnaean varieties. *Bot. J. Linn. Soc.* 101(4): 329–345.

Oost, E.H., Brandenburg, W.A., Reuling, G.T.M. & Jarvis, C.E. (1987). Lectotypification of *Brassica rapa* L., *B. campestris* L. and neotypification of *B. chinensis* L. (*Cruciferae*). *Taxon* 36(3): 625–634.

Oost, E.H. & Toxopeus, H. (1986). Scope and problems of cultivar-group formation as exemplified in *Brassica rapa* L. *Acta Hort.* 182: 117–123.

Pancho, J.V. (1959). Notes on cultivated aroids in the Phillipines 1. The edible species. *Baileya* 7: 63–70.

Pangalo, K.L. (1948). Novyje principy vnutrividovoj sistematiki kulturnych rastenij [New principles for the infraspecific systematics of cultivated plants]. *Bot. Zhurn.* (*Kiev*) 33: 151–155. [In Russian].

Paris, H.S. (1989). Historical records, origins and development of the edible cultivar groups of *Cucurbita pepo* (*Cucurbitaceae*). *Econ. Bot.* 43(4): 423–443.

Perrino, P., Pignone, D. & Hammer, K. (1992). The occurrence of a wild *Brassica* of the *oleracea* group (2n=18) in Calabria (Italy). *Euphytica* 59(2–3): 99–101.

Pickersgill, B. (1977). Taxonomy and the origin and evolution of cultivated plants in the New World. *Nature* 268 (5621): 591–595.

Pickersgill, B. (1981). Biosystematics of crop-weed complexes. *Kulturpflanze* 29: 377–388.

Pickersgill, B. (1986). Evolution of hierarchical variation patterns under domestication and their taxonomic treatment. In B.T. Styles (ed.). Infraspecific classification of wild and cultivated plants, pp. 191–209. Clarendon Press, Oxford.

Pickersgill B. (1997). Genetic resources and breeding of *Capsicum* spp. *Euphytica* 96: 129–133.

Pickersgill, B. & Heiser Jr., C.B. (1977). Origins and distribution plants domesticated in the New Tropics. In C.A. Reed (ed.). Origins of Agriculture, pp. 804–835. Moutons Publ., The Hague.

Pickersgill, B., Heiser Jr., C.B. & McNeill, J. (1979). Numerical taxonomic studies on variation and domestication in some species of *Capsicum*. In J.G. Hawkes, R.N. Lester and A.D. Skelding (eds). Biology and taxonomy of the *Solanaceae*, pp. 669–700. Academic Press, London.

Plucknett, D.L., De la Pena, R.S. & Obrero, F. (1970). Taro (*Colocasia esculenta*). *Field Crops Abstr.* 23: 413–426.

Prakash, S. & Hinata, K. (1980). Taxonomy, cytogenetics, and origin of crop Brassicas, a review. *Opera Bot.* 55: 1–57.

Prince, S.D. & Carter, R.N. (1977). Prickly Lettuce (*Lactuca serriola*) in Britain. *Watsonia* 11(4): 331–338.

Raamsdonk, L.W.D. van (1993). Wild and cultivated plants: the parallellism between evolution and domestication. *Evol. Trends Pl.* 7: 73–84.

Raamsdonk, L.W.D. van & Maesen, L.J.G. van der (1996). Crop-weed complexes: the complex relationship between crop plants and their wild relatives. *Acta Bot. Neerl.* 45(2): 135–155.

Schulz, O.E. (1919). *Cruciferae-Brassiceae.* 290 pp. Wilhelm Engelmann, Leipzich.

Schwanitz, F. (1967). Die Evolution der Kulturpflanzen. 463 pp. Bayerischer Landwirtschaftsverlag, München.

Small, E. (1978). A numerical taxonomic analysis of the *Daucus carota* complex. *Canad. J. Bot.*: 56(3): 248–276.

Small, E. (1984). Hybridization in the domesticated-Weed-Wild Complex. In W.F. Grant (ed.). Plant Biosystematics, pp. 195–210. Academic Press, Don Mills, Canada.

Snogerup, S., Gustafsson, M. & Bothmer, R. von (1990). *Brassica* sect. *Brassica* (*Brassicaceae*) I. Taxonomy and variation. *Willdenowia* 19(2): 271–365.

Stammers, M., Harris, J., Evans, G.M., Hayward, M.D. & Forster, J.W. (1995). Use of random PCR (RAPD) technology to analyse phylogenetic relationships in the *Lolium/Festuca* complex. *Heredity* 74(1): 19–27.

Styles, B.T. (ed.) (1986). Infraspecific classification of wild and cultivated plants. 432 pp. Clarendon Press, Oxford.

Swingle, W.T. (1946). The botany of *Citrus* and its wild relatives of the orange subfamily (Family *Rutaceae*, Subfamily *Aurantioideae*). In H.J. Webber & L.D. Batchelor (eds). The Citrus Industry. Vol. 1. History, botany and breeding, pp. 129–474. Berkeley, Los Angeles.

Tanaka, T. (1954). Species problem in *Citrus*. A critical study of wild and cultivated units of *Citrus*, based upon field studies in their native homes. Tokyo.

Toxopeus, H., Oost, E.H. & Veerman, I. (1986). Systematic description of the cultivated *Brassica* species – a translation of Johann Metzger's Systematische Beschreibung der kultivirten Kohlarten, with introduction. 65 pp. Wageningen.

Trehane, P., Brickell, C.D., Baum, B.R., Hetterscheid, W.L.A., Leslie, A.C., McNeill, J., Spongberg, S.A. & Vrugtman, F. (eds). (1995). International Code of Nomenclature for Cultivated Plants – 1995. 175 pp. Quarterjack Publishing, Wimborne.

Vries, I.M. de (1997). Origin and domestication of *Lactuca sativa* L. *Genet. Resources Crop Evol.* 44: 165–174.

Vries, I.M. de & Jarvis, C.E. (1987). Typification of seven Linnaean names in the genus *Lactuca* L. (*Compositae: Lactuceae*). *Taxon* 36(1): 142–154.

Wang, J.-K. (1983). Taro. A review of *Colocasia esculenta* and its potentials. 400 pp. University of Hawaii Press, Honolulu.

Warwick, S.I. (1997). Use of biosystematic data, including molecular phylogenies, for biosafety evaluation. In S. Matsui, S. Miyzaki & K. Kasamo (eds). The 3rd JIRCAS International Symposium: the 4th international symposium on the biosafety results of field tests of genetically modified plants and microorganisms, pp. 53–65. Tsukuba, Japan.

Warwick, S.I. & Black, L.D. (1991). Molecular systematics of *Brassica* and allied genera (subtribe *Brassicinae, Brassiceae*): chloroplast genome and cytodeme congruence. *Theor. App. Genet.* 82(1): 81–92.

Waters, T. (1998). A comment on the culton concept. *Hortax News* 1(4): 13–15.

Wet, J.M.J. de (1981). Species concepts and systematics of domesticated cereals. *Kulturpflanze* 29: 177–198.

Wet, J.M.J. de & Harlan, J.R. (1975). Weeds and domesticates: evolution in the man-made habitat. *Econ. Bot.* 29(2): 99–107.

Wijnheijmer, E.H.M., Brandenburg W.A., & Ter Borg, S.J. (1989). Interactions between wild and cultivated carrots (*Daucus carota* L.) in the Netherlands. *Euphytica* 40(1–2): 147–154.

Zeven, A.C. & Brandenburg, W.A. (1986). Use of paintings from the 16th to the 19th centuries to study the history of domesticated plants. *Econ. Bot.* 40(4): 397–408.

Zohary, D. (1984). Modes of speciation in plants under domestication. In W.F. Grant (ed.). Plant Biosystematics, pp. 579–586. Academic Press, Don Mills, Canada.

Pickersgill, B. & Karamura, D.A. (1999). Issues and options in the classification of cultivated bananas, with particular reference to the East African Highland bananas. In: S. Andrews, A.C. Leslie and C. Alexander (Editors). Taxonomy of Cultivated Plants: Third International Symposium, pp. 159–167. Royal Botanic Gardens, Kew.

ISSUES AND OPTIONS IN THE CLASSIFICATION OF CULTIVATED BANANAS, WITH PARTICULAR REFERENCE TO THE EAST AFRICAN HIGHLAND BANANAS

BARBARA PICKERSGILL[1] AND DEBORAH A. KARAMURA[2]

[1]Department of Agricultural Botany, School of Plant Sciences, The University of Reading, Whiteknights, PO Box 221, Reading RG6 6AS, UK
[2]Kawanda Agricultural Research Institute, Box 7065, Kampala, Uganda

Abstract

Various techniques of multivariate analysis have been used to classify the Ugandan Highland bananas into five clusters which conform to a considerable extent with local practice and also show some predictive value for scientists working on the crop. However, the Highland bananas represent just a small portion of the spectrum of variation among the cultivated bananas in *Musa* section *Musa*. The hierarchical nature of this variation makes it difficult to determine at what level the category of cultivar-group should be employed. At least two further categories, in addition to cultivar-group and cultivar, are necessary to portray adequately the variation in cultivated members of the section considered as a whole. In bananas, and also in other crops, we consider that one phenetic classification, based on many characters, may substitute for, or at least usefully complement, multiple special-purpose classifications based on one or a few characters.

Introduction

The foreword to the 1995 edition of the *International Code of Nomenclature for Cultivated Plants* or Cultivated Plant Code contains the statement that "Any code which has undergone such a major change in design will have its shortcomings and these will only be discovered when the Rules of the *Code* are put to the test" (Brickell 1995). Hard cases reputedly make bad law but, to remain credible, laws have to be tested on hard as well as easy cases. Some of the hardest cases in the classification of cultivated plants concern species in which different morphological or use-groups occur, for example *Brassica oleracea* L. (cabbage, kale, cauliflower, etc.), or *Musa* L. section *Musa* (dessert, cooking and beer bananas). Furthermore, most previous discussions of the classification of cultivated plants and the application of the Cultivated Plant Code have concerned plants grown in developed countries, with well-organised trades in harvested products, planting material or both, and often with International Registration Authorities to regulate the application of cultivar names. Hetterscheid & Brandenburg (1995), Hetterscheid & van den Berg (1996) and Hetterscheid *et al.* (1996) have argued forcefully that classification of cultivated plants should be user-driven, by which they seem to mean driven by those who grow, sell, consume or otherwise utilise the plants concerned. We wish to consider whether similar principles can, or should, be applied to a crop developed, grown and consumed in less-developed countries, namely the East African Highland bananas. However, the Highland bananas cannot be considered in isolation: account has also to be taken of other cultivated members of *Musa* section *Musa*.

Cultivated Bananas in *Musa* section *Musa*

A. Genome Groups and their Origins

Bananas were initially domesticated in South-East Asia. Some are derived from a single wild diploid species which carries the A genome and whose name is discussed below; others are hybrids between this species and a second wild diploid, *M. balbisiana* Colla, which carries the B genome. Some diploid bananas with the genomic formulae AA and AB are still grown today, but the most widespread and economically most important bananas are now the triploids. These can be classified into three distinct groups on the basis of their genome(s). The widely exported dessert bananas, and the East African Highland bananas, are autotriploids (genomic formula AAA). A second group contains some locally-valued but little-exported dessert bananas, the Polynesian cooking bananas, and the plantains, which are staple foods in the humid lowlands of West and Central Africa and parts of tropical America. This group has the genomic formula AAB. A third group of triploids with the genomic formula ABB includes some other cooking bananas and some beer bananas.

B. Variation within Genome Groups

All domesticated bananas, whether diploid or triploid, are virtually seedless and are propagated clonally. Somatic mutants are not uncommon, and have had the greatest opportunity to arise and become established in those bananas which are grown on the largest scale, namely the AAA dessert bananas, the East African Highland bananas and the plantains. The AAA dessert bananas and the plantains have each been subdivided by banana scientists and banana growers into clusters of related clones. Within the AAA dessert bananas there are the Cavendish, Gros Michel and other subgroups (Simmonds 1959, Stover & Simmonds 1987), while the plantains are subdivided into French, Horn and two intermediate clusters (Swennen *et al.* 1995). Each cluster of clones may be further subdivided, for example into Dwarf Cavendish, Giant Cavendish, etc. before one reaches the level of the clone or cultivar.

C. Classification of Cultivated Bananas in *Musa* section *Musa*

Article 1 of the 1995 Cultivated Plant Code (Trehane *et al.* 1995) states that "Cultivated plants are named in accordance with the International Code of Botanical Nomenclature ... to the level of species or below, if, and only in so far as, they are identifiable with botanical taxa in those ranks." The AAB and ABB triploid bananas certainly do not occur in the wild, and we have found no reports of wild-occurring AB diploid hybrids either. All cultivars which combine the A and B genomes, at whatever level of ploidy, should therefore presumably be assigned simply to *Musa* section *Musa*.

However, AA diploid bananas do occur in the wild as well as in cultivation. The accepted name for these is *M. acuminata* Colla. Cheesman (1948) pointed out that Colla's description refers to a sterile, hence cultivated, banana but Colla's name has been widely used by Cheesman and others for wild seedy bananas morphologically similar to Colla's cultigen. We consider the AAA autotriploids to belong to the same species as the AA cultivated and AA wild bananas. However, it would be anomalous to give the AAA triploids names in Latin form at species level while the AAB and ABB triploids have Latin names to genus and section only. We agree with Simmonds (1962a) that in cultivated bananas all Latin nomenclature below the generic level should be jettisoned.

Hetterscheid & Brandenburg (1995) also considered that the species level is not needed for naming cultivated plants. However, their further suggestion that a combination of generic name and cultivar epithet is sufficient for purposes of classification and identification does not fit the experience of those who have worked with bananas. Simmonds (1959) and Stover & Simmonds (1987) used the genome groups, as originally identified by Simmonds & Shepherd (1955), as a basis for further classification of the numerous clones of cultivated bananas in *Musa* section *Musa*. These genome groups are now well-established among banana scientists, though the genomic formula is often used as a prefix to a more descriptive name (without any indication of rank), which may distinguish a subgroup within a genome group, e.g. AAA dessert, AAA East African Highland. No terms have been consistently used to designate the clusters or subclusters of clones within these subgroups.

There have been no studies of variation within the East African Highland bananas comparable to those in the dessert bananas or plantains. We therefore investigated morphological variation within the East African Highland bananas of Uganda, to determine whether a hierarchical pattern similar to that in the dessert bananas and plantains could be detected, and if so, how this variation might be treated in accordance with the 1995 Cultivated Plant Code.

East African Highland Bananas in Uganda

A. Background

The Highland bananas are a major food in the Great Lakes region of Africa. Although the fruits can be eaten raw, they are usually cooked and mashed to form matooke, which is eaten at most meals. Bananas with fruits too bitter to be palatable either raw or cooked are brewed into beer. The morphological variation in these bananas is so great that De Langhe (1996) considered the Great Lakes region to be a secondary centre of banana diversity. This diversity is recognised also by those who grow and use the crop. Karamura & Karamura (1994) recorded over 220 different names for Highland bananas in Uganda alone. Some of these are undoubtedly synonyms in different tribal languages: in a much earlier study, Shepherd (1957) collected 29 different names for what he considered to be 8 widely grown clones. A first task in classifying the Highland bananas of Uganda was therefore to determine the number of discrete entities that exist and their delimitation. A second task was to group these entities into classes on the basis of shared attributes so that the range of, and discontinuities in, their variation might be more readily comprehended.

B. Phenetic Analyses of Morphological Variation

Karamura (1998) investigated morphological variation among 192 accessions of Highland bananas in the national germplasm collections maintained at Kawanda and Kabanyolo Agricultural Research Institutes. 150 different local names are used for these accessions, but Karamura (1998) decided that they probably represent only 79 distinct clones and that the redundant names are synonyms in different languages.

From her data on 61 characters (13 quantitative and 48 qualitative), she calculated correlation and distance matrices for the 192 accessions and subjected these to cluster analysis. Five more or less well-defined clusters appeared consistently in each phenogram, though a significant number of outlying accessions could not be assigned confidently to any particular cluster. Principal component analysis also indicated much overlap between the five clusters and much uncorrelated variation among the 61 characters. Even the first principal component accounted for only 11% of the variation.

Classificatory discriminant analysis with a reduced set of 16 characters, mainly those with high loadings on the first or second principal component, was then used to maximise the differences between groups defined *a priori* (i.e. the five clusters from the cluster analyses) and to assign each accession to the most appropriate group. Only one accession remained intermediate between two groups. This confirmed our feeling that, despite their overlaps, the five clusters did have some validity. The five clusters also conformed well with a subjective classification made by Karamura at the start of the study.

We needed next to determine whether accessions grown by farmers but not necessarily represented in the collections could also be placed in these five groups, or whether the addition of more accessions, grown under different conditions, would blur distinctions that were reasonably clear in the collections. Data on 102 representative accessions from the collections were combined with data on 46 accessions grown in farmers' fields. One of the phenograms resulting from cluster analysis of this data set is shown in Fig. 1, together with the assignment of these accessions to five groups by classificatory discriminant analysis.

One cluster consisted entirely of beer bananas, though not all bitter-fruited bananas were in this cluster. The two which were omitted (arrowed in Fig. 1) both clustered with the same subset of cooking bananas. Both were claimed by their growers to be recent mutants from cooking clones. One did indeed cluster with its putative progenitor, but the other differed so much that we consider the farmer's opinion on its origin is in error. Some farmers in Uganda are concerned that cooking clones are mutating frequently to beer bananas. If this is indeed the case, we would expect beer bananas to be scattered throughout one or more clusters of cooking bananas, not forming a discrete cluster of their own. Our data suggest that most beer clones resemble one another in many morphological characters in addition to their bitter fruits, and so form a coherent group that has probably been evolving separately from the cooking bananas for some considerable time. However, the beer group, defined morphologically, does not coincide perfectly with the beer group, defined by bitter fruits (i.e. by use), possibly because there are occasional mutations from non-bitter to bitter fruits.

The four clusters within the cooking bananas bring together clones which share characters important to farmers and consumers, though these characters were not used in the multivariate analyses. Thus, clones in the Nakitembe and Nakabululu clusters sucker profusely, mature quickly, and produce a soft-textured matooke. Most clones of the Nfuuka cluster are slow to produce suckers, take a long time to mature, and produce a hard-textured matooke. However, one subcluster within Nfuuka shares with the Nakabululu and Nakitembe clusters the characteristics of rapid maturity and production of soft-textured matooke. Nfuuka is more heterogeneous than the other clusters and overlaps with most of them. Its name means "I am changing" and reflects the farmers' perception that somatic mutants are particularly frequent in these clones. The final cluster, Musakala, contains the highest-yielding clones (e.g. 'Lumenyamagali' – I break bicycles), grown on a commercial scale to supply the urban markets. Clones in this cluster are intermediate with regard to time to maturity and quality of their matooke.

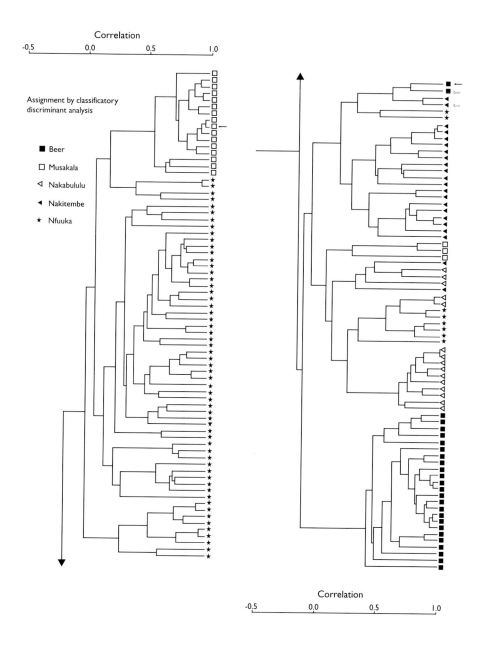

FIG. 1. Phenogram resulting from unweighted pair group arithmetic average (UPGMA) clustering of matrix of correlation coefficients comparing 102 accessions of East African Highland bananas from the Ugandan national collections and 46 accessions from farmers' fields (symbols represent results of classificatory discriminant analysis; arrows indicate two accessions of beer banana claimed by their growers to be recent mutants from the two accessions of cooking banana also indicated by arrows; form of arrow indicates claimed progenitor-derivative relationships).

Clones of Highland bananas in Uganda can therefore be sorted into morphological groups which coincide to a considerable extent, though not completely, with the groups recognised by banana growers and consumers and which parallel the groups or clusters of clones within the AAA dessert bananas and the plantains.

Discussion

A. Hierarchical Variation in Cultivated Bananas

Hetterscheid *et al.* (1996) claimed that the examples of hierarchical variation given by Pickersgill (1986) in some crops other than banana are ephemeral, due partly to random selection by man, and liable to be blurred by multiple or ongoing domestications. They argued against any hierarchy of categories for classification of cultivars. We consider that bananas provide a further example of a crop in which the variation is clearly hierarchically structured. The levels of variation represented by the genome groups (e.g. AAA), subgroups (e.g. AAA dessert or East African Highland bananas), and further subdivisions (e.g. Cavendish dessert bananas or Nakitembe Highland bananas) are not ephemeral and do not appear to result from random selection by man.

Furthermore, multiple or repeated origins have not undermined the validity of these hierarchical levels. AA cultivated diploids apparently contain genes from at least three wild subspecies of *M. acuminata* (Horry & Jay 1990); the AAA triploids also probably originated independently many times (Simmonds 1962b). This helps to explain why different groups may exist at the same hierarchical level, for example the AAA dessert bananas and the AAA East African Highland bananas, but it does not invalidate the hierarchy.

B. Taxonomic Treatment of Hierarchical Variation in Cultivated Bananas

Only one category, cultivar-group, is provided by the Cultivated Plant Code to treat variation above the level of cultivar. In trying to decide how and where this category should be applied in bananas, we looked for precedents in other cultivated species. *Brassica oleracea* is another cultivated species with essentially hierarchical variation: an initial division by part of the plant eaten into cabbage, kale, cauliflower, etc.; further subdivision, for example into white, red or Savoy cabbages, or curly, marrowstem or thousand-headed kale; possible finer distinction into spring cabbages and winter cabbages, etc. In a recent classification of *B. oleracea*, these first two levels of variation were collapsed so that cultivar-groups were considered to be, not Cabbage or Kale, but White Headed Cabbage or Marrowstem Kale (Jansen *et al.* 1994). It would not take much further elaboration to turn these cultivar-group names into something akin to the pre-Linnean phrase names, which were discarded in favour of binomials because they had become too cumbersome. Rather than lengthen cultivar-group names unduly, or resurrect hierarchies of unfamiliar terms which, as Hetterscheid *et al.* (1996) pointed out, will lead to a classification being ignored by most potential users, we prefer to supplement the single category above cultivar provided by the Cultivated Plant Code with some informal categories, conforming as closely as possible to categories developed by those who work actively with the crop, hence categories which have been tried by experience.

In bananas, Simmonds' (1959) genome groups are well-established, as are major subgroups such as the AAA dessert bananas or the plantains and also further subdivision of economically important and/or variable subgroups, for example distinction of the French and Horn plantains. The category of cultivar-group could presumably be used at any one of these three levels. Simmonds (1959) regarded his genome groups as corresponding in rank to cultivar-group, though he never used the term cultivar-group. However, this usage would produce cultivar-groups which have no name, only a genomic formula, hence would not conform to Article 19 of the Cultivated Plant Code (Trehane *et al.* 1995) which specifies that a cultivar-group epithet must be a word or phrase. It may therefore be preferable to use cultivar-group at the level of Simmonds' (1959) subgroups, e.g. Cavendish Group, East African Highland Group, Plantain Group. The genome group then becomes a useful informal category above cultivar-group.

We need also an informal category below cultivar-group to accommodate the clusters of clones that we recognise within the Highland bananas and that others have recognised within the AAA dessert bananas or the plantains. We do not wish to use the category subgroup for these, because of potential confusion with Simmonds' (1959) use of subgroup. Simmonds (1959) sometimes used the term type for these clusters of clones, but type has undesirable nomenclatural implications. We therefore designate these clusters "clone sets". We are thus using a hierarchy of four ranks: genome group, cultivar-group, clone set and cultivar. A given clone, for example 'Lumenyamagali' would then be classified in *Musa* section *Musa*, genome group AAA, East African Highland Group, clone set Musakala. Of course, these ranks are not all essential parts of the name of this clone, just as a species in a large and variable genus does not have to be referred to its correct series, section and subgenus every time its name is used. The additional ranks provide a means of locating a clone more precisely within the range of variation represented by the cultivated bananas when it is useful to do so.

C. Special-purpose versus Phenetic Classifications

Hetterscheid & Brandenburg (1995) considered it impossible to produce a natural classification of entities influenced by human selection. They advocated instead that, for each crop, there should be a series of user-driven special-purpose classifications into cultivar-groups. The cultivar-groups in each of these classifications were to be defined by different user-related characters. One cultivar could thus belong simultaneously to more than one cultivar-group. The example cited in the Code is the potato cultivar 'Desiree', which belongs to both the Red-skinned Group and the Maincrop Group.

There are two rather different groups of users of a classification of the East African Highland bananas. The first group consists of those who grow, sell and use the crop and who have generated a classification which suits their needs without reference to any Code of Nomenclature. Karamura (unpublished data) records that the classification of Ugandan farmers first distinguishes the East African Highland bananas from other bananas, then subdivides the Highland bananas into beer versus cooking bananas. Locally named clones are then recognised within both beer and cooking bananas, though farmers do not always agree over the identification of a particular clone and often explain that "you know these bananas change". After such a change, the old clonal name is often retained with a suffix to distinguish the new clone, e.g. Nakitembe-Nakamali (=Nakitembe the small one), Nakitembe-Nakawere (=Nakitembe the glossy one). This is akin to our classification into clone sets, while the highest category recognised by the Ugandan farmers, East African Highland versus other bananas, is akin to our use of cultivar-group.

The second group of users who require a classification of East African Highland bananas are the scientists who work on this crop. Bananas in East Africa are threatened by numerous pests and diseases. National collections of local clones have been assembled in many countries and national and international efforts are under way to locate useful resistances. Maintaining living collections of bananas is costly, evaluating them even more so. Curators of banana and other germplasm collections need classifications which provide a basis for establishing core collections (Brown 1995), to be used for training and given priority in evaluation. Breeders, crop protection scientists and extension workers need some way of selecting a maximally distinct set of accession from national or international germplasm collections for initial evaluation and then need a means of locating other accessions likely to have similar properties to any accessions found to be desirable. To conduct their work and communicate its results, this group of users needs a classification based on many, rather than few, attributes, which identifies "natural" groups which may have some predictive value, rather than the artificial groups assembled by special-purpose classifications which have no predictive value. This need is demonstrated by the number of papers reporting use of multivariate methods to classify genetic resource collections in Hodgkin *et al.* (1995).

Rather than producing multiple special-purpose classifications of the East African Highland bananas, we have combined classificatory discriminant analysis with other forms of multivariate analysis to produce a single classification which we feel reflects the practice of the local people and may also satisfy the needs of international scientists. This classification has already demonstrated some predictive value: breeders have found that most clones of Highland banana that retain slight female fertility belong to the Nfuuka and Nakabululu clone sets (Vuylsteke *et al.* 1996). However, time alone will tell whether this classification will indeed prove useful and acceptable to the different groups of people who, in their different ways, make their living from the East African Highland bananas.

Acknowledgements

The work reported here was supported by a grant to Deborah Karamura from the Rockefeller Foundation, administered by the International Institute of Tropical Agriculture (IITA). We are grateful also to colleagues at the National Agricultural Research Organisation, Kawanda, Uganda, and at IITA-ESARC for providing equipment, transport and other facilities and for helpful discussions about the data.

References

Brickell, C.D. (1995). Foreword. In P. Trehane, C.D. Brickell, B.L. Baum, W.L.A. Hetterscheid, A.C. Leslie, J. McNeill, S.A. Spongberg & F. Vrugtman. (eds). (1995). International Code of Nomenclature for Cultivated Plants — 1995. pp. vii–viii. Quarterjack Publishing, Wimborne.

Brown, A.H.D. (1995). The core collection at the crossroads. In T. Hodgkin, A.H.D. Brown, Th.J.L. van Hintum & E.A.V. Morales, (eds). Core collections of plant genetic resources. pp. 3–19. John Wiley, Chichester.

Cheesman, E.E. (1948). Classification of the bananas. III. Critical notes on species. b. *M. acuminata* Colla. *Kew Bull.* 3: 17–28.

De Langhe, E. (1996). Banana and plantain: the earliest fruit crops? *INIBAP Annual Report 1995*. pp. 6–8. International Network for the Improvement of Banana and Plantain. Montpellier, France.

Hetterscheid, W.L.A. & Brandenburg, W.A. (1995). Culton versus taxon: conceptual issues in cultivated plant systematics. *Taxon* 44(2): 161–175.

Hetterscheid, W.L.A. & Berg, R.G. van den (1996). Cultonomy of *Aster* L. *Acta Bot. Neerl.* 45(2): 173–181.

Hetterscheid, W.L.A., Berg, R.G. van den & Brandenburg, W.A. (1996). An annotated history of the principles of cultivated plant classification. *Acta Bot. Neerl.* 45(2): 123–134.

Hodgkin, T., Brown, A.H.D., Hintum, Th.J.L. van & Morales, E.A.V. (eds). Core collections of plant genetic resources. 269 pp. John Wiley, Chichester.

Horry, J.P. & Jay, M. (1990). An evolutionary background of bananas as deduced from flavonoids diversification. In R.E. Jarret (ed.). Identification of genetic diversity in the genus *Musa*. Proceedings of an International Workshop held at Los Baños, Philippines, 5–10 September 1988. pp. 41–55. INIBAP, Montpellier, France.

Jansen, P.C.M., Siemonsma, J.S. & Narciso, J.O. (1994). *Brassica oleracea* L. In J.S. Siemonsma & K. Piluek (eds). Plant resources of South-East Asia No. 8. Vegetables. pp. 108–111. Pudoc-DLO, Wageningen, Netherlands.

Karamura, D.A. (1998). Numerical taxonomic studies of the East African Highland bananas (*Musa* AAA-East Africa) in Uganda. Ph.D. thesis. The University of Reading, Reading, U.K.

Karamura, D.A. & Karamura, E.B. (1994). A provisional checklist of bananas in Uganda. 28 pp. INIBAP, Montpellier, France.

Pickersgill, B. (1986). Evolution of hierarchical variation patterns under domestication and their taxonomic treatment. In B.T. Styles (ed.). Infraspecific classification of wild and cultivated plants. pp. 191–209. Oxford University Press, Oxford. (Systematics Association Special Volume No. 29).

Shepherd, K. (1957). Banana cultivars in East Africa. *Trop. Agric. (Trinidad)* 34(4): 277–286.

Simmonds, N.W. (1959). Bananas. (Ed. 1). 466 pp. Longman, London. (Tropical Agriculture Series).

Simmonds, N.W. (1962a). The classification and nomenclature of the bananas and potatoes: some implications. *Proc. Linn. Soc. Lond.* 173: 111–113.

Simmonds, N.W. (1962b). The evolution of the bananas. 170 pp. Longman, London.

Simmonds, N.W. & Shepherd, K. (1955). The taxonomy and origins of cultivated bananas. *J. Linn. Soc. Bot.* 55: 302–312.

Stover, R.H. & Simmonds, N.W. (1987). Bananas. (Ed. 3). 468 pp. Longman, London.

Swennen, R., Vuylsteke, D. & Ortiz, R. (1995). Phenotypic diversity and patterns of variation in West and Central African plantains (*Musa* spp., AAB Group *Musaceae*). *Econ. Bot.* 49(3): 320–327.

Trehane, P., Brickell, C.D., Baum, B.R., Hetterscheid, W.L.A., Leslie, A.C., McNeill, J., Spongberg, S.A. & Vrugtman, F. (1995). International Code of Nomenclature for Cultivated Plants — 1995. 175 pp. Quarterjack Publishing, Wimborne.

Vuylsteke, D., Karamura, D., Ssebuliba, R.N. & Makumbi, D. (1996). Seed and pollen fertility in the East African Highland bananas. Mus*Africa* No. 10 pp. 13–14. (Abstract of paper presented at International Conference on Banana and Plantain for Africa, Kampala, Uganda, 14–18 October 1996).

THE CULTIVAR – DEFINITION AND RECOGNITION

6

Spencer, R.D. (1999). Cultivated plants and the codes of nomenclature – towards the resolution of a demarcation dispute. In: S. Andrews, A.C. Leslie and C. Alexander (Editors). Taxonomy of Cultivated Plants: Third International Symposium, pp. 171–181. Royal Botanic Gardens, Kew.

CULTIVATED PLANTS AND THE CODES OF NOMENCLATURE – TOWARDS THE RESOLUTION OF A DEMARCATION DISPUTE

ROGER D. SPENCER

Royal Botanic Gardens, Melbourne, Birdwood Ave, South Yarra, Victoria 3141, Australia

Abstract

The Botanical Code and Cultivated Plant Code deal not only with the names and ranks of plants but also give attention to their origins (whether wild or "cultivated"). Although the Botanical Code deals with the Latin names of all plants, the Cultivated Plant Code defines its scope in terms of the additional names of those plants whose origin or selection is primarily due to human intention. A clear indication of the origins of plants as expressed in their names would serve both Codes and have application in a future BioCode. L.H. Bailey's terms "cultigen" and "indigen" are examined and it is proposed that the term cultigen be redefined in line with present-day usage which is commensurate with the definition of the "cultivated plant" given in the Cultivated Plant Code. It is suggested that all cultigens (as newly defined) have a name component under the Cultivated Plant Code. A mechanism is presented for giving cultigen names to those cultigens which at present have names under the Botanical Code only. Cultigens would then be recognised by the presence of a cultigen epithet and indigens (wild plants) by the absence of a cultigen epithet. The use of the proposed terminology would permit the removal of the confusing expression "cultivated plant" from both Codes. A simple, flexible, and utilitarian classification system for cultigens is outlined and the proposal that the use of the term "culton" be used for cultigens (similar to "taxon" for indigens) is endorsed.

Mandates of the Botanical Code and Cultivated Plant Code – Names, Ranks and Origins

Both the Botanical Code (Greuter *et al.* 1994) and the Cultivated Plant Code (Trehane *et al.* 1995) deal with the names and ranks of plants. Principle 2 of the Cultivated Plant Code points out that the Botanical Code covers botanical names in Latin form for both wild and cultivated plants. In the sense that all botanical names have a Latin component (the genus and specific epithet), cultivated plants are under the jurisdiction of the Botanical Code but they may have, in addition, parts of the name governed by the Cultivated Plant Code – in most cases the cultivar epithet. Principle 2 defines the scope of the Cultivated Plant Code as being the "cultivated plant", loosely defined at first as one whose "... origin or selection is primarily due to the intentional actions of mankind." This part of Principle 2 is followed by a more detailed definition of the cultivated plant.

By definition, the Cultivated Plant Code is a separate Code because of the origin of the plants it covers: it acknowledges the existence of a distinct kind of plant, the cultivated plant, to be dealt with nomenclaturally in a special way. By implication, plants that are not cultivated must be wild. Thus the Codes are concerned not only with the names and ranks of plants but also their origins.

The scope, terminology and concepts relating to the two Codes would be greatly simplified if the Cultivated Plant Code dealt with all "cultivated plants" (it does not at present do so) and the distinction between wild and "cultivated" plants be immediately evident through the construction of their names. To achieve this clear demarcation requires a clarification of the concepts of wild and "cultivated" plants using simple terms, and a mechanism for giving "cultivated" plants treated under the Botanical Code additional epithets under the Cultivated Plant Code.

Wild and Domesticated Plants and Animals

The cornerstone of Darwin's Theory of Evolution is the idea of natural selection (Darwin 1859). Darwin derived his theory from his observation of selection processes operating on both wild and domesticated organisms, and came to the conclusion that these processes were importantly different from one-another. Now, more than ever, biological scientists are trying to unravel the story of evolution by defining "natural" groups in the attempt to determine "natural" lineages and to represent these in their classifications, a formal discipline known as phylogenetic systematics. In contrast to natural processes of selection there are the "artificial" processes performed by man. Included here would be the selection of horses, greyhounds and, of course, many of our garden plants.

McNeill (1998) supports Alexander (1997) in asserting that the distinction between natural and artificial taxa is itself artificial because man is part of the ecosystem. "The evolution of a wild orchid and the breeding of *Rosa* 'Peace' – or even the genetic engineering of transgenic Rt potato cultivars resistant to Colorado beetle ... – merely represent extremes of a continuum" (McNeill 1998). Continuum or not, the important point is that under the Code mandates mentioned above it is perfectly clear that the wild orchid would be given a binomial while the rose and potato cultivars would have additional parts to their names (probably cultivar epithets) that fall under the Cultivated Plant Code. [The correct name for the Peace rose is 'Madame A. Meilland' as Peace is a trade designation – author's note].

Thus we have a clear distinction between wild plants and animals subject to the forces of natural selection, and organisms that have arisen essentially under human intention by the process of artificial selection.

Cultigens and Indigens

Bailey (1918, 1923) in preparing his *Manual of Cultivated Plants* suggested that changes in botanical nomenclature would be desirable for people with "a scientific interest in cultivated plants". He was having problems naming domesticated plants using the principles of binomial nomenclature and Latin names according to the Linnaean tradition. Bailey wanted terms to define species and varieties that had arisen by the intentional activities of man and therefore, like Darwin, he was concerned with the demarcation between wild and domesticated taxa.

Bailey (1918) identified two classes of plants for which **species names** could be used:

1. Those that are discovered in the wild. These he termed "indigens". The most common present day casual term for these plants is "wild".
2. Those that arise in some way by human intention. These he termed "cultigens" and he later defined a cultigen (Bailey 1923) as " ... a species, or its equivalent, that has appeared under domestication, – the plant is cultigenous." These are species and hybrid species only; under Bailey's definition, cultivars are excluded.

There is no doubt that in his 1918 and 1923 articles Bailey used the term cultigen for plants at the rank of species but not below. Apart from his clear definition, he headed part 1 of his 1923 article "Various cultigens and cultivars". Neither Bailey's nor today's cultivars would be permitted as cultigens under his definition. As cultigens he cited, among others, the examples maize, oats, barley, lettuce, wheat, apple, and peach. The potato, *Solanum tuberosum* L., is another popular example.

Hortus Third (L.H. Bailey Hortorium 1976) defines a cultigen as: "... a plant or group of apparent specific rank, known only in cultivation, with no determined nativity, presumably having originated, in the form in which we know it, under domestication."

The 1995 Cultivated Plant Code enigmatically defines a cultigen negatively as: "A species believed not to have originated in the wild."

Bailey's meaning of the term cultigen, although completely clear, has been subject to (probably wishful) misinterpretation in the literature. It is easy to assume that by "cultigens" he meant all plants arising by human intention. *The New Royal Horticultural Society Dictionary of Gardening* (Huxley 1992) begins the description of a cultigen as "A plant found only in cultivation or in the wild having escaped from cultivation; included here are many hybrids and cultivars, ... ". Hetterscheid & Brandenburg (1995), refer to the cultivar as "the principle cultigen category" and "the principal fundamental subdivision of the cultigen" (Hetterscheid *et al.* 1996) although they point out that the term cultigen "has lost its original meaning and is now used with several redefinitions". The confusion is further compounded by the fact that the glossary of terms in Bailey's own completed *Manual of Cultivated Plants* (Bailey 1924), defines a cultigen as "Plant or group known only in cultivation; presumably originating under domestication; contrast with indigen", a definition that could include cultivars. All these later texts ignore Bailey's original and mutually exclusive distinction between a cultigen (at the rank of species) and a cultivar (as a botanical variety or rank below the level of species that has originated and persisted under cultivation).

It is proposed here that the term cultigen be formally accepted in its broad sense to refer to "cultivated plants" as defined in Principle 2 of the Cultivated Plant Code; this is the sense in which the term is generally used today. The terms cultigen and indigen and their definitions would be a valuable adjunct to the Preambles of both Codes and the Cultivated Plant Code could be more clearly named the Cultigen Code.

It is Bailey's cultigens (as he defined them) that require new names if the Cultivated Plant Code is to deal exclusively with cultigens as now defined. Indigens would then be easily recognised by the absence of cultigen epithets.

Cultivated Plants

Principle 1 and Article 1.1 of the Cultivated Plant Code have "cultivated plants" as the key concept. In Principle 2 cultivated plants covered by the Cultivated Plant Code are defined as plants " ... whose origin or selection is primarily due to the intentional action of mankind" and that "may arise by deliberate hybridisation or by accidental hybridisation in cultivation, by selection from existing cultivated stock, or may be a selection from variants within a wild population and maintained as a recognisable entity solely by continued propagation".

Unfortunately the general notion of a cultivated plant is not a clear one. It is probably incorrectly assumed by most scientific and other users that the Cultivated Plant Code deals with **all** plants whose origin or selection is primarily due to intentional human activity (cultigens in a broad sense) – as indeed the above definition would imply. This difficulty has been noted by Trehane (1993) who declaimed "Surely the Botanical Code should deal with wild taxa and the Cultivated Plant Code should deal with man's own plants". This thesis has also been proposed by Hetterscheid & Brandenburg (1994, 1995 and Hetterscheid *et al.* 1996). As we have already noted, Bailey's cultigens as originally defined are not catered for in the Cultivated Plant Code though the definition of "cultivated plant" implies that they should be.

There is a further difficulty. Most people regard a cultivated plant as one in cultivation. This is the popular notion of a cultivated plant. It is simply a plant being used in horticulture, agriculture or forestry. These are the plants implied by the book title *Manual of Cultivated Plants* (Bailey 1924); they are "cultivated" because they are deliberately grown, not because of their origin. They frequently do not have names under the Cultivated Plant Code and are simply indigens that have been brought into cultivation. This point is taken up by Article 28.1 of the Botanical Code which states "plants brought from the wild into cultivation retain the names that are applied to the same taxa growing in nature". This common usage notion of a cultivated plant should be alluded to either in the Principles or Preamble of the Cultivated Plant Code as this is probably the notion carried to the Cultivated Plant Code by most people before reading the definition of "cultivated plant" in Principle 2. Alternatively, and preferably, the expression "cultivated plant" could be omitted from the Codes.

It is clear that if cultigens (as originally defined by Bailey in 1923) are given names under the Cultivated Plant Code then the definition of a "cultivated plant" will be consistent with the origins of plants covered by both the Botanical Code and Cultivated Plant Code. There will also then be an equivalence of the notions of cultigen (in the broad sense) and cultivated plant as defined in Principle 2 of the Cultivated Plant Code. Most importantly, cultigens and indigens will be immediately apparent by the structure of their names. Once these distinctions have been clarified it is obvious that the term "cultigen" is a distinct improvement on the expression "cultivated plant".

Cultigens, Indigens and the Codes – Establishing a Clear Demarcation

We now have two clearly defined and useful terms, indigens for wild plants and cultigens for cultivated plants as defined in the Cultivated Plant Code: indigens originating by natural selection, cultigens essentially by human (artificial) selection.

The Botanical Code currently deals with two kinds of cultigen (as defined by Bailey 1923). Firstly, garden hybrids with binomials such as *Viburnum* × *juddii* Rehder, and secondly cultigenous species such as *Zea mays* L. and *Solanum tuberosum*, which, although not implied in their names, are assumed to have been altered from their wild relatives at some time by human activity. If these were given names under the Cultivated Plant Code then the scope of the two Codes would become clear.

One way to achieve this would be by placing some sort of flag or abbreviation next to the appropriate names, e.g. "cultig." next to all cultigens without names under the Cultivated Plant Code. The Cultivated Plant Code could then list these as an appendix, which would be a useful item in itself.

However, provision is already made within the Codes to achieve the required result. The difficulty lies not so much in the mechanism of achieving this but in persuading botanists to follow recommended practices when dealing with the names of cultigens.

A. Cultigenous Hybrids

Under the provisions of the Cultivated Plant Code, cultigens arising through hybridisation may be assigned to cultivars or cultivar-groups (Article 16.4).

Article 28.2 of the Botanical Code states "Hybrids, including those arising in cultivation, may receive names as provided in App. 1", and in Note 1 of this Article that "nothing precludes the use for cultivated plants of names published in accordance with the requirements of the present Code". The Botanical Code in its treatment of hybrids devotes considerable space to the complications of hybridisation in cultivated plants. Since the facility exists for cultigenous hybrids to be given names under the Cultivated Plant Code then it would surely be simpler if this facility were more stringently applied. This should be encouraged and actively pursued for all cultigenous plants. It could also be retrospective. Thus, for cultigens, botanists should be actively encouraged to use names under the Cultivated Plant Code. Note 2 of Article 28.2 of the Botanical Code then points out that "Epithets in names published in conformity with this Code may be used as cultivar epithets under the rules of the Cultivated Plant Code, when this is considered to be the appropriate status for the groups concerned," adding that names published after 1 Jan. 1959 must be names in a modern language. *Viburnum × juddii* would become *Viburnum* 'Juddii', and so on. The kinds of names chosen would be appropriate to the status of the groups concerned. It may, for example, be more appropriate to make the name under the Cultivated Plant Code a cultivar-group as *Viburnum* Juddii Group. It should be added that, as implied by Hetterscheid & Brandenburg (1995), it makes no sense to publish a nothospecific name for an artificial hybrid if it contains no cultivar and is so-to-say "empty".

There will, of course, be cases where the hybrid background is unknown. This need not cause undue problems or inconsistencies, the plants being named according to the current state of knowledge.

The same hybrids may arise in nature and in cultivation. Cultigenous novelties should, of course, be given names under the Cultivated Plant Code.

B. Cultigenous Species

Bailey's other cultigens were plants with binomials, such as *Zea mays* and *Solanum tuberosum*, dealt with by the Botanical Code. Many of these were given names before the Cultivated Plant Code came into existence. They were presumably given binomials for convenience to keep the nomenclatural process simple and in line with the Linnaean system, and because their histories were uncertain. Nevertheless, they are cultigens and if we knew their true histories they would surely have names under the Cultivated Plant Code.

Article 17.3 of the Cultivated Plant Code states that "A Latin epithet at the rank of species or below which is validly published (established) and otherwise legitimate (acceptable) in conformity with the Botanical Code for a taxon subsequently reclassified as a cultivar (and which is considered, upon re-classification, to be coextensive with that taxon as described in Art. 6.2), is to be retained as a cultivar epithet." Thus *Mahonia japonica* (Thunb. ex Murray) DC. becomes *Mahonia* 'Japonica', *Solanum tuberosum* would probably become *Solanum* Tuberosum Group and so on (see Cultivated Plant Code 16A.1, and Botanical Code (Article 28.2, Note 2) . See also the notes on hybrid epithets above.

It may be possible to use the abbreviation 'cultig.' effectively in the Cultivated Plant Code. Thus *Solanum* cultig. *tuberosum* could provide an effective name for the class of all potato cultivars. This might be an alternative to a name such as *Solanum* Tuberosum Group.

Transferring the names of extremely well known plants and many commercial crops in this way, though involving minimal change and disruption, is likely to be resisted as a costly and impractical process. If this is considered an obstacle then the transition could be made a gradual one encouraged by the two Codes.

Difficulties with the Notion of the Cultigen

There remain some difficulties in providing a clear definition of a cultigen. The difficulties are not, however, of sufficient strength to erode the value of the term but they do require attention. Some of these difficulties could be clarified by a simpler terminology although that is not addressed here.

A. Cultivated Indigens

Many plants in gardens are cultivated indigens. They are plants brought into cultivation from the wild that retain the names applied to the same taxa growing in nature (see Botanical Code Article 28.1, Cultivated Plant Code Article 16.1). The names of wild plants brought into cultivation are dealt with in Article 16 of the Cultivated Plant Code. Certainly it may be argued that "From the moment a plant is taken from nature (selected) and propagated or maintained under man-controlled circumstances, it is no longer exclusively subject to the forces of evolution" (Hetterscheid & Brandenburg 1995) and hence better regarded as a cultigen. However, as they generally do not differ in any way from plants growing in the wild there seems no point in giving them names under the Cultivated Plant Code. They are not novelties that have arisen under human selection or influence and are therefore indigens.

There are occasional circumstances for example when plants have large or brightly coloured flowers but still lie within the circumscription of the wild taxon named under the Botanical Code. Plants specially selected from the wild become cultigens because they are part of an intentional and conscious novelty selection process and are thus "cultivated plants" under Principle 2 of the Cultivated Plant Code. It should be noted that it has always been difficult to decide which Code is appropriate under such circumstances.

Hybrids may occur between naturalised cultivated indigens. In most cases these would be considered no different from similar hybrids that might arise from hybridisation of indigens in the wild.

B. Aliens

Increasing numbers of plants are escaping from cultivation to become naturalised in the wild. Of great environmental concern and receiving increasing attention from biologists and governments alike, they are generally referred to as environmental weeds or garden escapes. Many have had a devastating impact on native floras. They may be either cultivated indigens that have returned to the wild, or cultigens that have become naturalised. A further complication is that plants may pass back and forth between wild and cultivated environments, hybridising as they do so.

It seems reasonable to regard cultivated indigens that have returned to the wild as indigens unless some clear distinction can be shown that would warrant a name under the Cultivated Plant Code. Cultigens escaping into the wild would remain cultigens as their cultigenous origin is not in question.

It is important to note that a plant is a cultigen or indigen according to its origin, not the place where it is growing. Indigens may, through breeding or selection, give rise to cultigens. However, cultigens can never become indigens: they and their progeny are cultigenous.

C. Indigens under Human Influence
The influence of human activity, and its degree, is not always clear in nature. Perhaps new species arise in the wild as a result of industrial pollution, or human-induced climate change. Have these taxa arisen under human influence, and should they be termed cultigens? Is this a form of what might be termed unconscious selection - somewhere between natural and artificial selection?

Since such taxa have arisen by unintentional action they should remain as indigens; it is assumed that artificial selection must involve a deliberate act of choice.

It may also be argued that many cultigens have arisen by accident in cultivation and are therefore not the result of intention. Here it should be noted that it is in no way contradictory to point out that although these variants have arisen by accident, they have been selected by intention.

D. Plants of Unknown Origin
In many cases it is simply not known if a plant is cultigenous or not. The exact nature of human influence in the origin of many commercially important plants will probably remain unknown.

In cases like this where the garden plant cannot be clearly distinguished from its wild counterpart then it should be regarded as a cultivated indigen; if it has distinctive features resulting from its time in cultivation then it should be given a name under the Cultivated Plant Code. If a decision cannot be made, then the name would remain the same (see Cultivated Plant Code Art. 16.2, 16.3).

Natural and Artificial Classification Systems and Cultigens

Important aspects of natural and artificial classifications are summarised and discussed by Brandenburg (1986), Hetterscheid & Brandenburg (1995), and Hetterscheid *et al.* (1996). The following points are drawn largely from their analyses and emphasise the place of cultigens in such systems.

A. The Linnaean Hierarchy
There are many cultigens whose background may be exceedingly complex. In some instances this relationship may have involved hybridisation between wild and cultivated forms, backcrossing, increase of ploidy levels and so on; we shall probably never know their true origins. Many modern cultivars cannot be traced to one particular species.

The use of the Linnaean hierarchy of names, and other multilevel hierarchical modifications of this form of name presentation for cultigens (see e.g. Jeffrey 1968) are discussed by Hetterscheid *et al.* (1996).

Names like *Brassica oleracea* L. subsp. *oleracea* convar. *capitata* (L.) Alef. var. *capitata* f. *alba* DC. (even when the potential number of ranks is abbreviated) present many problems. They are unwieldy, unstable and, above all, extremely non-user-friendly, especially to the horticultural community. In many cases the detail of the name implies a precision which belies the fact that we have little idea of true relationships.

Names such as *Brassica* Capitata Group or *Brassica* Capitata Alba Group would be a simpler solution for the above name. There appears to be little merit in forcing cultigen names into an extended formal hierarchy (Linnaean or otherwise), although nested groups will remain. Cultivars can be grouped in any way that is useful, or in no way. The resulting system can be combined with Botanical Code Latin names at whatever level is appropriate: subspecies, species, genus, or notho-category. No new names for categories are needed.

Nevertheless, an appendix to the Cultivated Plant Code listing stable groups of economic plants would be extremely useful.

B. Ranking Cultivated Plants

In the Linnaean classification system taxa have a rank within a taxonomic hierarchy – each group being a subset of another in a boxes-within-boxes manner. Unfortunately ranks and taxa may be confused with one-another. While ranks are subjective concepts or ideas that can be defined, taxa, in contrast, exist in nature and are empirical circumscribed entities. In common usage these terms are often used interchangeably.

For practical purposes, if the substitution of the words "taxonomic unit" for "taxon" in a sentence makes sense, then there should be no confusion.

The categories of the Cultivated Plant Code (Trehane *et al.* 1995) are: denomination class; graft-chimaera; cultivar-group; cultivar; and selection/maintenance. Do these have rank in the same way as the ranking of plants used by the Botanical Code? How will we rank the genetically manipulated cultivars of the future? There is clearly a strong element of utility to these categories although there is certainly a hierarchy of decreasing comprehensiveness from cultivar-group to cultivar then to selection/maintenance. However, it should be noted that the same cultivars may be allocated to different cultivar-groups based on different criteria and, as has been noted before, they can be grouped in any way that is useful, or in no way. There is, however, no reason why cultivars should not be classified by descent, should that be useful.

In short, the ranking of cultigens in the Cultivated Plant Code is different from the traditional Linnaean method of ranking indigens.

C. Open and Closed Classification Systems

Cultivated plants have open classifications, while those of classical taxonomy are closed. For example in a group of dissected-leaved maples a purple-leaved group may be given a name under the Cultivated Plant Code but the others left undefined. In the Linnaean hierarchy all entities must occupy a rank. As already stated, cultivars can be grouped in any way that is useful, or in no way. The resulting system can be combined with Latin names of the Botanical Code at whatever level is appropriate: subspecies, species, genus, or notho-category. No new names for categories are needed.

Taxon vs. Culton

"Taxonomic groups of any rank will, in this Code, be referred to as taxa (singular: taxon)." (Article 1.1, Botanical Code). A taxonomic group can be defined as a group into which a number of similar individuals may be classified.

Hetterscheid & Brandenburg (1994, 1995) have argued that the taxonomic groupings of cultivated plants are not true taxa and that a different term is required for them, that term being culton (pl. culta). The culton is defined as: " ... a systematic group of cultivated plants based on one or more user-driven criteria." Hetterscheid &

Brandenburg (1994) note that though not inherent in the rules of nomenclature, taxa are nowadays viewed in an evolutionary context and that the term taxon should embrace only the products of natural (not artificial) processes. Cultivated plants are no longer exclusively subject to the forces of evolution within a natural population structure. A consequence of this is that these "taxa" are artificial and maintained in artificial habitats, and the classifications produced for them are also highly artificial. The classification system used for culta has been termed a teleological classification, although "utilitarian" would perhaps be a simpler and more apt term. The categories may be based on non-natural or unusual criteria and in this sense are multi-purpose.

It should be added that taxonomic botanists (except for those with a special charter) have largely ignored or deliberately avoided cultivated plants. Cultivars and cultivated indigens are almost invariably omitted from revisions and phylogenies precisely because they are not seen as part of the "natural" system; in herbaria they are generally clearly annotated or separated altogether from their wild counterparts. There is also a significant difference between types (Botanical Code) and standards (Cultivated Plant Code).

Although accepting that the notion of a culton has value, the definition is perhaps unnecessarily complex and could be pared to its simplest form – something like: "... a category of classification used for cultigens."

McNeill (1998) accepts the notion of the culton but argues that the key issue for deciding whether a plant is a taxon or culton is not whether human influence has been involved in its evolution, but rather the **purpose** for which recognition is being sought. Thus *Fragaria × ananassa* Duchesne and *Triticum aestivum* L. are acceptable as botanical taxa, then if we wish to recognise discrete cultivated variants then culton status is appropriate. The issue here, it seems to me, is over the mandates of the two Codes. These plants are examples of Bailey's cultigens and the recommendation presented in this paper is that, being artificial taxa and therefore cultigens, they should be given epithets according to the Cultivated Plant Code even though at present they are legitimately placed under the Botanical Code. I therefore support Hetterscheid and Brandenburg in considering human influence as critical.

McNeill also argues that the same plant may be both a taxon, *Erica tetralix* L. f. *alba* (Aiton) Braun-Blanq., in the wild and a culton, *E. tetralix* 'Alba', in gardens according to the **purpose** of recognition. In my view a cultivated indigen becomes a cultigen at the point where it is deliberately selected or altered permanently in some way that makes it distinct, whereupon it becomes a cultigen (see section on cultivated indigens). Certainly purpose and utility are important but in my view human selection is paramount; such cases have always been difficult.

Contrary to McNeill (1998), I do not accept that a culton escaping into the wild becomes a taxon. This is the crux of the matter. If it is a cultigen it is a "cultivated plant" under the Cultivated Plant Code and should have an appropriate epithet. As mentioned previously, human influence can be as simple as the selection of a flower colour variant from the wild. Thus the cultigen *Crocosmia × crocosmiiflora* (Burb. & Dean) N.E.Br. could be given the culton name *Crocosmia* Crocosmiiflora Group (or other appropriate name under the Cultivated Plant Code) which clearly indicates the plant's cultigenous origin. It will be a culton in the garden and in the wild. The decision as to whether it is a taxon or culton is not then made on the dubious factor of where it happens to be growing.

We must follow the definition of the taxon and its intention as presented in the Botanical Code and that does not accomodate what happens to cultivated plants. *The New Royal Horticultural Society Dictionary of Gardening* (1992) defines a taxon as "a general term for a taxonomic group of any category or rank." This definition would certainly allow the continued use of the term taxon for cultivated plants but in my view such use is surely a misrepresentation of the actual definition and its intention as used in the Botanical Code.

Botanists should not be expected to use the term taxon for the highly utilitarian artificial assemblages of plants of such diverse genealogy as cultivars. By the same token, the utilitarian and flexible system as envisaged here, and by other workers, is powerful and easy to use; it would provide horticulturists, agriculturists and foresters with a system that is simple to understand and free from the constraints of the Linnaean hierarchy.

It should be noted that "taxon" is a term that applies across the biological world. In the light of the proposed BioCode (see Greuter *et al.* 1998) would culton be a suitable term for domesticated "taxa" of biology in general?

TABLE 1. Proposed scheme of terminology for the Botanical and Cultivated Plant Codes

	CULTIVATED PLANT (as defined in the Cultivated Plant Code)	**WILD PLANT**
TERM	CULTIGEN	INDIGEN
CLASSIFICATION CATEGORY	CULTON (CULTA)	TAXON (TAXA)
GOVERNING CODE(s)	CULTIVATED PLANT CODE ('Cultigen Code') and BOTANICAL CODE ('Indigen Code')	BOTANICAL CODE ('Indigen Code')

The synthesis proposed here follows sentiments expressed since before the introduction of the Cultivated Plant Code. Stearn (1986) indicates that the German botanist Karl Koch as early as 1865 considered Latin names for garden forms a source of confusion. The early pioneering work was that of Liberty Hyde Bailey, in many ways the "father" of horticultural taxonomy. In more recent times the work of Hetterscheid, Brandenburg and others (see e.g. Hetterscheid *et al.* 1996) also subscribes to such a view.

An approach such as the one outlined here clarifies confusing concepts, frees botanists of the problems relating to cultigens, and allows horticulturists, agriculturists and foresters to claim and be responsible for their own plants, with a simple, powerful, flexible, utilitarian and user-friendly classification scheme. Such a system removes the necessity for a complex hierarchy and the observation of taxonomic strictures inappropriate for cultivated plants. The ideas presented here could be easily adapted for use in the proposed BioCode.

Acknowledgements

I would like to thank Peter Lumley and Ian Clarke for useful discussions of the ideas presented here.

References

Alexander, C. (1997). Do we need the term "culton"? *Hortax News* 1(3): 6–9.

Bailey, L.H. (1918). The Indigen and Cultigen. *Science* (New Series). 47: 306–8.

Bailey, L.H. (1923). Various cultigens, and transfers in nomenclature. *Gentes Herb.* 1(3): 113–136.

Bailey, L.H. (1924). Manual of cultivated plants most commonly grown in the continental United States and Canada. 851 pp. Macmillan, New York.

Brandenburg, W.A. (1986). Classification of cultivated plants. *Acta Hort.* 182: 109–115.

Darwin, C. (1859). On the origin of species by means of natural selection, or the preservation of favoured races in the struggle for life. 502 pp. Murray, London.

Greuter, W. *et al.* (eds). (1994). International Code of Botanical Nomenclature (Tokyo Code). 389 pp. Koeltz Scientific Books, Königstein. (Regnum Vegetabile 131).

Greuter, W. *et al.* (the IUBS/IUMS International Committee for Bionomenclature) (1998). Draft Biocode (1997): the prospective international rules for the scientific names of organisms. *Taxon* 47(1): 127–150. (see also http://www.rom.on.ca/ebuff/biocode1997.html).

Hetterscheid, W.L.A. & Brandenburg, W.A. (1994). The culton concept: setting the stage for an unambiguous taxonomy of cultivated plants. *Acta Hort.* 413: 29–34.

Hetterscheid, W.L.A. & Brandenburg, W.A. (1995). Culton versus taxon: conceptual issues in cultivated plant systematics. *Taxon* 44(1): 161–175.

Hetterscheid, W.L.A., Berg, R.G. van den & Brandenburg, W.A. (1996). An annotated history of the principles of cultivated plant classification. *Acta Bot. Neerl.* 45(2): 123–134.

Huxley, A. (ed. in chief). (1992). The New Royal Horticultural Society Dictionary of Gardening. 4 vols. Macmillan, London.

Jeffrey, C. (1968). Systematic categories for cultivated plants. *Taxon* 17(2): 109–114.

L.H. Bailey Hortorium. (1976). Hortus Third. 1290 pp. Macmillan, New York.

McNeill, J. (1998). Culton: a useful term, questionably argued. *Hortax* 1(4): 15–22.

Stearn, W.T. (1986). Historical survey of the naming of cultivated plants. *Acta Hort.* 182: 18–28.

Trehane, P. (1993). What is a cultivated plant? *Hortax News* 1(1): 1–2.

Trehane, P. *et al.* (1995). International code of nomenclature for cultivated plants — 1995. 175 pp. Quarterjack Publishing, Wimborne, UK. (Regnum Vegetabile 133).

Culham, A. & Grant, M.L. (1999). DNA markers for cultivar identification and classification. In: S. Andrews, A.C. Leslie and C. Alexander (Editors). Taxonomy of Cultivated Plants: Third International Symposium, pp. 183–198. Royal Botanic Gardens, Kew.

DNA MARKERS FOR CULTIVAR IDENTIFICATION AND CLASSIFICATION

ALASTAIR CULHAM[1] AND MIKE L. GRANT[2]

[1]Centre for Plant Diversity and Systematics, Department of Botany, The University of Reading, Whiteknights, Reading RG6 6AS, UK
[2]The Royal Horticultural Society's Garden, Wisley, Woking, Surrey GU23 6QB, UK

Abstract

Based on genetic criteria, there are seven basic types of cultivar. Molecular techniques for the identification and classification of cultivars are discussed. The use of DNA fingerprint markers in cultivar identification and classification has expanded rapidly. The many techniques from DNA sequencing, through RFLP to PCR-based fingerprinting which have been used to study cultivars are reviewed. The choice of approach has largely been technique-driven with new techniques coming to the fore before existing ones have been properly evaluated. The current application, and applicability, of these techniques to cultivar identification is reviewed. The definition of the cultivar is put into a genetic context and the cultivar concept discussed. Molecular techniques offer immediate help in cultivar identification and could ultimately lead to an improved classification.

Introduction

Horticultural taxonomy suffers from, among other things, the combined problems of two codes of nomenclature: the Botanical Code (ICBN, Greuter *et al.* 1994) and the Cultivated Plant Code (ICNCP, Trehane *et al.* 1995). The difficulties however result not just from attempting to marry the two codes but also from the genetic complexity of cultivated plants through direct or indirect human manipulation. In this paper we aim to discuss the influence of molecular genetic techniques on our understanding of what cultivars are and how they may be classified and identified.

Many plants in cultivation are little different from those found in the wild. They simply represent selections of 'garden-worthy' individuals from a wild population. Others in contrast, have been intensively selected by man and either represent genetic combinations which would not survive in the wild or are the result of hybridisation between taxa that would not naturally meet (Rice 1998). Indeed some plants have been in cultivation so long that it is not certain what their wild progenitors were or whether they survive.

The Current Definition of a Cultivar

Trehane *et al.* (1995), state that a cultivar should be distinct, uniform and stable but sensibly do not attempt to quantify these three key features. This is a conveniently generalised definition allowing the broad use of the cultivar category seen currently. It is in line with the similarly broad definition of species seen in the Botanical Code. A precise definition of cultivar can be as challenging as its identification because differences may not always be apparent. Reliable and repeatable identification, or at least the ability to distinguish, is vital if the application of cultivar names is to be consistent, as required by gardeners, growers, breeders and those testing new cultivars.

Current tests for distinctness, uniformity and stability (DUS testing) may involve field trials lasting some years, but the regulation of this varies from group to group and for many ornamentals there seems to be little testing at all. What can be distinguished? Article 2.18 of the Cultivated Code states that: 'all indistinguishable variants, irrespective of their origin, are treated as one cultivar'. It does not specify the means of distinction. Accepted practice includes not only morphological variation but seasonal or biochemical variation too.

There are few published keys to cultivars, which reflects the widespread use of dichotomous keys, which are often unsuitable for cultivars due to the non-hierarchical nature of variation often found in cultivar-groups, and also the difficulty of putting into simple words the subtle differences between them. A good example is a key to some 50 cultivars of flowering cherry (Alexander 1995), although this is limited in scope as it only describes those that are widely cultivated. In some groups there are too many cultivars for a single printed key, such as *Narcissus* with over 20,000 cultivar names recorded (Kington 1998). Identification thus often depends on 'expert' opinion and laborious comparison with photographs and specimens. The identification of *Hemerocallis* 'Inspired Word' or *H.* 'Pandora's Box', for instance requires such expert knowledge as over 40,000 cultivars have been recorded.

The lack of a 'type' system exacerbates the problem, as there is usually no definitive specimen for comparison as seen in botanical nomenclature. The 'Standard Portfolio' system proposed by Miller (1995) is the first attempt to make a permanent record of cultivars for a reference collection that does not require perpetual or periodic cultivation. Such a system is being pioneered at the RHS (Royal Horticultural Society) herbarium at Wisley where an active policy of maintaining portfolios is being followed (Miller & Grayer 1999). Living reference collections exist for some groups (Brogdale: apples, Henry Doubleday Research Association: heritage vegetables, NIAB (National Institute of Agricultural Botany), Cambridge: PBR (Plant Breeders' Rights) protected cultivars and the NCCPG (National Council for the Conservation of Plants and Gardens) oversees 600 reference collections of ornamental plants for conservation and scientific purposes (NCCPG 1998). Unfortunately, living collections are subject to change and loss. A study of botanic gardens has shown both extinction and hybridisation of species to be common, even in well-tended collections (Maunder 1997, Maunder *et al.* 1998).

What are Cultivars?

Not all cultivars are equivalent in terms of either propagation or genetic makeup. Cultivars can be split into seven basic types (Table 1).

The easiest to define is the clone. This is a genetically uniform group propagated by asexual means (cuttings, division, grafting, tissue culture). They may be selections from breeding programmes (e.g. *Bergenia* 'Abendglut'), chance seedlings (e.g. *Choisya ternata* 'Sundance') or collections from the wild (e.g. *Daphne bholua* 'Ghurka') chosen for vigour, flower-colour, fruit-size or even proximity to a road or path (e.g. *Corydalis flexuosa* 'China Blue'). Some clonal cultivars are mutants that have arisen in cultivation as a result of somaclonal variation (i.e. sports, Fig. 1). These are usually colour variants such as *Camellia japonica* 'Elegans Champagne' or variegated (Sombrero 1997) such as *Pelargonium* 'Mrs Parker' (Fig. 2). In all of these cases, the cultivar is a single genetic individual.

TABLE 1. Basic types of cultivar.

Clonally Propagated

Genetically Uniform	**Genetically Mixed Individuals**
a) Selections b) Sports	c) Chimaeras d) Viral or other microbial infections

**Sexually Reproducing
Genetically Mixed Populations**

e) Seed raised
f) Repeated selections
g) F1 hybrids

FIG. 1. A double-flowered sport of *Clematis* 'Nelly Moser'. Photo: M. Dixon.

FIG. 2. *Pelargonium* 'Mrs Parker', a variegated cultivar. Photo: Harry Smith Collection.

Several well-known and widely-planted cultivars are not single genetic individuals but a mixture of more than one set of genes that are not combined. Graft-hybrids are chimaeric mixtures of cell types from the original stock and scion. +*Laburnocytisus* 'Adamii', +*Crataegomespilus* 'Jules d'Asnières' and *Syringa* + 'Correlata' are examples. In each case, the two 'parent' plants are closely related. Other genetically mixed individuals are the combination of a plant with a microbe or virus, which alters its vigour or appearance. Many dwarf conifers contain a witches broom infection (e.g. *Pinus nigra* 'Hornibrookiana') and variegated plants may be a result of a viral infection (e.g. *Lonicera japonica* 'Aureoreticulata' – vein-clearing virus and *Abutilon pictum* 'Thompsonii' – mosaic virus). Like the graft-hybrid, the two genotypes are separate but sharing the same living entity. Both types of combination can be unstable and may be prone to reversion thus failing the DUS requirement for cultivar recognition.

The most complex category is numerically the most common, the sexually propagated cultivar, in which the individuals are genetically similar but not identical. Seed-raised inbred lines are the products of repeated inbreeding to achieve close to genetic uniformity. Examples such as *Tagetes patula* 'Mr Majestic' and *Aquilegia vulgaris* 'Nora Barlow' are typical. Many horticultural and agricultural field crops fall into this category.

Some cultivars are a complex mixture of genotypes giving for example a range of colour and flower attributes. Plants such as stocks (*Matthiola* R.Br.) are sold as self-selections where particular attributes such as double flowers, are selected at the seedling stage using seedling colour as a genetic marker.

The most uniform and vigorous cultivars are often F1 hybrids. These are the product of hybridisation between two inbred lines. Though in principle this should make them definable on the basis of the inbred lines, it is common practice to vary one parent to gain an increased seed yield without altering the cultivar name (Rice 1996).

DNA, the Genetic Key

The rapid growth of molecular biology since the publication of the PCR (polymerase chain reaction) technique in 1985, has led to the development of techniques with the potential to discriminate or even identify plant cultivars. The advantage of using DNA markers over morphological markers is that DNA is present and largely unchanged from zygote formation through to maturity, senescence and even death in the form of herbarium specimens and botanical artefacts. There is no 'identification season' which relies on a particular stage of the life-cycle, be it flowering, fruiting or seeding. This is an immediate advantage over conventional methods of identification. In addition, many molecular techniques may elucidate the hierarchy of relationship among cultivars, where there is one to be discovered.

Much of the work towards the development of molecular aids to cultivar identification has been done by plant breeders seeking to characterise cultivars in major crop-breeding programmes. Every major crop from soya to potatoes has received the attention of breeders seeking to link desirable traits to molecular markers through gene mapping. Many of the techniques for mapping have proved useful in providing DNA markers that can distinguish different individuals or cultivars. These are known as 'DNA fingerprinting' techniques. The use of such fingerprints for discriminating cultivars of ornamental plants has many benefits, but investment in this area has inevitably lagged behind that for crop species. Reviews of the application of these techniques are provided by Grant & Culham (1997a & b) and Handa (1998) but a brief and general review of these is given here in the context of cultivar identification and registration.

Molecular techniques involving DNA can be summarised in three general categories: A) fingerprinting, B) sequencing and C) specific markers.

A. DNA Fingerprinting
DNA fingerprinting has been applied to only about 20 genera of ornamental plants (Table 1) and in each case to only a few of the cultivars (Grant & Culham 1997a & b). The technique involves the generation of a series of bands (DNA fragments) which are separated by size. The bands are generally not characterised fully but are defined by the presence of a particular target sequence of DNA. These techniques vary in sensitivity, cost, reproducibility and type of data generated. Few have been widely applied (Table 2). Without doubt, the most widely applied fingerprinting technique so far is RAPD (random amplified polymorphic DNA) (Williams et al. 1990). It is fast, sensitive and, when the chemical and physical conditions for copying reactions are optimal, robust, but perhaps the most important attribute leading to its success is that it is, by molecular standards, relatively cheap and requires little equipment. The RAPD technique has been applied to a variety of ornamental plants (Table 3, Fig. 3 and Grant, Miller & Culham 1999), and to horticultural and agricultural crops. Laboratories using the technique are routinely producing convincing and useful results for cultivar identification (Table 3). The bands generated are a mixture of nuclear and non-nuclear organellar DNA with some estimates suggesting up to 30%

TABLE 2. Major techniques in DNA fingerprinting.
(* indicates the first application of a technique to plants)

1) Jeffreys *et al.*	1985	Humans	Minisatellite
2) Tautz *et al.*	1986	Humans	Microsatellite (SSR)
3) Rogstad *et al.*	1988*	Plants	Minisatellite
4) Ryskov *et al.*	1988	Plants, Animals, Microbes	M13
5) Williams *et al.*	1990	Plants	RAPD
6) Condit & Hubbell	1991*	Plants	Microsatellite (SSR)
7) Caetano-Anolles *et al.*	1991	Plants	DAF
8) Zabeau & Vos	1993	Plants, Animals, Microbes	AFLP
9) Zietkiewicz *et al.*	1994	Plants	Inter-SSR

contribution from plastid DNA. The plastid bands in the fingerprint will all be inherited from one of the parents while nuclear ones will be inherited from both. The bands are generally dominant in their inheritance pattern so do not give good estimates of allelic diversity or heterozygosity without progeny testing.

The inter-SSR technique (Zietkiewicz *et al.* 1994) offers a level of resolution similar to RAPD (i.e. it has a similar power to distinguish individuals) but generates more bands per primer used. It generates DNA fragments defined at each end by the possession of a particular microsatellite DNA sequence. It is more expensive but also more reproducible. The bands produced are all nuclear in origin and some (generally the short DNA fragments) are co-dominantly inherited.

Both these techniques have to some extent been pushed aside by AFLP (amplified fragment length polymorphism) (Zabeau & Vos 1993). This expensive and currently fashionable technique has the big advantage of automation. More than 50 bands can be generated by a single AFLP reaction, though resolution on a band-for-band comparison suggests discrimination power similar to RAPD and inter-SSR. DNA quality is important for this technique. The bands produced are generally dominant in inheritance although a small proportion are co-dominant. This technique is favoured for use among crop plants but has been little used for ornamentals.

The fingerprinting techniques offering the best reproducibility and highest discriminating power are mini- and micro-satellites. These involve discriminating repeat lengths of short regions of DNA that generally consist of 1–4 base sequences repeated 5–50 times (eg. ATATATATATAT). Their use has been limited by the high cost of initial development. This is because they use DNA sequences that are specific to species or small species groups and cannot be transferred to other plant groups without some redevelopment. These systems produce co-dominant markers that give a good measure of heterozygosity as well as general genetic similarity.

All fingerprinting techniques allow some form of similarity or distance calculation to be performed on banding patterns that can give hierarchical groupings. They therefore provide data for both discrimination and classification. Banding patterns are scored on shared presence or absence of a band. Addition of further samples can result in new bands being detected thus altering overall patterns of relationship calculated. Essentially the samples have to be run together on a gel for direct comparison, or run with an internal size marker in the case of automated AFLP. A direct comparison is however preferable in terms of sensitivity of the technique.

TABLE 3. DNA fingerprinting studies of ornamental plants.

Genus/Species	Number of Taxa Compared	Method	Author
Acer rubrum L., *A. saccharinum* L., *A.* × *freemanii* A.E. Murray	13 cultivars 5 seedlings	RAPD	Krahl *et al.* 1993
Camellia L.	10 cultivars	RAPD	Christie & Tiao-Jiang 1996
Camellia L.	15 cultivars	RAPD	Tiao-Jiang & Christie 1997
Chrysanthemum L.	21 cultivars	RAPD	Scott *et al.* 1996
Chrysanthemum L.	20 cultivars	RAPD inter-SSR RFLP GATA	Wolff *et al.* 1995
Chrysanthemum L. *Dendranthema* (DC.) Des Moul.	18 cultivars 13 species	RAPD	Wolff & Peters-Van Rijn 1993
Cornus florida L.	10 cultivars	DAF RAPD	Windham & Trigiano 1998
Cyclamen L. *Petunia* Juss.	5 cultivars 6 cultivars	RAPD	Zhang & McDonald 1996 McDonald 1996
Dianthus L.	9 cultivars	M13 R18.1 33.6 + 33.15 probes	Tzuri *et al.* 1991
Dianthus L	16 cultivars	33.6 + 33.15 probes	Vainstein *et al.* 1991
Epimedium L.	7 species	RAPD	Nakai *et al.* 1996
Euphorbia pulcherrima Willd. ex Klotzsch	9 cultivars	RAPD	Ling *et al.* 1997
Hamamelis L.	40 cultivars	RAPD	Marquard *et al.* 1997
Heliconia L.	16 cultivars	RAPD	Kumar *et al.* 1998
Lilium L.	21 cultivars 14 species	RAPD	Yamagishi 1995
Loropetalum chinense (R.Br.) Oliv.	16 cultivars	RAPD	Gawel *et al.* 1996
Malus Mill.	23 cultivars, 55 unnamed seedlings	M13 probe	Nybom 1990
Ozothamnus diosmifolius (Vent.) DC.	6 cultivars	RAPD	Ko *et al.* 1996

TABLE 3. continued

Pelargonium L'Hér. ex Aiton	8 cultivars	DAF & ASAP	Starman & Abbitt 1997
Pelargonium L'Hér. ex Aiton	10 cultivars	AFLP RAPD	Barcaccia *et al.* 1999
Petunia Juss.	5 species 10 cultivars	DAF	Cerny *et al.* 1996
Rhododendron L.	4 species	RAPD	Iqbal *et al.* 1994
Rhododendron L.	13 assorted taxa	RAPD	Iqbal *et al.* 1995
Rhododendron L.	23 cultivars	RAPD	Kobayashi *et al.* 1995
Rosa L.	5 cultivars	RAPD	Torres *et al.* 1993
Rosa L.	20 cultivars 1 species	M13 +33.6 probes	Ben-Meir & Vainstein 1994
Rosa L.	25 cultivars	RAPD	Gallego & Martinez 1996
Rosa L.	22 cultivars	RFLP & RAPD	Ballard *et al.* 1996
Rosa L.	3 cultivars 1 species	RAPD	Walker & Werner 1997
Syringa L.	2 species 4 hybrids	RAPD	Marsolais *et al.* 1993
Ulmus americana L.	6 cultivars, 16 un-named hybrids	RAPD	Kamalay & Carey 1995
Viburnum L.	3 cultivars, 4 unnamed hybrids	RAPD	Hoch *et al.* 1995

B. DNA Sequencing

DNA sequencing differs from fingerprinting in data generation (Figs. 4 & 5). Sequencing is essentially totally reproducible and generates large quantities of data very quickly. Sequencing of specific genomic regions ensures that only homologous sequences are compared. Large amounts of sequence data are publicly available through the EMBL database in Europe (U.K. link at http://srs.ebi.ac.uk/), Genbank in the USA and DDBJ in Japan. These databases roughly double in size every year and provide a very valuable resource. DNA regions can be chosen that evolve either slowly or quickly and therefore the resolution of this approach can be very coarse (families, genera and above) or very fine (individuals and cultivars). The choice of regions to sequence is increasing rapidly as our knowledge of the plant genome improves.

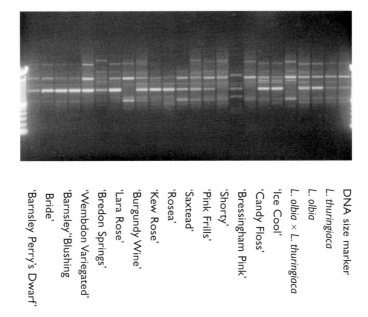

DNA size marker
L. thuringiaca
L. olbia
L. olbia × L. thuringiaca
'Ice Cool'
'Candy Floss'
'Bressingham Pink'
'Shorty'
'Pink Frills'
'Saxtead'
'Rosea'
'Kew Rose'
'Burgundy Wine'
'Lara Rose'
'Bredon Springs'
'Wembdon Variegated'
'Barnsley Blushing Bride'
'Barnsley Perry's Dwarf'

FIG. 3. RAPD banding patterns showing the differences between *Lavatera olbia*, *L. thuringiaca* and some cultivars derived from them. Image: M. Grant.

DNA sequencing offers the advantage that the data are absolute (i.e. the complete sequence of a gene is the same when recorded from other taxa). Comparison does not depend on running samples together or even in the same laboratory. Data can be exchanged between researchers as text information (the bases ACGT) and algorithms exist for searching large databases for similar sequences. This contrasts with DNA fingerprinting where the number of 'absent' bands is a function of the presence of these bands in other samples.

Sequence data offer the means to both distinguish and to classify and are amenable to a much greater variety of analytical approaches than fingerprint data. They can even offer direct measures of heterozygosity. The technique of cycle sequencing is fast and becoming cheaper. The use of DNA sequence data has generally not been applied below species level or to ornamental cultivars, largely due to cost and lack of knowledge of the plant genome. The tracing of cultivar ancestry may be possible in some groups using sequence approaches. Such an investigation into the origin of some *Pelargonium* L'Hér. ex Aiton cultivars is in progress (Culham, Miller *et al.* in prep.).

DNA sequence data have already been used successfully in the classification of several groups of ornamental species (Compton *et al.* 1998a & b, Bakker *et al.* 1998, Smith *et al.* 1998). It is likely that DNA sequence studies will continue to cause realignments of generic boundaries. This will have an impact on horticultural taxonomy and may alter the position of some cultivars.

Digitalis grandiflora

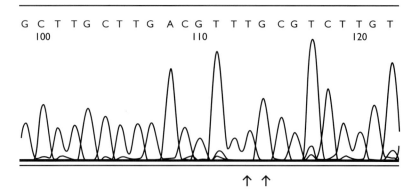

Digitalis thapsi

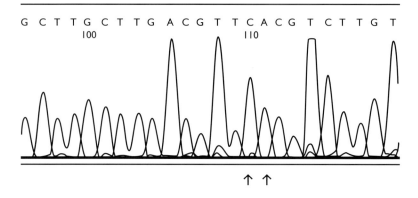

FIG 4. A sample of automated DNA sequence with arrows showing the difference in base sequence between two species of *Digitalis*. Photo: A. Culham.

C. Specific Markers

These are labels to particular sequences of known function or origin. The development of genetic markers is progressing in many agricultural crops and in some horticultural ones such as lettuce (Kesseli *et al.* 1996) but it is not really a cost-effective approach in most ornamentals although it has been applied to *Petunia* Juss. (Peltier *et al.* 1994) and *Rosa* L. (Ballard *et al.* 1996). Nevertheless the approach is likely to become quicker and more cost-effective in the near future. Many markers have been developed in plant breeding programmes, especially to traits of agronomic importance. Marker

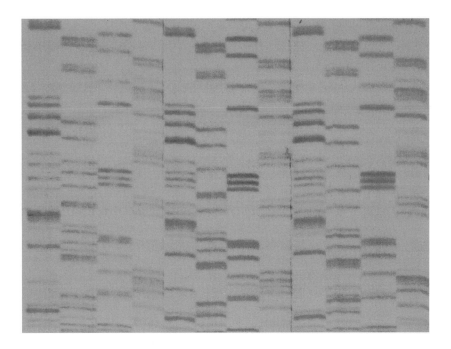

FIG 5. An autoradiogram of manual DNA sequencing reactions for *Digitalis grandiflora*, *D. ferruginea* and *D. thapsi* showing DNA sequence variation in the form of banding patterns. Photo: A. Culham.

technology has to some extent become linked to monitoring genetically modified organisms, though as yet there is little use in ornamentals. DNA markers are already in use for some agricultural crops. Markers can be both diagnostic and classificatory.

Which Markers for which Groups?

There is no simple answer to this. Marker technology has moved on very rapidly in the last 10 years and the latest system is always heralded by its inventors as better than earlier ones. In practice, comparison between the dominant marker fingerprint systems (RAPD, Inter-SSR, AFLP) suggests that all have similar resolution and it is a matter of balancing cost, time and available technology when deciding which to use. Where available, codominant microsatellite markers offer the highest resolution among fingerprinting techniques, if enough marker loci have been identified. Essentially any of these approaches could be used to distinguish most cultivars.

Specific genetic markers offer the chance to test for particular traits but are not yet really suited for cultivar classification and identification. DNA sequencing offers very high resolution with a series of highly variable regions being sequenced in parallel; at present this is still very expensive, though it is getting cheaper.

It is essential to recognise the fundamental difference between identification and classification. The process of classification involves grouping individuals, most often into a hierarchy. A successful classification has predictive value in terms of traits, and gives names to groups sharing those common traits. Identification involves distinguishing the different entities recognised in a classification by means such as keys and knowledge of the correct names to be applied.

Cultivar identification by conventional means requires specialist knowledge or references, and all too often a living collection. Plant material, which is often phenotypically plastic, has to show the correct appearance. The characters needed for identification may be seasonal: apples fruit in the autumn just as *Narcissus* L. flowers in the spring. Essentially, much cultivar identification is limited by season. In addition, the lack of diagnostic keys and a still embryonic reference system can result in considerable subjectivity in cultivar identification. A plant may gain a new cultivar name just by moving continent or losing popularity for a few years. The registration of 'new' cultivars derived unchanged from material in cultivation for many years is becoming commonplace. The lack of objective proof in demonstrating that a new cultivar is the same as an old one has allowed some people to profit from others' work and will further complicate identification in the future. In the case of *Loropetalum chinense* (R.Br.) Oliv. using RAPD (Gawel *et al.* 1996), showed that different cultivar names had been applied to the same cultivars and succeeded in providing a cultivar-name synonymy.

The use of DNA marker technology will not solve all the problems of cultivar identification, as many are matters of opinion or depend on interpretation of the cultivar concept. However, DNA markers offer a degree of objectivity in identification that should allow debate to be based on fact rather than opinion. DNA markers allow identification at all stages of growth and at any time of year. The generation of fingerprint or sequence data is becoming highly automated and reproducibility is high for many techniques. Computer algorithms for fingerprint comparison (developed for forensic science) and for DNA sequence comparison (developed for studies of the genome) and large databases of DNA sequence data are readily available. Analysis techniques are being improved by both geneticists and systematists, and the prospects for a more objective approach to cultivar identification seem good.

To produce a hierarchical classification of cultivars to parallel the classification of wild species may not be worthwhile unless there is a natural hierarchy among those cultivars. All plants, including those in horticulture, fit into the hierarchical system of Linnaean classification under the Botanical Code. Cultivars however, are also governed by the Cultivated Plant Code. It is generally accepted that there is an intrinsic hierarchy in wild taxa, that reflects the evolutionary relationship of those species. However, cultivated plants are known to be a complex mixture of species, hybrids and selected individuals that do not exist in a natural, nested, hierarchy. It is evident that any classification system that is to work must not be too rigid. Just as the species definition has been debated for many years with only the most pragmatic definition ('a species is what we choose to call a species') gaining any widespread popularity. Any attempt to formalise the cultivar more strictly in its many guises is unlikely to be successful.

In conclusion, though DNA techniques will not solve theoretical problems, they can offer practical mechanisms to improve objectivity, resolution, repeatability (with some exceptions) and identification. The results may cause difficulties through recognition of entities with no observable morphological differences. Molecular markers are not 'field' (garden) characters, and identification of cultivars using molecular techniques is still somewhat expensive compared with traditional morphological approaches.

Acknowledgements

The authors wish to thank James Compton for his critical reading of the manuscript, the two reviewers who made valuable suggestions and the organisers of the symposium for their support to AC at a 'busy' time.

References

Alexander, J.C.M. (1995). *Prunus*. In Cullen, J., Alexander, J.C.M., Brady, A., Brickell, C.D., Green, P.S., Heywood, V.H., Jörgensen, P-M., Jury, S.L., Knees, S.G., Leslie, A.C., Matthews, V.A., Robson, N.K.B., Walters, S.M., Wijnands, D.O. & Yeo, P.J. (eds). The European Garden Flora. Vol. 4. pp. 447–461. Cambridge University Press, Cambridge, U.K.

Bakker, F.T., Hellbrügge, D., Culham, A. and Gibby, M. (1998). Pylogenetic relationships within *Pelargonium* sect. *Peristera* (*Geraniaceae*) infered from nrDNA and cpDNA sequence comparisons. *Pl. Syst. Evol.* 211: 273–287.

Ballard, R., Rajapakse, S., Abbot, A. & Byrne, D.H. (1996). DNA markers in roses and their use for cultivar identification and genome mapping. *Acta Hort.* 424: 265–268.

Barcaccia, G., Albertinin, E. & Falcinelli, M. (1999). AFLP fingerprinting in *Pelargonium peltatum*: its development and potential in cultivar identification. *J. Hort. Sci. Biotechnol.* 74(2): 243–250.

Ben-Meir, H. & Vainstein, A.(1994). Assessment of genetic relatedness in roses by DNA fingerprint analysis. *Sci. Hort.* 58: 115–121.

Caetano-Anolles, G., Bassam, B.J. & Gresshoff, P.M. (1991). DNA amplification fingerprinting using very short arbitrary oligonucletide primers. *Bio/Technology* 9: 553–557.

Cerny, T.A., Caetano-Anolles, G., Trigiano, R.N. & Starman, T.W. (1996). Molecular phylogeny and DNA amplification fingerprinting of *Petunia* taxa. *Theor. App. Genet.* 92(8): 1009–1016.

Christie, C.B. & Tiao-Jiang, X. (1996). Paternity suits against 'Captain Rawes' and associates solved out of court. *New Zealand Camellia Bull.* 19: 25–29.

Compton, J.A., Culham, A., Gibbings, J.G. & Jury, S.L. (1998a). Phylogeny of *Actaea* including *Cimicifuga* (*Ranunculaceae*) inferred from nrDNA ITS sequence variation. *Biochem. Syst. & Ecol.* 26: 185–197.

Compton, J.A., Culham, A., & Jury, S.L. (1998b). Reclassification of *Actaea* to include *Cimicifuga* and *Souliea* (*Ranunculaceae*): phylogeny inferred from morphology, nrDNA ITS and cpDNA *trnL-F* sequence variation. *Taxon* 47(3): 593–634.

Condit, R. & Hubbell, S.P. (1991). Abundance and DNA sequence of two-base repeat regions in tropical tree genomes. *Genome* 34: 66–71.

Gallego, F.J. & Martinez, I. (1996). Molecular typing of rose cultivars using RAPDs. *J. Hort. Sci.* 71: 901–908.

Gawel, N.J., Johnson, G.R. & Sauve, R. (1996). Identification of genetic diversity among *Loropetalum chinense* var. *rubrum* introductions. *J. Environm. Hort.* 14(1): 38–41.

Grant, M.L. & Culham, A. (1997a). DNA fingerprinting and the identification of cultivars. *New Plantsman* 4(2): 85–87.

Grant, M.L. & Culham, A. (1997b). DNA fingerprinting and the identification of ornamental cultivars: part 2. *New Plantsman* 4(3): 157–168.

Grant, M.L., Miller, D.M. & Culham, A. (1999) A morphological and molecular investigation into the woody cultivars of *Lavatera* L. In S. Andrews, A.C. Leslie & C. Alexander (eds). Taxonomy of Cultivated Plants: Third Inernational Symposium, p. 347. Royal Botanic Gardens, Kew.

Greuter, W., Barrie, F.R., Burdet, H.M., Chaloner, W.G., Demoulin, V., Hawksworth, D.L., Jørgensen, P.M., Nicolson, D.H., Silva, P.C., Trehane, P. & McNeill, J. (eds). (1994). International Code of Botanical Nomenclature (Tokyo Code). 389 pp. Koeltz Scientific Books, Königstein. (*Regnum Vegetabile* 131).

Handa, T. (1998). Utilization of molecular markers for ornamental plants. *J. Jap. Soc. Hort. Sci.* 67: 1197–1199.

Hoch, W.A., Zeldin, J.N. & McCown, B.H. (1995). Generation and identification of new *Viburnum* hybrids. *J. Environm. Hort.* 13: 193–195.

Iqbal, M.J., Gray, L.E., Paden, D.W. & Rayburn, A.L. (1994). Research Note: feasibility of *Rhododendron* DNA profiling by RAPD analysis. *Pl. Var. Seeds* 7: 59–63.

Iqbal, M.J., Paden, D.W. & Rayburn, A.L. (1995). Assessment of genetic relationships among rhododendron species, varieties and hybrids by RAPD analysis. *Sci. Hort.* 63: 215–223.

Jeffreys, A.J., Wilson, V. & Thein, S.L. (1985). Hypervariable 'minisatellite' regions in human DNA. *Nature* 314: 67–73.

Kamalay, J.C. & Carey, D.W. (1995). Application of RAPD-PCR markers for identification and genetic analysis of American elm (*Ulmus americana* L.) selections. *J. Environm. Hort.* 13: 155–159.

Kesseli, R., Witsenboer, H., Hill, M., Zhang, Y-H., Chase, M. & Michelmore, R. (1996). Genome organisation and evolution of disease resistance genes in *Lactuca* spp. and the prospects for comparative mapping in the *Compositae*. In P.D.S. Caligari & D.J.N. Hind, (eds). *Compositae*: Biology and Utilization, Part 2, pp. 1–8. Royal Botanic Gardens, Kew.

Kington, S. (comp.). (1998). The international daffodil register and classified list 1998. 1166 pp. The Royal Horticultural Society, London.

Ko, H.L., Henry, R.J., Beal, P.R., Moisander, J.A. & Fisher, K.A. (1996). Distinction of *Ozothamnus diosmifolius* (Vent.) DC. genotypes using RAPD. *HortScience* 31: 858–861.

Kobayasha, N., Takeuchi, R., Handa, T. & Takayanagi, K. (1995). Cultivar identification of evergreen azalea with RAPD method. *J. Jap. Soc. Hort. Sci.* 64: 611–616.

Krahl, K.H., Dirr, M.A., Halward, T.M., Kochert, G.D., & Randle, W.M. (1993). Use of single-primer DNA amplifications for the identification of red maple (*Acer rubrum* L.) cultivars. *J. Environm. Hort.* 11(2): 89–92.

Kumar, P.P., Yau, J.C.K. & Goh, C.J. (1998). Genetic analyses of *Heliconia* species and cultivars with RAPD markers. *J. Amer. Soc. Hort. Sci.* 123(1): 91–97.

Ling, J.T., Sauve, R. & Gawel, N. (1997). Identification of poinsettia cultivars using RAPD markers. *HortScience* 32: 122–124.

Marquard, R.D., Davis, E.P. & Stowe, E.L. (1997). Genetic diversity among witchhazel cultivars based on RAPD markers. *J. Amer. Soc. Hort. Sci.* 122(4): 529–535.

Marsolais, J.V., Pringle, J.S. & White, B.N. (1993). Assessment of random amplified polymorphic DNA (RAPD) as genetic markers for determining the origin of interspecific lilac hybrids. *Taxon* 42(3): 531–537.

Maunder, M. (1997). Botanic garden response to the biodiversity crisis: implications for threatened species management. Unpublished PhD Thesis, University of Reading.

Maunder, M., Higgens, S. & Culham, A., (1998). Neither common nor garden: the garden as a refuge for threatened plant species. *Bot. Mag.* 15(2): 124–132.

McDonald, M.B. (1996). Genetic testing for seed vigor, uniformity. *Grower Talks* Winter 1996: 8–10.

Miller, D.M. (1995). Standard specimens for cultivated plants. *Acta Hort.* 413: 35–39.

Miller, D.M. & Grayer, S.R. (1999). Standard portfolios in the herbarium of the Royal Horticultural Society. In S. Andrews, A.C. Leslie & C. Alexander (eds.) Taxonomy of Cultivated Plants: Third International Symposium, pp. 397–399. Royal Botanic Gardens, Kew.

Nakai, R., Shoyama, Y. & Shiraishi, S. (1996). Genetic characterisation of *Epimedium* species using random amplified polymorphic DNA (RAPD) and PCR-restriction fragment length polymorphism (RFLP) diagnosis. *Biolo. Pharm. Bull.* 19(1): 67–70.

NCCPG. (1998). The National Plant Collections Directory 1998. 148 pp. NCCPG, Wisley.

Nybom, H. (1990). Genetic variation in ornamental apple trees and their seedlings revealed by DNA fingerprinting. *Hereditas* 113: 17–28.

Peltier, D., Farcy, E., Dulieu, H. & Bervillé, A. (1994). Origin, distribution and mapping of RAPD markers from wild *Petunia* species in *Petunia hybrida* Hórt. lines. *Theor. Appl. Genet.* 88(6–7): 637–645.

Rice, G. (1996). Regularising the names for annuals and bedding plants. *Hortax News* 1(2): 12–19.

Rice, G. (1998). Novelty value. *Garden (London)* 123(7): 508–511.

Rogstad, S.H., Patton, J.C. II & Schaal, B.A. (1988). M13 repeat probe detects DNA minisatellite-like sequences in gymnosperms and angiosperms. *Proc. Natl. Acad. Sci. U.S.A.* 85: 9176–9178.

Ryskov, A.P., Jincharadze, A.G., Prosnyak, P.L. & Limborska, S.A. (1988). M13 phage as a universal marker for DNA fingerprinting of animals, plants and microorganisms. *Fed. Eur. Biochem. Soc. Lett.* 233: 388–392.

Scott, M.C., Caetano-Anolles, G. & Trigiano, R.N. (1996). DNA amplification fingerprinting identifies closely related chrysanthemum cultivars. *J. Amer. Soc. Hort. Sci.* 121: 1043–1048.

Smith, J.F., Kresge, M.E., Möller, M. & Cronk, Q.C.B. (1998). A cladistic analyses of *ndhF* sequences from representative species of *Saintpaulia* and *Streptocarpus* subgenera *Streptocarpus* and *Streptocarpella* (*Gesneriaceae*). *Edinburgh J. Bot.* 55(1): 1–11.

Sombrero, C. (1997). Enigma of variegations. *New Plantsman* 3: 158–169.

Starman, T.W. & Abbitt, S. (1997). Evaluating genetic relationships of geranium using arbitrary signatures from amplification profiles. *HortScience* 32(7): 1288–1291.

Tautz, D., Trick, M. & Dover, G.A. (1986). Cryptic simplicity in DNA is a major source of genetic variation. *Nature* 322: 652–656.

Tiao-Jiang, X. & Christie, C.B. (1997). Identification of closely related camellia hybrids and mutants using molecular markers. *Int. Camellia J.* 29: 111–116.

Torres, A.M., Millán, T. & Cubero, J.I. (1993). Identifying rose cultivars using randomly amplified polymorphic DNA markers. *HortScience* 28: 333–334.

Trehane, R.P., Brickell, C.D., Baum, B.R., Hetterscheid, W.L.A., Leslie, A.C., McNeill, J., Spongberg, S.A. & Vrugtman, F. (eds). (1995). International Code of Nomenclature for Cultivated Plants – 1995. 175 pp. Quarterjack Publishing, Wimborne. (*Regnum Vegetabile* 133).

Tzuri, G., Hillel, J., Lavi, U., Haberfeld, A. & Vainstein, A. (1991). DNA fingerprint analysis of ornamental plants. *Pl. Sci. (Elsevier)* 76: 91–97.

Vainstein, A., Hillel, J., Lavi, U. & Tzuri, G. (1991). Assessment of genetic relatedness in carnation by genetic fingerprint analysis. *Euphytica* 56: 225–229.

Walker, C.A. & Werner, D.J. (1997). Isozyme and RAPD analyses of Cherokee rose and its putative hybrids 'Silver Moon' and 'Anemone'. *J. Amer. Soc. Hort. Sci.* 122(5) 659–664.

197

Williams J.G.K., Kubelic A.R., Livak K.J., Rafalski J.A. & Tingey S.V. (1990). DNA polymorphisms amplified by arbitrary primers that are useful as genetic markers. *Nucl. Acids Res.* 18: 6531–6535.

Windham, M.T. & Trigiana, R.N. (1998). Are 'Barton' and 'Cloud 9' the same cultivar of *Cornus florida* L.? *J. Environm. Hort.* 16(3): 163–166.

Wolff, K. & Peters-Van Rijn, J. (1993). Rapid detection of genetic variability in chrysanthemum (*Dendranthema grandiflora* Tzvelev) using random primers. *Heredity* 71: 335–341.

Wolff, K., Zietkiewicz, E. & Hofstra, H. (1995). Identification of chrysanthemum cultivars and stability of DNA fingerprint patterns. *Theor. Appl. Genet.* 91(3): 439–447.

Yamagishi, M. (1995). Detection of section-specific random amplified polymorphic DNA (RAPD) markers in *Lilium*. *Theor. Appl. Genet.* 91(6–7): 830–835.

Zabeau, M. & Vos, P. (1993). Selective restriction fragment amplification: a general method for DNA fingerprinting. EPO Patent No. 0534858A1.

Zhang, J. & McDonald, M.B. (1996). Abstract 057: Varietal identification of *Cyclamen* and *Petunia* seeds using RAPDs. *HortScience* 31: 574.

Zietkiewicz, E., Rafalski, A. & Labuda, D. (1994). Genome fingerprinting by simple sequence repeat (SSR)– anchored polymerase chain reaction amplification. *Genomics* 20(2): 176–183.

Borys, J. (1999). DUS testing of cultivars in Poland. In: S. Andrews, A.C. Leslie and C. Alexander (Editors). Taxonomy of Cultivated Plants: Third International Symposium, pp. 199–203. Royal Botanic Gardens, Kew.

DUS TESTING OF CULTIVARS IN POLAND

JULIA BORYS

The Research Centre for Cultivar Testing (COBORU), 63-022 Słupia Wielka, Poland

Abstract

The organisation and the methods used in Poland for testing distinctness, uniformity and stability (DUS) of cultivars before entry into the Register of Cultivars and granting of Plant Breeders' Rights are presented. All activities connected with plant variety testing, the maintenance of the Register of Cultivars and the Register of Plant Breeders' Rights, are provided by the Research Centre for Cultivar Testing (COBORU). Our DUS testing of cultivars is conducted according to the national guidelines prepared on the basis of UPOV guidelines.

Introduction

Poland, as a member of the International Union for the Protection of New Varieties of Plants (UPOV) since 1989 follows the UPOV Convention and adheres to UPOV regulations. The Polish Seed Industry Law of 1995 which has been in force since 20th January 1996, in the section concerning plant variety protection, is adjusted to the 1991 Act of UPOV Convention (UPOV 1991).

According to Polish law, a cultivar is a population of plants within a botanical systematic unit of the lowest known rank, which can be:

- Defined by the expression of the characteristics resulting from a given genotype or combination of genotypes.
- Distinguished from any other plant population by the expression of at least one of the characteristics.
- Remains unchanged after multiplication or at the end of the relevant cycle of multiplications or crosses.

One of the requirements for registration on the Register of Cultivars (National List) and/or for an award of Plant Breeders' Rights (PBR) in Poland is that the cultivar must be distinct, uniform and stable (DUS).

In Poland, there are cultivars of 302 taxa which are eligible for PBR protection and inclusion into the Register of Cultivars. These include: 122 agricultural, 63 vegetable, 92 ornamental and 25 orchard taxa. The legal base of the list of taxa is the Decree of the Minister of Agriculture and Food Economy of 15 April 1996.

Every cultivar must pass the whole DUS testing procedure before entering the Register of Cultivars and/or being granted protection (Borys 1997). In addition, 46 cultivars out of 302 taxa are only included in the Register of Cultivars on condition of satisfactory economic value (value for cultivation and use or VCU).

The administration of the statutory variety testing (DUS and VCU trials) and the maintenance of the Register of Cultivars and the Register of Plant Breeders' Rights are provided by the Research Centre for Cultivar Testing (COBORU) located in Słupia Wielka. This work includes the handling of applications, fees, preparing the methods of testing, naming procedures, registration and publication of the Register of Cultivars and Plant Breeders' Rights information. Cultivar testing is carried out at the Experimental Stations for Cultivars Testing (SDOO) which belongs to COBORU.

The Organisation and Methods of Testing

DUS testing is carried out in SDOO at one or two sites. The cultivars of some taxa are tested in co-operation with the Research Institute of Pomology and Floriculture (orchard cultivars) and the Czech, Hungarian and Slovakian test authorities (agricultural and vegetable cultivars).

DUS trials are carried out with at least two replications. Similar cultivars are placed close to each other. The cultivars can be in row trials and/or spaced plant trials. They are randomised within each block and subgroup. Within the same DUS trials, the seeds of two or three generations of the same cultivar are sown. In 1997, the following number of candidates (cultivars) were tested: 806 agricultural (out of 1891) in 85 taxa, 437 vegetable (1330) in 46 taxa, 167 ornamental (1486) in 47 taxa and 63 fruit (159) in 21 taxa (Table 1).

TABLE 1. Number of taxa and cultivars in DUS tests in Poland in 1997.

Group of plants	Number of taxa	Number of cultivars tested					
		Total	Domestic	Foreign	Registered NL and PBR	Candidates for NL and PBR	Other
Agricultural	85	1891	1165	726	797	806	288
Vegetables (total)	51	1330	586	744	865	437	28
– under cover	5	353	126	227	232	121	
– in the field	46	977	460	517	633	316	28
Ornamentals (total)	47	1486	500	485	534	451	501
– under cover	13	431	174	174	135	213	83
– in the field	27	850	310	133	304	139	407
– trees and shrubs	7	205	16	178	95	99	11
Orchard plants (total)	31	159	139	17	39	117	3
– fruit trees	25	121	104	14	28	90	3
– berry plants	6	38	35	3	11	27	
Total	214	4866	2390	1972	2235	1811	820

DUS tests of cultivars are conducted according to the national guidelines on the basis of Guidelines for the Conduct of Tests for Distinctness, Homogeneity (Uniformity) and Stability (Test Guidelines) prepared for different taxa by the UPOV Technical Working Parties who seek the views of the breeders' associations. At present about 160 UPOV Test Guidelines are available for agricultural and vegetable species as well as for ornamentals, trees and orchard (fruit) species. Among them are Test Guidelines helpful for the DUS testing of the cultivars of 302 taxa in Poland: 53 for agricultural, 42 for vegetable, 30 for ornamental and 20 for orchard taxa.

The general principles of conducting DUS tests have been incorporated into the general introduction of *Guidelines for the Conduct of Tests for Distinctness, Homogeneity and Stability of New Varieties of Plants* (UPOV 1979). Test Guidelines are then recommended to the member States of UPOV and provide a common basis for DUS tests and for establishing cultivar descriptions in standardised form.

Distinctness, Uniformity and Stability are judged on the basis of characteristics and their expression. The nature of the characteristics used in the technical procedure of DUS testing is an important element. Such characteristics may be morphological, physiological, biochemical or of another nature, but they must be capable of precise recognition and description. There are two main categories of characteristics:

- **Qualitative characteristics** which show discontinuous states with no arbitrary limit on the number of states, e.g. shape, colour, number of rows on barley ear (two or more). They are often controlled by one gene and are usually independent of environmental conditions. Therefore, they are good tools for distinguishing cultivars.
- **Quantitative characteristics** are those that are measurable on a one dimensional scale and show continuous variation from one extreme to the other, e.g. length of leaf. For the purpose of description they are divided into a maximum of 9 states, e.g. 1 (very short) to 9 (very long), with the middle state 5 (medium) and the other states formed correspondingly. These characteristics are mostly polygenic and they are more or less influenced by environmental conditions.
- Characteristics which are assessed separately may be combined, e.g. the length/width ratio. They are often less influenced by environment than the component characteristics.

Testing for Distinctness

According to Article 7 of the Convention, a cultivar must be clearly distinguishable from any other cultivar whose existence is a matter of common knowledge at the time of filling in the application.

Criteria of distinctness: two cultivars are considered distinct if the difference between them: (i) has been determined in at least one testing place, (ii) is clear, (iii) is consistent.

In the case of truly qualitative characteristics where the respective characteristics show expressions which fall into different states, the difference between two cultivars is considered clear. In the case of other qualitative characteristics an eventual fluctuation has to be taken into consideration in establishing distinctness. For measured quantitative characteristics the difference is considered clear if it occurs with 1% probability of error when using, for example, the Least Significant Difference (LSD) method.

The differences are consistent, if they show in the same way in two consecutive, or two out of three growing seasons. UPOV now recommends the use of a Combined Over Years (COYD) method, based on the principles of the analysis of variance and the interaction between cultivars and years.

The main statistical tool used in Poland for continuous characteristics is the analysis of variance. Following analysis of variance, other well-known methods are used for cultivar comparison.

For categorial data the χ^2 test is used for testing the differences between cultivars and the weighted regression method is used, where the levels of expression of characters are related to time. These two methods are used in Poland, e.g. for so-called dynamics of flowering in grasses.

Electrophoresis (SDS PAGE) is used as an additional method for the identification of barley and wheat cultivars.

Testing for Uniformity

According to Article 8 of the Convention, a cultivar shall be deemed uniform if, subject to the variation which may be expected from the particular features of its propagation, it is sufficiently uniform in its relevant characteristics.

For vegetatively propagated cultivars and truly self-pollinated cultivars the maximum acceptable number of off-types in samples of various sizes is fixed based on a population standard (often 1%) and an acceptance probability (usually 95%) (UPOV 1998). For cultivars which are not fully self-pollinated a higher tolerance is required — it is usually double that of vegetatively propagated cultivars.

The population standards are currently being established by crop experts. Usually standards of 1%, 2% and 5% are used, and are fixed for each species/crop based on the experience of technical experts. Standards are statistically based but the appropriate level is decided by DUS testers.

Open-pollinated cultivars normally exhibit wider variation than vegetatively propagated or self-pollinated cultivars and it is more difficult to distinguish off-types. Therefore relative tolerance limits are used through comparison with comparable cultivars that are already known. For measured characteristics the standard deviation or variance is used as the criterion for comparison. A cultivar is considered not uniform in a measured characteristic if its variance exceeds 1.6 times the average of the variance of the cultivars used for comparison. At present this criterion is being replaced by a combined over years uniformity (COYU) method mainly used for grass and fodder species.

Visually assessed characteristics are handled in the same way. The number of plants visually different from those of the cultivar should not significantly (5 per cent probability of error) exceed the number found in comparable cultivars that are already known.

In the case of hybrid cultivars the tolerance depends on the type of hybrid. Single cross hybrids are treated as mainly self-pollinated cultivars, but a tolerance is allowed for inbred plants. Normally the maximum number tolerated is given in the Test Guidelines. For other categories of hybrids, the segregation of certain characteristics is acceptable if it is in agreement with the breeding of the cultivar. These hybrids are very often treated as open-pollinated cultivars and the uniformity is compared with that of comparable known cultivars.

Electrophoresis (Acid PAGE) is useful in testing uniformity of wheat, barley and oat cultivars.

Testing for Stability

According to Article 9 of the Convention, a cultivar shall be deemed stable if its relevant characteristics remain unchanged after repeated propagation or, in the case of a particular cycle of propagation, at the end of each cycle.

It is not generally possible during a period of two to three years to perform tests for stability which lead to the same certainty as with testing of distinctness and uniformity.

In general, stability is the function of uniformity which means that heterogeneous cultivars may lack stability. When a submitted sample has been shown to be uniform, it can also be considered to be stable. During the testing of distinctness and uniformity, careful attention is paid to stability. The stability is tested by growing a further generation or new seed stock to verify that it shows the same characteristics as previous seed material of the cultivar.

Summary

In Poland DUS testing is conducted for all taxa for which applications are made to the Register for Cultivars and for the granting of Plant Breeders' Rights. Every year we receive an increasing number of candidates to test and also applications in new taxa. In future DUS testing will extend to all plant genera and species for which protection is requested.

The technical tests for distinctness and stability are very expensive. Trial fields, the use of modern methods, e.g. electrophoresis, DNA profiling, disease tests, statistical tools and image analysis as well as well-trained personnel may all have to be provided. For each species a seed reference collection and the comparison of a candidate with other cultivars, a database containing previously established descriptions and the development and improvement of Test Guidelines are necessary for modern DUS testing.

Therefore, UPOV is providing member States with an international system for cooperation in examination. Based on a UPOV model agreement, bilateral agreements have been established between Poland and the Czech Republic, Hungary and Slovakia.

There is a necessity for close harmonisation in the methods used in DUS testing of cultivars. A research programme named RING-TEST has helped in the exchange of expertise in DUS testing and harmonisation of the assessment and description of cultivars.

References

Borys, J. (1997). DUS testing of cultivars. Wiadomości Odmianoznawcze 65: 3–11. COBORU, Słupia Wielka. (In Polish).

UPOV. (1998). Testing of uniformity of self-fertilised and vegetatively propagated species using off-types. TC/34/5. International Union for the Protection of New Varieties of Plants, Geneva, Switzerland.

UPOV. (1991). International convention for the protection of new varieties of plants. International Union for the Protection of New Varieties of Plants, Geneva, Switzerland.

UPOV. (1979). Revised general introduction to the guidelines for the conduct of test for Distinctness, Homogeneity and Stability of new varieties of plants. TG/1/2. International Union for the Protection of New Varieties of Plants, Geneva, Switzerland.

DATABASES FOR COLLECTIONS,
NOMENCLATURE AND TAXONOMY

7

Wiersema, J.H. (1999). Nomenclature of world economic plants from the USDA's GRIN database. In: S. Andrews, A.C. Leslie and C. Alexander (Editors). Taxonomy of Cultivated Plants: Third International Symposium, pp. 207–208. Royal Botanic Gardens, Kew.

NOMENCLATURE OF WORLD ECONOMIC PLANTS FROM THE USDA'S GRIN DATABASE

JOHN H. WIERSEMA

United States Department of Agriculture/Agricultural Research Service, Systematic Botany and Mycology Laboratory, Bldg. 011A, Beltsville Agricultural Research Center (BARC-West), Beltsville, MD 20705-2350 USA

Abstract

The GRIN Database contains accurate information on plants of economic importance to US agriculture. A new publication: *World Economic Plants: a standard reference* provides substantive reference data for over 9,500 vascular plants that are important or of potential importance in world trade.

This project arose from the need to provide accurate information on economically important plants to the Germplasm Resources Information Network (GRIN) of the National Genetic Resources Program of United States Department of Agriculture's Agricultural Research Service (USDA/ARS). This need had been met previously through USDA/ARS Agricultural Handbook 505 *A checklist of names for 3,000 vascular plants of economic importance* (Terrell 1977, Terrell *et al.* 1986), which has been a valuable reference on economic plants to many scientists in agriculture. Both editions of this Handbook focused on plants important to U.S. agriculture and included only scientific and common names for those plants. Both the scope and content of this earlier Handbook have been expanded considerably in our project. GRIN and our new publication *World Economic Plants: a standard reference* (Wiersema & León 1999) now provide essential reference data for over 9,500 vascular plants important or potentially important in the global economy. These data include the accepted scientific name, about 2,500 important synonyms, 19,200 multilingual common names, 13,500 economic uses, and geographical distribution. Plants treated are used for food, spice, forage, fodder, fibre, wood, gum, oil, fuel, pesticide, both pharmaceutical and folk medicine, ornament, erosion control, green manure, shade, gene sources, nectar sources, beneficial invertebrate host, social purpose, and a variety of other uses. Also included are plants having a negative economic impact, such as weeds and poisonous plants.

Data on economic plants in GRIN have accumulated during more than two decades of literature and nomenclatural research on economic plants, with the incorporation of relevant data into GRIN and selective review by specialists to achieve the most current and accurate information. Taxonomic decisions follow those of recognised specialists for various plant groups whenever possible, with important alternative taxonomies represented in synonymy. For many cultivated taxa, alternative cultivar- group names have also been provided. Insofar as it is possible, all data categories conform to the internationally recognised standards for nomenclature and taxonomic databases of the Taxonomic Databases Working Group (TDWG), including Greuter *et al.* (1994) and

Trehane *et al.* (1995) for nomenclature, Brummitt & Powell (1992) for author abbreviations, Hollis & Brummitt (1992) for geographical distributions, and Cook (1995) for economic data. The data in GRIN will provide regular updates for future editions of the published reference, at <www.ars-grin.gov/npgs/tax/taxecon.html>.

References

Brummitt, R.K. & Powell, C.E. (eds). (1992). Authors of plant names: a list of authors of scientific names of plants, with recommended standard forms of their names, including abbreviations. 732 pp. Royal Botanic Gardens, Kew.

Cook, F.E.M. (1995). Economic botany data collection standard. 146 pp. Royal Botanic Gardens, Kew.

Greuter, W., Barrie, F.R., Burdet, H.M., Chaloner, W.G., Demoulin, V., Hawksworth, D.L., Jørgensen, P.M., Nicolson, D.H., Silva, P.C., Trehane, P. & McNeill, J. (1994). International Code of Botanical Nomenclature (Tokyo Code). (*Regnum Vegetabile* 131). 389 pp. Koeltz Scientific Books, Königstein.

Hollis, S. & Brummitt, R.K. (1992). World geographical scheme for recording plant distributions. 104 pp. Hunt Institute for Botanical Documentation, Pittsburgh.

Terrell. E.E. (1977). A checklist of names for 3,000 vascular plants of economic importance. Agric. Handb. (U.S.D.A.) 505, 201 pp.

Terrell. E.E., Hill, S.R., Wiersema, J.H. & Rice, W.E. (1986). A checklist of names for 3,000 vascular plants of economic importance. Agric. Handb. (U.S.D.A.) 505, 244 pp.

Trehane, P., Brickell, C.D., Baum, B.R., Hetterscheid, W.L.A., Leslie, A.C., McNeill, J., Spongberg, S.A., & Vrugtman, F. (1995). International Code of Nomenclature for Cultivated Plants – 1995. (*Regnum Vegetabile* 133). 175 pp. Quarterjack Publishing, Wimborne, U.K.

Wiersema, J.H. & León, B. (1999). World economic plants: a standard reference. xxxv + 749 pp. CRC Press, Inc., Boca Raton.

Clennett, C. (1999). Q-Collector: Development of a portable database for expeditions by the Royal Botanic Gardens, Kew. In: S. Andrews, A.C. Leslie and C. Alexander (Editors). Taxonomy of Cultivated Plants: Third International Symposium, pp. 209–213. Royal Botanic Gardens, Kew.

Q-COLLECTOR: DEVELOPMENT OF A PORTABLE DATABASE FOR EXPEDITIONS BY THE ROYAL BOTANIC GARDENS, KEW

CHRIS CLENNETT

Royal Botanic Gardens, Kew, Wakehurst Place, Ardingly, West Sussex RH17 6TN, UK

Abstract

Use of a portable field collection database for the Living Collections Department of the Royal Botanic Gardens, Kew, was driven by a need to accelerate incorporation of material into the Kew collections. Electronic data collection in the field eliminates much transcription and potential for delay and errors.

Q-Collector™ uses Microsoft Access™ to create a relational database, with entry and retrieval of data via forms and reports. Use of the table relationships reassembles entries into readable English in reports with provision for a variety of summaries and detailed outputs. Following feedback from expeditions, version 2.0 of the software incorporates greatly increased use of pick lists, with population and descriptive data collected for both seed banking and cultivation of collections.

Battery limitations, and size and weight of laptop computers, have shown that bulk editing is more efficient than attempting to record at the time of collection. Working quickly from a field collection book, or pocket recorder is easily achieved using pre-formatted notebook pages or prompt list generated by the system.

Delays of up to three months with manual transcription from field notebooks or tapes, before introduction of the system to Living Collections expeditions, have been replaced by field data made available throughout the institute within 1–2 weeks of return. This enables collections to be swiftly sown and fully curated in the Kew nurseries.

Introduction

The understandable desire of expedition members in the Living Collections Department of the Royal Botanic Gardens, Kew, to reduce their workload on return, and to accelerate incorporation of material into the Kew collections brought about the consideration of a portable field collection database, using laptop or palmtop computers. The need for an improvement was highlighted by an expedition in 1992 where three months elapsed between return and full incorporation into the Living Collections database. During this period, horticulturists had no information on which to base their efforts in germinating collections, and voucher material could not be processed.

The primary aim of development was therefore to accelerate the processes of accession, and to reduce the level of transcription caused in producing documents in varying formats based on the same hard copy data. To this end Q-Collector is seen as a data capture tool supporting the Kew Living Collections and Seed Bank databases.

209

Electronic collection of such data in the field which is subsequently output in the many formats required by a large institute eliminates much of the work of transcription and potential for delay and errors. Errors may not be eliminated, but many can be simply traced and amended.

Q-Collector is a relational database written in Microsoft Access™, with entry and retrieval of data via forms and reports. Most data is stored in a primary table, with a set of smaller tables holding data that is infrequently altered. Use of the table relationships re-assembles entries into readable English in the output reports, with provision for a variety of summaries and detailed outputs including customs declarations, an expedition report, lists and labels for herbarium vouchers, seed packet labels and updated lists for surplus material, pre-formatted accession forms for the Kew Living Collection (LCD) and Seed Bank databases, and distribution lists for material lodged within the gardens.

Since development of the system was initially driven by Living Collections, it is closely tailored to the LCD database, but with much of the data captured in separate fields. Currently, the information is re-assembled when output to fit the LCD database fields, but separate capture means that it can also be used for population analysis should the need arise. The author's combination of skills: managing one of Kew's nursery units; having participated in detailed survey and collection expeditions with the Cyclamen Society; managing Wakehurst Place plant records; and being computer literate, provides a balanced view of design and use of such a system.

A major problem encountered by the first expeditions using Q-Collector was in re-typing entries from record to record, particularly with complicated find spot and field note entries. Simple methods of repeating text were seen as vital in improving usability, and therefore enthusiasm for the system. Refinements following feedback from the LCD expeditions to the Russian Far East (1994), Pakistan (1995), China (1995 & 1997) and Japan (1995), have therefore incorporated in version 2.0 of the software, greatly increased use of pick lists. This allows easy and quick selection for entry of much data, minimises typing errors and ensures consistency between records.

The author's use of the system under field conditions was during a 2 week stay in Cyprus in October 1996. The primary aim of the trip was to study *Cyclamen cyprium* Kotchy in its habitat since this species is one of a group of related taxa studied for a dissertation for MSc Pure & Applied Plant & Fungal Taxonomy at the University of Reading (Clennett 1997), and the genus forms a long standing interest. The Cyprus Department of Forests were extremely helpful throughout, in allowing the collection of herbarium specimens, and seed of some other taxa, which proved very enlightening in using Q-Collector to record the information. The opportunity to test the system in the field, albeit on a small scale, resulted in additional data fields being incorporated.

Opening and Setting up the Database

When using Q-Collector from within Access or from a separate icon, an opening screen carrying information about terms of supply greets users, followed by the Main Menu, from which the functions of the database are accessed. Eight transactions (pre-expedition routines, add or edit collections, allocate material, notebook page generation, voucher edits, back-up of data, and exit) and four methods of enquiry (expedition based, specimen based, surplus, and statistics) are built into the Main Menu. Since "Add Collections" will begin loading collection records in the field this option is the default. However, in order to function correctly, the Pre-Expedition routines are used to set up the system prior to the expedition, so this option is the first transaction listed.

These routines consist essentially of four operations. Initially the geographic codes need to be loaded for the area to be examined, and here RBG Kew uses the TDWG code system (Taxonomic Databases Working Group, Hollis 1992), with a corresponding area name. Existing records can be viewed by using three navigation buttons to scroll through the underlying table, and new records added as necessary.

The Donor area of these routines allows detail of the members of the expedition to be entered. Again the existing expeditions can be scrolled through to allow editing. Detail of the donor and expedition code need to be added, and these are currently tied to the 4 letter codes used in the Kew Living Collections database. Adding the names of expedition participants allows the output reports to carry full information.

The Notebook option allows two formats of note page to be generated. Pages for a field notebook can be printed with tick boxes and spaces for each entry, or a simple prompt sheet can be created for using a pocket recorder. This requires reducing on a photocopier, but can be invaluable in remembering the data to note when in the field.

There is also an operation to remove old collection records from the main database table. Although not essential to the operation of the database, this speeds up field use and for data capture is therefore useful. Data is moved to a reserve table from which it can be retrieved at a later date.

Use in the Field

Adding new records to the database is the most frequently used option in the field, and it is here that most work has been done to simplify operations. The form can be set up to default to the chosen expedition and geographic code before use, but this is relatively straightforward. Information is selected from lists or typed into fields which can be selected using a mouse or pointer, but with a laptop moving from field to field is easiest using <tab>. The taxon information is recorded first, and here the genus field uses a list of all the taxa in the database, so as collections of new genera are made the list expands. The first of four screens also covers the collection number, date of collection, geography for the collection site, and site number.

Continuing with the second entry screen, detail is entered for findspot, then latitude and longitude. These two fields are checked for consistency, refusing numbers above 90 degrees lat. or 180 degrees long. Use of GPS (Global Positioning Systems) here has caused some difficulties in that the third level of accuracy is frequently recorded by GPS instruments as a decimal fraction of a minute, where the Kew Living Collections system requires seconds. Q-Collector is tied to the latter so conversions have to be made. (multiply by 0.6). Selection is then made for the provenance of collection, the material collected, the number of plants sampled, approximate number found, area of sampled population and percentage of plants fruiting.

The third entry screen covers habit and habitat information. The height and spread of the specimen are recorded, followed by selection for plant habit and the plant type. The latter is later used by the KewScape mapping (Billson *et al.* 1992) and TRAMS tree maintenance (Ruddy 1999) systems to key accessions in the gardens to symbols for their habit. The habitat field is selected from a list from the Living Collections database, with a brief explanation of terms used, and a qualifier allows more precise information to be added. Aspect, exposure and slope fields allow detail of the habitat to be selected. The parent rock and soil fields carry short statements on these site features, of great importance to horticulturists when growing on collections. Both fields allow new entries to be made, but also selection from all the previous terms used in the database.

The final screen for adding new collections allows selection of category for frequency, and typed text for field notes, conservation notes, photographic data and free text. The check box here for vouchers is also important since only records showing a voucher here will be accessible in voucher label outputs.

Generating Output

Most reports output from Q-Collector will be produced after return from the expedition, but generation of a customs declaration is seen as an important feature of the system, and is the only enquiry designed expressly to be used before return. The expedition to be queried is selected using a standard search box, and then automatically opens a report with the necessary data. With the expeditions that have so far used Q-Collector, access to a printer has been possible at the offices of local collaborators, allowing hard copy of this information to be made available to customs inspectors.

After the return of the expedition there are two main areas of editing to be completed before reports are output. Information on the distribution of duplicate vouchers is added as a list of standard abbreviations, drawn from existing records on the system, and the information is then output on the voucher labels and reports.

The second editing process is vital for living collections, and is very much tied to RBG practice. The five Living Collections Sections are listed, together with the Micropropagation Unit, RBG Millennium Seed Bank and surplus. The expedition is selected using a standard search box, and each collection can be displayed. Check buttons against each LCD Section, or seed bank or surplus can be clicked to show the location of material, and these boxes control the output lists generated for accessioning and distribution.

Having completed these edits, output of pre-formatted reports can be completed simply. The Expedition Report contains all the information recorded in the field. This can be the most time-consuming operation after an expedition if it has to be typed by hand from handwritten field notes, and the Q-Collector report is therefore formatted to become an integral part of all RBG expeditions, usually forming an appendix. It can be printed as hard copy or exported from Q-Collector to Microsoft Word™ for further editing.

Accessioning of collections to the Living Collections database is vitally important to allow the nurseries to sow and curate their collections. Speedy accessioning is an essential advantage of using Q-Collector, and is achieved using preformatted reports. A standard Kew accession request form is output for the target LCD Sections which can be printed and sent to Plant Records. Currently the information contained here is then re-typed to the LCD database, but as there is no check on taxon accuracy within Q-Collector, this stage allows controls to be introduced. Seed bank accessions are generated in exactly the same way as for LCD and use a different selection and combination of data. This is to mimic the standard form used in the RBG Millennium Seed Bank, and again the data is then manually transferred to the main database.

By clicking the surplus button when allocating material, a list giving brief detail of collections can be generated to distribute to interested gardens or organisations, and subsequently packet labels can be printed to simplify distribution. In a similar way, having updated the location of any duplicate vouchers, a list of vouchers taken by the expedition and their location can be produced. Pre-formatted labels can also be printed to accompany the voucher specimens as they are incorporated into the RBG Herbarium collections. Since the Living Collections expeditions use four letter codes, provision has been made to record the names of collectors alongside the code to avoid later confusion.

Simple statistics are also available for expeditions on return. The expedition can be selected and information output on the total collections made, the collections made for each day in the field, the uses made of collections at RBG Kew, and the collections listed according to plant habit categories. The system is set up to query all the records held, so data for several expeditions can be output together, querying by genus rather than expedition.

Conclusion

Successful use of Q-Collector has allowed accession numbers and field data to be made available throughout the institute within 1–2 weeks of return, enabling collections to be sown and fully curated in the Kew nurseries. This compares with delays of up to 3 months with manual transcription from field notebooks or tapes before introduction of the system to Living Collections expeditions — more than justifying the time invested by the author in refining the database. The flexible design, separating data into individual fields, allows construction of varied reports, combining data in differing ways to conform to its users needs.

Battery limitations, and size and weight of laptop computers, have shown that bulk editing in evenings (or even in the airport lounge!) is more efficient than attempting to record at the time of collection. Working quickly from a field collection book, or pocket recorder can be easily achieved using pre-formatted notebook pages or prompt list generated by the system.

The RBG Kew Corporate Plan (Prance *et al.* 1998) now expressly states that Q-Collector should be used on all Living Collections expeditions, and it is hoped that staff in the Seed Bank and Herbarium will be sufficiently impressed by the system to adopt it for some of their work. To this end, increasing the usability for herbarium staff in particular is seen as an important step to take in the near future. Ultimately the data collected will be downloaded to the Living Collections system directly, but checks for accuracy need to be strengthened to ensure errors are not imported to the main database.

The Royal Botanic Gardens, Kew have undertaken to distribute Q-Collector to interested institutions and individuals free of charge. However, RBG is unable to provide any software support and users are asked to agree to a small set of conditions of receipt prior to using the system. Users will require Microsoft Windows®, 3.1 and Access version 2.0 or later to be installed on their computer. Details and copies can be obtained at the author's address, or by e-mail at Q-Collector@rbgkew.org.uk. Information will also become available shortly on the Kew web site (www.rbgkew.org.uk).

References

Billson, C. *et al.* (1992). User documentation for KewScape data collection and field information system. Tangent Technology Design Associates, Loughborough.

Clennett, C. (1997). A taxonomic review of *Cyclamen* L. subgenus *Gyrophoebe* O. Schwartz. Unpublished MSc. Thesis. University of Reading

Hollis, S. & Brummitt, R.K. (1992). World geographical scheme for recording plant distributions. Plant Taxonomic Database Standards No. 2. Hunt Institute for Botanical Documentation, Pittsburgh (available electronically).

Prance, G.T. *et al.* (1998). Corporate Plan of the Royal Botanic Gardens Kew. Unpublished.

Ruddy, S. (1999). TRAMS99v1C. Unpublished.

Knüpffer, H. & Hammer, K. (1999). Agricultural biodiversity: a database for checklists of cultivated plant species. In: S. Andrews, A.C. Leslie and C. Alexander (Editors). Taxonomy of Cultivated Plants: Third International Symposium, pp. 215–224. Royal Botanic Gardens, Kew.

AGRICULTURAL BIODIVERSITY: A DATABASE FOR CHECKLISTS OF CULTIVATED PLANT SPECIES

H. KNÜPFFER AND K. HAMMER*

Institut für Pflanzengenetik und Kulturpflanzenforschung (IPK), Corrensstr. 3, D-06466 Gatersleben, Germany
*Present address: Universität Gesamthochschule Kassel, Fachbereich 11, Landwirtschaft, Internationale Agrarentwicklung und Ökologische Umweltsicherung, Fachgebiet Agrarbiodiversität, Steinstrasse 11, D-37213 Witzenhausen, Germany

Abstract

The database is designed for information about species of cultivated plants in selected countries. It contains taxonomic data, folk names, geographical information, plant uses and plant parts used, narrative text information and literature references. So far, data for Cuba (1029 spp.), Korea (578) and South Italy (521) have been recorded; work on Central and North Italy (550), Albania and East Asia (including China and Japan) is in progress. Information sources are publications, personal observations, and contributions of national specialists. Taxonomic information is being validated against other taxonomic databases. Various inventories of cultivated plant species have been published using the database. Such checklists are useful for plant explorers and collectors, but also as a starting point for compiling floras of cultivated plants. The information from the database will be made available in the WWW, together with a database derived from *Mansfeld's World Manual of Agriculture and Horticultural Crops*, which is being updated and prepared for the English edition by IPK, and which will contain *c.* 6,000 cultivated species, excluding ornamentals and forestry species.

Background

Of the estimated 250,000 higher plant species, 40% can be considered as useful genetic resources (see Table 1). According to the data provided in Table 2, about 7,000 species are cultivated in a narrow sense, excluding forestry and ornamental species.

Species checklists are a well-known and widely used tool in field research, e.g. in ethnobotany. However, few examples of checklists of cultivated plants exist.

Country checklists of agricultural and horticultural plant species have been developed by genebank staff of IPK since the 1980s, in cooperation with specialists

TABLE 1. Numbers of wild species of higher plants, plant genetic resources, and cultivated plants in Germany, in Europe and in the World (estimates, cf. Hammer 1995).

	Higher Plants	**Plant Genetic Resources**	**Cultivated Plants**
Germany	2,500	1,055	150
Europe	11,500	4,730	500
World	250,000	100,000	7,000

TABLE 2. Numbers of cultivated agricultural and horticultural plant species (after Hammer 1995).

Authors	Year	Number of Species		Reference
		Included	Estimated	
Mansfeld	1959	1430	1700–1800	Mansfeld 1959
Vul'f	before 1941	2288		Vul'f 1987
Vul'f & Maleeva	1969	2540		Vul'f & Maleeva 1969
Mansfeld 2nd ed.	1986	4800		Schultze-Motel 1986
Mansfeld 3rd ed.	in prep.		6000	
General estimation			7000	

from various countries. The first checklists were developed for Libya (Ghāt oases: Hammer & Perrino 1985, whole country: Hammer *et al.* 1988), followed by those for Korea (Baik *et al.* 1986 and additions), Cuba (complete list: Esquivel *et al.* 1992) and Italy (complete list for South Italy and Sicily: Hammer *et al.* 1992; Central and North Italy: Hammer *et al.* 1999).

The Database of Cultivated Plants was initially developed to support preparation and updating of checklists for Cuba and South Italy (Knüpffer 1992). Data for Korea was included in the database from our earlier publications, and meanwhile updated. The database is being extended for East Asia, to include also China and Japan. Inclusion of data for Albania is under preparation.

The compilation of cultivated plant species of the world initiated by Rudolf Mansfeld in 1959 and updated in 1986 (Schultze-Motel 1986) is being revised. The third edition will be published in English as *Mansfeld's World Manual of Agricultural and Horticultural Crops* (Hanelt in prep.). This approach is being complemented "bottom-up" by detailed country-specific studies through the checklists.

Use of Checklists

A country checklist of cultivated plant species is a valuable tool for exploration and collection of plant genetic resources (Hammer 1990). It enables the collector to ask farmers for rare or obsolete cultivated plant species. It allows one to quickly check whether a species found in cultivation was known from that area before. It also forms a good starting point for floras or catalogues of cultivated plant species (e.g. South Italy; Hammer *et al.* 1992), or monographs (e.g. Cuba; Hammer *et al.* 1992–1994).

The inclusion of widely used synonyms and those found in local literature, as well as vernacular names makes the checklists useful in studying older literature sources.

Structure of the Database

Initially the database served mainly to produce a checklist, including indices of synonyms and vernacular names. Therefore, the following data elements were included in the relational database:

- Nomenclature (accepted name: genus, species, family, infraspecific name, author, place of original publication), including important and regional synonyms.

- Vernacular names, including abbreviated language or region.
- Geographical information (origin, distribution, own collections/observations).
- Uses and plant parts used.
- Narrative text of any length, e.g. for information about the history, diversity, breeding and ethnobotanical aspects.
- Editorial remarks concerning a particular species, e.g. literature references yet to be checked.
- Literature references.

Nomenclatural information for each species is stored only once, in a relational group of tables forming the taxonomic-nomenclatural core. Errors and inconsistencies thus can be largely avoided. Country-specific information is stored separately for each country.

Outputs of the Database

The design of the output lists resembles the checklists published earlier (cf. Table 3) which, in turn, reflect the structure of Mansfeld's Manual. Indices for synonyms (Table 4), vernacular names (Table 5), plant families (Table 6) and uses can be

TABLE 3. Example for checklist format, from Cuba. The references quoted in this example are explained in Esquivel *et al.* (1992).

Matricaria recutita L., Sp. Pl. (1753) 891. – *Matricaria chamomilla* L., Fl. suec. ed. 2 (1755) 296 et Sp. Pl. ed. 2 (1763) *et auct. non* L. (1753). (*Compositae*).
Camomila, manzanilla alemana, manzanilla dulce
M. (fl.)
Origin: Europe-Siberia.
Ref.: Esquivel *et al.* 1989a, Roig 1974, Roig 1975

Medicago sativa L., Sp. Pl. (1753) 778. (*Leguminosae*).
Alfalfa
Fo., green manure
Origin: Central Asia, West Asia, Mediterranean coastal and adjacent regions, Europe-Siberia.
According to García (1910), alfalfa was brought by the Spaniards to America (Mexico, Peru, Chile), where it naturalized. From there it was introduced to California and Texas under the name alfalfa. With the colonization of the western part of North America by settlers from other European countries, it was brought in under the name lucerna, without much success, so this name did not become popular. The Cuban cultivated material originates from seeds from North America. The cultivars 'Arizona', 'Texas' and 'Americana' were grown.
Today 'Gilboa Africana' is also cultivated.
Ref.: Alain 1974, Anon. 1989, Calvino 1918a, Esquivel *et al.* 1989a, García 1910, Menéndez *et al.* s.a., Roig 1975

Melaleuca leucadendra (L.) L., Mant. Pl. 1 (1767) 105. – *Myrtus leucadendra* L., Amoen. Acad. 4 (1759) 120. (*Myrtaceae*).
Cayeput, melaleuca
M. (h., fr.)
Origin: Indochina-Indonesia.
Ref.: Esquivel *et al.* 1989a, Roig 1974, Roig 1975

Melia azedarach L., Sp. Pl. (1753) 384. (*Meliaceae*).
Arbol quitasol, cinamono, paraíso, paraíso de la India, ponciana, pulsiana
Shade tree, living fences, M. (ba., l., fr., r.)
Origin: Central Asia.
Cultivated already c. 1800 (Fernández *et al.* 1990).

Contd. over

TABLE 3 continued

M. azedarach var. *umbraculiformis* Berk. is reported with the folk name árbol quitasol. Ref.: Esquivel *et al.* 1989a, Esquivel *et al.* 1990b, Fernández *et al.* 1990, Hammer *et al.* 1991, Hermann 1951, Roig 1974, Roig 1975
Melica uniflora Retz., Obs. Bot. 1 (1779) 10. (*Gramineae*). Fo., in collections Origin: Europe-Siberia. Ref.: Hammer *et al.* 1992
Melicoccus bijugatus Jacq., Enum. syst. pl. (1760) 19. – *Melicocca bijuga* L., Sp. Pl. ed. 2 (1762) 495. (*Sapindaceae*). Anoncillo, mamoncillo Fr., M. (fr.) Origin: Central America. Ref.: Cañizares 1982, Esquivel *et al.* 1989a, Gómez de la Maza & Roig 1914, León & Alain 1953, Roig 1974, Roig 1975
Melilotus officinalis (L.) Pall., Reise russ. Reich 3 (1776) 537. – *Trifolium officinale* L., Sp. Pl. (1753) 765, p.p. max. (*Leguminosae*). Green manure Origin: Europe-Siberia. Ref.: Calvino 1918a

TABLE 4. Example for index of families, from Korea (Hoang *et al.* 1997).

Iridaceae (4) – *Belamcanda, Crocus, Iris* (2)
Juglandaceae (6) – *Carya, Juglans* (4), *Pterocarya*
Labiatae (26) – *Agastache, Dracocephalum, Elsholtzia, Lavandula, Leonurus* (2), *Lycopus, Melissa, Mentha* (3), *Nepeta, Ocimum* (2), *Perilla, Pogostemon, Prunella, Salvia* (4), *Schizonepeta, Scutellaria, Stachys, Teucrium, Thymus*
Lauraceae (2) – *Cinnamomum, Lindera*
Leguminosae (55) – *Albizia, Amorpha, Amphicarpaea, Arachis, Astragalus* (3), *Caesalpinia* (2), *Cajanus, Canavalia, Caragana* (2), *Cassia* (3), *Cercis, Gleditsia* (2), *Glycine, Glycyrrhiza* (2), *Lablab, Lespedeza* (2), *Lotus, Lupinus, Medicago* (3), *Melilotus* (3), *Phaseolus* (3), *Pisum, Psoralea, Pueraria, Robinia* (2), *Sophora, Trifolium* (3), *Trigonella, Vicia* (4), *Vigna* (4), *Wisteria*
Liliaceae (25) – *Allium* (9), *Anemarrhena, Asparagus* (2), *Convallaria, Fritillaria* (2), *Hemerocallis* (2), *Hosta* (2), *Lilium, Liriope, Ophiopogon, Polygonatum* (2), *Rohdea*
Linaceae – *Linum*
Lobeliaceae – *Lobelia*
Magnoliaceae (5) – *Magnolia* (5)
Malvaceae (11) – *Abelmoschus* (2), *Abutilon, Alcea, Althaea, Gossypium, Hibiscus* (3), *Malva* (2)
Meliaceae (2) – *Melia, Toona*
Moraceae (7) – *Broussonetia* (2), *Cannabis, Ficus, Humulus, Morus* (2)
Musaceae – *Musa*

TABLE 5. Example for index of synonyms, from South Italy and Sicily (Hammer *et al.* 1992).

Capparis ovata Desf. = **Capparis sicula**
Carduus marianus L. = **Silybum marianum**
Carica candimarcensis Hook.f. = **Carica pubescens**
Carica cundimarcensis Linden = **Carica pubescens**
Carissa grandiflora DC. = **Carissa macrocarpa**
Carya microcarpa Nutt. = **Carya ovalis**
Carya pecan Engl. & Graebn. = **Carya illinoinensis**
Cassia senna L. = **Cassia italica**
Casuarina equisetifolia L. = **Casuarina equisetifolia**
Casuarina quadrivalvis Labill. = **Casuarina stricta**
Celtis excelsa Salisb. = **Celtis australis**
Cepa prolifera Moench = **Allium × proliferum**
Cerasus avium (L.) Moench = **Prunus avium**
Cerasus mahaleb (L.) Mill. = **Prunus mahaleb**
Cerasus vulgaris Mill. = **Prunus cerasus**
Chamomilla nobilis (L.) Godr. = **Anthemis nobilis**
Cheiranthus cheiri L. = **Erysimum cheiri**

Chenopodium album L. subsp. *amaranticolor* Coste & Reynier = **Chenopodium giganteum**
Chenopodium amaranticolor (Coste & Reynier) Coste & Reynier = **Chenopodium giganteum**
Citrullus vulgaris Schrad. ex Eckl. & Zeyh. = **Citrullus lanatus**
Citrus aurantium L. var. *grandis* L. = **Citrus maxima**
Citrus aurantium L. var. *myrtifolia* Ker-Gawler = **Citrus myrtifolia**
Citrus aurantium L. var. *sinensis* L. = **Citrus sinensis**
Citrus bigaradia Lois. = **Citrus aurantium**
Citrus grandis Osbeck = **Citrus maxima**
Citrus japonica Thunb. = **Fortunella japonica**
Citrus limetta Risso var. *bergamia* Risso = **Citrus bergamia**
Citrus limonia Osbeck var. *limetta* (Risso) Aschers. & Graebn. = **Citrus limetta**
Citrus margarita Lour. = **Fortunella margarita**

TABLE 6. Example for index of vernacular names, from North and Central Italy (Hammer *et al.* 1999). The abbreviations in parentheses give the provinces where a particular name has been reported from.

melega (To.) = *Sorghum bicolor*
melega (Lo.) = *Zea mays*
melega da scope (Lo.) = *Sorghum bicolor*
meleghetta (Lo.) = *Sorghum bicolor*
meleghetta (To.) = *Sorghum bicolor*
melella (Ma.) = *Crataegus oxyacantha*
meless (Fr.) = *Sorbus aucuparia*
melessàr (Fr.) = *Sorbus aucuparia*
melester (Ve.) = *Sorbus aucuparia*
melestra (Ve.) = *Sorbus aucuparia*
melestri (Ve.) = *Sorbus aucuparia*
melga (Em.) = *Sorghum bicolor*
melga (Lo.) = *Sorghum bicolor*
melga (Em.) = *Zea mays*
melga (Lo.) = *Zea mays*
melga bianca (Lo.) = *Sorghum bicolor*
melga da garnè (Em.) = *Sorghum bicolor*
melga da granèd (Em.) = *Sorghum bicolor*
melga da poll (Em.) = *Sorghum bicolor*
melga da porch (Em.) = *Sorghum bicolor*
melga da scooli (Ve.) = *Sorghum bicolor*
melga da spazzadore (Ve.) = *Sorghum bicolor*
melga negra (Lo.) = *Sorghum bicolor*
melga rossa (Lo.) = *Sorghum bicolor*
melga selvadega (Lo.) = *Sorghum halepense*
melga spargolo (Lo.) = *Sorghum bicolor*
melgàsc (Lo.) = *Zea mays*

melgaster (Em.) = *Sorghum halepense*
melgastro (Ve.) = *Sorghum halepense*
melghèr (Lo.) = *Sorghum halepense*
melghèta (Lo.) = *Sorghum halepense*
melghetta (Lo.) = *Sorghum bicolor*
melghetta (Em.) = *Sorghum halepense*
melghetta (Lo.) = *Sorghum halepense*
melgòn (Lo.) = *Zea mays*
melia (Pi.) = *Zea mays*
melia d'le ramasse (Pi.) = *Sorghum bicolor*
melia d'ramasse (Pi.) = *Sorghum bicolor*
melia roussa (Pi.) = *Sorghum bicolor*
meliaca (To.) = *Prunus armeniaca*
meliach (To.) = *Melia azedarach*
meliaco (To.) = *Prunus armeniaca*
melica = *Sorghum bicolor*
melica (To.) = *Sorghum bicolor*
melieta (Pi.) = *Sorghum bicolor*
melifillo (To.) = *Melissa officinalis*
meliga = *Sorghum bicolor*
meliga (To.) = *Sorghum bicolor*
meliga (Li.) = *Zea mays*
meliloto = *Melilotus officinalis*
meliloto bianco = *Melilotus alba*
meliloto comune = *Melilotus officinalis*
meliloto odoroso = *Melilotus officinalis*
meliloto odoroso (To.) = *Melilotus officinalis*

generated. Output can be tailored to specific needs, using various selection and sorting criteria, such as plant families or uses. Such partial lists are sent for editing to specialists for the particular area. They are also used for publications on special aspects, e.g. the cultivated medicinal plants of a country (e.g. Hammer *et al.* 1997).

Sources of Information

For each country, information is being extracted from literature, and completed by observations made during expeditions. Later, printouts are checked and completed by specialists from the particular country or those for a particular group of cultivated plants. Searches in other databases accessible via the Internet may be useful to add relevant information.

Monitoring of Information Quality

The data compiled from different sources are often very heterogeneous. The information in the database is, therefore, verified with taxonomic standard literature, such as Mansfelds Verzeichnis (Schultze-Motel 1986), *Index Kewensis* and cross-checked against available databases, e.g. the taxonomic core of GRIN, the Genetic Resources Information Network, USA (cf. Wiersema 1999). Internet searching will be of use also here. Tools such as alphabetically sorted lists of authors of botanical names or literature references, or KWIC (key word in context) indices support the standardisation of abbreviations and spellings used.

Contents of the Database

The present content of the database is summarised in Tables 7–10. Table 7 gives the totals and country-specific figures for numbers of taxa, species, genera, families, synonyms, and vernacular names. Tables 8–10 show the most frequently occurring plant families, genera, and uses of plants, respectively.

TABLE 7. Summary of contents of the database (as of mid 1998). The years refer to the respective publications for Cuba (Esquivel *et al.* 1992), South Italy and Sicily (Hammer *et al.* 1992), Central and North Italy (Hammer *et al.* 1999) and Korea (Hoang *et al.* 1997). Figures for areas in preparation are still incomplete.

	Total	Cuba (1992)	S Italy (1992)	C&N Italy (1999)	Korea (1997)	E Asia (in prep.)	Albania (in prep.)
Taxa	2181	1044	540	568	605	949	433
Species	2102	1029	521	550	578	896	418
Genera	953	531	298	327	378	503	255
Families	177	117	86	92	111	140	82
Synonyms	1425	729	348	344	497	656	225
Vernacular names		1669	2981	10802	714	2614	264

TABLE 8. Plant families with the largest numbers of species. See Table 7.

Family	Total	Number of Species					
		Cuba (1992)	S Italy & Sicily (1992)	C & N Italy (1999)	Korea (1997)	SE Asia (in prep.)	Albania (in prep.)
Leguminosae	263	164	72	70	55	87	37
Gramineae	180	95	54	55	37	65	40
Compositae	119	44	32	38	38	55	38
Rutaceae	94	55	18	15	14	56	14
Rosaceae	93	11	41	41	43	53	35
Labiatae	66	26	27	35	26	26	19
Umbelliferae	53	14	16	19	28	40	12
Liliaceae	48	16	16	15	25	30	15
Solanaceae	45	35	18	16	15	18	10
Euphorbiaceae	42	33	1	3	6	11	2
Myrtaceae	41	41	7	3	1	1	3
Cruciferae	36	11	20	24	14	18	13
Moraceae	30	20	6	8	7	13	6
Cucurbitaceae	28	22	12	15	13	15	9
Malvaceae	28	22	6	6	11	12	5
Polygonaceae	25	5	7	12	12	14	2
Zingiberaceae	24	9	0	0	8	20	0
Rubiaceae	21	14	1	2	1	9	0
Verbenaceae	21	16	3	3	1	3	1

TABLE 9. Genera with the largest numbers of species. See Table 7.

Genus (Family)	Total	Number of Species					
		Cuba (1992)	S Italy & Sicily (1992)	C & N Italy (1999)	Korea (1997)	East Asia (in prep.)	Albania (in prep.)
Citrus (Rutaceae)	62	30	12	10	5	44	11
Prunus (Rosaceae)	24	3	13	13	15	15	9
Cassia (Leguminosae)	16	7	1	1	3	11	0
Artemisia (Compositae)	15	4	3	4	5	11	1
Allium (Liliaceae)	14	9	9	9	9	9	5
Desmodium (Leguminosae)	14	14	0	0	0	0	0
Angelica (Umbelliferae)	13	0	0	1	4	12	2
Erythrina (Leguminosae)	12	12	0	0	0	0	0
Vigna (Leguminosae)	12	12	3	2	4	4	1
Acacia (Leguminosae)	11	1	6	2	0	7	2
Annona (Annonaceae)	11	10	1	0	0	6	0
Festuca (Gramineae)	11	7	4	4	3	3	4
Mentha (Labiatae)	11	5	3	9	3	3	1
Pyrus (Rosaceae)	11	1	3	3	8	8	3
Agave (Agavaceae)	10	10	2	1	1	1	0
Bambusa (Gramineae)	10	1	0	0	0	9	0
Chrysanthemum (Compositae)	10	1	5	5	5	7	4

Contd. over

TABLE 9 continued

Datura (Solanaceae)	**10**	10	2	2	2	2	1
Passiflora (Passifloraceae)	**10**	10	2	1	0	0	0
Rosa (Rosaceae)	**10**	0	3	2	6	6	6
Solanum (Solanaceae)	**10**	10	3	2	4	4	2
Syzygium (Myrtaceae)	**10**	10	0	0	0	0	0

TABLE 10. Summary of plant uses: number of taxa for which a particular use is reported. See Table 7.

	Number of Species				
Use	**Cuba (1992)**	**S Italy & Sicily (1992)**	**C & N Italy (1999)**	**Korea (1997)**	**SE Asia (in prep.)**
Medicinal (M.)	432	121	140	427	587
Fruits (Fr.)	262	108	95	53	139
Forage (Fo.)	173	90	97	93	137
Vegetable (V.)	99	110	124	100	138
Spice, aromatic (Sp.)	60	60	71	23	54
Fibre (Fi.)	44	8	7	31	36
Industrial (I.)	41	50	51	59	101
Starch (St.)	31	7	9	6	10
Oil (Oi.)	24	14	15	48	109
Cereals (C.)	9	23	22	15	29
Total:	**1044**	**540**	**568**	**605**	**949**

Advantages of Database Technology, Additional Search Possibilities

Information from new sources can easily be added, and the lists and indices are updated automatically and consistently. Indices and sub-lists can be generated. The quality of the data is ensured by specific tools. The data can be used in more different ways and more efficiently than would be possible with a manual approach.

Uses of the Database

The Database of Cultivated Plants is useful for plant breeding and diversification of agriculture, since it documents the present and past diversity of cultivated plant species in particular countries. It can be used to search for obsolete, neglected, under-utilised, potential or new crops. Other possible uses of the database are the comparison of floras of cultivated plants of different countries, or a comparison of uses of the same species in different countries (cf. Table 11). The flora of cultivated plants is, in addition, an important aspect of the agro-biodiversity of a country. An overview of the flora of cultivated plants of a country is a good starting point for developing a national system of plant genetic resources.

TABLE 11. Comparison of uses of some cultivated plant species in Cuba, South Italy and Korea. For abbreviations of plant uses, see Table 10. Plant parts used are given in parentheses in abbreviated form: fl. = flowers, fr. = fruits, h. = herb, l. = leaves, r. = roots or rhizomes, s. = seeds.

Taxon	Cuba	South Italy	Korea
Abelmoschus esculentus	V. (fr.), Oi. (s.), M. (fr., l., fl.)	V. (fr.)	V. (fr.)
Anethum graveolens	M. (h., fr.), V. (l.)	Sp. (l.), M. (fr.)	M. (fr.)
Calendula officinalis	M. (h.), V. (fl.), Sp. (fl.), ornamental	M. (fl.)	M. (fl.)
Colocasia esculenta	St. (r.), ornamental	St. (r.)	V. (l.), St. (r.)
Coriandrum sativum	Sp. (l., s.), M. (fr.)	M. (fr.), Sp. (fr.)	I. (perfume), M (fr., h.), Sp. fr.)
Cucurbita maxima	Fo. (fr.), V. (fr.), Fr., M. (s., fr.)	V. (fr.), Fo. (fr.)	V. (fr.)
Cymbopogon citratus	M. (l.), aromatic, soil erosion control	Sp. (l.)	I. (h., perfume)
Eucalyptus globulus	M. (l.), wind break, soil erosion control and reforestation	M. (s.)	I., Oi. (l.)

Other countries for which information is already available, will be included in the database. Many requests for specific information could be satisfied so far by querying the database. For example, information about single species and their use or breeding activities has been provided for an IPK-IPGRI project on Neglected Crops (Hammer & Heller 1998).

The database will also form part of the "German Federal Information System on Genetic Resources" (BIG) which is under development (Becker 1999). BIG will also include a WWW database derived from the English version of *Mansfeld's World Manual of Agricultural and Horticultural Crops* (Hanelt in prep.) which will contain *c.* 6,000 cultivated species, excluding ornamentals and forestry species (Ochsmann *et al.* 1999).

References

Baik, M.-C., Hoang, H.-Dz. & Hammer, K. (1986). A check-list of the Korean cultivated plants. *Kulturpflanze* 34: 69–144.

Becker, B. (1999). Bundesinformationssystem Genetisch Ressourcen – Konzept und erste Ergebnisse. In F. Begemann, S. Harrer & J.D. Krause (eds). Dokumentation und Informationssysteme im Bereich pflanzengenetischer Ressourcen in Deutschland. Schriften zu Genetischen Ressourcen, Bd. 12, pp. 93–104. Zadi, Bonn.

Esquivel, M., Knüpffer, H. & Hammer, K. (1992). Inventory of the cultivated plants. In K. Hammer, M. Esquivel & H. Knüpffer (eds). "...y tienen faxones y fabas muy diversos de los nuestros..." — Origin, Evolution and Diversity of Cuban Plant Genetic Resources, vol. 2, pp. 213–454. IPK, Gatersleben.

Hammer, K. (1990). Botanical checklists prove useful in research programmes on cultivated plants. *Diversity* 6(3–4): 31–34.

Hammer, K. (1995). *Ex-situ-* und *In-situ-*Erhaltung pflanzengenetischer Ressourcen in Deutschland. IWU-Tagungsberichte. Die Erhaltung von genetischen Ressourcen von Bäumen und Sträuchern. pp. 17–32. Magdeburg.

Hammer, K., Esquivel, M. & Knüpffer, H. (eds). (1992–1994). "...y tienen faxones y fabas muy diversos de los nuestros..." — Origin, Evolution and Diversity of Cuban Plant Genetic Resources. 3 vols., 824 pp. IPK, Gatersleben.

Hammer, K. & Heller, J. (1998). Promoting the conservation and use of underutilized and neglected crops. *Schriften Genet. Ressourcen* 8: 223–227.

Hammer, K., Knüpffer, H. & Hoang, H.-Dz. (1997). Koreanische Heilpflanzen — eine Liste der kultivierten Arten. *Drogenreport* 10(16): 57–59, and special issue, 26 pp. (unnumbered).

Hammer, K., Knüpffer, H., Laghetti, G. & Perrino, P. (1992). Seeds from the past. A catalogue of crop germplasm in South Italy and Sicily. Ist. Germoplasma. 173 pp. Bari.

Hammer, K., Knüpffer, H., Laghetti, G. & Perrino, P. (1999). Seeds from the past. A catalogue of crop germplasm in Central and North Italy. 1st. Germoplasma. iv + 257 pp. Bari.

Hammer, K., Lehmann, C.O. & Perrino, P. (1988). A check-list of the Libyan cultivated plants including an inventory of the germplasm collected in the years 1981, 1982 and 1983. *Kulturpflanze* 36: 475–527.

Hammer, K. & Perrino, P. (1985). A check-list of the cultivated plants of the Ghāt oases. *Kulturpflanze* 33: 269–286.

Hoang, H.-Dz., Knüpffer, H. & Hammer, K. (1997). Additional notes to the checklist of Korean cultivated plants (5). Consolidated summary and indexes. *Genet. Resources Crop Evol.* 44(4): 349–391.

Knüpffer, H. (1992). The database of cultivated plants of Cuba. In K. Hammer, M. Esquivel & H. Knüpffer (eds). "...y tienen faxones y fabas muy diversos de los nuestros..." — Origin, Evolution and Diversity of Cuban Plant Genetic Resources, vol. 1, pp. 202–212. IPK, Gatersleben.

Mansfeld, R. (1959). Vorläufiges Verzeichnis landwirtschaftlich oder gärtnerisch kultivierter Pflanzenarten (mit Ausschluß von Zierpflanzen). *Kulturpflanze*, Beih. 2, 569 pp.

Ochsmann, J., Biermann, N., Knüpffer, H. & Bachmann, K. (1999). Aufbau einer WWW-Datenbank zu "Mansfeld's World Manual of Agricultural and Horticultural Crops". In F. Begemann, S. Harrer & J.D. Krause (eds). Dokumentation und Informationssysteme im Bereich pflanzengenetischer Ressourcen in Deutschland. Schriften zu Genetischen Ressourcen, Bd. 12, pp. 57–63. Zadi, Bonn.

Schultze-Motel, J. (ed.). (1986). Rudolf Mansfelds Verzeichnis landwirtschaftlicher und gärtnerischer Kulturpflanzen (ohne Zierpflanzen). 4 vols. Akademie-Verlag Berlin.

Vul'f, E.V. (1941). (publ. 1987). Kul'turnaya flora zemnogo shara [Cultivated plant flora of the world]. 326 pp. Leningrad. (In Russian).

Vul'f, E.V. & Maleeva, O.F. (1969). Mirovye resursy poleznych rastenij [World resources of useful plants]. 564 pp. Leningrad. (In Russian).

Wiersema, J. H., (1999). Nomenclature of world economic plants from the USDA's GRIN database. In S. Andrews, A.C. Leslie & C. Alexander. Taxonomy of Cultivated Plants: Third International Symposium. pp. 207–208. Royal Botanic Gardens, Kew.

Thornton-Wood, S.P. (1999). The RHS plans for its information on cultivated plants — reconciling good taxonomic practice with the needs of gardeners. In: S. Andrews, A.C. Leslie and C. Alexander (Editors). Taxonomy of Cultivated Plants: Third International Symposium, pp. 225–228. Royal Botanic Gardens, Kew.

THE RHS PLANS FOR ITS INFORMATION ON CULTIVATED PLANTS — RECONCILING GOOD TAXONOMIC PRACTICE WITH THE NEEDS OF GARDENERS

SIMON P. THORNTON-WOOD

Department of Botany, RHS Garden, Wisley, Woking, Surrey GU23 6QB, UK

Abstract

The recent integration of *The RHS Plant Finder* with the RHS Horticultural Database is an important step forward for the Society in gathering information on cultivated plants, both historical and up-to-date. The RHS is working to manage such information to serve its wide range of publications (now to include electronic media) and is having to reconcile (somewhat disparate) needs of populist books, specialist journals, garden visitor information, catalogues and reference databases, as emphasis shifts from collation to dissemination. A strategy for relating RHS information to that of other organisations is developing, increasing the potential for a widely-approved standard — or at least cross-reference — of cultivated plant nomenclature.

A possible future relationship for the Society, its publications and the imperatives of good taxonomic practice is presented, in the context of an expanding database resource of taxonomic information at Wisley.

Introduction

The Royal Horticultural Society's Botany Department at Wisley is a centre for horticultural taxonomy and maintains a database of plant names, with extensive coverage of plants in cultivation in the United Kingdom. The aim of this paper is to explain the sources of information in this database which reinforce its value to users.

The key to the success of this work is the relationship between the Society's integrated horticultural database (using *BG-BASE*™) and *The RHS Plant Finder*. *The RHS Plant Finder* is unique in the way in which other sources of botanical information reinforce the value of the plant name list in the book, through careful recording of other taxonomic activities of the Society within the database. It is important, when using such a reference work, to understand:

- Who compiled the work?
- How was it compiled?
- When (and how often) was it compiled?

The way in which confidence may be gauged in such a compiled reference is partly analogous to judging the value of a herbarium specimen label: knowledge of the person determining the identity of a plant affects one's opinion of their identification. Judging a "Plant Finder" is made more difficult, as users have many requirements of the publication, and the compilers try to maintain a balance between botanical accuracy and familiarity for users.

Is sufficient consideration given to the authorship of each of the systematic databases proliferating around the world? Whilst the authorship of the RHS Horticultural Database cannot be pinned to an individual, it reflects the work of the Botany Department at Wisley, which is engaged in a very wide range of activities in the field of horticultural taxonomy, and other selected specialists. This affects the development of the database from which *The RHS Plant Finder* is compiled. These activities are described here, and may be broadly classed within the themes of gathering, reviewing and disseminating information.

Gathering

The RHS Plant Finder was started (as *The Plant Finder*) in 1987 (Philip & Lord 1987) by Chris Philip (devisor and compiler) and W. Anthony Lord (editor), and has become established as one of the most useful references to cultivated plant nomenclature, partly due to its comprehensive annual updating. The eleventh edition of the book (Lord 1998) was the first to be produced directly from the Royal Horticultural Society's integrated database; a complicated procedure which was undertaken in the belief that the RHS database would enhance the value of an already highly-respected reference. One important benefit has been the direct contribution of RHS botanists to the editorial work.

The RHS Plant Finder imposes a discipline on information-gathering with its annual publication and demand for consistency of treatment within genera. It is a very good means of ensuring that the record of plant names in cultivation is kept up-to-date, and gives a measure of the rarity of each taxon in current nursery supply. To a certain extent, the book also keeps the database editors informed of the use of names by the nursery industry, from the small specialist supplier to the more major outlets. Electronic data collection is being introduced this year (1998) on a trial basis, to improve the way in which the book reflects each nursery's stock list for the year ahead. Whilst a great deal of interpretation of the information supplied by nurseries is made, the database retains the essential form of names as they are delivered for inclusion, cross-referenced to our interpretation of correct usage. This is the most visible part of the nomenclatural structure of the database, but other activities strengthen it.

The herbarium at Wisley (WSY) is one of the largest cultivated plant herbaria in the world. A volunteer programme been established by the Keeper of the Herbarium to gather voucher specimens from the extensive RHS garden collections, a rich and well-documented resource. The herbarium has been particularly active in the development of representative collections from RHS Trials — each a unique record of the range of cultivars within a group at the time of Trial. However, many other sources strengthen the collection, such as the RHS Members' Advisory Service, to which a wide range of specimens are sent for identification.

The RHS is the International Registration Authority for nine groups of plants of major horticultural significance (*Rhododendron* L., *Narcissus* L., *Dianthus* L., *Dahlia* Cav., *Delphinium* L., *Iris* L., orchids, conifers and *Clematis* L.), and continues to develop its approach in quest of best practice in the administration of such work, whilst addressing the differing needs of each group for which it is responsible. The Internet may become an additional resource of information on the naming of cultivars across the world. RHS Registration databases are maintained separately from the RHS Horticultural Database at this time, as management of the nomenclatural structure is rather different.

Reviewing

The RHS Plant Finder encourages a continual process of review across the range of taxa in cultivation. An RHS Advisory Committee on Nomenclature and Taxonomy, comprising a small group of active horticultural taxonomists, is regularly convened to address broad issues and establish a consensus for the treatment of taxa of contentious nomenclature. All this is reflected in the database from which the book is directly drawn, with some important issues highlighted in nomenclatural notes in the book.

A verification programme has been established for the RHS garden collections — currently concentrating on Wisley — which is carefully recorded in the horticultural database. Verifications are closely linked in the database to the "determinations" undertaken for related herbarium material (a subtle difference in the jargon!).

The RHS Members Advisory Service provides some input into the continual process of review; this gives constant contact with a sample of plants grown throughout the United Kingdom (and beyond) and helps relate the Department's work to the many questions and areas of confusion for gardeners.

A programme for gradually cataloguing the herbarium with *BG-BASE*™ is an important step toward fully integrating the specimen collection with other sources of information. A current programme to identify all "Standard Specimens" (the horticultural equivalent to to botanical type specimens) is in progress; and new material entering the collection is, where possible within constraints of space and time, fully catalogued.

A "Trials Module" in the RHS Horticultural Database promises greater opportunities for input from the Botany team to this work, whilst giving the Trials Administration team greater access to information on appropriate taxa for inclusion in the Trials Programme. RHS Trials are a valuable opportunity for the Botany Department to undertake taxonomic research, making use of the comparison of taxa side by side under fairly uniform conditions. Increasingly, this work is undertaken in collaboration with other academic institutions.

Every nomenclatural decision cascades through the system, affecting herbarium specimens and garden plant labels, publications and the worldwide web. It is these consequences which help maintain an editorial balance.

Disseminating

This structured basis for disseminating information has only just begun to show its value, beyond *The RHS Plant Finder*. Refinement of the book continues, with a view to maintaining its handy format whilst accommodating the full range of appropriate plant names listed as being in current supply. To this end, the 'Plant Deletions' section of the book — plants no longer listed by suppliers — has been reduced in the 1998/9 edition. A new, electronic format for *The RHS Plant Finder* database is now well established as the complete reference to all names associated with the book since its first publication in 1987.

Part of the impetus for future developments in information dissemination from the database is the Society's commitment to education. *The RHS Plant Finder* has educated people in the use of plant names: the database will enable its extension to other media, including the worldwide web.

RHS International Registers are similarly being produced directly from databases; most recently, *The Daffodil Register and Classified List 1998* (Kington 1998) was produced in this way. Electronic publication is familiar for orchid greges, and the Daffodil Register is now reflected in an Internet version on the RHS website (www.rhs.org.uk).

A start has been made on making the RHS living collections open to searching on-line with the Wisley On-line Garden Plant Catalogue, though concerns remain over the security implications of freely distributing precise details of valuable parts of the collection. Novel ways of introducing the collection to visitors are being explored through manned and, before too long, public access computer terminals.

The RHS website has a new section devoted to Trials, which have proven so popular with visitors to Wisley over the years. Details of Award plants and more general conclusions of the assessments are currently shown; the development of a fully-integrated Trials module in *BG-BASE*™ has more potential for immediate and in-depth access to such information for a wide readership.

RHS Trials give information on the past as well as the future introduction of plant novelties into cultivation; a small start has been made on capturing historical information on such activities of the Society since 1859/1860 from the Society's archives.

RHS publications stand to gain a great deal from ready access to a consensus nomenclature compiled at Wisley. A framework for dictionary-style publications may be established as the database is strengthened. A reciprocal benefit would be better consistency in the handling of information associated with taxa: horticultural descriptions, extent of synonymy citation and classification into horticultural groups.

Strategy for information links with other organisations

The first tentative steps are now being taken towards integrating the information resources of the Society with other organisations. The On-line Garden Plant Catalogue is part of a small group involved in a *BG-BASE*™ initiative to link collection catalogues electronically, though it is to be expected that, eventually, even dissimilar database types may be queried jointly and simultaneously across the Internet. Other initiatives include the linking of Standard Specimen catalogues and the development of a reference to collectors' codes used in horticulture.

Conclusion

The future strength of the Society's work in cultivated plant nomenclature clearly lies in further integration of its activities through its horticultural database. The cascading consequences of taxonomic changes affects our judgement on them in a very direct way. The traditional work of this long-established organisation finds new expression and a strengthened role in informing, advising and educating gardeners with the help of new electronic media, making us all more effective communicators.

References

Kington, S. (comp.). (1998). The International Daffodil Register and Classified List 1998. 1166 pp. The Royal Horticultural Society, London.

Lord, W.A. (ed.). (1998). The RHS Plant Finder 1998–99. 914 pp. Dorling Kindersley, London.

Philip, C. (comp.) & Lord, W.A. (ed.). (1987). The Plant Finder. 408 pp. Headmain, for The Hardy Plant Society, Whitbourne, Worcestershire.

MODERN TECHNIQUES IN
BREEDING AND TAXONOMY

8

Baum, R.B. (1999). DNA fingerprinting of cereal cultivars for intellectual property rights protection. In: S. Andrews, A.C. Leslie and C. Alexander (Editors). Taxonomy of Cultivated Plants: Third International Symposium, pp. 231–238. Royal Botanic Gardens, Kew.

DNA FINGERPRINTING OF CEREAL CULTIVARS FOR INTELLECTUAL PROPERTY RIGHTS PROTECTION

BERNARD R. BAUM

Eastern Cereal and Oilseed Research Centre, Agriculture & Agri-Food Canada, Research Branch, Neatby Building, Central Experimental Farm, Ottawa, Ontario K1A 0C6, Canada

Abstract

Various seed certification schemes have been established to ensure cultivar identity and purity in the market place. Certification of identity has relied primarily on morphology. Novelty of identity, i.e. a new cultivar, is as a rule based on distinctness, uniformity and stability (DUS), and still relies primarily on morphology. New cultivar material may be protected by intellectual property (IP) laws, whereas new cultivar names may be protected by inscription on national lists. IP rights may be obtained country by country. With the increasing number of new cultivars, especially those created by biotechnology, DUS based on morphological traits that are susceptible to environmental conditions becomes problematic. DNA fingerprinting has enormous potential to be used in addition to or instead of morphology to fulfil DUS and thus also IP requirements. Some DNA fingerprinting methods are briefly described. An identification scheme for 128 barley cultivars registered or previously registered in Canada, with special emphasis on 65 six-rowed barley cultivars, operating in the Eastern Cereal and Oilseed Research Centre (ECORC), Ottawa, is briefly described along with an interactive identification key based on DNA fingerprints.

Introduction

This paper aims to explain the potential role of DNA fingerprinting in cultivar identification and its importance in cultivar protection for trademark registration, for Plant Breeders' Rights registration and for cultivar name registration in the sense of the *International Code of Nomenclature for Cultivated Plants – 1995* (Cultivated Plant Code), (Trehane *et al.* 1995) in general, and to provide an example based on DNA fingerprinting with particular emphasis on barley cultivar identification in Canada. Some issues associated with DNA fingerprinting technologies will also be discussed. Since intellectual property protection varies between countries, some generalisations made here may not apply universally; for instance whole plants are not yet patented in Canada.

A. Past and Extant Cultivars

Late in the 19th century and more so in the 20th century, cultivars released by various sources lost their identity and their unique traits. This is because they found their way into different admixtures and also became known by different names. For instance

'Aurora' oats is also known as 'Appler' and 'Yellow Peruvian'; 'Alber' oats is also known as 'Dasix', a synonym, the name being valid for a different cultivar; 'Dawn' oats from Australia is different from 'Dawn' oats in the US (Baum 1972), and 'Gold' barley from Germany is different from Swedish 'Gold' barley, based on their different pedigrees (Baum *et al.* 1985). Because admixtures and cultivars were traded under different names, cultivar identification was deemed to be important early in the evolution of the seed industry. The International Crop Improvement Association (ICIA) was established in 1920 to prevent confusion in the market place. The aim of the ICIA was to unify and standardise the seed certification programmes that had been developed since the turn of the century in many countries. Certification ensures genetic purity and integrity of cultivars from multiplication of the original or breeder's material, through subsequent multiplication stages (in Canada called foundation), to registration. Through such schemes a grower obtaining certified seed can be confident that the material grown is of a specific quality, is true-to-type, and carries with it reliable information about origin, yield, growing conditions, disease resistance, etc.

The International Seed Testing Association (ISTA) was founded in 1924. One of its most significant developments was unification of the testing methods that have appeared in various editions of the Rules (ISTA 1996). In 1957, the Organization for Economic Cooperation and Development (OECD) undertook to develop and manage schemes for cultivar certification of seed lots traded internationally, (OECD 1967). The International Union for the Protection of New Varieties of Plants (UPOV), established in 1961, is concerned with breeders' rights. The issue of breeders rights, as well as the description of new cultivars, of concern to both ISTA and UPOV, will be briefly discussed later in this paper.

Certification of trueness to cultivar is carried out by crop and seed inspection and by laboratory tests and field trials. The requirements for laboratory tests are explained in the Association of Official Seed Analysts (AOSA) handbook (AOSA 1991). Laboratory identification tests must ideally be reproducible within and among different laboratories, be easy to perform or lend themselves to automation, be completed in a short time, and be inexpensive. Since identification of most cultivars is based on morphology, identification must rely on experts or depend on well-trained personnel who will also grow reference plants from breeder's seed. This does not guarantee identity because of environmental effects on morphology. An example is the problem in identifying faba bean cultivars based on morphology as described by Higgins *et al.* (1988). As the number of cultivars increases it becomes difficult to find enough morphological features to distinguish one from another, especially when cultivars are bred for physiological or agronomic traits such as yield or earliness, or when developed by single gene insertions using transgenic technologies. To overcome these problems, other approaches or techniques were attempted, including chemical assays such as the phenol test (Elekes 1980), protein electrophoresis (Gebre *et al.* 1986), and allozyme assays (Fedak 1974). A good summary of laboratory-based cultivar identification schemes, including DNA-RFLPs, for crops of major economic importance can be found in Smith & Smith (1992). These techniques however, based on methods developed in the 1980s, are no longer sufficient because in an increasing number of cases they fail to distinguish between cultivars. New fingerprinting technologies were recently developed that are less expensive, faster and more refined, and are known to detect considerably more polymorphism than isozyme electrophoresis or DNA-RFLPs. New DNA-based markers for recognising value-added traits are seriously being considered, especially for new cultivars produced by transgenic technology, but their reliability has not yet been established.

B. New Cultivars

Identification of cultivars is essential for plant cultivar protection, registration and patents, as well as for seed certification. A new cultivar must undergo DUS testing for registration or for protection as a new cultivar, at least in the major commercial crops. In practice many ornamental groups have no such testing of these traits. In some countries a new cultivar may be patented if it is novel, inventive and useful. Though new cultivars may be developed for agronomic characters, the latter suffer from the same shortcomings as morphological characters when used in cultivar characterisation. Thus, there is a growing need to use biochemical, or better, DNA-based methods that reflect genetic differences between cultivars.

Until recently in the United Kingdom, though publication of new wheat cultivars was based on detailed morphological description, they were also assessed on gliadin (protein) profiles (Parnell 1983); new cultivars with profiles similar to those of existing cultivars were not recommended. This was a shortcoming. In Canada for instance, 22 registered red, hard-spring wheat cultivars were found to have identical gliadin profiles (Lukow *pers. comm.*), even though they were distinct on agronomic features. Although allozyme profile data have recently been required in France for registration application of new maize cultivars, they are not considered sufficient alone for distinctness testing purposes. Likewise, allozyme data are accepted in plant cultivar protection and patent applications in the US (for example, patent No. 4594810 of an inbred corn line, by Troyer 1986). Canada follows the UPOV recommendations that molecular tests, including RFLPs and allozyme data, are used as criteria supplementary to the essentially morphologically-based DUS for Plant Breeders' Rights (PBR) protection.

The results of mental labour are called intellectual property (IP). New cultivars form such property and result from investment in research and breeding. Many countries provide incentives for further investment in the form of rights and protection for the owners of IP. These rights are based on federal patent, trademark, copyright laws and state trade secret laws. In general, **patents** protect inventions of tangible things; **copyrights** protect various forms of written and artistic expression, and **trademarks** protect a name or symbol that identifies the source of goods or services. (American Intellectual Property Law Association 1995). Brand names are synonymous with trademarks and service marks; they are important intellectual properties upon which the public learns to rely to identify a source and a standard of quality in the products or services it purchases. Trademarks and service marks may be words, phrases, designs, sounds or symbols. They are used on or in association with goods or with services to be performed. The public learns through experience, that goods or services bearing such a mark come from, or are controlled by, a single source and will meet an expectation of quality found in the past. This predictability provides the owner or user of the mark with the benefit of goodwill held by the public for the product or service on offer. This goodwill is often the cornerstone of the owner's business. (American Intellectual Property Law Association 1995). The symbol ™ is often used with unregistered trademarks to give notice that the user is staking a claim on the mark (or design) as a trademark, but such use is optional. In Canada, however, under the Plant Breeders Rights Act, a mark may not be a cultivar name and under the Trade Mark Act a cultivar name may not be used as a Trade Mark.

Plant cultivars may however be sold in association with trademarks. PBR are a form of IP protection where the breeder or discoverer has the exclusive right to sell his cultivar material and to produce reproductive material for sale. The breeder or discoverer may grant these rights to others for a fee. This creates a suitable environment to encourage further investment in plant improvement. PBR are granted to the originators, as with patents and trademarks, on the basis of DUS based on morphological tests or other proof. Protection of a new cultivar by PBR differs from protection by patents in that it allows protected cultivar material to be freely used in further research and development (breeder's exemption) and to be reproduced by farmers for their own use (farmer's privilege).

The Cultivated Plant Code (Trehane *et al.* 1995), on the other hand, is not concerned with trademarks (Article 7, Note 2), but provides guidelines for the choice and protection of new cultivar names (Trehane *et al.* 1995, Appendix VIII Quick guide for new cultivar names), but not for the protection of cultivar material which is covered by patents, including breeder's rights. Suggestions for procedures to differentiate between trademarks and cultivar names were made by Gioia (1995), however some form of registration is required in both cases. The implication of "registration" in trademarks is different from that in cultivars, and they are subject to different legislation.

In all cases, whether plant material is registered for patenting or whether a name is registered under PBR as a cultivar, DUS requirements apply, but in the former, patent criteria must be met, i.e. novelty must be demonstrated. Cultivar identification is of crucial importance for protection, registration, patenting and certification. With the development of techniques more refined and technically sophisticated than visual inspection, the amount of variation detected may be prone to abuse especially in the determination of novelty (Smith *et al.* 1991) and thus there is need for standardization.

Techniques
"DNA fingerprinting" was established by Jeffreys *et al.* (1985) for the detection of variable DNA loci by hybridisation of specific multilocus probes to electrophoretically separated restriction fragments for identifying human individuals. With the invention of the Polymerase Chain Reaction (PCR) (Saiki *et al.* 1985), several methods were developed to detect DNA polymorphisms, commonly known as "DNA fingerprinting", "DNA profiling" or "DNA typing".

A variety of techniques is now available for DNA fingerprinting of cultivars. The most recent include random amplified polymorphic DNA markers (RAPD), (Welsh & McClelland 1990, Williams *et al.* 1990), amplified fragment length polymorphisms (AFLP), (Zabeau & Vos 1993), cleaved amplified polymorphic sequences (CAPS), (Konieczny & Ausubel 1993), inter-repeat amplifications (IRA), (Sinnett *et al.* 1990) and simple sequence repeat polymorphisms or microsatellites (SSR), (Tautz 1989, Weber & May 1989). The polymorphisms resulting from AFLP and Restriction Fragment Length Polymorphisms (RFLP), a somewhat older technique, detect restriction size variation from a subset of the variation of the genome. The DNA profiles from RAPD, i.e. bands visualized on gels, result from differences in primer binding sites and DNA length (nucleotide base pairs) between the binding sites, obtained from PCR amplification. AFLPs also depend on binding sites and result from PCR amplification. SSR polymorphisms are detected with PCR using pairs of primers flanking simple sequence repeats of 2–4 units. RFLP is now rarely used for cultivar identification as it is time consuming and labour intensive (Powell *et al.* 1996). In CAPS, specific primers from a partially known sequence of interest are used to amplify it in different individuals, followed by restricting this sequence to identify RFLPs among the individuals. Sequence characterized amplified regions (SCARs), (Paran & Michelmore 1993) can be derived from the DNA sequence of specific RAPD markers. Primers for SCARs are usually longer (24mer) than those for RAPDs (usually 10mer), and are thus more specific. Other DNA fingerprinting techniques are variations or extensions of those described. Some, such as variable number of tandem repeats (VNTR), are especially useful for fingerprinting individuals. VNTRs are based on highly variable copies of very short repeats occurring in tandem in the genome, apparently characteristic in each individual. Cultivars, even though many are homozygous with respect to traditional DUS criteria, are often populations of individuals that may exhibit intra-cultivar DNA polymorphisms; this may depend on the species although it is also found among clones. The goal of fingerprinting is to concentrate on DNA polymorphisms among cultivars.

Example: An Identification Scheme for Six-rowed Barley Cultivars of Canada

When work on the identification scheme for barley using DNA fingerprinting began in 1995, there were 65 registered six-rowed barley cultivars in Canada. RAPD assays were carried out on breeder's seed material of single seedlings and subsequently with 30 replicate single seeds for each cultivar (Baum *et al.* 1998). An automated computerized identification key based on the diagnostic bands found was generated. A small portion of this key is shown here.

	Band	Cultivar
15(14).	GEN911.1 absent	29. Mingo
	GEN911.1 present	42. OAC Elmira
16(14).	OPAA04.1 absent	41. OAC Acton
	OPAA04.1 present	17
17(16).	GEN874.1 absent	61. Excel
	GEN874.1 present	30. Noble

SCARs were developed from the diagnostic DNA bands. The bands were sub-cloned and sequenced, and primer pairs were designed for each diagnostic band. The primer pairs are 24–26 bp long and very specific, and thus less prone to the problems with repeatability that some workers have noted with RAPDs (e.g. Devos & Gale 1992). From these SCARs a new computerized identification key was generated (Baum & Mechanda in prep.). The resulting key is not identical to the key based on RAPDs for molecular reasons not to be discussed here. A small portion of this key is shown below, with the names given to the primer pairs identical to the RAPD primers with which the RAPD diagnostic bands originated.

	Band	Cultivar
42(40).	OPAB11 absent	32. OAC Kippen
	OPAB11 present	61. Excel
43(39).	OP-G03 absent	44
	OP-G03 present	46
44(43).	GEN911 absent	65. AC Nadia
	GEN911 present	45

We have accomplished a similar identification scheme for the 53 two-rowed barley cultivars in Canada (Penner *et al.* 1998). Thus, all 128 barley cultivars can now be identified at the breeder's seed level by DNA fingerprints. The advantage of this scheme is that it allows accurate identification within 24 hours from single seeds or material from small leaf discs taken from a single plant. We have not yet investigated the performance of this key commercially. Impurities and admixtures may be expected at this level, and other possible unknown factors that could contribute some polymorphism. Should that be the case, then the concept of minimum distance and the definition of boundary of minimum distance (Smith & Smith 1992) could be applied to determine the level of tolerance of cultivar identity, or other statistical parameters might be sought. Although there are still some technological barriers to overcome, such as faster determination and quicker methods to determine transgenes, the technology to automate DNA-based identification can now be routinely used at the point of grain handling.

To determine novelty, i.e. new cultivar status of material or a new trademark, a small difference in DNA banding pattern is currently insufficient since novelty is still determined largely on morphological traits (DUS) and agronomic performance. DNA fingerprinting assists only in determining differences for characterizing identity of new cultivar status or trademark. DNA identity, such as fingerprinting, is certainly more objective for supplying proof of distinctness, uniformity and stability. For uniformity and stability of cultivars (or trademarks for that matter), criteria at the DNA level need to be investigated and assessed for the application of DUS for IP protection at the commercial level, if necessary using statistical means.

Acknowledgments

I thank the following colleagues for making beneficial comments on earlier drafts of this paper: Ms. Lisa James, Marketing and Development (ECORC); Ms. Louise Duke, Variety Registration Office and Ms. Valerie Sisson, Plant Breeders Rights Office, both of the Canadian Food Inspection Agency, Nepean, Ontario, Canada. I am grateful to Mrs. Joy Morrow, Intellectual Property firm of Smart & Biggar, Ottawa, Ontario, Canada, for reading a draft of the manuscript and for clarifying a number of points. Mr. A.B. Ednie, Associate Director Laboratory Services Division, Canadian Food Inspection Agency, Ottawa, made useful comments on the manuscript, and provided financial support for the work on barley cultivars. The Alberta Barley Commission, Calgary, Alberta, which provided a three-year research grant in support of the Canadian barley cultivar identification work in the Agriculture and Agri-Food Matching Industry Initiative scheme, is acknowledged.

References

American Intellectual Property Law Association. (1995). An Overview of Intellectual Property. What is a Patent, a Trademark, and a Copyright? http://www.aipla.org/html/whatis.html.

AOSA. (1991). Cultivar purity testing handbook. Association of Official Seed Analysts. 78 pp.

Baum, B.R. (1972). Material for an international oat register. 266 pp. Information Canada Cat. # A52-4772. Ottawa.

Baum, B.R., Bailey, L.G. & Thompson, B.K. (1985). Barley register. Publication 1783/B Supply and Services Canada, Cat. # A53-1783/1985.

Baum, B.R., Mechanda, S., Penner, G.A. & Ednie, A.B. (1998). Establishment of a scheme for the identification of Canadian barley (Hordeum vulgare L.) six-row cultivars using RAPD diagnostic bands. Seed Sci. Techn. 26: 449–462.

Devos, K.M. & Gale, M.D. (1992). The use of random amplified polymorphic DNA markers in wheat. Theor. Appl. Genet. 84(5–6): 567–572.

Elekes, P. (1980). The nature and improvement of the phenol reaction of wheat seeds. 19th ISTA Congress Preprint No 13-S.V.

Fedak, G. (1974). Allozymes as aids to Canadian barley cultivar identification. *Euphytica* 23: 166–173.

Gebre, H., Khan, H. & Foster, A.E. (1986). Barley cultivar identification by polyacrylamide gel electrophoresis of hordein proteins: Catalogue of cultivars. *Crop Sci. (Madison)* 26: 454–460.

Gioia, V.G. (1995). Using and registering plant names as trademarks. *Acta Hort.* 413: 19–25.

Higgins, J., Evans, J.L. & Law, J.R. (1988). A revised classification and description of faba beans cultivars (*Vicia faba* L.). *Pl. Var. Seeds* 1: 27–35.

ISTA. (1996). International rules for seed testing, Rules 1996. *Seed Sci. Techn.* 24: 1–335. Supplement.

Jeffreys, A.J., Wilson, V. & Thein, S.L. (1985). Hypervariable 'minisatellite' regions in human DNA. *Nature.* 314: 67–73.

Konieczny, A. & Ausubel, F.M. (1993). A procedure for mapping *Arabidopsis* mutations using co-dominant ecotype-specific PCR-based markers. *Pl. J.* 4: 403–410.

OECD. (1967). OECD scheme for the varietal certification of cereal seed moving in international trade. 35 pp. OECD.

Paran, I. & Michelmore, R.W. (1993). Development of reliable PCR-based markers linked to downy mildew resistance genes in lettuce. *Theor. Appl. Genet.* 85(8): 985–993.

Parnell, A. (1983). The identification of new wheat varieties using a standard electrophoresis method. *J. Natl. Inst. Agric. Bot.* 16: 183–188.

Penner, G.A., Zheng, Y. & Baum, B.R. (in press). Identification of DNA fingerprints capable of differentiating all two-row barley cultivars registered in Canada. *Canad. J. Pl. Sci.*

Powell, W., Morgante, M., Chaz, A., Hanafey, M., Vogel, J., Tingey, S. & Rafalski, A. (1996). The comparison of RFLP, RAPD, AFLP and SSR (microsatellite) markers for germplasm analysis. *Molec. Breed.* 2: 225–238.

Saiki, R.K., Scharf, S., Faloona, F., Mullis, K.B., Horn, G.T., Erlich, H.A. & Arnheim, N. (1985). Enzymatic amplification of ß-globin genomic sequences and restriction site analysis for diagnostic sickle cell anemia. *Science* 230: 1350–1354.

Sinnett, D., Deragon, J.M., Simard, L.R. & Labuda, D. (1990). Alumorphs - Human DNA polymorphisms detected by polymerase chain reaction using Alu-specific primers. *Genomics* 7: 331–334.

Smith, J.S.C. & Smith, O.S. (1992). Fingerprinting crop varieties. *Advances Agron.* 47: 85–140.

Smith, J.S.C., Smith, O.S., Bowen, S.L., Tenborg, R.A., & Walls, S.J. (1991). The description and assessment of distances between inbred lines of maize. III. A revised scheme for the testing of distinctiveness between inbred lines utilizing DNA RFLPs. *Maydica* 36: 213–226.

Tautz, D. (1989). Hypervariability of simple sequences as a general source for polymorphic DNA markers. *Nucl. Acids Res.* 17: 6463–6471.

Trehane, P., Brickell, C.D., Baum, B.R., Hetterscheid, W.L.A., Leslie, A.C., McNeill, J., Spongberg, S.A. & Vrugtman, F. (eds). (1995). The international code of nomenclature for cultivated plants — 1995. 175 pp. Quarterjack Publishing, Wimborne, U.K.

Troyer, A.F. (1986). United States Patent; Inbred corn line. Patent No. 4,594,810. US Patent Office, Washington, D.C.

Weber, J. & May, P.E. (1989). Abundant class of human DNA polymorphisms which can be typed using the polymerase chain reaction. *Amer. J. Human Genet.* 44: 388–396.

Welsh, J. & McClelland, M. (1990). Fingerprinting genomes using PCR with arbitrary primers. *Nucl. Acids Res.* 18: 7213–7218.

Williams, J.G.K., Kubelik, A.R., Livak, K.J., Rafalski, J.A. & Tingey, S.V. (1990). DNA polymorphisms amplified by arbitrary primers are useful as genetic markers. *Nucl. Acids Res.* 18: 6531–6535.

Zabeau, M. & Vos, P. (1993). Selective restriction fragment amplification: a general method for DNA fingerprinting. European Patent Application 92402629.7.

Bachmann, K., Blattner, F. & Dehmer, K. (1999). Molecular markers for characterisation and identification of genebank holdings. In: S. Andrews, A.C. Leslie and C. Alexander (Editors). Taxonomy of Cultivated Plants: Third International Symposium, pp. 239–252. Royal Botanic Gardens, Kew.

MOLECULAR MARKERS FOR CHARACTERISATION AND IDENTIFICATION OF GENEBANK HOLDINGS

KONRAD BACHMANN, FRANK BLATTNER AND KLAUS DEHMER

Institut für Pflanzengenetik und Kulturpflanzenforschung, IPK, Corrensstraße 3, D-06466 Gatersleben, Germany

Abstract

Genebanks (*ex situ* collections of plant genetic resources) are the repositories of the dwindling genetic variability of cultivated plants. Such collections are maintained as a resource for continued progress in plant breeding and their size makes it imperative to characterise the present holdings. This is necessary to ensure a balanced and representative sampling of global biodiversity in crop plants and to facilitate the introduction of this diversity into breeding programmes. Molecular marker methods permit large-scale surveys of genetic variability, but require standardisation of the experimental protocols. We suggest that standard methods adopted by all genebanks would greatly increase the efficiency of *ex situ* conservation world-wide and at the same time provide a reliable cumulative database for crop-plant taxonomy. Some of the many technical problems involved in such a standardisation are discussed beginning with the choice of a marker system. The development of crop-specific sets of PCR primers derived from different methods for the detection of DNA sequence polymorphism is recommended. Such sets could provide data for the analysis of variation from individual genotypes up to the taxonomic categories that link crops to their wild relatives. The need to standardise methods of data treatment and acquisition is emphasised. Plastid DNA phylogenies are recommended as basic data for plant taxonomy.

Introduction

Many domesticated plants have been selected for centuries or even millennia and their original ranges have spread far from their centres of origin. During this process they have accumulated modified genepools enriched for agriculturally relevant alleles in addition to the natural genepools of their wild relatives. The value of these genepools as a resource for plant breeding has been recognised since the end of the last century. At that time, international collection missions and the first *ex situ* germplasm collections were started in Russia (Vavilov 1992) and in the USA. A conference of the International Biological Programme in 1967 (Frankel & Bennett 1970) was instrumental in stimulating world-wide recognition of the value of genetic resources. Today, there are about 1300 genebanks world-wide (FAO 1996). Even in the face of rapid genetic erosion in crop species (Hammer *et al.* 1996), there is still a virtually unlimited number of genetically different populations in many species. As the facilities available for storage and regeneration begin filling up, there is an increasing need for quantitative and qualitative assessment of the diversity present and its relationship to the total diversity available. An ongoing programme of collection and maintenance depends on a clear idea of what is to be maintained and for what purpose. With this in mind, evaluation and characterisation of collections becomes increasingly important, and newly collected material should be screened to determine if it needs to be maintained.

Comparative data on agriculturally relevant traits are of primary importance to the potential users of germplasm collections. Such information will greatly enhance the value of the stored material. However, evaluation for agriculturally relevant characters cannot supply the data needed for rapid decisions on the long-term maintenance of samples.

(1) Evaluation is costly and time-consuming and usually involves raising plants in the field during a normal growing season, if possible in multiple replicates under different climatic conditions.

(2) Each relevant character should be studied in a separate trial. There are many varied characters of interest to breeders, and it is possible that in the future new characters may become important that cannot be foreseen.

(3) Direct evaluation can reveal only characters that are phenotypically expressed. Since recombination between accessions can generate character qualities that exceed the values of either parent, breeding successes can only partially be predicted from the parental phenotypes. For approaches such as hybrid breeding (Zhang *et al.* 1996) or recombinant backcross breeding (Tanksley & McCouch 1997), the overall genetic difference between the parents of a cross may be more important than the visible parental traits.

The Advantages of Molecular Markers

These considerations argue for the use of molecular marker techniques for the characterisation of germplasm collections (Phippen *et al.* 1997). At present this means detection of DNA sequence polymorphisms throughout the nuclear and organelle genomes of plants. The major advantages of this method are the following:

• Screening for genomic polymorphisms reveals heritable characters, i.e. characters susceptible to selection, even if there is no immediate or direct relationship between the polymorphisms detected and specific traits. Sequence polymorphisms are landmarks spread throughout the genome and therefore indicators for the genetic diversity of a sample of plants.

• Whereas many useful traits are sensitive to the environment, genomic polymorphisms can be determined in the laboratory at any time.

• Molecular screening techniques can be applied to small samples of material such as seeds or seedlings or small pieces of tissue obtained non-destructively from a plant.

• There are several basic types and endless variations of molecular marker protocols. However, the basic procedures are relatively independent of the material to be studied and can be generally applied. Molecular characterisation therefore does not require taxon-specific training and can be fully or partially automated.

• Sequence polymorphism data obtained by various methods in all kinds of organisms have a similar simple structure that is ideally suited to computer analysis. Nevertheless, it should be emphasised that automatic data processing does not obviate the necessity for intelligent interpretation of the results, and more important still, with such a wide variety of programmes for data-evaluation available, the choice of an appropriate one is crucial for extracting the relevant information from a data set.

- Any screening of sequence polymorphisms provides direct access for detailed molecular analysis. Even anonymous statistical surveys such as RAPDs (random amplified polymorphic DNA) or AFLPs (amplified fragment length polymorphism) can be followed up with a precise investigation of individual markers at the level of the DNA sequence.

With these advantages, it is not surprising that molecular methods for the characterisation of germplasm have rapidly become established, and the literature on the subject is growing steadily. With rare exceptions, the published studies deal with samples of less than 100 plants and address specific questions. Considering that there are between 100,000 and 1,000,000 accessions of some major crop species in the genebanks of the world (FAO 1996), and about 10,000 accessions of species in a single resource collection (Bachmann 1997b), approaches on a completely different scale are needed. Large-scale screening of germplasm collections requires methods that are reliable and reproducible, but also puts a premium on the efficiency of the method in terms of time and costs. In addition, the protocols should be structured so that different technicians working at different times can obtain data that are directly comparable to produce a cumulative dataset.

Efforts to establish routine molecular screening procedures are underway in several institutions. Published examples of larger surveys include studies of the allelic variation at four microsatellite loci in 207 accessions of wild and cultivated barley (Saghai Maroof *et al.* 1994), at 10 microsatellite loci in 238 accessions of landraces and cultivars of rice (Yang *et al.* 1994) and the study of 700 soybean genotypes with 115 RFLP (restriction fragment length polymorphism) probes (Powell *et al.* 1996b). Here, a few points chosen on the basis of experience at IPK Gatersleben will be discussed. Those already using molecular techniques will be familiar with these questions and with the suggested ways of dealing with them on the basis of presently available methods. Two main points will be further considered:

(1) As far as feasible, the development of methods for large-scale screening should be co-ordinated among the various institutions using such methods (Jones *et al.* 1997). If the protocols were standardised, this would not only increase their efficiency, but also allow us to develop a world-wide cumulative database of the results. The added value of a global approach to the biodiversity of cultivated plants would be immense.
(2) It is possible that such a standardised approach could help considerably in dealing with problems in the taxonomy of cultivated plants. Users of taxonomic information want to be able to identify a plant and may also want to know its (phylogenetic) relationship with other plants. The output of standardised and automated molecular data may be the best way of providing this information.

These suggestions, even if accepted as desirable goals, will not be easy to implement. A number of technical questions arise when evaluating protocols for the production of standardised data. A few of these will now be discussed. By necessity this discussion will start with an assessment of the available methods. Molecular techniques for the analysis of biodiversity have evolved rapidly, and it is likely that this trend will continue for some time. It is thus unlikely that a set of protocols based on present technology will survive unchanged for more than a few years. However, most of the points raised here should be valid regardless of which protocols are chosen.

The Need for Crop-specific, Standardised PCR (Polymerase Chain Reaction) Marker Sets

The various molecular methods will not be described or discussed in detail. However, a short list of some key references may be useful.

- Amplified Fragment Length Polymorphisms (AFLP): Vos *et al.* (1995), Breyne (1997).
- Inter-simple sequence repeat polymorphisms (ISSR): Zietkiewicz *et al.* (1994); Kantety *et al.* (1995), Charters *et al.* (1996), Nagaoka & Ogihara (1997).
- Microsatellite RFLPs (Variable Number of Tandem Repeats. VNTRs): Bruford *et al.* (1992).
- Polymerase Chain Reaction (PCR): Saiki *et al.* (1995).
- Random Amplified Polymorphic DNA (RAPD): Welsh & McClelland (1990), Williams *et al.* (1990, 1993), Caetano-Anolles (1994).
- Restriction Fragment Polymorphisms (RFLP) of PCR amplified fragments: Tsumura *et al.*(1995), Mes *et al.* (1997).
- Sequence Tagged (single-locus) Microsatellites (STMS): Beckmann & Soller (1990), Morgante & Olivieri (1993), Weising *et al.* (1998).

Comparative discussions can be found in Bachmann (1994; 1997a), Morell *et al.* (1995), Powell *et al.* (1995a), Weising *et al.* (1995), Russell *et al.* (1997), Milbourne *et al.* (1997) and Jones *et al.* (1997).

The criteria on which these methods may be compared include speed, cost, reproducibility, information content of data, and range of applicability. As a rule, reliable and informative data require more expensive and technically demanding methods, whereas technically simple and inexpensive methods produce less reliable data. Single-locus microsatellite markers identify strictly homologous sites in the genome but require an expensive and time-consuming period of development for each crop. Though RAPDs and AFLPs require little preliminary work, the homology of bands is not certain and estimates of band identity will include a margin of error that increases with the genetic diversity of the sample. While this margin of error can be estimated and is often acceptable in local comparisons, it could become prohibitive in a standardised large-scale application. In this case, an investment in the establishment of specific and reliable methods, including automation, would be essential, the costs being offset by the eventual high level of use.

These considerations argue for single-locus microsatellites as the method of choice. However, there are two serious limitations to their application:–

- A set of microsatellite primers usually works for only one species and, for some loci, its nearest relatives (Kijas *et al.* 1994; Röder *et al.* 1995; Peil *et al.* 1998; Westman & Kresovich 1998).
- Whereas the homology of the loci detected by this method is virtually certain, there is no proof that repeats of the same length are identical by descent. Orti *et al.* (1997) have shown this by using nucleotide variation in sequences flanking a specific microsatellite to reconstruct its evolutionary history. In large samples and especially in a taxonomically wider range including a crop plant and its nearest relatives, these features of microsatellites may become limiting.

Comparisons using RAPDs and AFLPs frequently show markers that are constant within species (Wachira *et al.* 1997) or even genera, in addition to many bands that are informative at any taxonomic level below the species. However, conclusions based on individual RAPD or AFLP bands require proof of their identity. The bands can be isolated and used as hybridisation probes (Rieseberg 1996, Roelofs & Bachmann 1997a & b) or they can be cloned and sequenced to assure band identity. The great advantage of universally applicable methods such as RAPDs or AFLPs is the ease with which a very large number of markers can be screened to detect those that are potentially diagnostic for groups of related plants or for specific characters (Michelmore *et al.* 1991, Polley *et al.* 1997), even if the validity of these markers has to be carefully evaluated. Specific PCR primers for individual markers can then be designed as Sequence Characterized Amplified Regions or SCARs (Paran & Michelmore 1993). As with single-locus microsatellites, this needs an initial investment to obtain well-characterised specific markers. Based on current methods, we recommend the development of crop-specific sets of PCR primers derived from different approaches so that a limited number of amplification reactions for each plant will produce a data set with which the plant could be individually identified and at the same time classified within the relevant breeding genepool. Such a set of primers could also be used sequentially in order of increasing specificity, which might eliminate the need to test each sample against the entire set.

The use of markers derived from various approaches to genome screening is based not only on the wish to identify individual genotypes but also to classify them into taxonomic categories with one standardised method. The various marker methods preferentially scan different fractions of nuclear DNA having different rates and modes of evolution. RAPDs, for instance, seem to arise more often from repetitive DNA than AFLPs do (authors' unpublished results). Some striking differences between results obtained with different marker systems for the same material have been found by Sharma *et al.* (1996), Powell *et al.* (1996a), Parsons *et al.* (1997) and Russell *et al.* (1997).

Applying Molecular Markers to Genetic Resource Collections

Molecular markers have been employed for the following purposes:–

- Identification of plants, including descrimination between accessions (or genotypes) and assessing the degree of similarity of closely related accessions.
- Assessing the genetic changes associated with reproduction of genebank material.
- Assessing genetic variability within a sample of plants.
- Elucidating relationships among accessions as a basis for a stable system of classification.
- Reconstructing history in cultivation and geographical distribution.
- Assessing variability of specific relevant phenotypic characters (Worland et *al.* in press) or of their distribution via genetically linked markers (Michelmore *et al.* 1991, Paran & Michelmore 1993, Fahima *et al.* 1998, Korzun *et al.* in press).

For drawing up a collecting strategy for *ex situ* collections, these functions range in importance roughly in the order in which they have been listed. Whereas identification and reduction of redundancy in a collection is one of the first and easiest tasks, linking molecular variation to the characters relevant for the breeder requires considerable additional investment. Below we comment on some of the advantages and problems related to the co-ordinated undertaking of these tasks on a large scale.

Identifying Duplicates, Controlling Sample Identity and Increasing Collection Diversity

Duplicates are most easily identified in inbred lines or in clonally propagated plants. Zeven *et al.* (1998) tested a collection of 64 vegetatively propagated perennial kale (*Brassica oleracea* Acephala Group) accessions from the Benelux countries and France, maintained in Wageningen and at the IPK Genebank in Germany. By applying nine different RAPD primers they found seven different patterns, one being present in 37 of the 64 accessions. Data on cytology, plant colour, flowering behaviour and morphology allowed a further subdivision. Nevertheless, up to 16 accessions appeared to belong to one and the same clone. It should be emphasised that reliance on the available molecular data alone would not have picked up all the differences. It is therefore frequently asked how detailed a molecular analysis has to be before we can be sure that two accessions represent the same genotype.

For analysing large-scale collections, this question misses the point. The important question here is, to which degree the diversity present in the collection represents the diversity of the base sample, or even that of the entire population of the crop throughout the world. If the collection contains a large number of very similar accessions, while there is poor representation of other components of the available diversity, there is no need to search exhaustively for possible differences among the near-identical plants. Of course, many agriculturally important characters are based on alleles of single loci, and it is quite possible that useful differences are overlooked by the molecular screening procedure. The use of molecular markers is based on the supposition that diversity of phenotypic traits is correlated with diversity of molecular markers and that overall, a collection based on an even distribution of molecular variability also constitutes an optimal sample of available phenotypic variability. This is the statistical reasoning that underlies the selection of core collections (Hotchkin *et al.* 1995). Whereas core collections are limited samples representative of the total variation already deposited in genebanks, the total content of all genebank collections should be representative of the global variation of the species as far as still available. Estimating the size of this global variation, especially the part not yet adequately represented, should be an integral part of any collection strategy.

The use of molecular markers for the identification of duplicate accessions is illustrated by an ongoing study at IPK (Dehmer unpublished data). Eighty-three accessions of 34 named lettuce (*Lactuca sativa* L.) cultivars obtained from various sources have been analysed with anonymous markers. Lettuce is self-pollinating, and accessions are expected to be uniform inbred lines. Only nine of the 34 cultivars were found to be uniform, and in another 10 there was minor variation among the accessions. In four cultivars the variation was appreciable, and in no fewer than 11, accessions from different sources had little in common. It is likely that these large discrepancies are due to errors in labelling. This underscores one of the most important aspects of molecular data in plant identification: identifying markers are inherent in the plant, and there can be no errors due to misplaced labels. It is difficult to estimate how much *ex situ* material is mislabelled. Indications are that it may be more serious than we suspect, and this is very difficult to control or correct. Of course, genebanks try to minimise such errors. The genebank at IPK maintains reference collections of seeds, and herbarium specimens of all of its samples so that a detailed phenotypic comparison is possible (Bachmann 1997b). On an international scale, the deposition of genetic fingerprint patterns based on crop-specific standard PCR reactions might be easier to realise and could be as informative, especially when cross-referenced with relevant morphological information as text and/or image databanks.

Sampling and Data Evaluation

General agreement on a standard set of protocols to obtain data on identity and variation in crop plants will have to be accompanied by general agreement on procedures for sampling and data evaluation. There is possibly more variation in the methods of data processing than there is in the methods of data gathering, and it would be difficult to find two publications that have followed exactly the same procedure even when the final presentation of results looks very similar.

The problem of sampling is not limited to the number of accessions that are to be analysed. Outcrossing species in which there is genetic variation within each accession raise the question of how to sample this variation and thus to characterise the accession (Morell *et al.* 1995). There are several ways of dealing with such accessions. Data can be obtained from bulked samples (Virk *et al.* 1995, Kraft *et al.* 1997); if properly done this will reveal all the alleles in the sample as present or absent without regard to frequency. Alleles may however go undetected if not included in the bulked sample or if very infrequent (Dulson *et al.* 1998). Alternatively, individual plants may be sampled. This will provide information on allele frequency and combinations if the number of plants per accession is sufficiently large. An adequate population sample should consist of dozens of plants, and this will multiply the cost of the survey. There are arguments for and against bulk versus single-plant analysis. The choice depends very much on the type of marker system used (Kraft *et al.* 1997) and effect not only the reliability of the data, but also the method of data evaluation.

Del Rio *et al.* (1997a) have compared populations one generation apart and sister populations generated from a common source for the two North American potato species, *Solanum jamesii* Torr., a diploid (2n=24) outcrossing species, and *S. fendleri* A. Gray, an inbreeding disomic tetraploid (2n=48). DNA was extracted from 15–20 plants of each population, and the samples were pooled. The RAPD banding patterns from the pooled samples were then compared with the Sneath & Sokal (1973) matching coefficient. For *S. jamesii*, single individuals were also analysed and frequencies of RAPD alleles calculated according to Lynch & Milligan (1994). In general, minimal loss or change of genetic diversity due to seed increase was found. In contrast, a parallel study (Del Rio *et al.* 1997b) showed considerable differences between samples preserved in genebanks and new collections from the original wild populations.

This example indicates some of the many aspects of data evaluation that would require standardisation. Using a standardised set of crop-specific microsatellite and SCAR primers, doubts about the reproducibility of RAPDs and the method of estimating allele frequencies from RAPD patterns would be avoided. While such methods are acceptable in individual case studies (such as the one cited), they would introduce too much uncertainty into a programme co-ordinated among several laboratories. Aside from the choice of marker system, a comparative study such as the one described requires consideration of the following questions:

(1) Should bulk samples or individual plants be analysed?
(2) In bulk samples, should DNA from individual plants be bulked or can DNA be isolated from a bulked collection of plant material such as leaf cuttings?
(3) Should allele frequencies be calculated or merely the presence or absence of alleles be scored?
(4) In each case, how many plants per accession should be sampled?
(5) Which coefficient for genetic similarity or genetic difference should be used?
(6) Once all the similarities have been calculated, how should the data matrix be evaluated?

The importance of the first four points is rarely appreciated, and all too often, points (5) and (6) are considered only when the data are available.

All coefficients of genetic similarity or difference are based on the four possible relationships between alleles of two genotypes when the presence of an allele (or band, scored as "1") is compared with its absence ("0"). These are a (1,1), b (1,0), c (0,1) and d (0,0). There are at least 10 coefficients that are used more or less regularly (Weising *et al.* 1998). For instance, the simple matching coefficient (Sneath & Sokal 1973) cited above is the fraction of common character states (absent or present in both):

$$S= (a+d)/(a+b+c+d)$$

In contrast the frequently cited Jaccard coefficient assigns no informative value to similarity of absence (d) and is defined as:

$$S= a/(a+b+c)$$

Whatever the reasons for the choice of coefficient, quantitative values based on different coefficients cannot be directly compared. It is important to remember that the quantitative value for "genetic similarity" depends not only on the kind of marker used but also on the definition of the similarity coefficient. The fact that such coefficients can be calculated very accurately is no proof that they describe the overall relatedness of the plants with similar accuracy. In fact it can be occasionally shown that they do not (Manninen & Nissila 1997).

There are also about 10 different methods for matrix evaluation in current use. Two frequently used means of analysing data are principal component analysis (PCA) and similarity dendrograms. PCA summarises the genetic distances among accessions in a two- or three-dimensional graph and provides a simple visual representation of the even or clustered distribution of diversity. Similarity dendrograms are hierarchical trees of similarity among the various accessions. An obvious danger of similarity dendrograms is the suggestion that they represent a family tree of the accessions. Under certain conditions, similarity dendrograms of molecular data may indeed represent the most probable reconstruction of the evolutionary history of the sample. Unfortunately, the literature abounds with cases in which such a congruence is at least implicitly suggested even though quite unwarranted. The most obvious cases are those in which diploid or allopolyploid hybrids are present in the sample. Representation of data by PCA or similar "neutral" ways of illustrating similarity avoids false interpretation of data. These methods, however, depend on the variability in the sample and the graphic representation can change profoundly when more accessions are added and new components of diversity (geographical or taxonomic) are added to the sample. Apart from the final evaluation, a standardised method of data acquisition would at least guarantee comparable similarity indices and permit the cumulative assembly of large data matrices.

Molecular Markers and the Taxonomy of Cultivated Plants

It is obvious that the aim of the calculations discussed above is more or less the same as a taxonomic analysis of the material. The common aim is the identification of accessions, the determination of their genetic similarities and the identification of more or less well-circumscribed clusters of similar accessions. In crop plants especially, both approaches also face the same problems: neither (infraspecific) taxonomic categories nor similarity clusters of marker patterns are necessarily reflections of phylogenetic relationship. They may also represent similar character constellations that have arisen independently. Also, the family history of cultivated plant accessions at species level and below is likely to involve so much reticulate evolution through (artificial) hybridisation and introgression that it cannot possibly be reflected in a nested hierarchy. In this respect, cultivated plants are an extreme illustration of the broad and fuzzy transition from panmictic populations (with randomised allele associations from a universally shared genepool) to genetically isolated genepools that evolve by lineage splitting. The transition from recombination within a shared genepool to evolution of genetically separate lines underlies practically all definitions of the species as a taxonomic category. In crop species, genetically isolated taxa below species level are common and the evolution of crops typically involves gene exchange among related species. As a result, not even the species is a solid basic category, and the taxonomy of cultivated plants must be open to flexible, non-hierarchical and non-exclusive groupings. Marker methods can provide some urgently needed objective criteria to identify and characterise useful and biologically reasonable groupings.

This could be approached from two directions:

- A search for diagnostic markers could provide simple and unequivocal definitions of taxa already described.
- Analysis of molecular datasets could indicate the most natural way to define taxa.

In practice, a combination of both methods will be the most efficient approach. A search for diagnostic markers will implicitly also be a test of a proposed taxon. If it is a natural group based on common ancestry and a shared genepool, diagnostic markers will be found in a survey of RAPDs or AFLPs. Occasionally, molecular marker surveys reveal infrasubspecific taxa as ad hoc assemblies of unrelated plants sharing one or very few striking or important characters. In that case, molecular markers may indicate an alternative taxonomic treatment that reflects overall genetic relationship.

Molecular markers provide the simplest data to determine how precisely circumscribed a taxon is. Surprisingly there has apparently not been a single study of the fuzziness of taxon boundaries in higher plants, i.e. the degree of congruence of diagnostic characters (Van Regenmortel 1997). The identification and description of "good" taxa is still mostly a matter of educated guesswork. It might be very illuminating to obtain precise quantitative data for some selected taxa. Of course, such data would accumulate as a by-product of the large-scale characterisation of genebank holdings.

Most of the taxonomy of cultivated plants takes place in the difficult transition zone between free gene exchange and complete genetic isolation, in taxonomic terms from the individual to the genus. To connect the man-made and largely non-hierarchical taxa of cultivated plants to the traditional taxonomy of wild species, it would be necessary to refer cultivated plant taxa to the nearest clades among their natural relatives, i.e. to identify the inclusive group that contains the genepool that can or has been used in breeding the cultivated taxa. This could be anything from a subspecies or variety from which cultivars have been derived to a set of related genera that have contributed genes to the breeding of a crop. Here again, molecular data are the simplest tools. Above, we suggest including SCAR markers for infraspecific taxa or species in the crop-specific set of PCR primers to provide a constant background for the markers that identify individual genotypes. Such markers will also provide a connection with wild plant taxonomy.

So far, we have almost exclusively considered markers for the nuclear genome. Since nuclear polymorphisms are markers for traits as well as for relationship, they are of primary importance for resource characterisation. However, there are important reasons why polymorphisms in the plastid genome (usually chloroplast DNA , cpDNA) should also be considered. Variation in plastid DNA plays a crucial role for two reasons: (1) Plastid DNA in general evolves very slowly (Wolfe *et al.* 1987) and is therefore informative at higher taxonomic levels, and (2) cpDNA is usually transmitted uniparentally, in most crops through the maternal line of descent, and almost without recombination (Clegg & Zurawski 1992). Plastid DNA haplotypes therefore evolve in a strictly clonal way by accumulating mutations. The algorithms of cladistic analysis are applicable to chloroplasts at any taxonomic level at which there are sequence polymorphisms. When reticulate relationships among diploid genotypes become as complex as they are in many crops, a statistically reliable phylogeny of plastids can provide one solid line of evidence for the reconstruction of their relationships and a clear link between crop taxonomy and the taxonomy of their wild relatives.

The identification of chloroplast haplotype variation among closely related species or even within species requires an exploration of the limits of cpDNA variation. There are microsatellite loci in cpDNA and these can be informative for resource characterisation (Powell *et al.* 1995b). A relatively quick and general way to detect cpDNA variation makes use of the observation that there are several relatively variable intergenic spacers between conserved coding sequences so that primers for the flanking coding sequences can be used to amplify variable spacer DNAs from a wide variety of plants (Taberlet *et al.* 1991). The only variation in cpDNA of tea (*Camellia sinensis* (L.) Kuntze) and related species was found in one of four such fragments (Wachira *et al.* 1997). Amplification products can be characterised by the restriction fragments generated with a series of restriction enzymes. Mes *et al.* (1997) illustrate this method for the subgeneric classification of *Allium* L. and give details of the amplification of fragments adding up to about 20 kb of cpDNA.

Conclusion

Now that molecular methods for the identification and classification of accessions in genebanks are available, the time has come for a standardised and generally agreed set of protocols that would result in world-wide characterisation of resource variation and better use of resource collections. We have listed and discussed some of the major theoretical and technical problems that will have to be dealt with if such an undertaking is to be carried out. The number of problems to which there seems to be no single best solution may suggest that a general standardisation of methods, however desirable, is unlikely to be a realistic goal. This is not what we want to suggest. On the contrary, such a project is likely to stimulate and provide a general focus for research on cultivated plant taxonomy and the conservation of genetic resources, to generate documentation for the maintenance of resource collections and thus added value for research and collections.

References

Bachmann, K. (1994). Molecular markers in plant ecology. *New Phytol.* 126(3): 403–418.

Bachmann, K. (1997a). Nuclear DNA markers in plant biosystematic research. *Opera Bot.* 132: 137–148.

Bachmann, K. (1997b). The genebank and genetic resources at the IPK, Gatersleben, Germany. *Pl. Var. Seeds* 10: 173–184.

Beckmann, J.S. & Soller, M. (1990). Towards a unified approach to the genetic mapping of eukaryotes based on sequence-tagged microsatellite sites. *Bio/Technology* 8: 930–932.

Breyne, P., Boerjan, W., Gerats, T., Montagu, M. van & Gysel, A. van (1997). Applications of AFLP in plant breeding, molecular biology and genetics. *Belg. J. Bot.* 129: 107–117.

Bruford, M.W., Hanotte O., Brookfield, J.F.Y. & Burke, T. (1992). Single locus and multi locus DNA fingerprinting. In A.R. Hoelzel (ed.). Molecular Genetic Analysis of Populations. pp. 225–269. Oxford University Press, New York.

Caetano-Anolles, G. (1994). MAAP – a versatile and universal tool for genome analysis. *Pl. Molec. Biol.* 25: 1011–1026.

Charters, Y.M., Robertson, A., Wilkinson, M.J. & Ramsay, G. (1996). PCR analysis of oilseed rape cultivars (*Brassica napus* L. ssp. *oleifera*) using 5'-anchored single-sequence repeat (SSR) primers. *Theor. Appl. Genet.* 92(3–4): 442–447.

Clegg, M.T. & Zurawski, G. (1992). Chloroplast DNA and the study of plant phylogeny. In P.S. Soltis, D.E. Soltis & J.J. Doyle (eds). Molecular Systematics of Plants. pp. 1–13. Chapman and Hall, New York.

Del Rio, A.H., Bamberg, J.B. & Huaman, Z. (1997a). Assessing changes in the genetic diversity of potato gene banks. 1. Effect of seed increase. *Theor. Appl. Genet.* 95(1–2): 191–198.

Del Rio, A.H., Bamberg, J.B., Huaman, Z., Salas, A. & Vega, S.E. (1997b). Assessing changes in the genetic diversity of potato gene banks. 2. *In situ* vs. *ex situ*. *Theor. Appl. Genet.* 95(1–2): 199–204.

Dulson, J., Kott, L.S. & Ripley, V.L. (1998). Efficacy of bulked DNA samples for RAPD DNA fingerprinting of genetically complex *Brassica napus* cultivars. *Euphytica* 102: 65–70.

Fahima, T., Röder, M.S., Grama, A. & Nevo, E. (1998). Microsatellite DNA polymorphism divergence in *Triticum dicoccoides* accessions highly resistant to yellow rust. *Theor. Appl. Genet.* 96(2): 187–195.

FAO, (1996). The state of the world's plant genetic resources for food and agriculture. 336 pp. FAO, Rome.

Frankel, O. & Bennett, E. (eds). (1970). Genetic resources in plants – their exploration and conservation. 544 pp. IBP Handbook N. 11. Oxford.

Hammer, K., Knüpffer, H., Xhuveli, L. & Perrino, P. (1996). Estimating genetic erosion in landraces – two case studies. *Genet. Resources Crop Evol.* 43: 329–336.

Hotchkin, T., Brown, A.H.D., Hintum, T.H.J.L. van & Morales, E.A.V. (eds). (1995). Core collections of plant genetic resources. 269 pp. J. Wiley and Sons, Chichester.

Jones, C.J., Edwards, K.J., Castiglione, S., Winfield, M.O., Sala, F., Wiel, C. van de, Bredemeijer, G., Vosman, B., Matthes, M., Daly, A., Brettschneider, R., Bettini, P., Buiatti, M., Maestri, E., Malcevschi, A., Marmiroli, N., Aert, R., Volkaert, G., Rueda, J., Linacero, R., Vazquez, A. & Karp, A. (1997). Reproducibility testing of RAPD, AFLP and SSR markers in plants by a network of European laboratories. *Molec. Breed.* 3: 381–390.

Kantety, R.V., Zeng, X., Bennetzen, J.L. & Zehr, B.E. (1995). Assessment of genetic diversity in dent and popcorn (*Zea mays* L.) inbred lines using inter-simple sequence repeat (ISSR) amplification. *Molec. Breed.* 1: 365–373.

Kijas, J.M.H., Fowler, J.C.S. & Thomas, M.R. (1994). An evaluation of sequence tagged microsatellite site markers for genetic analysis within *Citrus* and related species. *Genome* 38(2): 349–355.

Korzun, V., Röder, M.S., Ganal, M.W., Worland, A.J. & Law, C.N. (in press). Genetic analysis of the dwarfing gene (*Rht8*) in wheat. I. Molecular mapping of the *Rht8* locus on the short arm of chromosome 2D of bread wheat (*Triticum aestivum* L.). *Theor. Appl. Genet.*

Kraft, T., Fridlund, B., Hjerdin, A., Säll, T., Tuvesson, S. & Halldén, Ch. (1997). Estimating genetic variation in sugar beets and wildbeets using pools of individuals. *Genome* 40(4): 527–533.

Lynch, M. & Milligan, B.G. (1994). Analysis of population genetic structure with RAPD markers. *Molec. Ecol.* 3: 91–99.

Manninen, O. & Nissila, E. (1997). Genetic diversity among Finnish six-rowed barley cultivars based on pedigree information and DNA markers. *Hereditas* 126: 87–93.

Mes, T.H.M., Friesen, N., Fritsch, R.M., Klaas, M. & Bachmann K. (1997). Criteria for sampling in *Allium* based on chloroplast DNA PCR-RFLP'S. *Syst. Bot.* 22(4): 701–712.

Michelmore, R.W., Paran, I. & Kesseli, R.V. (1991). Identification of markers linked to disease resistance genes by bulked segregant analysis: a rapid method to detect markers in specific genomic regions by using segregating populations. *Proc. Natl. Acad. Sci. U.S.A.* 88: 9828–9832.

Milbourne, D., Meyer, R., Bradshaw, J.E., Baird, E., Bonar, N. Provan, J., Powell. W. & Waugh, R. (1997). Comparison of PCR-based marker systems for the analysis of genetic relationships in cultivated potato. *Molec. Breed.* 3: 127–136.

Morgante, M. & Olivieri, A.M. (1993). PCR-amplified microsatellites as markers in plant genetics. *Pl. J.* 3: 175–182.

Morell, M.K., Peakall, R., Appels, R., Preston, L.R. & Loyd. H.L. (1995). DNA profiling techniques for plant variety identification. *Austral. J. Exp. Agric.* 35: 807–819.

Nagaoka, T. & Ogihara, Y. (1997). Applicability of inter-simple sequence repeat polymorphisms in wheat for use as DNA markers in comparison to RFLP and RAPD markers. *Theor. Appl. Genet.* 94(5): 597–602.

Orti, G., Pearse, D.E. & Avise, J.C. (1997). Phylogenetic assessment of length variation at a microsatellite locus. *Proc. Natl. Acad. Sci. U.S.A.* 94: 10745–10749.

Paran, I. & Michelmore, R.W. (1993). Identification of reliable PCR-based markers linked to downy mildew resistance genes in lettuce. *Theor. Appl. Genet.* 85(8): 985–993.

Parsons, B.J., Newbury, H.J., Jackson, M.T. & Ford-Lloyd, B.V. (1997). Contrasting genetic diversity relationships are revealed in rice (*Oryza sativa* L.) using different marker types. *Molec. Breed.* 3: 115–125.

Peil, A., Korzun, V., Schubert, I., Schumann, E., Weber, W.E. & Röder, M.S. (1998). The application of wheat microsatellites to disomic *Triticum aestivum-Aegilops markgrafii* addition lines. *Theor. Appl. Genet.* 96(1): 138–146.

Phippen, W.B., Kresovich, S., Candelas, F.G. & McFerson, J.R. (1997). Molecular characterisation can quantify and partition variation among genebank holdings: a case study with phenotypically similar accessions of *Brassica oleracea* var. *capitata* L. (cabbage) 'Golden Acre'. *Theor. Appl. Genet.* 94(2): 227–234.

Polley, A., Seigner, E. & Ganal, M.W. (1997). Identification of sex in hop (*Humulus lupulus*) using molecular markers. *Genome* 40(3): 357–361.

Powell, W., Orozco-Castillo, C., Chalmers, K.J., Provan, J. & Waugh R. (1995a). Polymerase chain reaction-based assays for the characterisation of plant genetic resources. *Electrophoresis* 16: 1726–1730.

Powell, W., Morgante, M., Andre, C., McNicol, J.W., Machray, G.C., Doyle, J.J., Tingey S.V. & Rafalski, J.A. (1995b). Hypervariable microsatellites provide a general source of polymorphic DNA markers for the chloroplast genome. *Curr. Biol.* 5: 1023–1029.

Powell, W., Morgante, M., Andre, C., Hanafey, M., Vogel, J., Tingey, S. & Rafalski, A. (1996a). The comparison of RFLP, RAPD, AFLP and SSR (microsatellite) markers for germplasm analysis. *Molec. Breed.* 2: 225–238.

Powell, W., Morgante, M., Doyle, J.J., McNicol, J.W., Tingey, S.V. & Rafalski, A.J. (1996b). Genepool variation in genus *Glycine* subgenus *Soja* revealed by polymorphic nuclear and chloroplast microsatellites. *Genetics* 144: 793–803.

Regenmortel, M.H.V. van (1997). Viral species. In M.F. Claridge, H.A., Dawah & M.R. Wilson (eds). Species. The Units of Biodiversity. pp. 17–24. Chapman & Hall, London.

Rieseberg, L.H. (1996). Homology among RAPD fragments in interspecific comparisons. *Molec. Ecol.* 5: 99–105.

Röder, M.S., Plaschke, J., König, S.U., Börner, A., Sorrells, M.E., Tanksley, S.D. & Ganal, M.W. (1995). Abundance, variability and chromosomal location of microsatellites in wheat. *Molec. Gen. Genet.* 146: 327–333.

Roelofs, D. & Bachmann, K. (1997a). Comparison of chloroplast and nuclear phylogeny in the annual *Microseris douglasii* (*Asteraceae, Lactuceae*). *Pl. Syst. Evol.* 204: 49–63.

Roelofs, D. & Bachmann, K. (1997b). Genetic analysis of a *Microseris douglasii* (*Asteraceae*) population polymorphic for an alien chloroplast type. *Pl. Syst. Evol.* 206: 273–284.

Russell, J.R., Fuller, J.D., Macaulay, M., Hatz, B.G., Jahoor, A., Powell, W. & Waugh, R. (1997). Direct comparison of levels of genetic variation among barley accessions detected by RFLPs, AFLPs, SSRs and RAPDs. *Theor. Appl. Genet.* 95(4): 714–722.

Saghai Maroof, M.A., Biyashev, R.M., Yang, G.P., Zhang, Q. & Allard, R.W. (1994). Extraordinarily polymorphic microsatellite DNA in barley: species diversity, chromosomal locations, and population dynamics. *Proc. Natl. Acad. Sci. U.S.A.* 91: 5466–5470.

Saiki, R.K., Scharf, S., Faloona, F., Mullis, K.B., Horn, G.T., Erlich, H.A. & Arnheim, N. (1985). Enzymatic amplification of b-globin genomic sequences and restriction site analysis for diagnosis of sickle cell anemia. *Science* 230: 1350–1354.

Sharma, S.K., Knox, M.R. & Ellis, T.H.N. (1996). AFLP analysis of the diversity and phylogeny of *Lens* and its comparison with RAPD analysis. *Theor. Appl. Genet.* 93(5–6): 751–758.

Sneath, P.H.A. & Sokal, R.R. (1973) Numerical taxonomy. 573 pp. Freeman, San Francisco.

Taberlet, P., Gielly, L. & Bouvet, J. (1991). Universal primers for the amplification of three non-coding regions of chloroplast DNA. *Pl. Molec. Biol.* 17: 1105–1109.

Tanksley, S.D. & McCouch, S. (1997). Seed banks and molecular maps: unlocking genetic potential from the wild. *Science* 277: 1063–1066.

Tsumura, Y., Yoshimura, K., Tomaru, N. & Ohba, K. (1995). Molecular phylogeny of conifers using RFLP analysis of PCR-amplified specific chloroplast genes. *Theor. Appl. Genet.* 91(8): 1222–1236.

Vavilov, N.I. (1992). Origin and geography of cultivated plants. 498 pp. Cambridge University Press, Cambridge, UK.

Virk, P.S., Ford-Loyd, B., Jackson, M. & Newbury, H.J. (1995). Use of RAPD for the study of diversity in germplasm collections. *Heredity* 74: 170–179.

Vos, P., Hogers, R., Bleeker, M., Reijans, M., Lee, T. van der, Hornes, M., Freijters, A., Pot, J., Peleman, J., Kuiper, M. & Zabeau, M. (1995). AFLP: a new technique for DNA fingerprinting. *Nucl. Acids Res.* 23: 4407–4414.

Wachira, F.N., Powell, W. & Waugh, R. (1997). An assessment of genetic diversity among *Camellia sinensis* L. (cultivated tea) and its wild relatives based on randomly amplified polymorphic DNA and organelle-specific STS. *Heredity* 78: 603–611.

Weising, K., Nybom, H., Wolff, K. & Meyer, W. (1995). DNA fingerprinting in plants and fungi. 322 pp. CRC Press, Boca Raton.

Weising, K., Winter, P., Hüttel, B. & Kahl, G. (1998). Microsatellite markers for molecular breeding. *J. Crop Prod.* 1: 113–143.

Welsh, S.L. & McClelland, M. (1990). Fingerprinting genomes using PCR with arbitrary primers. *Nucl. Acids Res.* 18: 7213–7218.

Westman, A.L. & Kresovich, S. (1998). The potential for cross-taxa simple-sequence repeat (SSR) amplification between *Arabidopsis thaliana* L. and crop brassica. *Theor. Appl. Genet.* 96(2): 272–281.

Williams, J.G.K., Kubelik, A.R., Livak, K.J., Rafalski, J.A. & Tingey, S.V. (1990). DNA polymorphisms amplified by arbitrary primers are useful as genetic markers. *Nucl. Acids Res.* 18: 6531–6535.

Williams, J.G.K., Hanafey, M.K., Rafalski, J.A. & Tingey, S.V. (1993). Genetic analysis using random amplified polymorphic DNA markers. *Meth. Enzymol.* 218: 704–740.

Wolfe, K.H., Li, W.-H. & Sharp, P. (1987). Rates of nucleotide substitution vary greatly among plant mitochondria, chloroplast, and nuclear DNAs. *Proc. Natl. Acad. Sci. U.S.A.* 84: 9054–9058.

Worland, A.J., Korzun, V., Röder, M.S., Ganal, M.W. & Law, C.N. (in press). Genetic analysis of the dwarfing gene (*Rht8*) in wheat. II. The distribution and adaptive significance of allelic variants at the *Rht8* locus of wheat as revealed by microsatellite screening. *Theor. Appl. Genet.*

Yang, G.P., Saghai Maroof, M.A., Xu, C.G., Zhang, Q. & Biyashev, R.M. (1994). Comparative analysis of microsatellite DNA polymorphism in landraces and cultivars of rice. *Molec. Gen. Genet.* 245: 187–194.

Zeven, A.C., Dehmer, K.J., Gladis, T., Hammer, K. & Lux, H. (1998). Are the duplicates of perennial kale (*Brassica oleracea* L. var. *ramosa* DC.) true duplicates as determined by RAPD analysis? *Genet. Resources Crop. Evol.* 45: 105–111.

Zhang, Q., Zhou, Z.Q., Yang, G.P., Xu, C.G., Liu, K.D. & Saghai Maroof, M.A. (1996). Molecular marker heterozygosity and hybrid performance in *indica* and *japonica* rice. *Theor. Appl. Genet.* 93(8): 1218–1224.

Zietkiewicz, E., Rafalski, A. & Labuda, D. (1994). Genome fingerprinting by simple sequence repeat (SSR)- anchored polymerase chain reaction amplification. *Genomics* 20(2): 176–183.

Möller, M. & Cronk, Q.C.B. (1999). New approaches to the systematics of *Saintpaulia* and *Streptocarpus*. In: S. Andrews, A.C. Leslie and C. Alexander (Editors). Taxonomy of Cultivated Plants: Third International Symposium, pp. 253–264. Royal Botanic Gardens, Kew.

NEW APPROACHES TO THE SYSTEMATICS OF *SAINTPAULIA* AND *STREPTOCARPUS*

MICHAEL MÖLLER[1] AND QUENTIN C.B. CRONK[2]

[1]**Royal Botanic Garden Edinburgh, 20A Inverleith Row, Edinburgh EH3 5LR, UK**
[2]**Institute of Cell and Molecular Biology, The University of Edinburgh, Kings Buildings, Mayfield Road, Edinburgh EH9 3JH, UK**

Abstract

Molecular phylogenies of the genus *Streptocarpus* based on the ribosomal DNA internal transcribed spacer (ITS) sequences show that the genus is paraphyletic; other African genera *Saintpaulia*, *Schizoboea* and *Linnaeopsis* are nested within *Streptocarpus*. The genus *Saintpaulia* has evolved from a caulescent African *Streptocarpus* species in subgenus *Streptocarpella*. This has been confirmed by sequence phylogenies of three additional genes; the chloroplast *trnL* (UAA) intron and the spacer between the *trnL* (UAA) 3' exon and *trnF* (GAA), the chloroplast *ndhF* gene and the putative single copy developmental gene *GCYC*. The ITS phylogenies of *Saintpaulia* reflects its biogeographic distribution. A group of Usambara species, the '*ionantha*-complex' show minimal ITS genetic differentiation. The *Streptocarpus* ITS phylogeny is also congruent with the two base chromosome numbers and reflects the existing subgeneric division, with a few phylogenetically interesting exceptions.

Introduction

The flowering plant genera *Saintpaulia* H. Wendl. and *Streptocarpus* Lindl. (*Gesneriaceae* Dumort., subfamily *Cyrtandroideae* Endl., tribe *Didymocarpeae* Endl.) are important horticultural plants with a multi-million pound trade-value world-wide. Although numerous in households, many species are endangered in the wild (Walter & Gillett 1998), and hybrid cultivars of *Saintpaulia* are more common on windowsills than some species are in their natural habitat. *Saintpaulia teitensis* B.L. Burtt, for instance, only occurs as a few hundred specimens in one population in the Teita Hills in Kenya. Despite being well known in horticulture, surprisingly little was understood until recently about the evolutionary relationships between *Saintpaulia* and *Streptocarpus*, and among species within these two genera.

In 1958 B.L. Burtt recognised 19 species of *Saintpaulia* based on morphology, particularly the leaf hairs (Fig. 1) of living and herbarium material. However, *S. amaniensis* E.P. Roberts was later included in *S. magungensis* E.P. Roberts as a subspecies (Burtt 1964), and re-examination of existing and new material resulted in an increase of the number of species to 20 with the recognition of *S. brevipilosa* B.L. Burtt and *S. rupicola* B.L. Burtt (Burtt 1964). The whole genus has a very small area of distribution and occurs only in southern Kenya and northeastern parts of Tanzania (Fig. 2). Since the early work little taxonomic revision has been carried out and the pioneering treatment by Burtt is still the standard work on *Saintpaulia* systematics.

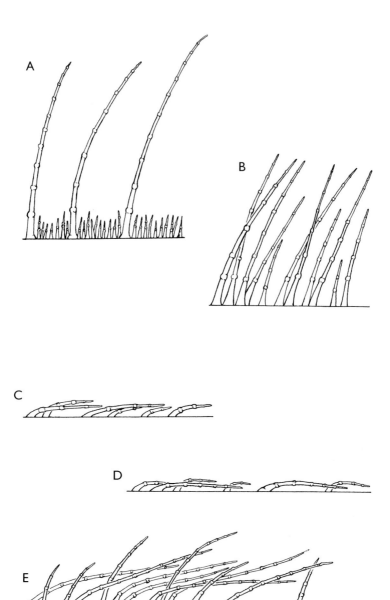

FIG. 1. Indumentum on the upper leaf surface of *Saintpaulia* species. A) *S. diplotricha* B.L. Burtt long erect + short erect. B) *S. ionantha* H. Wendl. long erect. C) *S. nitida* B.L. Burtt short appressed. D) *S. orbicularis* B.L. Burtt long appressed + short appressed. E) *S. intermedia* B.L. Burtt long appressed (modified after Burtt 1958).

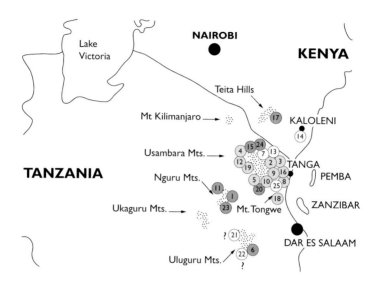

FIG. 2. Guide to the geographical distribution of *Saintpaulia* species. The position of each circle is an indication of the (very approximate) general distribution of the species represented by the number. ⬤ – species belonging to the '*ionantha*-complex'. Species are numbered as in Fig. 5; in addition no. 21 = *Saintpaulia inconspicua* B.L. Burtt, no. 22 = *S. pusilla* Engl.; *S. goetzeana* Engl. and *S. inconspicua* also occur in the Nguru Mts. and, the latter further south in the Ukaguru Mts.

The genus *Streptocarpus* is distributed in tropical and southern Africa (87 species) and Madagascar and the Comoro Islands (41 species). Another four are described from Asia; these are thought not to be closely related to the African or Madagascan taxa (Hilliard & Burtt 1971). The species are currently divided into two subgenera, mainly on vegetative morphological characters (Hilliard & Burtt 1971). Subgenus *Streptocarpella* Fritsch (44 species) contains mainly caulescent herbs with 'normal' shoot development, but a few unusual types occur in Madagascar, such as taxa with long-petioled leaves in a basal rosette (reminiscent of African *Saintpaulia*), and shrubby, woody caulescent species. Subgenus *Streptocarpus* Fritsch (88 species) comprises taxa without a shoot apical meristem, and form (in the extreme case) plants with a single, sometimes hugely expanded, cotyledon as the only aerial vegetative organ (Fig. 3B) Inflorescences are formed at the base of the lamina or on a shoot-like petiole, the 'petiolode'. Lamina and petiolode together form the 'phyllomorph'. Re-iteration of the phyllomorphic growth from a meristem on the petiolode results in rosulate forms (Fig. 3C, D) (Jong 1973, 1978). Rosette-type cultivars grown for commercial purposes belong to this type. Since Hilliard and Burtt's monograph (1971), based on morphology, the genus has not attracted much systematic study. Phylogenetic work based on morphological characters alone may be hampered, or at least be incomplete when comparing unifoliate with caulescent forms, as the former lack true leaves. However, the African members of tribe *Didymocarpeae* are relatively uniform when compared with the very diverse morphology seen in Asian members. For this reason it has been suggested that the African genera of *Didymocarpeae* are more closely related to each other than to Asian taxa (Hilliard & Burtt 1971).

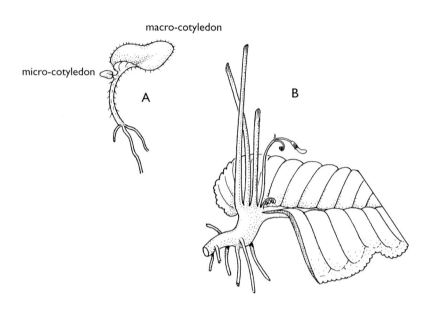

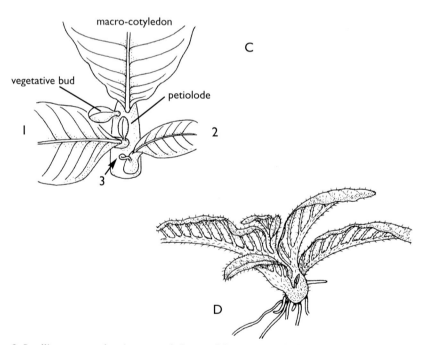

FIG. 3. Seedling stage and various growth forms of *Streptocarpus*. A) Seedling showing micro- and macro-cotyledon. B) *S. wendlandii* Spreng., a unifoliate form. C) Schematic sketch of a 'typical' rosulate, showing the formation of successive phyllomorphs (1, 2 & 3) on the petiolode of the macro-cotyledon. D) *S. primulifolius* Gand., a rosulate form (modified after Hilliard & Burtt, 1971 and Jong 1978).

The present work summarises the results of recent molecular work at the Royal Botanic Garden Edinburgh on African taxa of *Gesneriaceae*, with particular reference to *Streptocarpus* and *Saintpaulia*, complemented by newer findings. The genes chosen for various aspects of this research involve the chloroplast *trnL* (UAA) intron and the spacer between the *trnL* (UAA)3' and *trnF* (GAA) exons (Fig. 4C), the internal transcribed spacer (ITS) region of nuclear ribosomal DNA (Fig. 4B), and a putative *Gesneriaceae* homologue of the *Antirrhinum* L. single copy developmental gene *CYCLOIDEA* (Fig. 4A), which we have called *GESNERIACEAE CYCLOIDEA* (*GCYC*) (Möller *et al.* 1999) involved in the expression of flower symmetry (Cronk & Möller 1997). Previous work has also used the chloroplast gene *ndhF* (Smith *et al.* 1998). The chloroplast DNA (cpDNA) sequences are relatively conserved and thus suitable for phylogenetic reconstruction at genus level. In contrast, ITS sequences are about 5 times faster evolving than cpDNA in *Gesneriaceae*, and are suitable for resolution at the species level. The nuclear developmental gene *GCYC* has an intermediate substitution rate about three times faster than the chloroplast intron/spacer region (Möller *et al.* 1999).

Phylogenetic Relationships between *Saintpaulia* and *Streptocarpus*

A. The Phylogenetic Origin of *Saintpaulia*

Initially, *Saintpaulia* species were included in the molecular work to serve as potential outgroup taxa for studies in *Streptocarpus*. However, the ITS phylogenies indicated that *Streptocarpus* is not monophyletic, but rather that a monophyletic *Saintpaulia* is nested within a paraphyletic *Streptocarpus* (Fig. 6) (Möller & Cronk 1997a). The results clearly showed that *Saintpaulia* has evolved from within the genus *Streptocarpus* subgenus *Streptocarpella*. This supports Hilliard and Burtt's (1971) suggestion of a close relationship among African genera of *Gesneriaceae*, tribe *Didymocarpeae*.

The apparent evolution of *Saintpaulia* from within *Streptocarpus* was surprising, and provoked the study of additional genes to test the validity of the phylogeny based on the multicopy nuclear ribosomal DNA sequences. However, all additional phylogenies, inferred from *trnL+F* cpDNA sequences (Fig. 4C) and *GCYC* nucleotides and amino acid data (Fig. 4Ac) showed the same *Streptocarpus–Saintpaulia* relationships as those inferred from ITS data (Möller *et al.* 1999); in all cases the sister group of *Saintpaulia* comprises caulescent *Streptocarpus* species. This has also been confirmed independently in a cladistic analysis using sequences from another cpDNA gene, *ndhF* (Smith *et al.* 1998). The close relationship between *Saintpaulia* and *Streptocarpus* subgenus *Streptocarpella* is strengthened by shared morphological and cytological features; both have verruculose seeds and the same base chromosome number (x=15), while the majority of species in subgenus *Streptocarpus* have reticulate seeds and a base chromosome number of x=16 (Ratter 1975).

The major differences between typical *Streptocarpus* subgenus *Streptocarpella* and *Saintpaulia* are a reduction of the aerial stem (*Saintpaulia* mainly being rosette herbs) and the absence of a marked corolla tube in the latter. These features may be related to habitat, as many chasmophytes (e.g. *Saxifraga* L.) growing on wet cliffs (the main habitat of many *Saintpaulia* species) are typically rosette plants. The absence of a marked corolla tube may be associated with the loss of specialist long-tongued pollinators and a switch to generalist short-tongued insects (Cronk & Möller 1997). The two large exserted yellow anthers of *Saintpaulia* species also imply that they evolved pollen flowers rather than retaining the nectar flowers of *Streptocarpus*, further supporting a pollinator switch theory. A third significant difference between

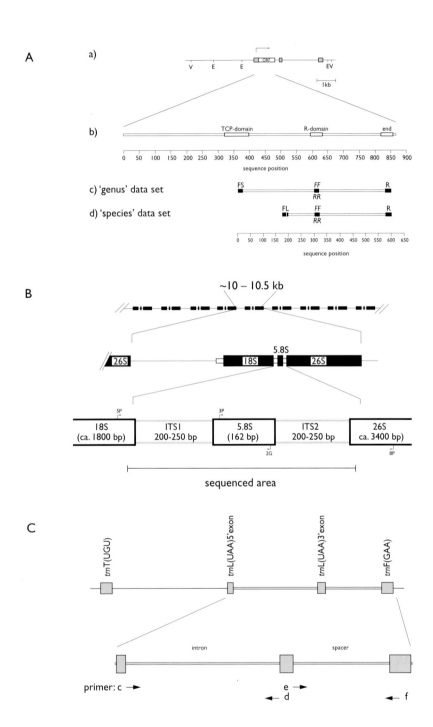

Saintpaulia and *Streptocarpus* is the absence of a twisted fruit in *Saintpaulia*. It is the main generic character of *Streptocarpus*, but is also found in other *Gesneriaceae* taxa, such as *Boea* Juss., *Dichiloboea* Stapf, *Ornithoboea* C.B. Clarke, *Paraboea* Ridl., *Rhabdothamnopsis* Hemsl. and *Trisepalum* C.B. Clarke. It is interesting that fruits of *Streptocarpus capuronii* Humbert and *S. tanala* Humbert have only a very slight or no twist, indicating that this trait is relatively easily lost; this has apparently happened several times independently in *Streptocarpus*, as indicated by an extended ITS phylogenetic analysis (Möller & Cronk in prep.) of the position of *Schizoboea* B.L. Burtt and *Linnaeopsis* Engl., two other genera with straight fruits having apparently evolved from within *Streptocarpus*. The evolutionary function of the twisted fruit may be related to an extension of the period of seed dispersal (Hilliard & Burtt 1971).

B. Limitations of ITS for the Phylogenetic Reconstruction in *Saintpaulia*

The ITS of ribosomal DNA in *Saintpaulia* has evolved both by base substitutions and by insertion/deletion events, with a maximum sequence divergence of 15.8% between *S. goetzeana* Engl. and *S. nitida* B.L. Burtt (Möller & Cronk 1997b). However, *Saintpaulia* is unusual in that a large group (12 accessions in the '*ionantha*-complex', Figs. 5 & 2) was found to have sequence divergences too low for clear phylogenetic resolution, and in an extended ITS analysis several accessions had identical sequences (Fig. 5) (Möller & Cronk 1997b, Möller *et al.* in prep.). To resolve further the relationships between species within the '*ionantha*-complex', *GCYC* sequences (Fig. 4Ad) were obtained for 22 species and subspecies of *Saintpaulia*, representing all areas of geographical distribution. We have shown that the evolutionary mutation rate of *GCYC* is higher than in ITS of closely related species (Möller *et al.* 1999). However, despite the higher divergence rate for *GCYC* at low divergence level, the phylogenetic resolution found was not higher when compared to ITS. Other gene sequences and molecular techniques are currently being evaluated at the Royal Botanic Garden Edinburgh with the aim of further characterising and differentiating taxa within the '*ionantha*-complex'. These include DNA fingerprinting using interspersed short sequence repeats (inter SSRs; Tsumura *et al.* 1996); the results of these studies are so far inconclusive.

The Evolution of *Streptocarpus*

A. The Phylogenetic Relationships within *Streptocarpus*

The genus *Streptocarpus* is, at least in vegetative morphology, one of the most diverse in the plant kingdom. Phylogenetic analyses based on ITS sequences clearly separate the species analysed into the two subgenera; the unifoliate/rosulate group (subgenus *Streptocarpus*) and the caulescent forms from Africa and Madagascar (subgenus *Streptocarpella*) (Fig. 6). Moreover, the two growth forms within subgenus *Streptocarpus*, the unifoliate (*S. dunnii* Hook.f., *S. eylesii* S. Moore and *S. wittei* De Wild.), and the

FIG. 4. Structures and maps of the genes sequenced. A)a) *CYCLOIDEA* (*CYC*) locus. Exons and predicted open reading frame (ORF) are indicated in rectangles; the arrow indicates the direction of transcription; restriction enzyme sites: E – *Eco*RI, V – *Eco*RV, (modified after Luo *et al.* 1996); b) The *CYC* ORF and the location of conserved regions (open boxes). Fragment of *CYC* sequenced for c) the 'genus' and d) for the 'species' data set and the PCR primer positions (closed boxes). B) ITS of ribosomal DNA, illustrating primer position and regions sequenced. C) cpDNA, illustrating the *trnL* (UAA) intron and the spacer between the *trnL* (UAA) 3' exon and *trnF* (GAA) sequenced.

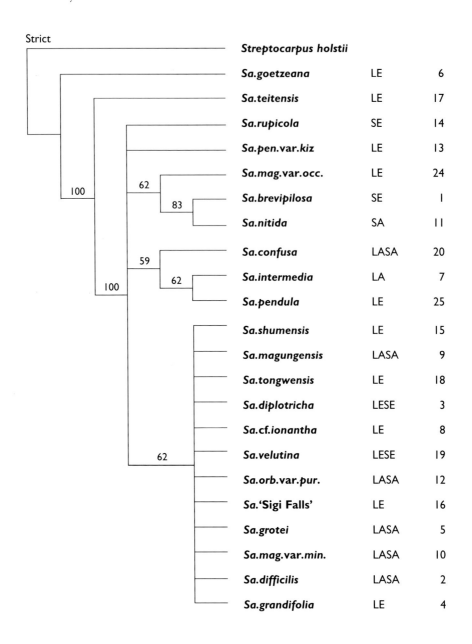

FIG. 5. Strict consensus tree of eight most parsimonious trees of 126 steps based on combined ITS 1 and ITS 2 sequence data (CI: 0.913; RC: 0.779), and the respective leaf hair type of 22 *Saintpaulia* accessions. Numbers above branches are heuristic bootstrap values of 1000 replicates. *Sa.* – *Saintpaulia*, *Sa.mag.*var.*occ.* – *Saintpaulia magungensis* E.P. Roberts var. *occidentalis* B.L. Burtt, *Sa.orb.*var.*pur.* – *Saintpaulia orbicularis* B.L. Burtt var. *purpurea* B.L. Burtt, *Sa.mag.*var.*min.* – *Saintpaulia magungensis* E.P. Roberts var. *minima* B.L. Burtt, *Sa.pen.*var.*kiz.* – *Saintpaulia pendula* B.L. Burtt. var. *kizarae* B.L. Burtt; L. – long hairs on upper surface of leaves, S – short, E – erect, A – appressed. Numbers beside hair types refer to the general distribution of the species in Fig. 2.

rosulate forms, were clearly separated into two distinct clades (Fig. 6). However, in an extended ITS analysis the subgeneric division is no longer clear cut (Möller & Cronk in prep.). Caulescent species such as *S. macropodus* B.L. Burtt, *S. papangae* Humbert and *S. schliebenii* Mansf. are placed in the subgenus *Streptocarpus* (acaulescent) clade in this analysis, which may indicate that the ancestor of unifoliate/rosulate taxa was caulescent. In our first ITS analysis (Fig. 6), the unifoliate taxa analysed were characterised by a large deletion in ITS2, but in the extended analysis a group of 17 taxa share this character (Möller & Cronk in prep.). This group is made up of unifoliate, plurifoliate and rosulate taxa. The rosulate habit appears in more than one clade, one of which consists of typical rosulate species of the *S. rexii* aggregate, including the most horticulturally important species. This indicates that the rosulate habit evolved independently several times.

Streptocarpus subgenus *Streptocarpella* includes 19 species in tropical mainland Africa, and 21 species from Madagascar. The molecular phylogenies indicate independent evolution of 'true rosette' growth forms (with long-petiolate leaves and axillary inflorescences) from caulescent ancestors in both Africa and Madagascar (*Saintpaulia* on the African mainland is one example). The 'true rosette' type from Madagascar, such as *Streptocarpus andohahelensis* Humbert and *S. beampingaratrensis* Humbert, (with vegetative morphology reminiscent of *Saintpaulia*), is the result of independent evolution from Madagascan caulescent species (Möller & Cronk in prep.). Fruit twisting is a diagnostic character of the genus *Streptocarpus*. This has resulted in the exclusion from *Streptocarpus* of African taxa which lack this character, including *Saintpaulia*. Molecular ITS phylogenies, however, indicate that at least two other genera have evolved from within *Streptocarpus* (Möller & Cronk in prep.). *Schizoboea*, originally part of *Didymocarpus* Engl., has straight fruits, but is otherwise very similar to caulescent *Streptocarpus* taxa (Burtt 1974). *Linnaeopsis* (another genus with straight fruits, but with flowers closely resembling those of rosulate *Streptocarpus* taxa) is also in this category.

B. The Cytology of *Streptocarpus*

The cytology of the genus is relatively uniform, subgenus *Streptocarpus* having $2n=32$ chromosomes while subgenus *Streptocarpella* has $2n=30$ (Ratter 1975, Jong & Möller in prep.). This is clearly reflected in the ITS topology (Fig. 6). The only exception appears to be confined to a clade of characteristically polyploid Madagascan species, with *S. perrieri* Humbert being tetraploid and *S. hildebrandtii* Vatke being octoploid. A third taxon belonging to this group (also characterised by branched veins ascending from the leaf base), *S. variabilis* Humbert, is hexaploid (Jong & Möller in prep.). The morphology different from other rosulate species and the tendency to form high polyploids, set this group apart from other *Streptocarpus* species. The close affinity of species within this group is also substantiated by molecular data (Möller & Cronk in prep.).

The fact that *S. papangae* has a base chromosome number of x=16 (typical of subgenus *Streptocarpus*) (Jong & Möller in prep.), but has caulescent morphology typical of subgenus *Streptocarpella* indicates that the caulescent growth form may be ancestral with regard to the subgenus *Streptocarpus*. This is reflected in the basal position of *S. papangae* in the clade containing most *Streptocarpus* in the extended ITS phylogeny (Möller & Cronk in prep.).

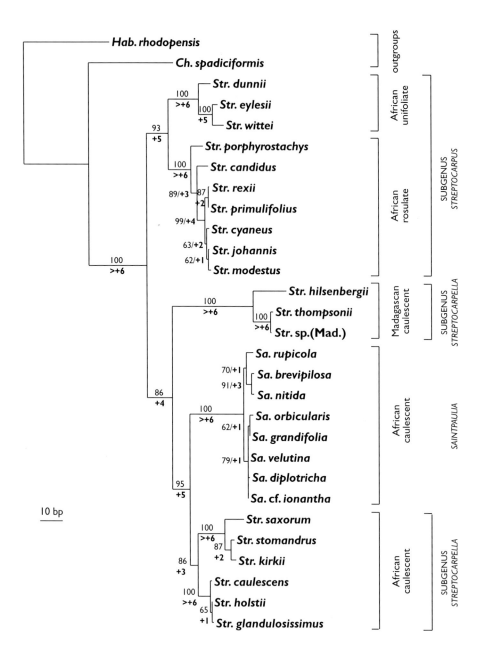

FIG. 6. Single most parsimonious tree for *Streptocarpus* and *Saintpaulia* of 419 steps length based on parsimony analysis of the combined ITS 1 and ITS 2 sequence data plus the alignment gap matrix (CI: 0.778; RC: 0.711). Upper numbers are boostrap values of 1000 replicates. Lower (boldface) numbers are decay indices (the numbers of steps necessary to cause collapse of monophyletic groups) modified after Möller & Cronk (1997a). Hab. = *Haberlea*, Ch. = *Chirita*.

Conclusions

The evolution of the genus *Saintpaulia* from *Streptocarpus* is now undisputed. The shared characters of chromosome number and seed coat structure combined with molecular evidence from four independent genes form a strong base for a close relationship between *Saintpaulia* and taxa belonging to *Streptocarpus* subgenus *Streptocarpella*. This might suggest that *Saintpaulia* species should be transferred to *Streptocarpus* or that *Streptocarpus* should be dismembered, if strictly monophyletic groups are considered desirable. However, the morphological differences between the two taxa are very clear and consistent. Furthermore, two other genera were found also to be nested within the genus *Streptocarpus*, which would then also need to be sunk into the genus *Streptocarpus*. This would make the diagnostic character of fruit twisting an unsuitable character for generic delimitation. Any revision of generic boundaries is premature, however, as further African and Madagascan genera of the tribe *Didymocarpeae* (*Acanthonema* Hook.f., *Trachystigma* C.B. Clarke, *Colpogyne* B.L. Burtt, *Hovanella* A. Weber & B.L. Burtt), which are possibly closely related to *Streptocarpus* have not yet been available for study, as they are rare in the wild and mostly not in cultivation. Until the relationships of these genera have been assessed by further morphological and molecular study, the *status quo* should be maintained.

Acknowledgements

The authors thank B.L. Burtt for stimulating discussions; H. Sluiman, J. Preston and C. Guihal for their technical support; U. Gregory, S. Scott, D. Mitchell and J. Main for the maintenance and expansion of the Edinburgh *Gesneriaceae* collection, and the Regius Keeper, and staff at the Royal Botanic Garden Edinburgh for research facilities. We thank A. Andrianjafy, G. Rafamontanantsoa and S. Irapanarivo from PBZT for logistic support during plant collections in Madagascar. We further thank the Institute of Cell and Molecular Biology, University of Edinburgh, for access to sequencing facilities and N. Preston for assistance. The receipt of a Leverhulme Trust Award, No. F/771/B, a research grant from the Systematics Association and the African Violet Society of America, and expedition funds from the Davis Expedition Fund of the University of Edinburgh, the Carnegie Trust for the Universities of Scotland and the Percy Sladen Memorial Fund, are gratefully acknowledged.

References

Bawa, K.S. (1994). Pollinators of tropical dioecious angiosperms: a reassessment? No, not yet. *Amer. J. Bot.* 81(4): 456–460.

Burtt, B.L. (1958). Studies in the *Gesneriaceae* of the Old World XV: The genus *Saintpaulia*. *Notes Roy. Bot. Gard. Edinburgh* 22(6): 547–568.

Burtt, B.L. (1964). Studies in the *Gesneriaceae* of the Old World XXV: Additional notes on *Saintpaulia*. *Notes Roy. Bot. Gard. Edinburgh* 25(3): 191–195.

Burtt, B.L. (1974). Studies in the *Gesneriaceae* of the Old World XXXVII: *Schizoboea*, the erstwhile African *Didymocarpus*. *Notes Roy. Bot. Gard. Edinburgh* 33(2): 265–267.

Cronk, Q.C.B. & Möller, M. (1997). Genetics of floral symmetry revealed. *Trends Ecol. Evol.* 12: 85–86.

Hilliard, O.M. & Burtt, B.L. (1971). *Streptocarpus*: an African plant study. 410 pp. University of Natal Press, Pietermaritzburg.

Jong, K. (1973). *Streptocarpus* (*Gesneriaceae*) and the phyllomorph concept (abstr.). *Acta Bot. Neerl.* 22(3): 244.

Jong, K. (1978). Phyllomorphic organisation in rosulate *Streptocarpus*. *Notes Roy. Bot. Gard. Edinburgh* 36(2): 369–396.

Luo, D., Carpenter, R., Vincent, C., Copsey, L. & Coen, E. (1996). Origin of floral asymmetry in *Antirrhinum. Nature*, 383: 794–799.

Möller, M. & Cronk, Q.C.B. (1997a). Origin and relationships of *Saintpaulia* H. Wendl. (*Gesneriaceae*) based on ribosomal DNA internal transcribed spacer (ITS) sequences. *Amer. J. Bot.* 84(7): 956–965.

Möller, M. & Cronk, Q.C.B. (1997b). Phylogeny and disjunct distribution: evolution of *Saintpaulia* (*Gesneriaceae*). *Proc. Roy. Soc. London, Series B, Biol. Sci.* 264: 1827–1836.

Möller, M., Clokie, M., Cubas, P. & Cronk, Q.C.B. (1999). Integrating molecular and developmental genetics: a *Gesneriaceae* case study. In P.M. Hollingsworth, R.M. Bateman & R.J. Gornall (eds). Molecular Systematics and Plant Evolution, pp. 375–402. Taylor & Francis, London.

Ratter, J.A. (1975). A survey of chromosome numbers in the *Gesneriaceae* of the Old World. *Notes Roy. Bot. Gard. Edinburgh* 33(3): 527–543.

Smith, J.F., Kresge, M.E., Möller, M. & Cronk, Q.C.B. (1998). A cladistic analysis of *ndhF* sequences from representative species of *Saintpaulia* and *Streptocarpus* sections *Streptocarpus* and *Streptocarpella* (*Gesneriaceae*). *Edinburgh J. Bot.* 55(1): 1–11.

Tsumura, Y., Ohba, K. & Strauss, S.H. (1996). Diversity and inheritance of inter-simple sequence repeat polymorphisms in Douglas-fir (*Pseudotsuga menziesii*) and sugi (*Cryptomeria japonica*). *Theor. Appl. Genet.* 92: 40–45.

Walter, K.S. & Gillett, H.J. (eds). (1998). 1997 IUCN Red list of threatened plants. 862 pp. IUCN, The World Conservation Union, Gland, Switzerland and Cambridge, UK.

Zhang, D., Dirr, M.A. & Price, R.A. (1999). Classification of cultivated *Cephalotaxus* species based on *rbcL* sequences. In: S. Andrews, A.C. Leslie and C. Alexander (Editors). Taxonomy of Cultivated Plants: Third International Symposium, pp. 265–275. Royal Botanic Gardens, Kew.

CLASSIFICATION OF CULTIVATED *CEPHALOTAXUS* SPECIES BASED ON *rbcL* SEQUENCES

DONGLIN ZHANG, MICHAEL A. DIRR AND ROBERT A. PRICE

Landscape Horticulture, University of Maine, Orono, ME 04469-5722, USA
Department of Horticulture, University of Georgia, Athens, GA 30602-7273, USA
Department of Botany, University of Georgia, Athens, GA 30602-7271, USA

Abstract

The nomenclature and taxonomy of *Cephalotaxus* in the nursery trade are frequently confused due to morphological similarities. Recently, molecular techniques especially DNA sequencing have been employed in taxonomic studies. This paper reports the sequencing of the chloroplast gene *rbcL*, and its use in determining relationship in 11 accessions of *Cephalotaxus* in cultivation. Seven 'species' in the commercial trade were examined and only three distinct sequences were found. The accessions named *C. harringtonii*, *C. drupacea*, *C. wilsoniana*, and *C. koreana* share an identical sequence and may represent a single species, for which the correct name is *C. harringtonii*. Material named *C. fortunei* and *C. sinensis* shares another identical sequence, and may represent a single complex species, though the latter may be a hybrid derivative. Accessions named *C. oliveri* have a third unique sequence, in keeping with its greater morphological distinctness. These sequence data, combined with morphological and horticultural characteristics, suggest that the seven species currently listed in the horticultural trade should be reduced to three, thus reducing the confusion in cultivated *Cephalotaxus* nomenclature. The use of *rbcL* sequencing in other morphologically homogeneous species of landscape plants should be considered.

Introduction

Cephalotaxus Siebold & Zucc. (plum yew) is a small genus of evergreen trees and shrubs native to eastern Asia. It is easily distinguished from *Taxus* L. (yew) by its naked ovules borne on small scales in compound cones. Only one or a few fleshy-coated seeds actually mature in each cone. The leaves are arranged in two ranks with two whitish bands on the lower leaf surface. In the most recent taxonomic revision, Fu (1984) recognized nine species, of which five are endemic to China, three are distributed in China and adjacent countries, and one, *C. harringtonii* (Knight ex J. Forbes) K. Koch, is native to Japan and Korea. At generic level and below, a variety of taxonomic interpretations have been published, resulting in numerous nomenclatural changes (Table 1). Since *Cephalotaxus* was established in 1842, 32 species and at least six varieties and one forma have been published (*Index Kewensis* 1895–   ). Even now there is no clear agreement on the delimitation of species. After critical revisions, Fu (1984) accepted nine species and one cultivar, while Silba (1984) accepted only four species and one variety but later added four more varieties (1990). Other species have been published by Lee (1935 & 1973), Li (1954), Hu (1964), Dallimore & Jackson (1967), Cheng & Fu (1978), Ohwi (1984) and Krüssmann (1985).

TABLE 1. Taxa of *Cephalotaxus* accepted by Fu (1984) and their habitat.

Taxon	Synonym and Basionym (names published by other authors)	Habitat
C. alpina (H.L. Li) L.K. Fu	*C. fortunei* Hook. var. *alpina* H.L. Li., *C. fortunei* Hook. var. *brevifolia* Dallim. & A.B. Jacks.	Endemic to China (northwestern Yunnan and southern Gansu); 2300–3700 m, mixed with broadleaf evergreen forests.
C. fortunei Hook.	*C. kaempferi* K. Koch, *C. fortunei* Hook. var. *concolor* Franch., *C. fortunei* Hook. var. *longifolia* Hort. ex Dallim & A.B. Jacks., *C. fortunei* Hook. var. *globosa* S.Y. Hu	Endemic to China (southern, eastern, western, and central China); 200–2000 m, mixed with broadleaf and needle-broadleaf mixed evergreen forests.
C. harringtonii (Knight ex J. Forbes) K. Koch	*Taxus harringtonii* Knight ex J. Forbes, *C. drupacea* Siebold & Zucc., *C. pedunculata* Siebold & Zucc., *C. drupacea* Siebold & Zucc. var. *pedunculata* (Siebold & Zucc.) Miq.	Korea and Japan; under 900 m, mixed with broadleaf forests.
C. lanceolata K.M. Feng	*C. fortunei* Hook. var. *lanceolata* (K.M. Feng) Silba	Endemic to China (only northwestern Yunnan); 1900 m, mixed with broadleaf evergreen forest.
C. latifolia (K.M. Feng) L.K. Fu	*C. sinensis* Rehder & E.H. Wilson var. *latifolia* K.M. Feng	Endemic to China (southern and western China); 900–2400 m, mixed forests.
C. mannii Hook.f.	*C. griffithii* Hook.f., *C. hainanensis* H.L. Li	China (west and south China), India, Burma, Thailand, Laos; 600–2900 m, mixed forests.
C. oliveri Mast.	None	China (central, southern, and western China) and Vietnam; 300–1800 m, mixed with broadleaf and needle evergreen forests.
C. sinensis Rehder & E.H. Wilson	*C. drupacea* Siebold & Zucc. var. *sinensis* Rehder & E.H. Wilson, *C. drupacea* Siebold & Zucc. var. *sinensis* (Rehder & E.H. Wilson f. *globosa* Rehder & E.H. Wilson, *C. harringtonii* (Knight ex J. Forbes) K. Koch var. *sinensis* (Rehder & E.H. Wilson) Rehder, *C. sinensis* Rehder & E.H. Wilson f. *globosa* (Rehder & E.H. Wilson) H.L. Li	Endemic to China (southern, eastern, western, and central China); 600–2300 m (–3200), mixed with evergreen broadleaf forests.
C. wilsoniana Hayata	*C. harringtonii* (Knight ex J. Forbes) K. Koch var. *wilsoniana* (Hayata) Kitam.	Endemic to China (Taiwan only); 1400–3000 m, mixed forests.

At species level, delimitation of taxa has been based primarily on the characters of the leaves, seeds, and cones. *Cephalotaxus oliveri* Mast. has distinctive leaf characters and is easy to recognise. Other species, such as *C. harringtonii*, *C. drupacea* Siebold & Zucc., *C. koreana* Nakai, and *C. wilsoniana* Hayata, are morphologically less distinct and are separated only on the basis of more or less constant, but not usually pronounced characters (Li 1954). When Li (1954) described *C. hainanensis* H.L. Li, he stated that it differed from *C. sinensis* Rehder & E.H. Wilson in being a shrub with shorter and broader leaves that did not taper at the apex, and that it differed from *C. drupacea* in having shorter, broader, and much thinner leaves with less conspicuous midribs. In general, differences between the species are slight; Dirr (1990) noted that species in the genus are hard to distinguish, especially *C. harringtonii* and *C. drupacea*. From a horticultural perspective also, there seems to be little difference between species and wide variation within them.

Micromorphological studies were conducted by Phillips (1941), Hu (1984), and Xi (1993) to delimit the species but with little success. Using pollen structure (Xi 1993) and leaf microstructure (Hu 1984), the authors concluded that only *C. oliveri* was easily distinguishable from the other putative species. Based on overall morphological distinctness, *C. oliveri* has been placed by itself in Section *Pectinata* L.K. Fu (Fu 1984), while the other eight putative species have been placed in Section *Cephalotaxus* (Fu 1984).

Cephalotaxus harringtonii was introduced to the U.S.A. as an ornamental shrub in 1830 (Dirr 1990). Fruitland Nursery (established in 1856) in Augusta, Georgia listed *C. harringtonii*, *C. drupacea*, and *C. harringtonii* 'Fastigiata' in a catalogue of the late 1800s; *C. fortunei* Hook. was added to the catalogue in 1900. Specimens of *C. fortunei* taller than 5 m (16.5 ft) occur in Augusta, Georgia and Aiken, South Carolina. Recently, *Cephalotaxus* has become popular because of several desirable attibutes including sun and shade tolerance (a good substitute for junipers), resistance to deer browsing (a problem in *Taxus*), tolerance to diseases or insects, and cold and heat adaptability in USDA Zones (5)6-9 (Dirr 1992). In 1994, *C. harringtonii* 'Prostrata', a low-growing form, was awarded a Gold Medal by the Georgia Green Industry Association (Harlass 1994). Unfortunately, *Cephalotaxus* nomenclature in the literature and nursery trade is highly confused due to morphological similarities and the complicated history of *Cephalotaxus* taxonomy. Plants given different names in horticulture often represent the same morphological taxon. Conversely, in the process of collecting material for this study, six accessions named *C. harringtonii* var. *drupacea* (Siebold & Zucc.) Koidz. were all morphologically different. It has been impossible clearly to distinguish most putative species of *Cephalotaxus* by morphological characters (Dirr 1990) and molecular techniques seemed worth employing.

Molecular data can be useful in horticultural taxonomic studies (Staub & Meglic 1993). Rose cultivars have been distinguished by nuclear restriction fragment length polymorphisms (RFLP) (Hubbard *et al.* 1992). Random amplified polymorphic DNA (RAPD) has been shown to be a reliable and easy method for identification of clonal red maples (Krahl *et al.* 1993) and American elms (Kamalay & Carey 1995). At species level and above, the chloroplast (cp) genome is well-suited to phylogenetic study because it is an abundant component of plant total DNA, thus facilitating extraction, amplification, and analysis (Clegg & Zurawski 1992). Its genes are present in only one distinct copy and its rate of nucleotide substitution is moderate (Palmer *et al.* 1988, Olmstead & Palmer 1994). From complete cpDNA sequences of *Pinus thunbergii* Parl. (Wakasugi *et al.* 1994), *Nicotiana tabacum* L. (Shinozaki *et al.* 1986), and *Oryza sativa* L. (Hiratsuka *et al.* 1989), comparisons of gene content, nucleotide substitution, and rate of gene evolution have been greatly facilitated. Taxonomists have used cpDNA sequences to help elucidate the phylogeny of genera and species. Xiang *et al.* (1993) successfully grouped the species of *Cornus* L. (*sensu stricto*) as a distinct clade and also separated each species based on *rbcL* data. The *rbcL* gene encodes for the large subunit of ribulose-1,5-biphosphate carboxylase/oxygenase (RUBISCO), the enzyme that catalyses CO_2 fixation during photosynthesis (Hallick & Bottomly 1983). It is a conservative and slow evolving gene. Brunsfeld *et al.* (1994) found that sequence evolution of *rbcL* in long-lived members of the *Taxodiaceae* Warm. and *Cupressaceae* Rich. ex Bartl. is extremely slow and the rate of silent nucleotide substitution is approximately 13 times slower than that estimated for short-lived monocotyledons. *Cephalotaxus* species are long-lived, with fossil distributions in western Europe, eastern North America, and eastern Asia dating from the Tertiary Period (Florin 1963). Nucleotide substitution of its *rbcL* gene is extremely slow and any base change should be a significant molecular character for species identification (Florin 1963). In this study, the sequence of the chloroplast gene *rbcL* has been used to determine relationships in cultivated material of *Cephalotaxus*.

Materials and Methods

Eleven accessions of cultivated *Cephalotaxus* (Table 2) were grown at the University of Georgia campus. Total genomic DNA was isolated from leaves following the acidic extraction protocol (modified from Guillemaut & Marechal-Drouard 1992). Details of the modifications and of reaction conditions for the subsequence PCR amplification of *rbcL* are available from the first author. The amplification primers used were *rbcL*5' and *rbcL*3' (Table 3).

TABLE 2. Plant source for *Cephalotaxus rbcL* gene.

Plant	Source
C. drupacea Siebold & Zucc.	R. Price, Univ. of California-Berkeley Botanical Garden, CA
C. fortunei Hook.	1846–80-A, Arnold Arboretum, Jamaica Plain, MA
C. koreana Nakai	1699–77-B, Arnold Arboretum, Jamaica Plain, MA.
C. harringtonii (Knight ex J. Forbes) K. Koch	1994–1497A, RBG Edinburgh, Scotland
C. harringtonii (Knight ex J. Forbes) K. Koch	800273, Univ. of Georgia Botanical Garden, Athens, GA
C. harringtonii (Knight ex J. Forbes) K. Koch	26–64, Arnold Arboretum, Jamaica Plain, MA
C. oliveri Mast.	D. Zhang 0233, Piroche Plants, Pitt Meadows, B.C., Canada
C. oliveri Mast.	AA 400–95, Arnold Arboretum, Jamaica Plain, MA (ex Nanjing, China)
C. sinensis Rehder & E.H. Wilson	1889–80, Arnold Arboretum, Jamaica Plain, MA
C. sinensis Rehder & E.H. Wilson	AA 777–94, Arnold Arboretum, Jamaica Plain, MA (ex Nanjing, China)
C. wilsoniana Hayata	1993–4074B, RBG Edinburgh, Scotland

TABLE 3. Primer sequences for *Cephalotaxus rbcL* gene amplification (PCR) and sequence.

Primer	Sequence	Source
rbcL 5'	TCA CCA CAA ACA GAA ACT AAA GC	Z-1, Gerard Zurawski
rbcL 3'	ATT TGA TCT CCT TCC ATA CTT CAC AAG	R. Price
A1010	GTA GGT AAA CTT GAA GG	R. Price
691R	CCC TTA ATT TCA CCC GTC TC	R. Price
1091R	AGA GAC CCA ATC TTG AGT GA	R. Price
1373R	ATT TGA TCT CCT TCC ATA CTT CAC AAG	R. Price

Five primers were used for sequencing each DNA sample:– *rbcL* 5', A1010, 691R, 1091R, and 1373R (Table 3), allowing the gene to be read from both strands for about 1000 bp out of 1347. To avoid PCR error and to confirm the accuracy of sequences within species, some were sequenced from two or three different accessions and *rbcL* sequences for each species were read at least twice (Table 2).

Results and Discussion

Three distinct *rbcL* sequences were found among the eleven accessions sampled. Group I consists of material labelled *C. harringtonii* (three accessions), *C. drupacea* (one accession), *C. koreana* (one accession), and *C. wilsoniana* (one accession). Group II contains *C. fortunei* (one accession) and *C. sinensis* (two accessions). Group III contains only *C. oliveri* (two accessions). (See Tables 2, 4 & 5). If we regard any base change of the *Cephalotaxus rbcL* gene as a good molecular character for species identification (Clegg & Zurawski 1992), there would be at least three potential species in the accessions sampled. Whether the accessions sharing a sequence represent a species or a group of species will be discussed in conjunction with morphological data.

A total of 11 nucleotide substitutions were found. Among them, six are synonymous changes and five are non-synonymous changes. Among groups, there is a six base-pairs (bp) difference between Group I and II (three synonymous and three non-synonymous), nine bp between Group I and III (five synonymous and four non-synonymous), and seven bp between Group II and III (four synonymous and three non-synonymous) (Tables 4 and 5). Since the synonymous changes do not affect amino acid composition, it is understandable that synonymous substitutions are more common than non-synonymous substitutions.

The number of nucleotide substitutions in the *rbcL* gene varies among plant taxa. The sequence divergence among ten species of *Cornus* is from five base substitutions (between *C. mas* L. and *C. officinalis* Siebold & Zucc.) to 52 base substitutions (*C. canadensis* L. from *C. mas* and *C. kousa* Hance) (Xiang *et al.* 1993). In gymnosperms, the rate of base substitutions for chloroplast genes, such as *rbcL*, is generally much slower. For example, *Thuja occidentalis* L. differs from *T. plicata* Donn ex D.Don by only two base substitutions (Genbank), and *Taxus × media* Rehder from *Amentotaxus argotaenia* (Hance) Pilg. (different genera) by only 43 base substitutions. Since three significantly different *rbcL* sequences were found in the accessions sampled here and no differences were detected within the species as labelled, it is likely that at least three species should be accepted.

TABLE 4. The *rbcL* sequence of *Cephalotaxus harringtonii*.

Sequence					Sequence Number
NNNNNNNNNN	NNNNNNNNNN	NNNNNNNAGT	GTCGGATTCA	AAGCTGGTGT	50
TAAAGATTAC	AGATTAACTT	ATTATACTCC	GGAATATAAG	ACCAAAGATA	100
CTGATATCTT	GGCAGCATTC	CGAGTAACTC	CTCAACCTGG	AGTGCCCCCT	150
GAGGAAGCAG	GAGCAGCAGT	AGCTGCCGAA	TCTTCCACTG	GTACATGGAC	200
TACTGTTTGG	ACCGATGGAC	TTACGAGTCT	TGATCGTTAC	AAGGGACGAT	250
GCTATGATAT	TGAACCCGTT	CCTGGAGAGG	AAAGTCAATT	TATTGCCTAT	300
GTAGCTTACC	CCTTAGATCT	TTTTGAAGAA	GGTTCTGTTA	CTAACCTGTT	350
CACTTCCATT	GTAGGTAATG	TATTTGGATT	CAAAGCCCTA	CGAGCTCTAC	400
GTCTGGAAGA	TCTGCGAATT	CCTCCTGCTT	ATTCAAAAAC	TTTCCAAGGC	450
CCACCACATG	GTATCCAAGT	GGAAAGAGAT	AAACTAAATA	AATATGGTCG	500
TCCTTTGTTG	GGATGTACAA	TCAAACCAAA	ATTGGGTCTA	TCTGCCAAGA	550
ATTATGGTAG	AGCGGTTTAC	GAATGTCTCC	GCGGTGGACT	TGATTTTACC	600
AAGGATGATG	AAAATGTGAA	TTCCCAACCA	TTCATGCGCT	GGAGAGATCG	650
TTTCTGCTTC	TGTGCAGAAG	CACTTTATAA	AGCTCAGGCT	GAGACGGGTG	700
AGATTAAGGG	ACATTACTTG	AATGCTACTG	CAGGTACATG	TGAAGAAATG	750
ATGAAAAGAG	CAGTATTCGC	CAGAGAATTG	GGAGTTCCTA	TAGTCATGCA	800
TGACTATTTG	ACTGGAGGTT	TTACCGCAAA	TACTTCGTTG	GCTCATTATT	850
GCCGAGACAA	CGGCCTACTT	CTCCACATTC	ATCGTGCAAT	GCATGCAGTT	900
ATTGACAGAC	AAAGAAATCA	TGGTATGCAC	TTCCGTGTAC	TGGCTAAAGC	950
ACTGCGTATG	TCTGGTGGAG	ATCATATTCA	CGCTGGTACT	GTAGTAGGTA	1000
AACTTGAAGG	AGAACGAGAA	GTCACTTTGG	GTTTTGTTGA	TCTATTGCGT	1050
GATGATTTTA	TTGAAAAAGA	CCGAAGTCGT	GGTATTTATT	TCACTCAAGA	1100
TTGGGTCTCT	ATGCCGGGTG	TCCTGCCTGT	AGCTTCAGGA	GGTATTCACG	1150
TTTGGCATAT	GCCTGCTTTG	ACCGAAATCT	TTGGTGATGA	TTCTGTATTA	1200
CAGTTTGGTG	GAGGGACTTT	GGGACACCCT	TGGGGAAATG	CACCAGGTGC	1250
AGTAGCTAAT	CGGGTTGCTT	TAGAAGCTTG	TGTACAAGCT	CGTAATGAAG	1300
GACGTGATCT	TGCTCGTGAA	GGTAATGAAG	TGATCCGAGA	AGCCACTNNN	1350
NNNNNNNNNN	NNNNNNNNNN	NNNNNNNNNN	NNNNNNNNNN	NNNNNNNNNN	1400
NNNNNNNNNN	NNNNNNNNNN	NNNNNNNN			1428

Without morphological data, it is unacceptable to draw conclusions on species identification and base phylogeny reconstruction on molecular evidence alone (Donoghue & Sanderson 1992). Historically, the nomenclature of *Cephalotaxus* was confused due to morphological similarities among species.

In Group I, *C. drupacea* is now regarded as a synonym of *C. harringtonii*, which was published four years earlier; this is supported by a number of modern authors (Bailey & Bailey 1976, Cheng & Fu 1978, Fu 1984, Krüssmann 1985). Tripp (1995) presented detailed leaf descriptions and concluded that the differences between *C. drupacea* and *C. harringtonii* were not stable and were inadequate to separate them. In Rehder's *Manual of cultivated trees and shrubs* (1954), this species is incorrectly called *C. drupacea*.

Cephalotaxus koreana was described by Nakai in 1930 in *Bot. Mag.* (*Tokyo*), based on plants collected from Korea. However, the species is not accepted by most modern authors (Fu 1984, Hu 1964, den Ouden & Boom 1965, Rehder 1954) because of morphological similarities to *C. harringtonii*. In cultivation, the plant retains its

TABLE 5. *Cephalotaxus rbcL* gene base differences among *C. harringtonii, C. fortunei* and *C. oliveri.*

Base Comparison			
	84	120	138	284
C. harringtonii (Knight ex J. Forbes) K. Koch	A	C	T	G ......
C. fortunei Hook.	A	C	T	C ......
C. oliveri Mast.	C	T	C	C......
	429	484	660	838
C. harringtonii (Knight ex J. Forbes) K. Koch	T	C	C	T ......
C. fortunei Hook.	G	T	T	C ......
C. oliveri Mast.	T	T	T	T ......
	873	896	976	
C. harringtonii (Knight ex J. Forbes) K. Koch	C	C	A	
C. fortunei Hook.	T	C	A	
C. oliveri Mast.	T	G	G	

remarkable black-green foliage throughout the year and has dense branching and foliage which are most effective for mass planting (Tripp 1995). However, the morphological characters have not generally been accepted as separating *C. koreana* from *C. harringtonii*. Rehder (1954) and Dallimore & Jackson (1967) treated it as *C. harringtonii* var. *koreana* (Nakai) Rehder, which is a more reasonable arrangement.

Cephalotaxus wilsoniana was described by Hayata (1914) based on specimens collected from Taiwan, and separated from *C. harringtonii* by its shorter male strobilus peduncle, 3 mm rather than 5 mm. However, Kitamura (1974) regarded this difference as inadequate, and reduced it to *C. harringtonii* var. *wilsoniana* (Hayata) Kitam. However, several modern authors (Cheng & Fu 1978, Fu 1984, Farjon *et al.* 1993) do not agree. Morphological and molecular data suggest that Group I (*C. harringtonii, C. drupacea, C. koreana,* and *C. wilsoniana*) might be combined into a single species, though it is appreciated that in other groups, well-differentiated morphological species may show identical or very similar *rbcL* sequences.

In Group II, *C. sinensis* and *C. fortunei* are widely accepted at species level by modern authors (Bailey & Bailey 1976, Cheng & Fu 1978; Dallimore & Jackson 1967, Fu 1984, Hu 1964; Krüssmann 1985, Li 1954 & Tripp 1995). Rehder (1954) treated the former species as *C. harringtonii* var. *sinensis* (Rehder & E.H. Wilson) Rehder. Morphologically, the key differences between the two species are that *C. sinensis* has rounded or rounded-cuneate leaf bases compared with the cuneate or widely cuneate leaf bases of *C. fortunei* (Fu 1984). In leaf length, *C. sinensis* is intermediate between *C. fortunei* and *C. harringtonii* (Tripp 1995). Comparisons of plants grown at our Georgia site show that leaf-shape in *C. fortunei* and *C. sinensis* are similar but those of the latter are $^1/_3$ to $^1/_2$ shorter. According to K. Tripp (*pers. comm.*), the Chinese sample may be a hybrid between *C. fortunei* and *C. harringtonii*. The *rbcL* sequences are identical to those of *C. fortunei* and *C. sinensis* from the Arnold Arboretum (Table 2), suggesting that if this accession of *C. sinensis* is of hybrid origin, it received its chloroplast genome from *C. fortunei*. The two taxa are endemic to China and share the same geographical distribution (Table 1).

Cephalotaxus oliveri is the only cultivated species that is clearly distinct without argument among historical and modern taxonomists. DNA sequences of the *rbcL* gene also strongly support that *C. oliveri* is a good species.

The use of *rbcL* sequence data in conjunction with morphological data forms the basis of a new approach to the classification of *Cephalotaxus* material in cultivation. Since morphological differences between the groups were slight, we propose reducing the seven species to three:–

1. *Cephalotaxus harringtonii* (Knight ex Forbes) K. Koch, Japanese plum yew (including *C. drupacea*, *C. koreana* & *C. wilsoniana*): Shrub or small tree to 10 m tall, with spreading branches; bark grey, fissured into narrow detachable strips. Leaves 2-ranked, forming a V-shaped trough on branchlets, linear, 3–5 cm × 2–3.5 mm, rounded or obtuse at the base, acuminate or cuspidate (abruptly mucronulate) at apex, dark green above with a prominent midrib and two pale glaucous bands beneath. Male strobili with an 8 mm stalk. Distributed in Japan, South Korea (Southern portion) and China (Taiwan).

Since this species was introduced into cultivation in 1830 (Dirr 1990), many cultivars have been selected. Among the 36 varieties and cultivars grown at the University of Georgia, wide variations occur in habit (tree, shrub, spreading, and weeping), leaf arrangement (2-ranked, whorled, and semi-whorled), leaf colours (dark green, light green to yellow), and growth rate from 20 to 310 cm (total shoot growth per plant per year). The faster-growing, more attractive clones should be given cultivar names and released to growers.

2. *Cephalotaxus fortunei* Hook. f., Chinese plum yew (including *C. sinensis*): Tree to 16 m tall, usually with several stems; bark reddish brown, peeling off in large flakes leaving light brown markings. Leaves 2-ranked, spreading horizontally from the branchlets, linear-lanceolate, 3–14 cm × 3–4.5 mm, cuneate or widely cuneate at the base with a short twisted petiole, narrowly acute at apex, usually tapering to a fine point, lustrous green above with a strongly raised midrib in the groove and two pale glaucous bands beneath. Male strobili short-stalked. Endemic to China (Yunnan, Sichuan, Guizhou, Hubei, Hunan, Guangdong, Guangxi, Jiangxi, Fujiang, Zhenjiang, Jiangsu, Anhui, Henan, Shaanxi and Gansu).

Only a few cultivars are available in the nursery industry. This species has the longest leaves in the genus, especially *C. fortunei* 'Grandis' (up to 15 cm). Four varieties and cultivars have been collected at University of Georgia for continued evaluation.

3. *Cephalotaxus oliveri* Mast., Oliver's plum yew: Shrub, to 4 m tall, usually with multiple stems and stiff spreading branches. Leaves dense (margins nearly touching) and curving slightly, 1.5–2.5 cm × 2–3.5 mm, sessile, obliquely cordate at base, obtuse or cuspidate at apex, lustrous green above with two pale glaucous bands beneath. Distributed in Vietnam, China (Yunnan, Sichuan, Guizhou, Hubei, Hunan, Guangdong and Jiangxi).

Key to the Species

1. Leaves arranged spirally in two ranks or appearing whorled, separate from each other, 3–15 cm long, rounded, obtuse, or cuneate at base · · · · · · · · · · · · · 2
   Leaves arranged closely in two ranks, margins nearly touching, 1.5–2.5 cm long, obliquely cordate at base · 3. *C. oliveri*

2. Leaves forming a more or less V-shaped trough, or whorled, 3–5 cm long, linear, rounded or obtuse at base · 1. *C. harringtonii*
Leaves spreading horizontally, not whorled, 3–14 cm long, linear-lanceolate (gradually tapering toward apex), cuneate or widely cuneate at base
· 2. *C. fortunei*

Conclusions

Currently, cultivated material of *Cephalotaxus* is sold under seven species names. Based on the *rbcL* sequences and morphological comparison, only three species may be present. Material grown as *C. sinensis* shares the *rbcL* sequences of that grown as *C. fortunei*, and may be a hybrid between the latter as male parent and *C. harringtonii*, which it more closely resembles in leaf length. Further studies will be carried out on taxa below species level, including varieties and cultivars.

References

Brunsfeld, S.J., Soltis, P.S., Soltis, D.E., Gadek, P.A., Quinn, C.J., Strenge, D.D. & Ranker, T.A. (1994). Phylogenetic relationships among the genera of *Taxodiaceae* and *Cupressaceae*: evidence from *rbcL* sequences. *Syst. Bot.* 19: 253–262.

Cheng, W.J. & Fu, L.K. (1978). Flora reipublicae popularis sinicae. Vol. 7 (Gymnospermae). 542 pp. Science Press, Beijing.

Clegg, M.T. & Zurawski, G. (1992). Chloroplast DNA and the study of plant phylogeny: present status and future prospects. In P.S. Soltis, D.E. Soltis, & J.J. Doyle (eds). Molecular systematics of plants. pp. 1–13. Chapman & Hall, New York.

Dallimore, W. & Jackson, A.B. (1967). A handbook of *Coniferae* and *Ginkgoaceae*. 729 pp. St. Martin's Press, New York.

Dirr, M.A. (1990). Manual of woody landscape plants: their identification, ornamental characteristics, culture, propagation and uses. (Ed. 4). 1007 pp. Stipes Publishing Company, Champaign.

Dirr, M.A. (1992). *Cephalotaxus harringtonii*, the Japanese plum yew: superbly tolerant of heat, drought, sun, and cold dipping to –15° to –20°. *Nursery Manager* 8(4): 24-25.

Donoghue, M.A. & Sanderson, M.J. (1992). The suitability of molecular and morphological evidence in reconstructing plant phylogeny. In P.S. Soltis, D.E. Soltis, & J.J. Doyle (eds). Molecular systematics of plants. pp. 30–368. Chapman & Hall, New York.

Farjon, A., Page, C.N. & Schellevis, N. (1993). A preliminary world list of threatened conifer taxa. *Biodiversity & Conserv.* 2: 304–326.

Florin, R. (1963). The distribution of conifer and taxad genera in time and space. *Acta Hort. Berg.* 20(4): 1–258.

Fu, L.K. (1984). A study on the genus *Cephalotaxus* Sieb. et Zucc. *Acta Phytotax. Sin.* 22(4): 277–288.

Greuter, W., Barrie, F.R., Burdet, H.M., Chaloner, W.G., Demoulin, V., Hawksworth, D.L., Jørgensen, P.M., Nicholson, D.H., Silva, P.C., Trehane, P. & McNeill, J. (eds). (1994). International code of botanical nomenclature (Tokyo Code). 389 pp. Koeltz Scientific Books, Königstein.

Guillemaut, P. & Marechal-Drouard, L. (1992). Isolation of plant DNA: a fast, inexpensive, and reliable method. *Pl. Mol. Biol. Reporter.* 10: 60–65.

Hallick, R.B. & Bottomly, W. (1983). Proposals for the naming of chloroplast genes. *Pl. Mol. Biol. Reporter.* 1: 38–43.

Harlass, S. (1994). Georgia names 4 outstanding plants. *Greenh. Manager* 13(1): 79–82.

Hayata, B. (1914). Icones plantarum formosanarum IV. 263 pp. & 35 pl. Taihoku, Taipei.

Hiratsuka, J., Shimada, H., Whittier, R., Ishibashi, T., Sakamoto, M., Mori, M., Kondo, C., Honji, Y., Sun, C.-R., Meng, B.-Y., Li, Y.-Q., Kano, A., Nishizawa, Y., Hirai, A., Shinozaki, K. & Sugiura, M. (1989). The complete sequence of the rice (*Oryza sativa*) chloroplast genome: intermolecular recombination between distinct tRNA genes accounts for a major plastid DNA inversion during the evolution of the cereals. *Mol. Gen. Genet.* 217: 185–194.

Hu, S.-Y. (1964). Notes on the flora of China IV. *Taiwania* 10: 13–62.

Hu, S.-Y. (1984). Comparative anatomy of the leaves of *Cephalotaxus* (*Cephalotaxaceae*). *Acta Phytotax. Sin.* 22(4): 289–296.

Hubbard, M., Kelly, J., Rajapakse, S., Abbott, A. & Ballard, R. (1992). Restriction fragment length polymorphisms in rose and their use for cultivar identification. *HortScience* 27: 172–173.

Index Kewensis. (1893–). Oxford University Press, Oxford: Royal Botanic Gardens, Kew, UK.

Kamalay, J.C. & Carey, D.W. (1995). Application of RAPD-PCR markers for identification and genetic analysis of American elm (*Ulmus americana* L.) selections. *J. Environm. Hort.* 13: 155–159.

Kitamura, S. (1974). Short reports of Japanese plants. *Acta Phytotax. Geobot.* 26(1–2): 1–15.

Krahl, K.H., Dirr, M.A., Halward, T.M., Kochert, G.D. & Randle, W.M. (1993). Use of single-primer DNA amplification for the identification of red maple (*Acer rubrum* L.) cultivars. *J. Environm. Hort.* 11: 89–92.

Krüssmann, G. (1985). Manual of cultivated conifers. 361 pp. Timber Press, Portland, Oregon.

Lee, S.C. (1935). Forest botany of China. 991 pp. The Commercial Press Ltd, Shanghai.

Lee, S.C. (1973). Forest botany of China supplement. 477 pp. Chinese Forestry Association Publisher, Taipei.

Li, H.L. (1954). New species and varieties in *Cephalotaxus*. *Lloydia* 16(3): 162–64.

Liberty Hyde Bailey Hortorium. (1976). Hortus Third. 1290 pp. MacMillan Publishing Co., Inc. New York.

Ohwi, J. (1984). Flora of Japan. 1067 pp. Smithsonian Institute Publisher, Washington, D.C.

Olmstead, R.G. & Palmer, J.D. (1994). Chloroplast DNA systematics: a review of methods and data analysis. *Amer. J. Bot.* 81: 1205–1224.

Ouden, P. den & Boom, B.K. (1965). Manual of cultivated conifers hardy in the cold- and warm-temperate zone. 526 pp. Martinus Nijhoff, The Hague.

Palmer, J.D., Jansen, R.K., Michaels, H.J., Chase, M.W. & Manhart, J.R. (1988). Chloroplast DNA variation and plant phylogeny. *Ann. Missouri Bot. Gard.* 75(3–4): 1180–1206.

Phillips, E.W.J. (1941). The identification of coniferous woods by their microscopic structure. *J. Linn. Soc., Bot.* 52 no. 343: 259–320.

Rehder, A. (1954). Manual of cultivated trees and shrubs. 996 pp. MacMillan Company, New York.

Shinozaki, K., Ohme, M., Tanaka, M., Wakasugi, T., Haysida, N., Matsubayashi, T., Zaita, N., Chungwongse, J., Obokata, J., Yamaguchi-Shinozaki, K., Ohto, C., Torazawa, K., Meng, B.Y., Sugita, M., Deno, H., Kamogashira, T., Yamada, K., Kusuda, J., Takaiwa, F., Kata, A., Tohdoh, N., Shimada, H. & Sugiura, M. (1986). The complete nucleotide sequence of the tobacco chloroplast genome. *Pl. Molec. Biol. Reporter.* 4: 110–147.

Silba, J. (1984). Phytologia memoirs VII: An international census of the *Coniferae*, I. 78 pp. Moldenke Publisher, Plainfield, New Jersey.

Silba, J. (1990). A supplement to the international census of the *Coniferae*, II. *Phytologia* 68(1): 7–78.

Staub, J.E. &. Meglic, V. (1993). Molecular genetic markers and their legal relevance for cultivar discrimination: a case study in cucumber. *HortTechnology* 3: 291–300.

Tripp, K.E. (1995). *Cephalotaxus*: The plum yew. *Arnoldia* 55(1): 24–39.

Wakasugi, T.,Tsudzuki, J., Ito, S., Nakashima, K., Tsudzuki, T. & Sugiura, M. (1994). Loss of all nfh genes as determined by sequencing the entire chloroplast genome of the black pine, *Pinus thunbergii*. *Proc. Natl. Acad. Sci. U.S.A.* 91: 9794–9798.

Xi, Y.Z. (1993). Studies on pollen morphology and exine ultrastructure in *Cephalotaxaceae*. *Acta Phytotax. Sin.* 31(5): 425–431.

Xiang, Q.Y., Soltis, D.E., Morgan, D.R. & Soltis, P.S. (1993). Phylogenetic relationships of *Cornus* L. *sensu lato* and putative relatives inferred from *rbcL* sequence data. *Ann. Missouri Bot. Gard.* 80: 723-734.

Cooper, W. & MacLeod, J. (1999). Genetically modified crops: the current status. In: S. Andrews, A.C. Leslie and C. Alexander (Editors). Taxonomy of Cultivated Plants: Third International Symposium, pp. 277–283. Royal Botanic Gardens, Kew.

GENETICALLY MODIFIED CROPS: THE CURRENT STATUS

WENDY COOPER[1] AND JOHN MACLEOD[2]

[1]51 Longdell Hills, Contessey, Norwich NR5 OPB, UK
[2]National Institute of Agricultural Botany (NIAB), Huntingdon Road,
Cambridge CB3 0LE, UK

Abstract

Genetic engineering techniques allow a specific trait, such as disease-resistance or herbicide-tolerance, to be conferred, either by manipulating the expression of genes within the plant's own genome, or via the introduction of a gene, or genes from another organism. This has potential advantages over conventional plant breeding methods, which can be slow and inefficient. However, despite the obvious benefits that may be realised by the use of genetic engineering, there are also concerns associated with the development of genetically modified organisms (GMOs), which have led to the development of rigorous policy concerning environmental releases of genetically modified (GM) cultivars. This paper outlines the current developments in GM crops in terms of experimental releases and commercialisation and reviews the regulatory procedures involved.

Introduction

Currently there are no GM crops on the market in Britain although increasing numbers are undergoing trials and commercialisation of the first GM cultivar is expected in the next year or two. The situation elsewhere in the world is very different and growth of GM plants in the USA is well established. By 1998 the acreage of GM crop plants is predicted to reach 73 million acres, principally grown in the USA, Canada, Argentina, China and Australia (Table 1).

The first field trials of transgenic crops were conducted in the USA and France in 1986. By the end of 1995 over 3,600 field trials of GM crops had been carried out in 34 countries (Table 1) with at least 56 different crop species (Table 2).

To date the UK has approved 135 releases, but for research purposes only; no consents for release for commercial purposes have yet been granted. Since each release approval can cover a research programme comprising a number of sites, transformation with a number of gene inserts and a number of years, there is now a substantive body of data and experience pertaining to GM cultivars. The grant of consent for marketing on a wide-scale is however very slow in the EC and only three consents were issued up to January 1999, while another eleven applications have been delayed within the Commission or by objection from other Member States.

The state of GM crop development in the UK can be summarised as follows. Oilseed rape is nearest to the market place, with the first anticipated commercialisation in 2001. Modifications include cultivars tolerant to glufosinate ammonium (Challenge) or glyphosate (Roundup), and cultivars with a high lauric acid content.

TABLE 1. 1998 releases of GMOs per country as a percentage of the world total.

Country	%	Country	%
USA	70.45	Sweden	0.37
Canada	11.83	New Zealand	0.34
France	4.72	Denmark	0.31
Belgium	2.02	Brazil	0.28
UK	1.84	South Africa	0.17
Italy	1.71	Finland	0.11
Holland	1.47	Portugal	0.06
Spain	1.20	Russia	0.06
Japan	1.17	Bulgaria	0.05
Germany	0.89	Austria	0.03
Australia	0.88	Switzerland	0.03

There is currently a wide range of GM crops in experimental trial, including sugar beet (herbicide tolerant and altered carbohydrate metabolism), potato (altered carbohydrate, virus resistant), spring wheat (disease resistant), and maize (herbicide tolerant and insect resistant).

Cultivars of spring and winter oilseed rape, sugar beet, fodder beet and forage maize are currently being assessed in the statutory National List (NL) trials. Inclusion of a cultivar in the NL and the EC common catalogue is an essential precursor to commercialisation. To put the state of development into perspective there are about 1000 cultivars in the current NL trials of which around 20 are genetically modified.

Future Crops

Current research interests are likely to dictate the types of genetically modified cultivars to be developed in the future. One of the most notable developments is the improvement in techniques for transforming cereals, previously a difficult group of crops to modify. Other current research developments and strategies include the large-scale initiatives to sequence the genomes of entire organisms such as *Arabidopsis* Heynh. and rice. Such approaches will provide invaluable information concerning new gene combinations and their function. Gene isolation using a variety of approaches continues to be a successful approach in conjunction with efforts to determine the function of these newly discovered genes.

Overall the emphasis is likely to remain on the production of cultivars with improved pest and disease resistance as well as production of plants compatible with effective weed control and environmentally friendly farming methods. However there is also likely to be an increasing emphasis on the development of cultivars bred for industrial purposes such as the production of novel oils, starches and high value pharmaceutical compounds such as vaccines, or the development of crops with tolerance to salinity, drought or frost.

Genetic engineering has enabled higher yielding hybrid systems to be introduced by creating male sterile plants. Most cultivated species are bisexual and capable of fertilising themselves. However, by crossing two individuals, hybrid offspring may be

TABLE 2. Genetically Modified Plant Species (OECD figures, 1998).
Major releases are indicated by the relevant percentage of releases within the OECD.

Actinidia deliciosa (A. Chev.) Liang & A.R. Ferguson var. *deliciosa* (kiwi fruit)
Agrostis stolonifera L. (creeping bentgrass)
Allium cepa L. (onion)
Ananas comosus (L.) Merr. (pineapple)
Arabidopsis thaliana (L.) Heynh. (thale cress)
Arachis hypogaea L. (peanut)
Asparagus officinalis L. (asparagus)
Atropa belladonna L. (belladonna)
Avena sativa L. (oat)
Beta vulgaris L. (beet)
Beta vulgaris L. (sugar beet) (2%)
Betula pendula Roth (silver birch)
Brassica carinata A. Braun (Ethiopian mustard)
Brassica juncea (L.) Czerniak. (mustard)
Brassica napus L. (oilseed rape) (13%)
Brassica nigra (L.) Koch (brown mustard)
Brassica oleracea L. (broccoli, cauliflower and cabbage)
Brassica oleracea L. var. *acephala* (DC.) Alef. (forage rape)
Brassica rapa L. (turnip rape)
Capsicum annuum L. (pepper)
Carica papaya L. (papaya)
Castanea sativa Mill. (European chestnut)
Castanea dentata (Marshall) Borkh. (American chestnut)
Chrysanthemum × *grandiflorum*(Ramat.) Kitam. (chrysanthemum)
Cichorium intybus L. (chicory)
Citrullus lanatus (Thunb.) Matsumae & Nakai (watermelon)
Citrus × *aurantium* L. (orange)
Cucumis melo L. (melon)
Cucumis sativus L. (cucumber)
Cucurbita pepo L. (marrow squash, courgette, etc.)
Cucurbita texana Scheele
Cyphomandra betacea Cav. (tamarillo)
Daucus carota L. (carrot)
Dianthus caryophyllus L. (carnation)

Eucalyptus camaldulensis Dehnh. (eucalyptus)
Eustoma grandiflorum (lisianthus)
Gladiolus sp. (gladiolus)
Gossypium hirsutum L. (cotton) (7%)
Helianthus annuus L. (sunflower)
Hordeum vulgare L. (barley)
Ipomoea batatas (L.) Lam. (sweet potato)
Juglans sp. (walnut)
Lactuca sativa L. (lettuce)
Linum usitatissium L. (flax)
Liquidambar sp. (sweetgum)
Lupinus angustifolius L. (lupin)
Lycopersicon esculentum Mill. (tomato) (10%)
Malus domestica Borkh. (apple)
Medicago sativa L. (alfalfa)
Nicotiana benthamiana Domin (tobacco)
Nicotiana tabacum L. (tobacco) (5%)
Oryza sativa L. (rice)
Pelargonium sp. ('geranium')
Picea abies (L.) Karst. (Norway spruce)
Picea sp. (spruce)
Pinus sp. (pine)
Pisum sativum L. (pea)
Poa pratensis L. (Kentucky bluegrass)
Populus sp. (poplar)
Rosa hybrida sensu auct. (rose)
Rubus idaeus L. (currant)
Saccharum officinarum L. (sugar cane)
Saintpaulia ionantha H. Wendl. (African violet)
Sinapis alba L. (white mustard)
Solanum melongena L. (eggplant)
Solanum tuberosum L. (potato) (12%)
Sorghum bicolor (L.) Moench. (sorghum)
Tagetes sp. (marigold)
Trifolium subterraneum L. (clover)
Triticum aestivum L. (wheat)
Vaccinium oxycoccus L. (cranberry, European)
Vitis vinifera L. (grape)
Zea mays L. (maize) (38%)

obtained that are stronger, more vigorous and often higher yielding. Cross-pollination is ensured by the creation of a castrated plant known as a male sterile, which is fertilised with pollen from another plant. Creating male-sterility by traditional breeding methods has been, at best, difficult and in some species impossible. Emasculation has traditionally used physical (e.g. detasselling in maize) or chemical means. More recently genetic approaches involving the modification of gene expression have successfully inhibited development of functional male organs.

Traditional breeding techniques offer only limited potential in dealing with the difficult genetics of the potato, and development of new cultivars has been relatively slow. However the potato is relatively easy to transform and can be reproduced by cloning making it amenable to genetic modification. Modified cultivars are likely to include those expressing quality trait improvements such as increased starch production and reduced uptake of oil when cooking chips and crisps, a potential health benefit. Changes to the starch composition are also possible and desirable as different starches are required for different commercial processes.

Quality traits such as improved shelf-life of fruit have been produced. The best known of these is the Flavr Savr tomato sold in the USA where the rate of fruit ripening has been retarded to produce fruit with a longer shelf life. Normal ripening occurs but softening is slower; this improves fruit quality in terms of higher soluble solids content and increased viscosity, some of the desired characteristics for better processing of tomato paste. By remaining firmer for longer, the GM fruits can be harvested at a later stage of ripening and are less likely to become infected by bacteria and fungi. A similar GM tomato has been used in the UK in the production of tomato purée, which went onto the supermarket shelves approximately two years ago.

Speciality crops are likely to be an important future development in providing a substitute for fossil fuels or other non-renewable sources from which we derive many industrial oils used in the manufacture of plastics, detergents, inks and lubricants. The particular type of oil produced by a plant depends on which enzymes the plant contains but there is good potential for developing plant-based alternatives to fossil fuels. By isolating the gene responsible for an enzyme essential to the formation of a target fatty-acid and inserting it into a widely-grown oilseed crop like oilseed rape, economically viable quantities of oil can be produced. An example of this is the isolation of the genes encoding the enzymes responsible for the synthesis of petroselinic acid, a fatty-acid with potential for use in making detergents and nylon polymers. The gene is obtained from coriander and has already been inserted into oilseed rape. Similar approaches are being taken to increase the content of erucic acid, which is used in lubricants, cosmetics and coatings for plastics. Oilseed rape is also one of the crops being modified to produce proteins or enzymes used in the production of pharmaceuticals for human and veterinary medicine.

The Concerns

Opponents of genetic engineering are concerned that the technology is unproven and that not enough is known about the potential risks to human or environmental health. Once genetically modified plants are released into the environment it will be very difficult to repair any damage caused by their release. One reason for concern is the insertion of marker genes conferring antibiotic resistance into a number of genetically modified cultivars. The reason for the presence of such genes is related entirely to the experimental development of these GM plants and the difficulties associated with identifying plant tissue that has been successfully modified.

Transformed plants are able to grow in the presence of the antibiotic whereas non-transformed plants are killed. Two commonly used antibiotics are kanamycin and hygromycin. Though this is a very quick visual way of identifying successfully transformed tissue, it has caused a great deal of concern about the potential for antibiotic resistant strains of bacteria developing via gene transfer in the gut of animals or even humans. Biotechnology companies are responding to public pressure by developing GM cultivars that no longer contain antibiotic resistance markers

The proponents of genetic engineering argue that its application, which is still a relatively new technology, is controlled by strict regulations concerning the release of GM cultivars into the environment.

Regulations surrounding the Release of GM Crops in the EC and the UK

Genetically modified crops are regulated by both EC Directive 90/220 and the UK Environmental Protection Act 90 and 92, and are also subject to other relevant legislation. In the UK these include: Control of Pesticide Regulation 1986, Plant Varieties Act 1997, Food Safety Act 1990, Agriculture Act 1970 and Feeding Stuff Regulations 1995. (Further information can be obtained from MAFF). The objective of the regulations is to ensure that the deliberate release of genetically modified plants does not cause risk to the environment or human health. Two levels of consent for release applications must be obtained before a GM cultivar can be released. For research, authority rests with each individual member state; for commercialisation, authority rests with the EC and approval requires the consent of all Member States or a qualified majority. In addition to approval under EC Directive 90/220, all GM cultivars must satisfy the existing EC seeds and plant varieties legislation which covers National List testing and inclusion in the Common Catalogue and demonstrates that a new cultivar, whether GM or conventional, has commercial potential.

Applications for release of GM crops must include a full **Risk Assessment** carried out by the applicant, which is subject to evaluation by legislative bodies. The consent may include conditions for release such as isolation distances from neighbouring crops and requirements for general duty of care. The deliberate release of genetically modified crops, whether on a small scale for research purposes or for general marketing, is subject to clearly defined legislation across the EC.

A. Regulation in the UK: Consideration of Applications
All applications for deliberate release of GM cultivars are considered by the Advisory Committee for Releases to the Environment (ACRE), an independent committee with expertise in genetics, molecular biology, agronomy, human health and the environment. ACRE advises the Secretary of State for the Environment who can issue approval for a release for research purposes within the UK. For consent to release for marketing purposes the Secretary of State for the Environment is required to give informed opinion to the EC. All Member States are subsequently required to give approval for marketing release whether or not the release is scheduled to take place within their own country.

Any application involving food use, either directly or indirectly after processing, must also be considered and approved by the Advisory Committee on Novel Foods and Processes (ACNFP), a similar independent committee with expertise in all aspects of food and food products. MAFF expert committees may also consider the use of any genetically modified material in animal feed.

To be granted approval, even for research purposes, for release to the environment and inclusion in food or feed, an application must be cleared through all these processes. The consent granted by the Secretary of State for the Environment collates and includes all these separate considerations, and incorporates any condition or monitoring requirement imposed on the applicant. The applicant has to supply extensive data on both the inserted genetic material and the characteristics of the new genetically modified plant. The source of the insert, details of the modification, means of insertion, level of interaction, level of expression, interaction with the environment, effect on human health, management of the site and the release must all be considered and summarised by the applicant in a **Risk Assessment**. For crop plants ACRE's consideration of the **Risk Assessment** focuses primarily on whether:–

- Inserted genes may make the modified crop more persistent
- Inserted genes may make the modified crop more invasive
- Inserted genes may make the modified crop more undesirable to living organisms or the environment
- Inserted genes may be transferred to other organisms

In the UK the process sets out to be as open as possible. A summary of every application is placed on the Public Register, releases must be advertised locally and the consent, approval and any conditions of release are also placed on the Public Register.

In summary, the Department for the Environment, Transport and the Regions (DETR) as the lead department, in consultation with all the other relevant UK departments, has outlined very specifically the requirements for anyone wishing to grow and market GM cultivars in the UK:–

1. The applicant must notify the Health and Safety Executive (HSE) of their intention to initiate research work at the laboratory stage and produce a risk assessment of any risks to the environment or human health.
2. The HSE will then consult the Advisory Committee on Genetic Modification (ACGM).
3. If the cultivar is to be grown outside for field trials, a detailed application and environmental risk assessment must be completed by the applicant and submitted to the DETR.
4. The DETR will then seek advice and guidance from ACRE.
5. Consent from the Secretary of State for the Environment may be given allowing the cultivars to be grown outside for experimental purposes.
6. If the experimental release field trials are considered successful the Advisory Committee on Novel Foods and Processes (ACNFP) may consider the new cultivar for food safety purposes under the EC Novel Foods regulation.
7. The ACNFP may then seek advice from the Food Advisory Committee (FAC), Committee on Toxicity of Chemicals (COT) and the Committee on Medical Aspects of Food Policy (COMA).
8. Marketing consent is also required before the crops can be grown commercially, which requires another detailed submission to the DETR for consideration by ACRE.
9. Once an application is approved by these committees their assessment is passed to other EU member states for consideration.
10. If no objections are raised, the product can be marketed (provided it is labelled in compliance with the new EU laws).

B. Regulation in the USA and Canada

The USA and Canada have adopted a different approach from that taken by the EC where the regulations relate to the process of modification. In the USA the regulation relates to the product and not the process, while in Canada the regulation relates to any plant with novel traits; this may include other than genetically modified approaches.

The product-driven approach in North America means that once a product is proven safe it is no longer regulated. Similar releases are then subject to notification only resulting in a fast track for proven modifications. The criteria used to determine safety to the environment and human health are equally rigorous to those considered in Europe, but the product-driven approach leads to faster approval and commercialisation. In the EC the case-by-case process-driven approach, linked to approval by the EC for marketing, is prone to long delay.

Conclusion

Genetic modification and modern biotechnology can not only introduce novel and potentially valuable new characters into food crops but can also speed up the process of plant breeding. Conventional breeding involves mixing all the genetic material from both parents and produces progeny with both the targeted desirable characters and also undesirable characters, which have then to be minimised by repeated selection. Typically it takes eight to ten years from the initial crossing to produce a cultivar ready to enter the National List trials. Using genetic modification methods, the ability to identify, isolate and move desirable characters from one plant into another applies equally within and between species and can significantly reduce the time taken. New genetically modified characters must, however, be linked to the best available cultivars in terms of yield, quality and agronomic characters, as they must not only satisfy the environmental and safety criteria of the Deliberate Release legislation, but also in the UK the performance criteria of the National List and the more stringent criteria of the Recommended List before they will be accepted into wide-scale agricultural production. The production of new cultivars using molecular breeding methods also poses a number of important questions for the taxonomist including:–

- Should GMOs be named in the same manner as other cultivars?
- Should they be specifically labelled?
- If so, should the modification be declared as part of the cultivar name?
- What effect would multiple modifications have on the name?

Sources of Information

MAFF website; http://www.maff.gov.uk
ACRE website; http://www.shef.ac.uk

CASE STUDIES IN THE TAXONOMY
OF CULTIVATED PLANTS

9

Plovanich-Jones, A.E., Coombes, A. & Hoa, D.T. (1999). Cultivars of *Quercus cerris* × *Quercus suber*: *Q.* × *hispanica*, the Lucombe oak and inter-simple sequence repeats (ISSRs). In: S. Andrews, A.C. Leslie and C. Alexander (Editors). Taxonomy of Cultivated Plants, pp. 287–296. Royal Botanic Gardens, Kew.

CULTIVARS OF *QUERCUS CERRIS* × *QUERCUS SUBER*: *Q.* × *HISPANICA*, THE LUCOMBE OAK AND INTER-SIMPLE SEQUENCE REPEATS (ISSRs)

ANNE E. PLOVANICH-JONES[1], ALLEN COOMBES[2] AND DIEM T.HOA[3]

[1]Department of Botany and W.J. Beal Botanical Garden, Michigan State University, East Lansing, MI 48824-1312, USA
[2]The Sir Harold Hillier Gardens and Arboretum, Jermyns Lane, Ampfield, Romsey, Hampshire SO51 0QA, UK
[3]Medical Technology Program, College of Natural Sciences, Michigan State University, East Lansing, MI 48824-1312, USA

Abstract

Morphological and chromosomal evidence for the identification of cultivars of *Quercus* × *hispanica*, the Lucombe oak has often been inconclusive. The use of DNA fingerprinting techniques has proved helpful for such identifications. Using inter-simple sequence repeat (ISSRs) primers, 66 *Q.* × *hispanica* (and related) samples were analysed and compared for band pattern differences to establish identities. DNA evidence revealed that samples from some trees said to be the same cultivar had identical banding patterns, which then aided in identification of other named or unnamed trees with the same DNA banding patterns. *Quercus* 'Lucombeana', with samples from nine different grafted trees, showed no two had identical banding patterns, while two grafted trees on the grounds of the University of Exeter, and dating in age approximately to the time of Lucombe, did have the same banding pattern and may indeed be of the original clonally propagated 'Lucombeana'.

Introduction

From the time that William Lucombe of Exeter discovered a different looking oak seedling in his *Quercus cerris* L. seedbed (probably from a *Q. cerris* pollinated by *Q. suber* L.) in 1763 (Elwes & Henry 1906), there has been confusion about the name and identity of that oak. Called *Q. lucombeana* by Sweet (1826), it has subsequently been called the Lucombe oak, the Lucombe oak, *Quercus* × *hispanica* Lam. var. *lucombeana* (Loudon) Rehder, and the Exeter oak among other names. Within a short time Lucombe clonally propagated thousands of ramets of this oak by grafting onto *Q. cerris* rootstock and distributed them, while the original may have been cut down for Lucombe's coffin boards when the tree was about twenty years old (Elwes & Henry 1906). Lucombe's son named a group of seedlings from the original tree, which are now regarded as cultivars: 'Crispa', 'Suberosa', 'Incisa', 'Heterophylla', and 'Dentata'. Subsequent cultivars originating from other crosses of *Q. cerris* and *Q. suber* (*Q.* × *hispanica*) include 'Cana Major' of 1849 from the Hammersmith Nursery, 'Diversifolia' from the Smith Nursery, and perhaps the most famous other than that of Lucombe, 'Fulhamensis', described by Loudon (1838), from the Whitley & Osborne Nursery at Fulham, England (Bean 1976).

Morphological characteristics, such as corky bark formation, leaf form, leaf retention, height at maturity, acorn size, etc. are quite variable. Studies of the chromosome number and morphology of the chromosomes showed no variation among the forms examined (Caldwell 1953). As a result of these difficulties in identification, it is likely that many plants in arboreta and botanic gardens are inappropriately named. Mitchell (1994) described the history of many of the cultivars mentioned here and considered that 'Lucombeana', the best known and most widely used cultivar epithet, was only common in the Exeter area of Devon, England.

It was decided to use DNA fingerprinting on a group of *Q. × hispanica* cultivars in an attempt to shed light upon the confusion surrounding the Lucombe oak. Were the many trees named 'Lucombeana' vegetatively identical, and was it possible to establish their lineage as the same as that original clonal (grafted) material of W. Lucombe? Although many fingerprinting techniques have been successfully used on plant material (Weising 1995) and on oaks specifically (Dow *et al.* 1995, 1996), it was decided to use ISSRs as they have been used to good purpose in population studies of oaks (Marquardt *pers. comm.*). ISSR markers are generated from single-primer PCR reactions where the primer is designed from di- or trinucleotide repeat motifs with anchoring sequences of one to three nucleotides. (Wolfe *et al.* 1998). Such repeat motifs occur in all eukaryotes. Using polymerase chain reaction (PCR) primers created to bind to these repeated regions, the intervening areas are amplified allowing variation to be detected between samples. Plant cultivars that have been vegetatively propagated should all have an identical banding pattern (barring somatic mutation). All of the clonally propagated oaks from Lucombe's nursery and all those that were grafted later will exhibit an identical, unique banding pattern for each primer used. It should be noted that seed propagated cultivars (e.g. oilseed rape), will have a wide range of patterns, as can individuals within the same species.

Materials and Methods

Sixty-four different plant samples of *Q. × hispanica* and related materials were collected, see Table 1. Other genera from the *Fagaceae* Dumort. (*Castanea* Mill. and *Fagus* L.) and one outgroup (*Rhododendron* L.) were used to evaluate various extraction methods and choice of primers (data not shown). Samples from various sources in the UK (see Table 1 for specific locations and accession numbers) were mailed to Michigan State University, Michigan, USA in plastic bags through regular postal channels. One sample, *Q. hinckleyi* C.H. Muell., was obtained from a herbarium voucher (*Plovanich-Jones* 002) at MSC. DNA was extracted using the method of Scott *et al.* (1996) that was developed for use on 'rain forest' plant species but which has often proved useful for samples high in polysaccharides and secondary metabolites. The posted samples proved difficult to extract with other methods because of long transit storage conditions.

PCR conditions: 30ul PCR reactions were performed with final concentrations of 1X PCR buffer, 12% sucrose, 0.2nM cresol red, 200uM each dNTP, 0.07U/ul Amplitaq, 30 ng. total DNA, 1.8uM UBC primer. An MJ Research, Inc. PTC 100 thermocycler was used with the following conditions: denaturation at 94°C for 2 min. (using a hot start, 3.5 mM MgCl$_2$ was added when temperature exceeded 80°C) 29 cycles at 92°C, 30 sec., 52°C for 45 sec., and 72°C for 2 min. A 2% agarose gel was used for resolution of the PCR products. 25μl was loaded into each well (the addition of the Cresol red dye to the PCR reaction obviated the need for any further loading or running dye) and run at 50 V for 16 hours using a cooling, recirculating tank maintained at 12°C.

TABLE 1. Plant samples used for DNA ISSR analysis. Accession numbers for some samples are those of the Arboretum providing the plant material.

Name	Accession #	Name	Accession #
Castanea dentata (Marsh.) Borkh. (MSU)		*Q.* × *hispanica* 'Ambrozyana'	80.0280A
Castanea sativa Mill. (Hillier's)		*Q.* × *hispanica* 'Ambrozyana'	82.0296A
Q. cerris 'Laciniata'	44.0004	*Q.* × *hispanica* 'Ambrozyana'	77.4402
Q. cerris L.	14.0079	*Q.* × *hispanica* 'Ambrozyana'	77.2213
Q. cerris L.	75.0105	*Q.* × *hispanica* 'Cana Major'	27.0057
Q. cerris 'Wodan'	89.2595	*Q.* × *hispanica* 'Cana Major'	27.0056
Exeter lower		*Q.* × *hispanica* 'Cana Major'	27.0055
Exeter root stock (*Q. robur* L.)		*Q.* × *hispanica* 'Cana Major'	43.0457
Exeter middle		*Q.* × *hispanica* 'Cana Major'	43.0455
Fagus sylvatica L. (MSU)		*Q.* × *hispanica* 'Cana Major'	27.0058
Kensington A West Ham Park		*Q.* × *hispanica* 'Crispa'	43.0463
Kensington B West Ham Park		*Q.* × *hispanica* 'Crispa'	43.0464
Kensington C Kensington Gardens		*Q.* × *hispanica* 'Crispa'	43.0465
Kensington D Chiswick House		*Q.* × *hispanica* 'Diversifolia'	87.2194
Q. × *hispanica* Lam.	95.0694B	*Q.* × *hispanica* 'Diversifolia'	08.0320
Q. × *hispanica* Lam.	95.0694A	*Q.* × *hispanica* 'Fulhamensis'	86.2228
Q. × *hispanica* Lam.	04.0434	*Q.* × *hispanica* 'Fulhamensis'	82.0142
Q. × *hispanica* Lam.	43.0472	*Q.* × *hispanica* 'Fulhamensis'	78.1942
Q. × *hispanica* Lam.	77.5535	*Q.* × *hispanica* 'Hemelrijk'	93.0030
Q. × *hispanica* Lam.	44.0403	*Q.* × *hispanica* 'Hemelrijk'	93.0030B
Q. × *hispanica* Lam.	77.1306	*Q.* × *hispanica* 'Heterophylla'	95.0454
Q. × *hispanica* Lam.	09.0047	*Q.* × *hispanica* 'Lucombeana'	77.5412
Q. × *hispanica* Lam.	77.1306	*Q.* × *hispanica* 'Lucombeana'	14.0133
Q. laceyi Small (Texas)		*Q.* × *hispanica* 'Lucombeana'	56.0039
Q. hinckleyi C.H. Muell. (Texas)		*Q.* × *hispanica* 'Lucombeana'	43.0384
Q. robur L. (MSU)		*Q.* × *hispanica* 'Lucombeana'	77.7380
Rhododendron sp.		*Q.* × *hispanica* 'Lucombeana'	88.0329
Q. suber L.	79.9326	*Q.* × *hispanica* 'Lucombeana'	44.0037
Q. suber L.	77.4783	*Q.* × *hispanica* 'Lucombeana'	17.0048
Q. suber L.	43.174	*Q.* × *hispanica* 'Lucombeana'	41.0425
Q. suber L.	78.3131	*Q.* × *hispanica* 'Suberosa'	95.1629
Q. suber L.	76.9326	*Q.* × *hispanica* 'Waasland'	97.0075
		Q. × *hispanica* 'Wageningen'	89.2597
		Q. × *hispanica* 'Wageningen'	89.2596

Bands were visualized by staining gels with ethidium bromide for thirty minutes and destaining in 0.5X TAE for an additional thirty minutes. DNA Proscan, (Nashville, Tennessee) was used for scoring of bands and evaluation of results. All samples were evaluated with a minimum of two gels for each UBC primer (Table 2). Four UBC primers (UBC 807, 811, 834 and 841) were used for each sample. Although UBC primer 807 can produce a number of bands that will also appear using primer 834 likewise for 811 and 841, will produce many unique bands. The four primers will produce different banding patterns for each clonally unique cultivar. Samples of 'Lucombeana' from twenty different PCR reactions and the four primers were repeated as the results were somewhat surprising and this was the focal point of the study. Results for all gels were consistent and repeatable.

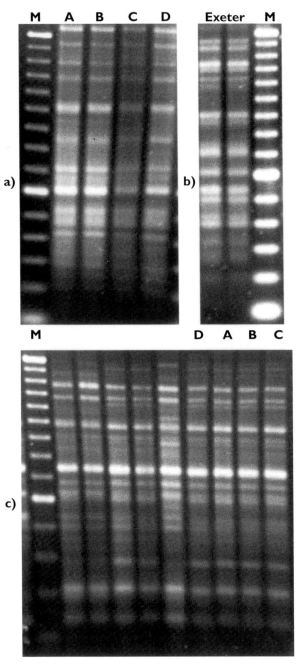

FIG. 1. Banding pattern UBC 834: a) 'Ambrozyana' Lane A 80.0280A, Lane B 77.4402, Lane C 82.0296A and Lane D 77.2263. M= Gibco 100 bp molecular weight marker. b) Exeter oaks, lower and middle. c) Lanes A and B, Exeter lower and middle, Lane C West Ham Park, Lane D West Ham Park, Lane E Kensington Gardens, Lane F Chiswick House, Lane G 'Fulhamensis' 82.0142, Lane H 'Fulhamensis' 78.1942, and Lane I 'Fulhamensis' 86.2228.

TABLE 2. DNA ProScan Molecular weight data for Fig. 2. Lanes 2, 3 and 4 correspond to Fig. 2a; Lanes 5, 6 and 7 correspond to Fig. 2b.

Lane 1	Lane 2	Lane 3	Lane 4	Lane 5	Lane 6	Lane 7	Lane 8
..	..	..	..	..	..	1622	..
1500	..	..	..	..	..	1560	1500
1400	..	1469	..	..	..	..	1400
1300	..	..	..	..	..	..	1300
1200	..	..	..	..	..	..	1200
1100	..	1049	..	1100	..	..	1100
1000	..	..	..	..	1021	..	1000
900	946	..	..	..	..	..	900
800	..	..	..	..	822	..	800
..	756	..	..	..	..	..	..
700	..	..	..	..	..	..	700
600	..	..	..	..	600	..	600
..	528	..	..	..	..	..	..
500	..	..	..	..	..	..	500
..	..	..	..	..	..	435	..
400	349	..	..	..	..	..	400
..	..	..	319	..	..	..	..
300	..	..	..	..	..	..	300
..	..	..	..	279	280	..	..
200	..	..	..	..	..	..	200
100	..	..	..	..	..	..	100

Results

The ISSR banding patterns of 'Ambrozyana' and 'Fulhamensis' and the two trees from the University of Exeter were identical within each sample group. (Fig. 1). Banding patterns are identical if the same banding phenotype is obtained with all primers used. Panel a shows four samples of 'Ambrozyana' using UBC 834, the bands are identical for all four samples. 'Ambrozyana' band patterns for the other UBC primers were consistent and identical within the four. This data is not shown. This was the case with the two samples from the University of Exeter shown in Panel b. Using UBC primer 834 (Panel c), Lanes A and B, the trees from Exeter show different banding patterns from GHI which are three samples of 'Fulhamensis' and CDF, three of the trees from Kensington which match the 'Fulhamensis'. Lane E from a tree at Kensington with similar morphology does not match the banding pattern of any other samples. The three trees from Kensington can be considered identical to the 'Fulhamensis'.

The samples labeled 'Crispa' and 'Cana Major' were different from each other and furthermore no 'Crispa' was the same as the other two, (Fig. 2) and no 'Cana Major' identical to the other five (Fig. 3). 'Lucombeana' showed that none of the nine different samples had identical banding patterns (Fig. 4). It should be noted that a different banding pattern always had more than two bands different from another sample being considered with the same name. There were no instances of samples being different by only one band. 'Fulhamensis' bands matched those of three of the samples from London and that of 'Lucombeana' #88.0329 (Fig. 5).

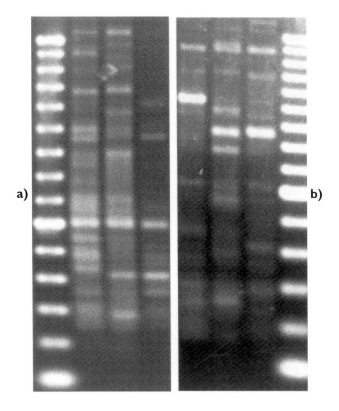

FIG. 2. Banding pattern a)UBC 811. b) UBC 841 of three samples of 'Crispa' 43.0463, 43.0464, 43.0465.

The two accessions of 'Diversifolia' were identical to each other, as were those of 'Hemelrijk' and 'Wageningen'. These latter two cultivars, often referred to as *Q.* × *hispanica* differed from each other as well as the third Dutch cultivar 'Waasland'. The ten samples identified only as *Q.* × *hispanica* showed no similarities to each other nor to any of the named cultivars. This data is not shown.

Discussion

It can be concluded from gel evidence that in the case where banding patterns are identical in multiple different samples of the same named cultivar (four or five minimum) that the given taxonomic designation is correct. Other unnamed samples with the same ISSR profile as the accepted cultivar then can be named. In the case of 'Ambrozyana' accessions 80.0280A, 82.0296A, 77.4402 and 77.2213, the different samples are identical and these samples are almost certainly correctly named. The three 'Fulhamensis' accessions 86.2228, 82.0142 and 78.1942 are identical and the unnamed samples from West Ham Park and Chiswick House that match them can be considered the same clone.

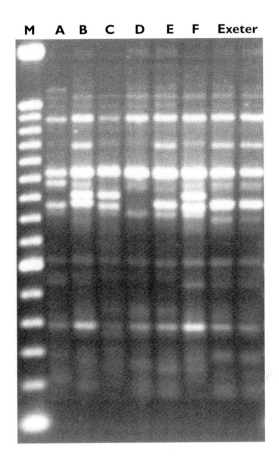

FIG. 3. Banding pattern UBC 834 for six samples of 'Cana Major', Lanes A–F: Accessions 27.0055, 43.0457, 27.0056, 27.0057, 27.0058 and 43.0455.

Certain cultivars remain problematical, 'Cana Major', 'Crispa' and 'Lucombeana' revealed no banding pattern that was identical within each sample group of the same name. Cultivars that had no more than one or two samples could be said to be not identical to other cultivars if their banding patterns using the same primer (and others) did not match banding patterns of any other cultivar. These include 'Laciniata', 'Suberosa', 'Diversifolia' and *Q. cerris* 'Wodan' which are different from each other and other named cultivars. The ten samples received only as *Q. × hispanica* did not match any named samples and could possibly have been grown from acorns (seed propagated), rather than have been vegetatively propagated.

The case of the Lucombe oak is particularly interesting. As no consensus was found from the samples received as 'Lucombeana', we would offer the following suggestion. A much larger sample set of 'Lucombeana' should be collected including those from historical collections (such as RBG, Kew, Cambridge Botanic Gardens, etc.) which date

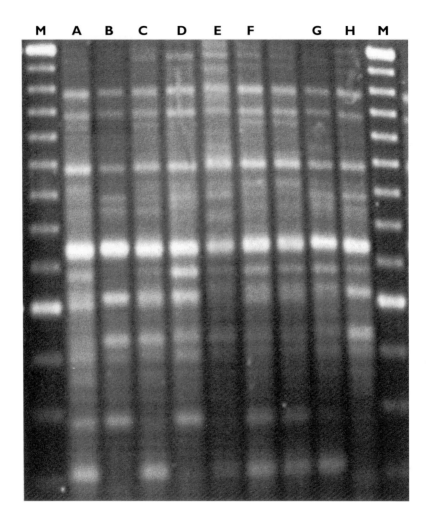

FIG. 4. Banding pattern UBC 834 for nine samples of 'Lucombeana', Lane A 88.0329, Lane B
43.0384, Lane C 56.0039, Lane D 14.0133, Lane E 17.0048, Lane F 77.5412, Lane G 41.0425,
and Lane H 44.0327. One lane of 'Lucombeana' is unlabeled due to a confusion on the label
of the posted sample.

from the time of Lucombe and are vegetatively propagated; a graft union being visibly
obvious. The remaining tree from Exeter should be evaluated and included in this
collection. Additional primers (7 to 10) should be used to evaluate the larger
'Lucombeana' sample set. A minimum number of samples with identical banding
patterns for all primers used should be established as a criterion for the identification
of a cultivar. When such a number is determined, then one could establish what is
validly a Lucombe oak. Such a proposal is not of course our decision and until that
time, we will still be in search of the Lucombe oak.

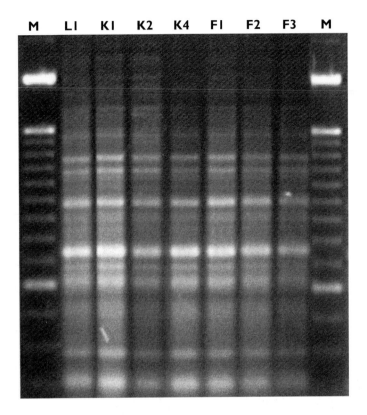

Fig. 5. Lanes 1 and 9 are molecular weight standards, Lane 2 is banding pattern of UBC 841 and 'Lucombeana' 88.0329, Lane 3, 4, 5 are unnamed oaks from the London area, Lanes 6, 7 and 8 are three samples of 'Fulhamensis'.

Acknowledgments

The following persons aided in the collection of samples, photographs, DNA extractions or PCR ISSR evaluations: Frank W. Telewski, Beal Botanical Garden, Michigan State University, USA; Hugh C. Angus, Westonbirt Arboretum, UK; Steven Scarr, University of Exeter, UK; Elinor Wiltshire, Carroll House, London, UK; Glen T. Jones and Jan Rademaker, Michigan State University, USA; Paula Marquardt, USDA, University of Wisconsin, Rhinelander, USA; Bryan Epperson, Department of Forestry and the MSU US DOE Sequencing Facility, Michigan State University, USA and M.V. Ashley, University of Illinois at Chicago, USA.

References

Bean, W.J. (1976). *Quercus*. In trees and shrubs hardy in the British Isles. Vol. 3. pp. 456–521. John Murray, London.

Caldwell, J. & Wilkinson, J. (1953). The Exeter Oak – *Quercus lucumbeana*. *Trans. Devonshire Assoc.*: 35–40.

DNA Proscan. (1997). DNA Pro Scan, Inc., Nashville, TN 37212.

Dow, B.D., Ashley, M.V. & Howe, H.F. (1995). Characterization of highly variable (GA/CT)$_n$ microsatellites in the bur oak, *Quercus macrocarpa. Theor. Appl. Genet.* 91(1): 137–141.

Dow, B.D. & Ashley, M.V. (1996). Microsatellite analysis of seed dispersal and parentage of saplings in bur oak, *Quercus macrocarpa. Mol. Ecol.* 5(5): 615–627.

Elwes, H.J. & Henry, A. (1906). *Quercus lucombeana* Lucombe oak. In The trees of Great Britain and Ireland. Vol. 5. pp. 1259–1267. Privately printed, Edinburgh.

Loudon, J.C. (1838). *Quercus cerris* var. *fulhamensis* Loudon. In Arboretum et Fruticetum Britannicum. Vol. 3. pp. 1850–1851. Longman, Orme, Brown, Green & Longmans, London.

Mitchell, A. (1994). The Lucombe Oaks. *The Plantsman*, 15(4): 216–224.

Scott, K.D. & Playford, J. (1996). DNA extraction technique for PCR in rain forest plant species. *Biotechniques* 20: 974–978.

Sweet, R.S. (1826). *Quercus lucombeana*. In Sweet's Hortus Britannicus or, a catalogue of plants cultivated in the gardens of Great Britain. (Ed. 1) p. 370. James Ridgeway, Piccadilly.

Wang, Z., Weber, J.L., Zhong, G. & Tanksley, S.D. (1994). Survey of plant short tandem DNA repeats. *Theor. Appl. Genet.* 88(1): 1–6.

Weising, K., Nybom, H., Wolff, K. & Meter, W. (1995). DNA fingerprinting in plants and fungi. 332 pp. CRC, London.

Wolfe, A.D., Xiang, Q.-Y., & Kephart, S.R. (1998). Assessing hybridization in natural populations of *Penstemon* (*Scrophulariaceae*) using hypervariable intersimple sequence repeat (ISSR) bands. *Mol. Ecol.* 7(9): 1107–1125.

Zietkiewicz, E., Rafalski, A. & Labuda, D. (1994). Genome fingerprinting by simple sequence repeat (SSR) – anchored polymerase chain reaction amplification. *Genomics* 20(2): 176–183.

McLewin, W. (1999). Fundamental taxonomic problems in and arising from the genus *Helleborus*. In: S. Andrews, A.C. Leslie and C. Alexander (Editors). Taxonomy of Cultivated Plants: Third International Symposium, pp. 297–304. Royal Botanic Gardens, Kew.

FUNDAMENTAL TAXONOMIC PROBLEMS IN AND ARISING FROM THE GENUS *HELLEBORUS*

W. McLewin

Phedar Research & Experimental Nursery, Bunkers Hill, Romiley, Stockport, Cheshire SK6 3DS, UK

Abstract

The conventional criteria for classification, i.e. flower morphology at generic and higher levels, and the addition of vegetative morphology at species levels have substantial shortcomings in the case of *Helleborus* (*Ranunculaceae*).

The genus contains a few very diverse species with extremely disparate characteristics and propagation requirements. This would not matter if hellebores were of only academic interest but it causes widespread confusion among gardeners and horticulturists.

Within *Helleborus* the degree of speciation varies enormously. There are some very well-defined species and a complex of overlapping species. In most wild colonies, especially of the acaulescent species, there is wide variation in all but the most basic morphological characters. This is another cause of inaccuracy and confusion.

These two distinct but interacting problems suggest that a change of emphasis and perhaps attitude in taxonomic practice would be beneficial, at least for those who live and work with the consequences of taxonomic decisions.

Introduction

Hellebores are fairly familiar plants, distributed, rather unevenly, over most of Europe and the Caucasus, particularly in the former Jugoslavia. In recent years they have become very popular as hardy, trouble-free, late-winter-flowering perennials; the true wild species less so than the larger-flowered more showy hybrids.

As a whole the genus has an irritating property — it contains plants sufficiently diverse to make it almost impossible to make a useful simple statement about hellebores that does not need to be qualified by a list of exceptions. The one definite statement that can be made is that the genus seems to be well defined and there is no doubt or dispute about which plants are included — except that *Helleborus vesicarius* Aucher is so unlike all the others it would be better in a genus by itself (*pers. obs.*).

Within *Helleborus* L. there are some very clearly defined species as well as a confusing group of less distinct taxa with weak differentiation. A useful division of the taxa can be made into two groups: caulescent and acaulescent. The caulescent group consists of three/four species that are clearly distinct in the sense that: they are unmistakable in the wild; the wild populations are fairly homogeneous; and either each does not interbreed with any other species or if they do albeit with help, (*H. niger* L. and *H. argutifolius* Viv., for example), they produce infertile offspring. In contrast the acaulescent species, e.g. *H. odorus* Waldst. & Kit. (see Figs. 4a, 4b), *H. purpurascens* Waldst. & Kit., *H. torquatus* Archer-Hind, *H. multifidus* Vis., etc. occur in wild colonies which are markedly heterogeneous (see Fig. 1) and they interbreed readily producing fertile offspring.

Thus, one fundamental problem in the taxonomic characteristics is relating phenetic characteristics with phylogeny.

The Acaulescent Species

The focus here is on the acaulescent species. Their geographical distribution is so wide and their appearance is so different that it is completely impracticable and unhelpful to lump them all together. The question then is how to distinguish them and how to describe them. Although there are differences in root and rhizome characteristics they are not sufficiently definitive and in any case there is no accepted taxonomic language for describing them. So one is left with flowers and leaves.

Consider as an example *H. atrorubens* Waldst. & Kit. This species is relatively well defined geographically in quite a small area (perhaps 2000 km^2) but the flower colour varies from green to purple inside and out, with all possible combinations in between, including stripes and dark veins, and even internal spotting, although the last has never been mentioned in the literature (Mathew 1985). The leaves of *H. atrorubens* do have a rudimentary morphological consistency — they are basically palmate with apparently about nine to eleven main leaflets (see Figs. 2, 3). However, the leaflets are not all attached to the supporting stalk in the same way and one can argue that there are only three, or alternatively only five, some of which divide at the base, or near the base, or elsewhere; the leaflets can be finely or very coarsely toothed, they can be very narrow or very wide and overlapping or they can be essentially undivided, partly so or all divided into smaller lobes; and of course their individual shape varies. Any formal description which encompasses all these possible combinations will encompass varying proportions of the foliage of all the other acaulescent species.

It is important to emphasise that the variation in flower colour and in leaf morphology is extensive, and also apparently continuous and random. Plants with widely different leaves and flower colours can be found in all wild populations, growing together with apparently no correlation to immediate environmental conditions. It is not the case that there are different 'forms' or entities in the usual sense, although some authors have succumbed to this temptation (Martinis 1973). In wild colonies of *H. atrorubens* leaves could be selected to represent most of the recognised acaulescent species.

In spite of this, *H. atrorubens* is a well defined and unambiguous taxon and these remarks apply with equal force to most other acaulescent species. It is not easy to explain this apparent contradiction — perhaps it is just a case of familiarity after a long period of study/struggle. Certainly the location of wild plants plays a part, also other characters not normally used, like ease of division of the rhizome for propagation purposes.

The leaf silhouettes from four separate sites in Figs. 1–4b do not illustrate different forms; they are examples of a continuous spectrum of leaf shapes.

FIG. 1. Leaf silhouettes of *Helleborus multifidus* subsp. *multifidus* showing variation. Specimens collected in NW Croatia at site number WM9635.

FIG. 2. Leaf silhouettes of *Helleborus atrorubens* showing variation. Specimens collected in SE Slovenia at site number WM9803.

FIG. 3. Leaf silhouettes of *Helleborus atrorubens* showing variation. Specimens collected in SE Slovenia at site number WM9616.

FIG. 4a. Leaf silhouettes of *Helleborus odorus* showing variation. Specimens collected in SW Hungary at site number WM9640.

Conclusions

As a result of combining intensive horticultural activity with hellebores and regular fieldwork over many years, the civil war in former Jugoslavia notwithstanding, I find myself with several conclusions.

The Linnean concept of a genus subdivided into species, in which each species is determined by a description of morphological features, seems inappropriate and is certainly inadequate in the light of today's knowledge of the genus. Thus, almost all references to acaulescent hellebores in horticulture are inaccurate.

It would have been better if the phylogenic element in speciation had been made more precise. If, in a batch of mixed cultivated hybrid seedlings, there is one which looks like *H. atrorubens*, is it actually *H. atrorubens*? I am not sure that taxonomists are clear about this question, so it is no wonder that in horticulture there is widespread abuse of species names.

FIG. 4b. Leaf silhouettes of *Helleborus odorus* showing variation. Specimens collected in SW Hungary at site number WM9640.

It would have been better if provenance was used more systematically and given more weight. Of course this would have the consequence that a plant could not be regarded as an example of a particular species without provenance, i.e. simply looking more or less right is not enough. This approach seems a necessity with hellebore species.

It would have been much more satisfactory if it were standard practice when preparing herbarium specimens always to have an accompanying assessment of the homogeneity of the parent population; and much better if a representative range of specimens (when appropriate) was regarded as necessary. It would also be desirable if formal descriptions of species included such an analysis and if published botanical drawings of species followed this approach.

It would have been better if the principle that every wild plant in a certain genus is in a particular species was abandoned and instead some plants, and indeed some colonies, where the morphology was variable and intermediate, were simply accepted as intermediates between well-defined species.

References

Martinis, Z. (1973). *Helleborus*. In S. Horvatic (ed.). Analitika Fl. Jugoslav. Vol. 1(2). pp. 231–243.

Mathew, B. (1989). *Helleborus*. In S.M. Walters *et al.* (eds). The European garden flora Vol. 3. pp. 328–331. Cambridge University Press, Cambridge.

Batdorf, L.R. (1999). Reducing new synonyms of infraspecific nomenclature through utilization of the world wide web with particular reference to *Buxus* (*Buxaceae*). In: S. Andrews, A.C. Leslie and C. Alexander (Editors). Taxonomy of Cultivated Plants: Third International Symposium, pp. 305–310. Royal Botanic Gardens, Kew.

REDUCING NEW SYNONYMS OF INFRASPECIFIC NOMENCLATURE THROUGH UTILIZATION OF THE WORLD WIDE WEB WITH PARTICULAR REFERENCE TO *BUXUS* (*BUXACEAE*)

LYNN R. BATDORF

International Registration Authority for *Buxus*, U.S. National Arboretum, 3501 New York Avenue NE, Washington DC 20002, USA

Abstract

An enumeration of *Buxus* taxa reveals 161 valid cultivars. However, in Europe and North America there are approximately 470 valid and invalid cultivar names. This discrepancy is due largely to synonyms that occur for species, cultivar and vernacular names. The checklists and registration lists of cultivar names published by International Registration Authorities have a limited impact on reducing the use of synonyms and validating cultivars. There is a need for a uniform and widely accessible world wide web site where the correct nomenclatural information is available to interested parties.

Introduction

Names that are synonyms occur in several situations. The first involves valid cultivar epithets that are mistakenly assigned to the wrong species. For example, *Buxus harlandii* Hance is somewhat obscure when compared to *B. sempervirens* L. Due to the popularity of *B. sempervirens* and its cultivars, *Buxus* L. has been perceived as a monotypic genus. Thus, when confronted with the binomial *B. harlandii*, the name is changed to *B. sempervirens* 'Harlandii'. At other times the one species is often confused with another because of shared similar characters, e.g. the valid name *B. harlandii* 'Richard' is changed to *B. microphylla* 'Richard' or 'Richardii', or *B. microphylla* Sieb. & Zucc. var. *japonica* 'Richardii'. When confronted with a less well-known species, the typical conclusion is that the specific epithet must be in error and belongs to a more common taxon. Thus, valid cultivar names are assigned to erroneous species.

It is possible for the same plant to have both a trademark name and a cultivar name. For example, *Buxus* 'Glencoe' is a hybrid boxwood with the trademark name of Chicagoland Green™. This has caused confusion in the nursery industry, which has removed the single quotation marks around the cultivar name and the trademark symbol. Nurserymen are generally more comfortable with common names, and often eliminate or confuse cultivar and trademark names. Thus, the names Chicagoland Green™ and *B.* 'Glencoe' have been replaced by the common name Chicagoland Boxwood, which has become accepted in the nursery trade. Similar trademark and cultivar name changes occur with other boxwoods.

Multiple synonyms can exist for the same clone. Perhaps the best example of this is *B. sempervirens* 'Suffruticosa', or English Boxwood. Cultivar synonyms include: 'Fruticosa', 'Humilis', 'Mt. Vernon', 'Nana', 'Rosmarinifolia Fruticosa', 'Rosmarinifolia Minor', 'Suffruticosa Nana', and 'Truedwarf'. In addition, botanical synonyms include: *B. humilis* Dod., *B. sempervirens* var. *nana* Hort., *B. sempervirens* var. *suffruticosa* Hort. and *B. suffruticosa* Mill. Further English vernacular synonyms include: Dwarf Box, Dwarf English, Edging Box, English Boxwood, English Dwarf Box, Ground Box, Old English Boxwood, True Edging Box and Truedwarf Boxwood. There are at least seven vernacular names in French, three in German and four in Dutch. There are probably additional names in other languages. As shown by this example this one clone has over 35 names.

Confusion over names also occurs between taxa. For example: *B. sinica* (Rehder & E.H. Wilson) Cheng var. *insularis* 'Winter Gem' is a Korean cultivar. Its vernacular name is Large Leaf Asiatic Boxwood. Synonyms include: *B. microphylla* 'Asiatic Winter Gem', *B. microphylla* var. *asiatica* 'Winter Gem', *B. microphylla* 'Wintergem', *B. sinica* var. *insularis* 'Large Leaf Asiatic', and simply, *B. sinica* var. *insularis* (Nakai) M. Cheng. The various synonyms occur because this plant is of Asian origin, has a relatively large leaf, maintains a dark green colour during the winter, and is a gem. The synonyms were created in an attempt by the nursery industry to describe and promote the various unique characteristics of this attractive boxwood.

There are other events that result in the creation of synonyms and invalid cultivar names. For example, if a cultivar name is lost or confused in the trade, the industry is often quick to create a new name to ensure the sale of the plant. Competing nurseries must respond to market pressures and are anxious to add new and exciting plants. In an effort to improve the marketability of plants, unexciting cultivar names like 'Graham Blandy' are changed to flashier names like 'Greenpeace' to increase their appeal and stimulate sales. Foreign cultivar names that seem difficult or awkward can have a negative impact on sales. On occasion the trade has modified or completely changed these names to make them seem more appropriate for the market area. Also at play are regional influences and landscape uses which give rise to the creation of new vernacular names. Sometimes, hybridizers and those naming new selections are unaware of the whole registration process and thus assign improper names.

Proposal

Many of the complexities surrounding infraspecific nomenclatural synonyms largely centre around correct information dissemination. Typically, registration and checklists (e.g. Batdorf 1995) produced by International Registration Authorities (IRAs) only reach a specialized audience. Arboreta, botanic gardens, research institutions, and plant societies which are primarily responsible for breeding, evaluating, the increasing of stock, naming, and the initial distribution of the plant, have excellent access to this information. However, several other large and important groups are not included in this distribution of information. They comprise the nursery industry, professional gardeners, propagators, researchers, and plant collectors, who are primarily responsible for the distribution of plants. Here, the need for proper names and descriptions to identify the plant and provide the correct nomenclature is at its greatest.

The challenge for the IRA is to communicate effectively with these various groups. One avenue is through the world wide web (WWW). Each IRA could establish a web site under the guidance and instruction of the ISHS (International Society for

Horticultural Science) Commission for Nomenclature and Registration which would improve the distribution of correct nomenclatural information. The traditional format is to list the valid name with its earliest bibliographic reference and any synonyms, if applicable. This web site would greatly promote the use of correct cultivar names and would also reduce the use of synonyms. While this format is exceptionally useful to informed audiences, it is difficult to access and has limited application for others. For example, verifying that a particular plant has the correct name is not possible. A different approach is required to include information concerning plant characteristics and to keep this information user-friendly. Each genus has its own unique characteristics with which the IRA is familiar. For example, *Buxus* at the cultivar rank is best differentiated by habit and leaf morphology. For *Hemerocallis* L., a range of floral characteristics are used to differentiate the thousands of hybrids. Brief descriptions in lay terms would permit a broader understanding of these characteristics and allow groups, and possibly individual cultivars to be distinguished. Synonyms and invalid names are often perpetuated because many are unaware of the resources available to obtain a definitive identification at the cultivar level.

Discussion

Providing correct information to a wide range of interested audiences has always been one of the greatest challenges for an IRA. A recent development has been the advent of electronic databases for plant nomenclature that can be accessed through the WWW. This proposed forum is unique in that it is constructed by the individual IRA, who has specialized knowledge regarding the cultivars of their respective genus. The common search engines available in the WWW do not permit quick or easy access to the ISHS Commissions' site. Access to the ISHS Nomenclature and Registration site could be improved by adding generic key words such as: plant registration, taxonomy, nomenclature, and cultivar. Additionally, the use of keywords such as the genus and its respective common name (i.e. *Nymphaea* and Water Lily) would properly guide the user to the Directory of International Registration Authorities at the ISHS site.

There are distinct advantages to this proposal. Increasing access by adding key words that allow search engines to locate the IRA site would improve information dissemination and would permit queries from a vast audience to access online information regarding a specific group of cultivars. The information on the site would include nomenclature, registration, synonyms, cultivar characteristics, etc. It would permit easy, accurate and quick access to interested parties such as hybridizers, the nursery industry, and institutions. The ability to contact the appropriate IRA when naming a plant will assist the IRA in processing cultivar and cultivar-group epithets and in maintaining more accurate records. The plant industry will benefit by having properly named plants. The IRAs will be able to reduce the use of synonyms and maintain accurate, comprehensive and up to date lists of cultivars within their respective fields.

The WWW has already established itself as an important forum. Indeed the ISHS has a web site that is regularly and often updated. Twelve main topics are presented on the home page of the ISHS and two topics are of interest in this discussion. The first is: "Links to other horticultural pages". The creation of other links to specialized horticultural areas within the interests of the ISHS would broaden resources. For example, IRAs and those seeking links to botanical nomenclature and resources may be interested in some of the following sites, see Table 1.

TABLE 1: Web sites relating to botanical nomenclature and resources.

Resource	Address
Aquatic & Wetland Database	http://aquat1.ifas.ufl.edu/database.html
Atlas Flora Europaeae	http://www.helsinki.fi/kmus/afe/database.html.
California State Univ. Bio. Sci. Web.	http://130.17.2.215/
Carnivorous Plants Database	http://www.hpl.hp.com/botany/public_html/cp/html/actualcp.htm
Checklists & Floras, Tax. Databases	http://bgbm3.bgbm.fu-berlin.de/botflora.html
Checklist of Floras of U.S.	http://trident.ftc.nrcs.usda.gov/plants/staselec.html
Chromosome number index	gopher://cissus.mobot.org/77/.Chromo/.index/chromo
Classification of flowering plants	http://www.systbot.uu.se/classification/overview.html
CropSEARCH	http://www.hort.purdue.edu/newcrop/CropSEARCH
CyperFlora California	http://www.csd.tamu.edu/FLORA/calflora/calflora.htm
Families & Genera of Vascular Plt.	http://www.ars-grin.gov/npgs/tax/taxfam.html
Flora Europea, RBG, Edinburgh	http://www.rbge.org.uk/forms.fe
Flora North America	http://www.fna.org/index1.html
Flora North America Online Search	http://www.fna.org/Libraries/plib/WWW/online.html
FLORIN Taxonomy	http://mitia.florin.ru/florin/brief/b_tax.htm
FlowerBase	http://www.flowerbase.com/
FlowerWeb	http://www.flowerweb.nl/
Germplasm Resources Info. Network	http://www.ars-grin.gov/npgs/tax/index.html
Global Plant Checklist Int'l Org.	http://iopi.csu.edu.au/iopi
Gray Herbarium Card Index	http://herbaria.harvard.edu:80/Data/Gray/gray.html
Index Herbariorum USA	http://www.nybg.org/bsci/ih/ih.html
Index Nominum Genericorum (ING)	http://www.nmnh.si.edu/ing/
Index Virum	http://life.anu.edu.su/viruses/lctv/index/html
Indices Nominum Supragenericorum	http://waffle.nal.usda.gov.lagdb/ind_nm-shtml
Int'l Assoc. Plant Taxonomy	http://bgbm3.bgbm.fu-berlin.de/IAPT/default.htm
Int'l Code of Bot. Nom. -Tokyo Code	http://www.bgbm.fu-berlin.de/iapt/nomenclature/
Int'l Org. of Palaeobotany Plt. Fossil	http://ibs.uel.ac.uk/ibs/palaeo/pfr2/pfr.htmcode/tokyo-e/default.htm
Int'l Society for Horticultural Sci.	http://www.ishs.org/
Internet Biodiversity Service	http://ibs.uel.ac.uk/ibs/
Internet Directory for Botany	http://www.helsinki.fi/kmus/botmenu.html

TABLE 1 continued

Resource	Address
Integrated Tax. Info. System (ITIS)	http://trident.ftc.nrcs.usda.gov/itis/
Links to Lower Plant Taxa	http://www.helsinki.fi/kmus/botcryp.html
Links to Vascular Plant Taxa	http://www.helsinki.fi/kmus/botvasc.html
Links to Fossil Taxa & Palynology	http://www.helsinki.fi/kmus/botpale.html
Names in use for extant plant genera	http://www.bgbm.fu-berlin.de/iapt/ncu/genera
Nat'l Biological Info. Infrastructure	http://www.nbii.gov/
Nat'l Center for Biotechnology Info.	http://www3.ncbi.nlm.nih.gov/Taxonomy/taxonomyhome.html
Nat'l Wildflower Research Center	http://www.wildflower.org/
New York Botanical Garden	http://www.nybg.org/bsci/hcol/hcol.html
New World Grass Checklist	http://www.mobot.org/MOBOT/tropics/Poa/agfnames.html
Noxious Weeds of USDA in GRIN	http://www.ars-grin.gov/cgi-bin/npgs/html/taxweed.pl
NOAA/Paleoclimatology Pollen Page	http://www.ngdc.noaa.gov/paleo/pollen.html
Peter's Carnivorous Plant Page	http://www.flytrap.demon.co.uk/
Phylogenetic Resources	http://www.ucmp.berkeley.edu/subway/phylogen.html
The Plant Kingdom	http://www.geocities.com/RainForest/6243/diversity4.html#Plant
Plant Chromosome Numbers Database	gopher://cissus.mobot.org/77/chromo/index/chromo
PLANTS Database	http://plants.usda.gov/
Plant Systematics & Evolution - Links	http://www.csdl.tamu.edu/FLORA/tfp/links.html
Plant Trivia Timeline	http://www.Huntington.org/BotanicalDiv/Timeline.html
Publications Database	http://www.herbaria.harvard.edu/Data/Publications/publications.html
Resources for Systematics Research	http://141.211.110.91/tool_dir.htm
Royal Bot. Garden, Kew	http://www.rbgkew.org.uk/web.dbs/webdbsintro.html
The Tree of Life	http://phylogeny.arizona.edu/tree/phylogeny.html
TreeBase	http://www.herbaria.harvard.edu/treebase/index.html
TROPICOS	http://mobot.mobot.org/Pick/Search/pick.html
Weed Images & Descriptions	http://www.rce.rutgers.edu/weeddocuments/index.htm
World Economic Plants in GRIN	http://www.ars-grin.gov/npgs/tax/taxecon.html
World Species Lists	http://www.envirolink.org/species/

The second main topic of interest on the home page of the ISHS is the International Registration Authorities List. This directory is also regularly updated giving a complete listing of IRAs in alphabetical order according to their respective taxa. An examination of this list (updated 12 May '98) reveals 127 registrars with mailing addresses, 68 with fax numbers, and 49 with e-mail addresses. The e-mail addresses provide a link to the respective registrar.

By providing an additional link to a page created by the registrar, the user could assess specific information regarding a particular cultivar-group, which could include links in the directory to all the IRAs for registration applications, registration lists, nomenclature, synonyms, plant characteristics and identification, commercial sources, or any other related topic to the appropriate IRA.

Conclusion

Use of the world wide web, assigning appropriate keywords so that interested parties could more easily link to and locate information on the proposed registrar sites, would accomplish several goals. The broad dissemination of information would reduce the use of synonyms and invalid names for all taxa. The nursery industry, botanic gardens, professional gardeners, propagators, researchers and other interested groups would have an authoritative resource that is easily accessible.

Registrars would be able to maintain a current registration list at this site. With improved access IRAs would have an additional resource to locate new taxa and new synonyms and name combinations would come to the attention of the IRA more quickly. The trade would be encouraged to submit registrations, thereby reducing invalid nomenclature.

Reference

Batdorf, L.R. (1995). Boxwood Handbook, a practical guide to knowing and growing boxwood. 99 pp. American Boxwood Society.

Stirton, C.H. (1999). The naturalised *Lantana camara* L. (*Lantaneae-Verbenaceae*) complex in KwaZulu-Natal, South Africa: a dilemma for the culton concept. In: S. Andrews, A.C. Leslie and C. Alexander (Editors). Taxonomy of Cultivated Plants, pp. 311–324. Royal Botanic Gardens, Kew.

THE NATURALISED *LANTANA CAMARA* L. (*LANTANEAE-VERBENACEAE*) COMPLEX IN KWAZULU-NATAL, SOUTH AFRICA: A DILEMMA FOR THE CULTON CONCEPT

CHARLES H. STIRTON

The National Botanic Garden of Wales, Middleton Hall, Llanarthne, Carmarthenshire SA32 8HW, Wales, UK

Abstract

Lantana camara is a widespread pantropical weed exhibiting enormous morphological variation across its range. A taxonomic study of the naturalised complex in Natal identified 18 taxa using floral and vegetative characters. Nine of the taxa are widespread in KwaZulu-Natal whereas most of the other taxa are either regionally localised or centred around urban areas. Attempts to provide names for the variants has proved difficult as the different taxa may be found both in cultivation and as naturalised invasive weeds. The study showed that the application of the culton and taxon concepts provided only a partial solution towards providing names that were of practical use to a wide range of users.

Introduction

Lantana L. is predominantly a New World genus of some 150 species (Mabberley 1987). It has been cultivated in temperate regions for over 300 years, having been introduced into Europe from the Neotropics as early as 1636 by Prince Johan Maurits van Nassau (Howard 1969). The history of how more than 650 cultivated varieties were distributed around the globe during the nineteenth century to become, in many areas, major weeds, is given in Howard (1969) and Stirton (1977). The most invasive taxon is *Lantana camara* L., which is the subject of this paper (Fig. 1).

Lantana camara is widely accepted as a major weed of tropical countries. Stirton (1977) provided a map of the earliest recorded dates of introduction into different tropical countries. The weed is noted for its toxicity to livestock (Morton 1994) and for its rapid invasion of natural veld, waste ground, forest margins and derelict or cultivated lands (Stirton 1977, Sharma *et al.* 1988, Munir 1996).

L. camara sensu lato is a polyploid aggregate species complex with a basic chromosome number of x=11, having diploid, triploid, tetraploid, pentaploid, and hexaploid representatives (Spies & Stirton 1982a, Spies 1984, Natarajan & Ahuja 1957). It represents throughout its naturalised range a formidable taxonomic problem (Stirton 1977).

Howard (1969) conceded that the genus "*Lantana* is not an easy one to consider taxonomically". Moldenke (1971) later said that ".. probably a lifetime project could be made of a thorough and intensive study of the *Lantana camara* Group" and he

311

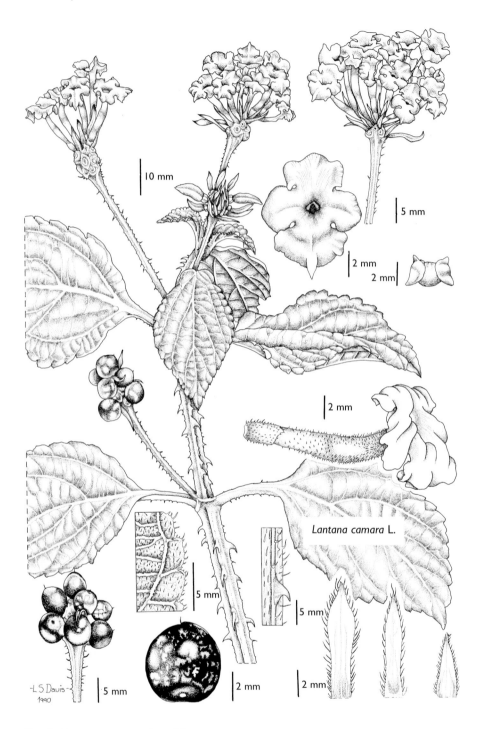

FIG. 1. *Lantana camara* collected by *L. Davis* s.n. (NU), Manderston Road, Pietermaritzburg, South Africa.

expressed serious reservations as to "how many taxa can really be distinguished, how they can be keyed out, and what is their relationship to each other. How greatly is hybridity involved here?" Munir (1996) provides an overview of the diverse opinions on the intractability of defining taxa within the complex. He concluded that "So far, no one has come up with a satisfactory answer to these questions."

This paper presents some of the results of two decades of study of the complex in southern Africa. It summarises the nature of the taxonomic difficulty, recognises that this difficulty is not the same in all regions of the world where the species has naturalised and that the naming of variants in such a complex is not made easy by the existing botanical and cultivated Codes of Nomenclature. Finally, the culton concept is examined and found too narrow to be of use in dealing with actively evolving weed complexes of horticultural value. A pragmatic naming solution is presented.

The Nature of the Taxonomic Problem

There has been little critical study of the *L. camara* complex across its native and naturalised ranges and only a handful of regional taxonomic studies have been made. Smith & Smith's (1982) review of the naturalised Australian taxa is the only detailed infraspecific analysis to date.

A promising start was made on the complex in the Caribbean by Sanders (1987a & b, 1989) but his work was abandoned as no further funding was available. The only other ongoing research that I am aware of is the molecular systematic investigation of Australian and South Pacific taxa by the Alan Fletcher Research Station in Australia (Martin Hannan-Jones *pers. comm.*).

The genus *Lantana* is based on at least two numbers, x=11 and x=12, with polyploid complexes in each of the lines (Sanders 1987b). In the Neotropics the juxtaposition of native taxa and naturalised cultivars has led to complex morphological variation that has confused the taxonomy of the genus there (Sanders 1987a).

There are strong differences between the taxa in the Caribbean and in Africa. In the Caribbean there are no meiotic irregularities in the diploids (Sanders 1987a) whereas, there are in South Africa (Spies 1984). Sanders has suggested that the southern African naturalised diploids are segmental allodiploids, with the degree of chromosomal homology among the plants being directly related to the history of hybridisation in the plants' parentage (Sanders 1987a).

Ploidy level and morphological correlation in section *Lantana* are weak in the Old World but strong in the New World. In the Caribbean Sanders (1987a) has shown that diploids and subspecies are morphologically distinctive, are part of the natural flora and have endemic distributions. This contrasts with the polyploids which are coarser, more variable morphologically, occupy a wider range of habitats and often overlap with the diploids in the expression of characters. In the Caribbean Sanders (1987a) has recommended that chromosomal data can augment morphological criteria of species concepts in *Lantana*. These should "rely first on discontinuities among diploid taxa, next on discontinuities between diploid taxa and even polyploid taxa with regular meiosis, and last on discontinuities among even-polyploids with regular meiosis. Triploids should be treated as hybrids." He goes on to say that "tetraploid morphological types with irregular meiosis are likely to be naturalised cultivars, probably of hybrid ancestry". He recommends that these latter tetraploids should be treated as intermediates or included in the polyploid taxa to which they are most familiar.

H. Moldenke published a series of fragmented notes based on his field studies and morphological investigations between 1940–1971 (Moldenke 1971). Unfortunately he caused more nomenclatural confusion than resolution. By 1971 he had concluded that *L. camara* was variable and polymorphic with many races differing in size and shape of leaves, presence or absence of prickles, amount of pubescence and the size and colour of flowers. Over this thirty year period Moldenke changed his mind many times describing several species and infraspecific taxa. But, because the delimiting characters he used were variable and inconsistent, he was forced to redefine his taxa to occupy species, variety, subspecies and forma categories for the same entity. Eventually he had used up all the available Linnean categories from species to forma. It would have been a great help to later workers if he done an overall summary and provided a key to section *Camara* Cham. Munir (1996) in reviewing Moldenke's work concluded that "..the value of infraspecific categories in *L. camara* seems questionable" and he gave a detailed summary of the treatment of the complex in Australia and SE Asia.

Smith & Smith (1982) recognised 29 forms or taxa in Australia. Unfortunately, L.S. Smith, the principal author, died before he could write up his investigations into the origin of the Australian taxa. In his revision, completed by his widow Doris A. Smith, he noted that the majority of the naturalised taxa in Australia were not conspecific with material he had seen in Central and South America. This led him to believe that these taxa were introduced into Australia as cultivars. This was supported by my historical analysis of introductions (Stirton 1977).

Smith & Smith (1982) did not record extensive hybridisation in the field but did note that some individuals are subject to branch mutations including "reversion shoots" (throwbacks). They recorded that "A mutant branch may be produced with characters sufficiently distinct from the remainder of the plant to be regarded as a different taxon. Seeds from mutant branches have been found to produce plants similar to the mutant branch and in this way a new taxon may become established". This has been confirmed in Australia by Parsons & Cuthbertson (1992) and the present study in South Africa.

The Smiths found that "..with variation of such complexity it was impossible to name each taxon according to the *International Code of Botanical Nomenclature* (ICBN) particularly since the literature on these plants is so confused". They therefore used the "field names" coined by L.S. Smith. In their paper they noted that these are probably best regarded as cultivar names as defined by the *International Code of Nomenclature for Cultivated Plants* – 1980 (ICNCP) (Brickell *et al.* 1980).

In South Africa it was anticipated that the complex would present a number of unusual problems since the lengthy period of biological control studies by C.J. Cilliers (1977) had shown preferential feeding by potential biocontrol agents on a number of different flower colour forms in the wild. A countrywide investigation of these forms was undertaken between 1976–1983. Despite detailed embryological and cytogenetic investigations the complex remains unresolved (Spies & Stirton 1982a, b, c). While it is not yet possible to produce a natural classification of the complex in southern Africa as a whole (Spies 1984), it has been possible to delimit and produce an artificial key to the major variants in Natal. This is because although there is extensive hybridisation and polyploidisation in the old Transvaal Province (mainly in the new Gauteng and Mapumalanga Provinces), there is only a limited amount of hybridisation taking place in the wild in Natal. It has not been possible to use the existing Linnean categories with great accuracy, to describe the reticulate relationships of what is primarily an extensive hybrid swarm (spontaneous and introgressive) across many parts of its distribution range in southern Africa.

Spies (1984) concluded his cytological investigations of the South African taxa by saying that ".. Cytogenetic data demonstrates that *Lantana camara* in southern Africa is in an active speciation phase and because most plants are intermediates in a transitional stage of evolution no attempt to recognise infraspecific entities will succeed." He estimated that thousands of variants could be generated within the South African hybrid swarms.

Fig. 2 summarises the position as I see it globulally. To differentiate the invasive variants of *L. camara* from all others, Bailey's (1923) concept of cultigens (groups of plants in cultivation) and indigens (groups of plants in the wild) was used. As the previous discussion, and what follows, indicates, there are a raft of variants which fit neither into the indigen nor the cultigen concept. For the purpose of the overview in Fig. 2 a device called an "invason", i.e. a group of plants neither naturally wild nor cultivated but alien and persistent in nature, has been used. This is not a proposal to set up a new system of classification for this neglected entity, but merely a device to either complement the culton concept or to explore a bridging concept between taxon and culton. It is clear from an inspection of Fig. 2 that the complex shows considerable variability in its development on each of the continents analysed. This has important implications for biological control programmes and chemical control strategies. It implies that biological control strategies that work in one region will not necessarily work elsewhere. So how is one to develop a practical identification and classification scheme in southern Africa?

Lantana camara and the Culton Concept

Up to this point I have presented the history of the *L. camara* species complex in taxon terms, within the Linnean Code, and under the rules of the ICBN. This is because all those who have previously attempted to deal with the complex have done so strictly within the taxon domain.

No attempt has been made by earlier students of *Lantana* to either try to extend beyond the categories of form, variety, subspecies and species or to make use of other available divisions as proposed by authors such as Bailey (1918), Jirásek (1958, 1961), and Mansfield (1953, 1954). Nor has anyone used non taxon-based systems of classification to solve the classification problem of *L. camara* (Hetterscheid & Brandenburg's cultonomy 1995b). The latter authors believe that their classification of cultivated plants is fundamentally different from the classification of wild plants and is better suited to the classification of cultivated plants. It is necessary therefore to explore whether a cultonomic classification of *L. camara* might be more effective.

Hetterscheid *et al.* (1996) provide a detailed annotated history of the principles of cultivated plant classification. They note that throughout history there has been a division in approach among taxonomists between classifying cultivated plants and plants as found in nature. Their review shows that no matter what scheme was developed within the Linnean hierarchy, it nearly always ended up in inflated names, unstable nomenclatures, and unsatisfactory classifications. They argue that their systematic theory emphasises the clear distinctions between the goals of classifying wild and cultivated plants. The new general term they use for systematic categories of cultivated plants is the culton as opposed to the taxon used in the Linnean hierarchy.

It seem that most of the examples cited by Hetterscheid & Brandenburg (1995a & b), Hetterscheid *et al.* (1996) and other writers on the culton concept refer to vegetable, forage and cereal crops. Brandenburg (1999) argues that the culton concept provides a better approach to solving weed-crop complexes.

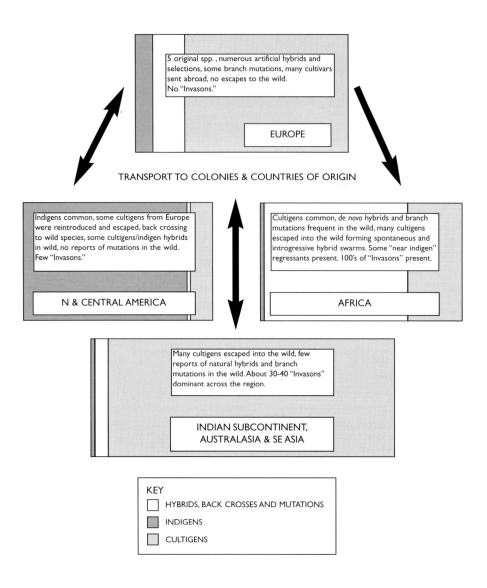

FIG. 2. The origins and dispersal of indigens and cultigens of the pantropical weed *Lantana camara* L. around the world.

Fig. 3 depicts the complicated derivative pathways that have and are happening in the evolution of the *L. camara* complex. These patterns are more complex than those depicted by Brandenburg. However, not many culton-based studies relate to horticultural species or weed species with horticulturally important congeners. It is vital therefore that the definition of a cultivar is examined more closely, as this is the basis of the culton approach.

Hetterscheid & Brandenburg (1995b) define a cultivated plant as one .."whose origin or selection is due to the activities of mankind" and that "Cultivars result from a mixture of natural and man-made processes, the influence of both differing from one cultivar to another."

Are cultonomic and taxonomic classifications really so different? Could it just be that the definition of a culton as being restricted to cultivars being "industrial products" is just too narrow, metaphorically or not? As was shown above, in *L. camara* the co-occurrence of cultigens and indigens contemporaneously makes no sense of such a restrictive definition. I agree with Alexander (1997) and McNeill (1998) when they question whether the processes of domestication and natural selection are so different. The *L. camara* example would suggest that where you have within the same species complex periodic bursts of spontaneous natural hybridisation, polyploidisation, coupled with secondary and tertiary contact with introgressive hybridisation, as well as "industrial selection", you are bound to end up with an assemblage where it will be very complex to define what is an "industrial product" and what is not.

My other concern focuses on the reconstitution criterion of cultivars as opposed to wild species. My own research into *Rubus* L. (Stirton unpublished data), *Eriosema* (DC.) Rchb. (Stirton 1981a & b, 1994), *Psoralea* L. and *Otholobium* C.H. Stirt. (Stirton 1989), and *L. camara* hybrid swarms (Spies 1984, Stirton 1979) show that spontaneous hybrids may reconstitute regularly within a hybridising habitat and can be similarly reconstituted within the greenhouse. A similar point has been made by Alexander (1997).

The above two concerns have been answered in part by Hetterscheid (1998). However, in focusing on some of the explanatory background to cultonomy there is a danger that the main point of the exercise might be lost. At the heart of Hetterscheid's arguments lies the need to create a simple, easily useable, consistent, nomenclaturally proofed classification for cultivated plants independent of the ICBN. I support such an initiative, where it is applicable.

The term culton has an instinctive clarity about it even to the point that it is more easily intelligible than the word taxon. The definition of culton is very succinct and unambiguous – "a systematic group of cultivated plants based on one or more user driven criteria." (Hetterscheid & Brandenburg 1995b). I agree with McNeill (1998) who says: "I believe that the key issue as to whether a group of plants should be recognised under the ICBN or the ICNCP ... is not whether human influence has been involved in its evolution, but rather the **purpose** for which recognition is being sought."

The two codes of nomenclature will always clash when dealing with taxonomic complexes such as *L. camara*. What becomes important is defining clearly why one is producing a classification and if both natural and artificial classifications need to be produced then they should. There seems to be a "tidy syndrome" among taxonomists that demands a universal classification for any taxonomic situation so as to meet all the needs of users, but only rarely enquiring as to what those needs might be.

With these few comments on a continuing lively debate on culta and cultonomy, I now wish to explore how the culton concept might be applied to the *L. camara* complex as it is found in KwaZulu-Natal, Republic of South Africa.

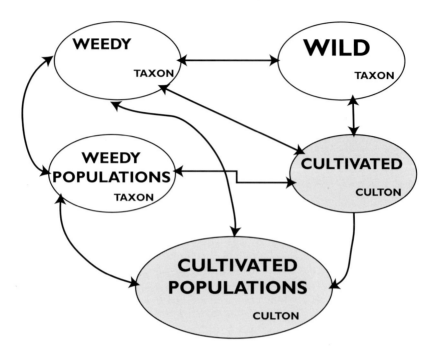

Fig. 3. Derivation pathways from a taxon-culton perspective in *Lantana camara*.

A South African Solution

How does the *L. camara* complex in southern Africa satisfy the Hetterscheid & Brandenburg (1995b) and Hetterscheid (1998) culton criteria?

Criterion 1: Cultivated plant — one ..."whose origin or selection is due to the activities of mankind"

Only some of the variants in the complex satisfy the definition of a cultivated plant (the original selections from the wild in the Neotropics, numerous early selections made in Europe, garden and nursery plants currently available, selections from naturalised populations in Australia, Africa, India and the Neotropics). Excluded are the wild species which make up the parentage and the hundreds of naturalised variants extant in the wild including their country of origin). Two classifications would be needed, one taxonomic, one cultonomic, but with some variants falling outside the scope of both. Thus Criterion 1 is only partly fulfilled.

Criterion 2: Cultivar-group — device to assemble named cultivars on the basis of a user-oriented classification

Although there are hundreds of named cultivars of *L. camara*, the majority do not fulfil the criteria of the ICNCP (Articles 2.2 and 17.1) (Trehane *et al.* 1995). Of the numerous "cultivars" causing problems around the globe very few have been formally named or would meet all the criteria of distinctness, uniformity and stability. It would be impractical to assign formal cultivar names for these variants, although this might be done for the most aggressive weeds, but those fell out under Criterion 1. It is therefore very difficult to form cultivar-groups. Criterion 2 is impractical to implement.

Criterion 3: Culton — "... systematic group of cultivated plants based on one or more user criteria", "A culton must have a name according to the rules of the International Code of Nomenclature for Cultivated Plants"

This is a difficult criterion to judge as the only names available are taxonomic. Most of the cultivar names available for *L. camara* variants do not fulfil Code criteria for the definition of cultivars or have not been formally delimited. As it would be an immense task to do this, they would be difficult to include within a culton approach, but if the effort and will were there, they could presumably be so defined. For practical purposes therefore they are difficult to deal with.

In southern Africa I recognise 5 "races" based broadly but not exactly on Moldenke's subspecies: Race 1 – **Aculeata** (pink-flowered), Race 2 – **Camara** (orange-flowered), Race 3 – **Nivea** (white-flowered), Race 4 – **Mista** (red-flowered) and Race 5 – **Flava** (yellow-flowered). Within each race there are a number of distinct variants which differ in habit, stem colour, armature, leaf texture, pubescence, flower colour transitions, fruit set, and in the size and shape of the flowers and flower bracts. A number of taxa do not always fall easily into any of these five races and the members of some races may in reality belong to another race, e.g. Race 3 – **Nivea** may contain albinos derived from any of the other races. However, for practical purposes the classification has been derived for ease of identification. There has, however, been a surprisingly high amount of natural clustering of the variants. These two points are important for the principal users of the classification: farmers who need to know whether to apply biological, mechanical or chemical control; and plant pathologists and entomologists who need to be able to run effective and repeatable biocontrol screening programmes against potential   control agents. Thus Criterion 3 is only partly fulfilled.

Treatment of Naturalised Variants from KwaZulu-Natal

I have not given any of the variants in Natal formal ranking as it has been very difficult to construct reliable keys. They do however have a distinct morphological fascies and cytogenetic profile and by a careful selection of certain field characters I have been able to develop character formulae that easily designate most of the variants. A detailed account of the Natal variants with keys, maps and illustrations will be given in Stirton (in press).

Each variant in southern Africa has been referred to a   race, as above, and designated by a numeral which refers to a character formula based on the presence of a number of fairly stable field characteristics (Stirton in press). For example, Mista 1 refers to variant 1 of the red-flowered race Mista. This variant is called Red 1 and has a character formula A1 B2 C1 D2 E5 F29 G9 H2 J1 K4. In this paper however I will not refer variants to named races but just refer each to one of six flower colour groups: pink, red, orange, orange-pink, pink-gold and white. So Mista 1 is referred in this paper to Red 1 and is given a colloquial name Vuma, which means fire in Zulu.

The field names and character formulae for the Natal taxa are given in Table 1 and are derived by scoring against the character list in Table 2.

TABLE 1. Field names and character formulae of naturalised variants of *Lantana camara* in KwaZulu-Natal.

Pink 1	A1 B2 C3 D2 E6 F6 G15 H3 J3 K4
Pink 2	A1 B1 C3 D2 E14 F5 G14 H3 J2 K4
Pink 3	A1 B2 C2 D2 E11 F6 G15 H2 J2 K3
Pink 4	A1 B2 C2 D2 E11 F6 G6 H2 J2 K3
Pink 5	A1 B2 C2 D2 E13 F5 G15 H3 J3 K4
Pink 6	A1 B2 C1 D1 E11 F6 <lighter scoop> G13 H2 J2 K2
Pink 7	A1 B2 C3 D4 E15 F27 G15 <SALMON> H2 J2 K4
Pink 8	A1 B2 C2 D2 E14 F19 G15 H3 J3 K4
Red 1	A1 B2 C1 D2 E5 F29 G9 H2 J1 K4
Red 2	A1 B2 C2 D2 E6 F21 G9 H2 J2 K3
Orange 1	A1 B2 C2 D2 E16 F7 G11/12 H3 J2 K3
Orange 2	A1 B2 C2 D1/2 E2 F7 G5 H2 J2 K3
Orange-pink 1	A1 B2 C3 D2 E5 F20 G7 H2 J2 K4
Orange-pink 2	A1 B1/2 C2 D2 E5 F30 G14 H3 J3 K2
Pink-gold 1	A1 B1 C3 D4 E2 F16 G7 H3 J2 K3/4
White 1	A1 B1 C2 D2 E3 F1 G16 H* J* K3
White 2	A1 B1 C3 D2 E3 F1 G1 H3 J3 K4

In this way new taxa or variants of existing ones can easily be incorporated into the current scheme when they are found or arise *de novo*. This simple system will provide field workers with an easily applicable characterisation of the variants they encounter.

During the early course of this study, I made use of a broader range of some 70 characters for taximetric analysis. These did not prove useful for natural clustering and key-making for most populations and regional accounts in South Africa, so I abandoned them for a shorter list of field characters (Table 2). However, as there are a number of widespread core variants (10–15) in the complex in southern Africa, the taximetric data was rerun and it was possible to produce detailed descriptions of the major invasive variants (Stirton in prep.). These will be given "invason/culton/taxon/field" names, e.g. Mamba. I am not sure yet precisely what these are and under which code they should be treated. The remaining 70–80 non-core variants will be identified by formulae and linked as subsets of the core taxa, e.g. A1 B2 C1 D2 E5 F29 G9 H2 J1 K4 – Vuma- Red 1.

TABLE 2. Characters and character states used for determining the character formulae of naturalised variants of *Lantana camara* in KwaZulu-Natal.

A. HABIT
    1. erect or scandent shrub
    2. prostrate spreading shrub

B. STEM COLOUR
    1. green
    2. green/purple
    3. purple

C. STEM COVERING
    1. pubescent or finely acicular
    2. covered in small scattered prickles
    3. heavily armed with large prickles

D. LEAF TEXTURE
    1. smooth, silky to the touch
    2. rough, raspy to the touch

E. OPENING COLOURS OF FLOWERS
    1. dark yellow
    2. gold
    3. white/gold
    4. pale yellow/gold
    5. yellow/gold
    6. dark yellow/gold
    7. pinkish-yellow/gold
    8. orange-yellow/gold
    9. red-edged/gold
    10. white/yellow/gold
    11. cream/yellow/gold
    12. pale yellow/dark yellow/gold
    13. pink/cream/dark yellow
    14. pink/cream/yellow/gold
    15. pink/yellow/gold
    16. orange
    17. reddish/yellowish-orange

F. CLOSING COLOURS OF FLOWERS
    1. white
    2. dark yellow
    3. gold
    4. pale pink
    5. dark pink
    6. mauvey-pink
    7. dark orange
    8. pinkish-orange
    9. reddish-orange
    10. reddish-orange with bleached edges
    11. bright red
    12. crimson
    13. pale pink/dark pink
    14. pale pink/yellowish-orange
    15. pale pinkish-orange/orange

    16. pale pinkish-orange/dark pink
    17. pale pinkish-orange/reddish-orange
    18. dark pink/very dark pink
    19. dark pink/mauve
    20. dark pink/orange
    21. dark pink/orange
    22. dark orange/reddish-orange
    23. crimson/yellowish-orange
    24. cream/yellow/gold
    25. pale pink/dark pink/pinkish-orange
    26. pale pink/yellow/orange
    27. pink/orange-red
    28. yellowish-orange/crimson
    29. crimson with pink lower and lateral limbs
    30. bleached yellowish-salmon/pink/orange

G. BUD COLOUR
    1. pale cream
    2. cream or pale yellow
    3. dark yellow
    4. gold
    5. orange
    6. yellowish-orange
    7. pinkish-orange
    8. reddish-orange
    9. dark pink
    10. reddish-pink
    11. crimson
    12. orangey-crimson
    13. pink becoming orangey-yellow
    14. pinkish-cream with pink flushes
    15. pale pink
    16. white

H. SHAPE OF LOWERMOST BRACT
    1. shortly triangular
    2. narrowly triangular to linear
    3. lanceolate <spear shaped>

I. BRACT/FLOWER LENGTH RATIO
    1. less than one third
    2. between one and two thirds
    3. more than two thirds

J. NUMBER OF FRUITS PER INFLORESCENCE
    1. nil
    2. poor, 1–3
    3. good, 4–9
    4. excellent, more than 10

K. INTERMEDIATE FLORAL COLOUR CHANGES
    1. present
    2. absent

Conclusion

The culton concept is an interesting paradigm when it is restricted to those groups where it works well. We still need to address the "grey areas" raised by Trehane (1997) and by this paper.

Cultonomy undoubtedly could play an important role in stabilising and simplifying the nomenclature of cultivated plants. Whether it has anything to contribute to the taxonomic unravelling of aggressive weedy cultivated species complexes has yet to be explored. The jury remains open on this.

I hope that this paper has highlighted the difficulties that are faced by taxonomists who deal with species and species complexes which become global in their distribution and continue to evolve rapidly.

Finally, it is important that the participants in the current culton/taxon debate always relate back to the customers their products ultimately serve. Hetterscheid and Brandenburg will have made a lasting contribution to the taxonomy of cultivated plants if we heed their core message – that our work should be relevant and useful to our customers.

Acknowledgements

I am grateful to the Director and staff of the Botanical Research Institute (now the National Botanical Institute), Pretoria, South Africa, especially my collaborator on *Lantana* Dr Johann Spies (University of the Orange Free State). Dr Carina Cilliers (Plant Protection Research Institute) supported this study from the beginning and provided critical support and field data. Dr Danie Joubert (Plant Protection Research Institute) made significant collections of *L. camara* in Natal and accompanied me on field trips. I owe a special debt of thanks to the late Drs David Annecke and Rob Kluge for their encouragement and support. They are both much missed. Ms Linda S. Thomas (née Davis) kindly did the excellent illustration of *L. camara*.

References

Alexander, C. (1997). Do we need the term "Culton"? *Hortax News* 1(3): 6–9.

Bailey, L.H. (1918). The indigen and cultigen. *Science* 47: 306–309.

Bailey, L.H. (1923). IV. Various cultigens, and transfers in nomenclature. *Gentes Herb.* 1(3): 113-115.

Brandenburg, W.A. (1999). Crop-weed complexes and the culton concept. In S. Andrews, A.C. Leslie & C. Alexander (eds). Taxonomy of Cultivated Plants: Third International Symposium. pp. 145–157. Royal Botanic Gardens, Kew.

Brickell, C.D. *et al.* (eds) (1980). International code of nomenclature for cultivated plants — 1980. 32 pp. W. Junk, The Hague.

Cilliers, C.J. (1977). On the biological control of *Lantana camara* in South Africa. pp. 341–344, In Proceedings 2nd National Weeds Conference, Stellenbosch. Balkema, Cape Town.

Hetterscheid, W.L.A. (1998). Yes, we need the term "Culton"! *Hortax News* 1(4): 9–12.

Hetterscheid, W.L.A. & Brandenburg, W.A. (1995a). Culton versus taxon: conceptual issues in cultivated plant systematics. *Taxon* 44(2): 161–175.

Hetterscheid, W.L.A. & Brandenburg, W.A. (1995b). The culton concept: setting the stage for an unambiguous taxonomy of cultivated plants. *Acta Hort.* 413: 29–34.

Hetterscheid, W.L.A., Berg, R.G. van den & Brandenburg, W.A. (1996). An annotated history of the principles of the cultivated plant classification. *Acta Bot. Neerl.* 45(2): 123–134.

Howard, R.A. (1969). A check list of cultivar names used in the genus *Lantana. Arnoldia* 29(11): 73–109.

Jirásek, V. (1958). Taxonomische Katagorien der Kulturpflanzen. *Index Sem. Hort. Bot. Univ. Carol. Praha*: 9–16.

Jirásek, V. (1961). Evolution of the proposals of taxonomic categories for the classification of cultivated plants. *Taxon* 10: 34–45.

Mabberley, D.J. (1987). The Plant-Book. 706 pp. Cambridge University Press, Cambridge.

Mansfield, R. (1953). Zur allgemeinen Systematik der Kulturpflanzen I. *Kulturpflanze* 1: 138–155.

Mansfield, R. (1954). Zur allgemeinen Systematik der Kulturpflanzen II. *Kulturpflanze* 2: 130–142.

McNeill, J. (1998). Culton, a useful term, questionably argued. *Hortax News* 1(4): 15–22.

Moldenke, H. (1971). A fifth summary of the *Verbenaceae, Avicenniaceae, Stilbaceae, Dicrastylidaceae, Symphomaceae, Nyctanthaceae,* and *Eriocaulaceae* of the world as to valid taxa, geographic distribution and synonymy. Moldenke, Wayne.

Morton, J.F. (1994). Lantana, or red Sage (*Lantana camara* L., [*Verbenaceae*]), notorious weed and popular garden flower; some cases of poisoning in Florida. *Econ. Bot.* 48(3): 259–270.

Munir, A.A. (1996). A taxonomic review of *Lantana camara* L. and *L. montevidensis* (Spreng.) Briq. (*Verbenaceae*) in Australia. *J. Adelaide Bot. Gard.* 17: 1–27.

Natarajan, A.T. & Ahuja, M.R. (1957). Cytotaxonomical studies in the genus *Lantana. J. Indian Bot. Soc.* 36(1): 35–45.

Parsons, W.T. & Cuthbertson, E.G. (1992). *Verbenaceae.* Noxious Weeds of Australia. pp. 625–637. Inkata Press, Melbourne.

Sanders, R.W. (1987a). Taxonomic significance of chromosome observations in Caribbean species of *Lantana* (*Verbenaceae*). *Amer. J. Bot.* 74(6): 914–920.

Sanders, R.W. (1987b). Identity of *Lantana depressa* and *L. ovatifolia* (*Verbenaceae*) of Florida and the Bahamas. *Syst. Bot.* 12(1): 44–60.

Sanders, R.W. (1989). *Lantana camara* sect. *Camara* in Hispaniola: novelties and notes. *Moscosoa* 5: 202–215.

Sharma, O.M, Makkar, H.P.S. & Dawra, R.K. (1988). A review of the noxious plant *Lantana camara. Toxicon* 26(11): 975–987.

Smith, L.S. & Smith, D.A.. (1982). The naturalised *Lantana camara* complex in eastern Australia. *Queensland Bot. Bull.* 1: 1–27.

Spies, J.J. (1984). A cytotaxonomic study of *Lantana camara* (*Verbenaceae*) from South Africa. *S. Afr. J. Bot.* 3(4): 231–250.

Spies, J.J. & Stirton, C.H. (1982a). Chromosome numbers of southern African plants. 1. *J. S. Afr. Bot.* 48: 21–22.

Spies, J.J. & Stirton, C.H. (1982b). Meiotic studies of some South African cultivars of *Lantana camara* (*Verbenaceae*). *Bothalia* 14(1): 101–111.

Spies, J.J. & Stirton, C.H. (1982c). Embryo sac development in some South African cultivars of *Lantana camara. Bothalia* 14(1): 113–117.

Stirton, C.H. (1977). Some thoughts on the polyploid complex *Lantana camara* L. (*Verbenaceae*). pp. 321–340. In Proceedings 2nd National Weeds Conference, Stellenbosch. Balkema, Cape Town.

Stirton, C.H. (1979). Taxonomic problems associated with invasive alien trees and shrubs in South Africa. In G. Kunkel (ed.). Taxonomic aspects of African Botany. pp. 218–229. Excmo, Las Palmas.

Stirton, C.H. (1981a). Natural hybridisation in the genus *Eriosema* (*Leguminosae*) in South Africa. *Bothalìa* 13(3–4): 307–315.

Stirton, C.H. (1981b). The *Eriosema cordatum* complex. II. The *Eriosema cordatum* and *E. nutans* groups. *Bothalia* 13(3–4): 281–306.

Stirton, C.H. (1989). A Revision of the genus *Otholobium* C.H. Stirt. (*Psoraleeae, Leguminosae*). Ph.D. University of Cape Town.

Stirton, C.H. (1994). Hybridisation in the genus *Eriosema* (*Leguminosae-Phaseoleae*) in Natal, South Africa. *Kew Bull.* 49(3): 529–535.

Stirton, C.H. (in press). The naturalised *Lantana camara* L. (*Verbenaceae*) complex in KwaZulu-Natal, South Africa. *S. Afr. J. Bot.*

Trehane, P. (1997). The *Code* – contentious issues to be resolved. *Hortax News* 1(3): 10–14.

Trehane, P., Brickell, C.D., Baum, B.R., Hetterscheid, W.L.A., Leslie, A.C., McNeill, J., Spongberg, S.A. & Vrughtman, F. (eds). (1995). International Code of Nomenclature for Cultivated Plants — 1995. 175 pp. Quarterjack Publishing, Wimborne.

Fantz, P.R., Rouse, R.J., & Bilderback, T.E. (1999). Cultivar-groups in Japanese cedar (*Cryptomeria japonica*). In: S. Andrews, A.C. Leslie and C. Alexander (Editors). Taxonomy of Cultivated Plants: Third International Symposium, pp. 325–334. Royal Botanic Gardens, Kew.

CULTIVAR-GROUPS IN JAPANESE CEDAR
(*CRYPTOMERIA JAPONICA*)

PAUL R. FANTZ[1], ROBERT J. ROUSE[2] AND TED E. BILDERBACK[1]

[1]Department of Horticultural Science, Box 7609, North Carolina State University, Raleigh, NC 27695-7609, USA
[2]Staff Arborist, National Arborist Association, Inc., The Meeting Place Mall, Route 101, P.O. Box 1094, Amherst, NH 03031-1094, USA

Abstract

Japanese cedar (*Cryptomeria japonica*) is regarded as one of the best gymnosperms adapted to the eastern United States, and is increasing in popularity as a landscape plant. A taxonomic study of cultivars grown in the eastern United States resulted in 45 cultivars recognized. Cultivar synonymy is provided in a table. Eleven cultivar-groups are established, ten newly described, based upon standard morphological terminology, except for leaves (juvenile state described as linear, adult state newly defined as saber-like). Warm and cool seasonal growth of shoots is an important character state newly defined. Four cultivars are not assigned to any cultivar-group. Forty one cultivars are assigned to one of the 11 cultivar-groups. A taxonomic key is presented to the cultivar-groups based on eastern USA data.

Cultivar-groups in Japanese Cedar (*Cryptomeria japonica*)

Japanese cedar, *Cryptomeria japonica* (Thunb. ex L.f.) D. Don [*Cupressaceae* Rich. ex Bart., formerly *Taxodiaceae* Warm.] is indigenous to Japan. This is one of the best gymnosperms adapted for the eastern United States, and is increasing in popularity as a landscape plant (Tripp 1993). Dirr (1990) noted that Japanese cedar is under-utilized in the southern states and has potential for use in any landscape situation. Cultivars of *C. japonica* performed well in heavy red clay soils during both prolonged dry and wet periods, exhibited remarkable tolerance to hot, humid, summer conditions, and are regarded as nearly pest free (Tripp 1993).

Trehane *et al.* (1995) cited several publications that included a checklist of ornamental cultivars. Rouse *et al.* (1997) provided a table of nearly 225 cultivar names reported in the literature. Tripp (1993) noted that several new cultivars introduced into the United States lacked descriptions.

Rouse *et al.* (1997) established that Japanese cedar in the eastern United States included a large number of named cultivars that needed an organized inventory of taxa, correct scientific names and synonyms, quantitative descriptions, taxonomic keys for segregation of taxa, and documentation of existing germplasm. The objective of this paper is to define groups of cultivars based upon a similarity of morphological characters, provide a key to these groups, and assign those cultivars being grown and used in the eastern United States to a cultivar-group.

Materials and Methods

Data and vouchers were collected by Rouse from mature taxa of Japanese cedar, grown primarily at the Atlanta Botanical Garden (Atlanta, Ga.), Morris Arboretum (Philadelphia, Penn.), the J.C. Raulston Arboretum of North Carolina State University (Raleigh, N.C.), and the U. S. National Arboretum (Washington, D.C.). Standardized data sheets were prepared listing various morphological characters appropriate for taxodiaceous plants. Morphological terminology used followed standard taxonomic references (Eiselt 1960, Harris & Harris 1994, Vidaković 1991). Data were obtained from three to six plants per cultivar. Many cultivars of Japanese cedar are rare in the landscape/nursery industries, and are found at only one or two sites in the United States, limiting sampling to only one or two plants. All plants sampled were labelled with cultivar names and were at least 4 years old.

Data on branch growth patterns were recorded from six to eight different primary branches per individual with measurements recorded from 12 to 15 different locations per individual. Leaf data were recorded from 10 to 20 leaves at 5 to 8 different locations per individual. Strobili data were recorded from 12 to 15 strobili per individual, when present. However, some cultivars lacked male and/or female strobili. Quantitative descriptions were prepared for each cultivar and compared to determine the groups of cultivars with similar morphological characters.

Results

Analysis of shoots of Japanese cedar required modification of standarized morphological terminology. Shoots of *Cryptomeria* cultivars exhibited distinct seasonal growth periods that could be characterized, but were lacking in literature descriptions. *Warm seasonal growth* exhibited longer internode length, larger leaves, a broader shoot width, and a longer shoot length. *Cool seasonal growth* exhibited short internode length, smaller leaves, a narrower shoot width, and a shorter shoot length.

Literature descriptions of cultivars available describe leaves as awns, needles, awl-shaped, acicular or subulate (Krüssmann 1985, den Ouden & Boom 1978, Tripp 1993, Welch 1991 & 1993). Different authors describing the same cultivar often use a different term for the leaves. In our opinion, the leaves of Japanese cedar do not agree with any of these types previously mentioned. We recognize four leaf types defined herein. *Linear leaves* are flat, straight to recurved, 2–4 mm wide, attached at an angle of 45° to 95°, with the apex acute to acuminate, straight or inflexed. Glaucous bands may be present or lacking on both surfaces. These leaves are similar to cotyledon leaves and may represent a juvenile trait.

Adult foliage is quadrangular or dorsiventrally compressed, and similar in appearance to a saber. Three saber leaf types are recognized. *Saber leaves* are incurved, 7–24 mm long, 1.5–3 mm wide, attached at an angle of 30° to 60°, with the apex acute to acuminate, reflexed or inflexed. *Straight saber leaves* are similar, except that the leaf is straight with an apex straight or weakly inflexed. Both are associated commonly with warm seasonal growth. *Short saber leaves* are keeled, straight to slightly incurved, 1–11 mm long, 1–4 mm wide, attached at an angle of 1° to 45°, with the apex acute and straight or inflexed. The leaf shaft is the portion of the leaf unattached to the twig.

An inventory by Rouse produced nearly a hundred cultivar names being used in the eastern United States. Forty five cultivars are recognized, with the other names reduced to synonymy. Article 19 of *The International Code of Nomenclature for Cultivated Plants – 1995* (Trehane *et al.* 1995) established guidelines for cultivar-groups, a method of

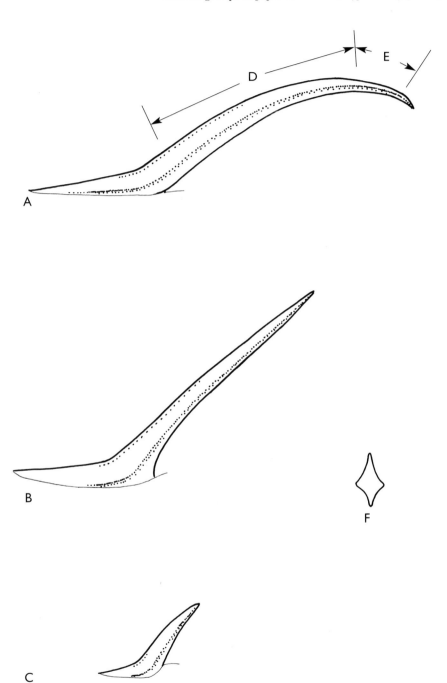

FIG. 1. Saber leaf types in *Cryptomeria japonica*. A. *Saber leaf*, lateral view. B. *Straight saber leaf*, lateral view. C. *Short saber leaf*, lateral view. D. Leaf shaft. E. Leaf apex. F. cross-section of saber leaf. Note that the stippeled area = the adaxial lateral leaf surface. Drawn by Sally Dawson.

classification for organization of large numbers of named cultivars. Comparison of morphological descriptions of these 45 cultivars led to the establishment of 11 cultivar-groups. Four cultivars ('Buckiscope', 'Green Pencil', 'Kukamiga' and 'Littleworth Dwarf') were not assigned to a cultivar group as data sets were limited. The remaining cultivars were assigned to a cultivar-group (Table 1).

A. Key to Cultivar-groups:

1a) Primary shoots with secondary shoots lacking or secondary shoots clustered near the apex and appearing "tuft-like"; quarternary shoots lacking; female strobili lacking; warm seasonal leaves saber-like ········· **2. Araucarioides Group**

1b) Primary shoot with secondary shoots distributed along shoot axis; quarternary shoots 1–8 or occasionally lacking; female strobili present or occasionally lacking; warm seasonal leaves linear, saber-like, short saber-like, or straight saber-like ···································· 2

2a) Warm and cool seasonal leaves linear; male strobili conical, simple, 5–14 mm long ····································· **7. Linear-leaf Group**

2b) Warm and cool seasonal leaves of a saber type, non linear; male strobili oblong or ovoid, simple or compound with two conelets, or fused into strobili complex, individual strobili 2–10 mm long ······················· 3

3a) Secondary and tertiary shoots with fasciations; trunk irregularly twisted ·· **6. Fasciate Group**

3b) Secondary and tertiary shoots lacking fasciations; trunk straight ·········· 4

4a) Majority of growth cool seasonal; warm seasonal leaves short saber-like, 3–7(–9) mm long, leaf shaft straight; cool seasonal leaves short saber-like, 1–3 mm long ······························ **5. Diminutive Saber-leaf Group**

4b) Majority of growth warm seasonal; warm seasonal leaves saber-like, short saber-like, or straight saber-like, (2–)6–28 mm long, shaft twisted spirally or lacking; cool seasonal leaves short saber-like, 2–6 mm long ············ 5

5a) Warm seasonal leaves straight saber-like; male strobili appearing fused at base arranged into a strobilus complex, or simple and arranged in spike-like clusters ······························· **11. Straight Saber-leaf Group**

5b) Warm seasonal leaves saber-like or short like; male strobili simple, or compound with one primary strobilus and 1 to 2 conelets fused at the base ········ 6

6a) Majority of warm and cool seasonal leaves with leaf shaft spirally twisted halfway around twig ····························· **10. Spiral-leaf Group**

6b) Majority of warm and cool seasonal leaves lacking spirally twisted leaf shaft ·· 7

7a) Warm seasonal leaves short saber-like, (4–)6–11 mm long; male strobilus oblong or ovoid, simple or occasional compound strobilus with a primary cone and 1 to 2 conelets fused at the base ············· **9. Short Saber-leaf Group**

7b) Warm seasonal leaves saber-like, 6–24 mm long; male cones ovoid, simple ··· 8

8a) Cool seasonal growth dominant with 20–80 secondary shoots; warm seasonal growth with (0–)8–20 secondary shoots (ratio 3.5–4); crown shape irregular ·· **8. Monstrose Group**

8b) Cool and warm seasonal growth more equally distributed in secondary shoots (ratio 1.5–3); crown shape regular ··························· 9

9a) Cool seasonal growth 0.3–1.6 cm long; warm seasonal growth 0.5–10.5 cm long
· **4. Diminitive Cool-seasonal Group**
9b) Cool seasonal growth (0.4–)1.5–7 cm long; warm seasonal growth to 36 cm long
· 10

10a) Cool seasonal leaves appressed to the stem; cool seasonal growth 2–4 mm wide;
dwarf or compact shrub · · · · · · · · · · · · · · · · · · · **1. Appressed-leaf Group**
10b) Cool seasonal leaves ascending; cool seasonal growth 2–7 mm wide; tree
· **3. Conical Tree Group**

B. Cultivar-groups

1. Appressed-leaf Group. Dwarf or compact shrubs. Warm seasonal growth dominant; cool seasonal growth 2–4 mm wide. Warm seasonal leaves saber-like, 6–24 mm long; cool seasonal leaves appressed to twig. Male strobili simple, ovoid. Standard cultivar: 'Bloomers Witches Broom' (standard specimen: *R.J. Rouse* 247, conserved at NCSC).

Members are distinguished easily by twig-like cool seasonal growth in which the saber-like leaves are appressed. Two cultivars — 'Bloomers Witches Broom' and 'Little Diamond' are recognized.

2. Araucarioides Group. Primary shoots unbranched, generally either with secondary shoots, or secondary shoots occurring profusely at the apex of primary shoot giving a "tuft-like" appearance. Quarternary shoots lacking. Warm seasonal leaves are saber-like. Female strobili lacking. Standard cultivar: 'Araucarioides' (standard specimen: *R.J. Rouse* 267, conserved at NCSC).

Members were classified historically (from 1865 onwards) as f. *araucarioides* (Siebold) Henkle & W. Hochst., then recognized as a cultivar-group (Welch 1979) characterized by the "rope-like" or "snake -like" foliage, bearing strong primary shoots and a paucity of secondary (lateral) shoots. Two cultivars — 'Araucarioides' and 'Dacrydioides' are recognized. The other seven cultivars cited by Welch (1979) as belonging to this cultivar-group were not located in the United States.

3. Conical Tree Group. Conical tree. Warm seasonal growth dominant; cool seasonal growth 2–7 mm wide. Warm seasonal leaves saber-like, 6–24 mm long; cool seasonal leaves saber-like, ascending. Male strobili simple, ovoid. Standard cultivar: 'Lobbii' (standard specimen: *R.J. Rouse* 249, conserved at NCSC).

Members are conical trees with a narrow crown and more profuse shoot growth than typical selections of Japanese cedar. The "tufting" of secondary shoots is more pronounced in consistently colder climates. Three cultivars —'Benjamin Franklin', 'Lobbii' and 'Yoshino' are recognized.

4. Diminutive Cool-seasonal Group. Dwarf or compact shrubs. Warm seasonal growth dominant, to 105 mm long; cool seasonal growth diminutive, to 16 mm long. Warm seasonal leaves 6–24 mm long. Male strobili ovoid, simple. Standard cultivar: 'Globosa' (standard specimen: *R.J. Rouse* 252, conserved at NCSC).

Members of this group are distinquished by the extremely short cool seasonal growth exhibited, and the simple male strobili. Five cultivars — 'Black Dragon', 'Giokumo', 'Globosa', 'Globosa Nana' and 'Tenzan' are recognized.

5. Diminutive Saber-leaf Group. Trees and compact shrubs. Cool seasonal growth dominant. Leaves short saber-like; warm seasonal leaves 3–6 mm long, leaf shaft straight; cool seasonal leaves 1–3 mm long. Male strobils rare. Standard cultivar: 'Gracilis' (standard specimen: *R.J. Rouse* 278, conserved at NCSC).

Members distinguished quickly by the dominant, shorter saber-like leaves and sparse warm seasonal growth. Four cultivars — 'Gracilis', 'Ikari', 'Knaptonensis' and 'Yataduta' are recognized. 'Knaptonensis' is confused often with 'Nana Albospicata' (which has longer warm seasonal leaves, 6–11 mm long) in the southeastern United States.

6. Fasciate Group. Secondary and tertiary shoots with periodic fasciations ("cockscombs") of flattened, broader branchlets composed of multiple meristems from shoots fusing and growing together. Trunks often twisted irregularly. Warm and cool seasonal leaves saber-like. Standard cultivar: 'Cristata' (standard specimen: *R.J. Rouse* 284, conserved at NCSC).

The twisted or knotted trunks that develop are unique to this cultivar-group, and may be a result of the grafting process used for these cultivars. Two cultivars — 'Cristata' and 'Kilmacurragh' are recognized, with smaller fasciations in the latter.

7. Linear-leaf Group. Warm seasonal leaves linear, cool seasonal leaves linear, rarely saber-like. Male strobili conical, simple, 5–14 mm long. Standard cultivar: 'Elegans' (standard specimen: *R.J. Rouse* 186, conserved at NCSC).

The foliage is soft and plants generally are vigorous in growth. The linear leaves are cotyledon-like, possibly representing a case of fixed juvenility. 'Compressa' is the only cultivar with cool season leaves short saber-like. Seven cultivars — 'Compressa', 'Elegans', 'Elegans Aurea', 'Elegans Nana', 'Elegans Viridis', 'Globes' and 'Pomona' are recognized, most bearing the word "Elegans" in the name. However, 'Elegans Compacta' does not belong here, but to the Straight Saber-leaf Group.

8. Monstrose Group. Crown shape irregular. Seasonal growth ratio 3.5–4; cool seasonal growth dominant with 20–80 secondary shoots; warm seasonal growth with up to 20 secondary shoots. warm seasonal leaves 6–24 mm long. Male strobili ovoid, simple. Standard cultivar: 'Monstrosa' (standard specimen: *R.J. Rouse* 190, conserved at NCSC).

Members of this group are distinguished by their irregular growth forms and unusual foliage appearance (*monstrous* = Latin for abnormal) as compared to typical Japanese cedars. The ratio (number of secondary shoots produced in season/length of seasonal growth) is high. Crown growth is regular and the ratio is 1.5–3 in cultivars of Japanese cedar assigned to other groups.

9. Short Saber-leaf Group. Warm seasonal growth dominant. Warm seasonal leaves short saber-like, 4–11 mm long. Male strobili oblong or ovoid, simple or rarely compound with 1–2 strobili fused basally to a primary strobilis. Standard cultivar: 'Vilmoriniana' (standard specimen: *R.J. Rouse* 271, conserved at NCSC).

Members of this group are distinguished easily from other cultivars with saber-like leaves, by the narrower shoots resulting from the dominant warm seasonal growth bearing short saber-like leaves. Five cultivars — 'Jindai-sugi', 'Lobbii Nana Aurea', 'Nana Albospicata', 'Taisho-tama' and 'Vilmoriniana' are recognized.

TABLE 1. Cultivar-groups of *Cryptomeria japonica*

Cultivar-group	Cultivars Recognized	Cultivar Synonyms
Appressed-leaf	'Bloomers Witches Broom' 'Little Diamond'	'National Arboretum Witches Broom', 'Reins Dense Jade'
Araucarioides	'Araucarioides' 'Dacrydioides'	'Enko-sugi', 'Yenko-sugi'
Conical Tree	'Benjamin Franklin' 'Lobbii' 'Yoshino'	'Bennies Best'
Diminutive Cool-seasonal	'Black Dragon' 'Giokumo' 'Globosa' 'Globosa Nana' 'Tenzan'	'Gyokruyu' 'Lobbii Nana' 'Tenzan Yatsabusa', 'Yatsubusa'
Diminutive Saber-leaf	'Gracilis' 'Ikari' 'Knaptonensis' 'Yataduta'	'Hime-ikari-sugi', 'Ikan-sugi', 'Ikar-sugi'
Fasciate	'Cristata' 'Kilmacurragh'	'Sekka-sugi', 'Sekkwia-sugi'
Linear-leaf	'Compressa' 'Elegans' 'Elegans Aurea'	'Birodo-sugi' 'Hime-sugi'

Table 1 continued

Cultivar-group	Cultivars Recognized	Cultivar Synonyms
	'Elegans Nana'	'Elegans Gracilis'
	'Elegans Viridis'	
	'Globes'	'Globus'
	'Pomona'	
Monstrose	'Bandai-sugi'	'Ito-sugi'
	'Monstrosa'	
	'Monstrosa Nana'	'Manhismi-sugi', 'Mankichi',
	'Yokohama'	'Yatsubusa'
Short Saber-leaf	'Jindai-sugi'	'Majiro-sugi', 'Mejero-sugi', 'Mejiro-sugi', 'Okina-sugi'
	'Lobbii Nana Aurea'	'Ito-sugi', 'Taisho-sugi', 'Taishotama-sugi',
	'Nana Albospicata'	'Taisho-tamasugi'
	'Taisho-tama'	
	'Vilmoriniana'	'Hino-sugi', 'Osaka-tama', 'Osaka-tama-sugi', 'Pygmaea'
Spiral-leaf	'Spiralis'	'Grannys Ringlets', 'Rasen-sugi' *pro parte*
	'Spiralis Falcata'	'Kusari-sugi', 'Spiralis Elongata', 'Yore-sugi',
		'Rasen-sugi' *pro parte*
Straight Saber-leaf	'Aurea'	'Ogon-sugi'
	'Elegans Compacta'	
	'Nana'	'Elegans Nana'
	'Sekkan'	'Sekkan-sugi'
	'Tansu'	'Yatsabusa'

10. Spiral-leaf Group. Warm seasonal growth dominant. Warm and cool seasonal leaves of a saber type, shaft spirally twisted to halfway around twig on a majority of the leaves. Male strobili rare. Standard cultivar: 'Spiralis' (standard specimen: *R.J. Rouse* 258, conserved at NCSC).

Members are distinguished quickly by the unique, twisted leaves. Two cultivars — 'Spiralis' and 'Spiralis Falcata' are recognized and commonly confused with each other. 'Spiralis' is a small tree with warm seasonal growth short and broad, whereas 'Spiralis Falcata' is a compact shrub with more elongated, but narrower warm seasonal growth.

11. Straight Saber-leaf Group. Warm seasonal growth dominant. Warm seasonal leaves saber-like, shaft straight; cool seasonal leaves saber-like, short, 2–6 mm long. Male strobili simple, arranged in a spike-like cluster, or appearing fused at the base into a strobili complex. Standard cultivar: 'Aurea' (standard specimen: *R.J. Rouse* 272, conserved at NCSC).

Members are characterized by most warm seasonal leaves being straight, not incurved nor inflexed near the apex, and the clustering of male strobili. Five cultivars — 'Aurea', 'Elegans Compacta', 'Nana', 'Sekkan' and 'Tansu' are recognized. 'Aurea' and 'Sekkan' are trees with male strobili in spike-like clusters. The other three cultivars are compact or dwarf shrubs and bear strobili complexes of multiple strobili fused basally.

Conclusion

Eleven cultivar-groups were established for the 41 cultivars of Japanese cedar (*Cryptomeria japonica*) being grown in the eastern United States, based upon morphological characters and segregated with a taxonomic key. Each cultivar-group was described, typified, and inventoried with cultivars currently grown. The establishment of these cultivar-groups will provide a foundation for classification of a large number of cultivars, both presently and in the future.

References

Dirr, M.A. (1990). Manual of woody landscape plants: their identification, ornamental characteristics, culture, propagation and uses. 1007 pp. Stipes Publishing Company, Champaign, Ill.

Eiselt, M.G. (1960). Nadelgeholze. 366 pp. [Radebeul] Neumann, Leipzig, Germany.

Harris, J.G. & Harris, M.W. (1994). Plant identification terminology, an illustrated glossary. 197 pp. Spring Lake Publishing, Spring Lake, Utah.

Krüssmann, G. (1985). Manual of cultivated conifers. 361 pp. Timber Press, Portland, Oregon.

Ouden, P. den & Boom, B.K. (1978). Manual of cultivated conifers, hardy in the cold – and warm – temperate zone. 526 pp. Martinus Nijhoff, The Hague.

Rouse, R.J., Fantz, P.R. & Bilderback, T.E. (1997). Problems identifying Japanese cedar cultivated in the United States. *Hort. Technology* 7(2): 129–133.

Trehane, P., Brickell, C.D., Baum, B.R., Hetterscheid, W.L.A., Leslie, A.C., McNeill, J., Spongberg, S.A. and Vrugtman, F. (1995). International Code of Nomenclature for Cultivated Plants — 1995. 175 pp. Quarterjack Publishing, Wimborne.

Tripp, K.E. (1993). Sugi: The ancient Japanese cedar finds new life in a profusion of outstanding cultivar forms. *Amer. Nurseryman* 178(7): 26–39.

Vidaković, M. (1991). Conifers morphology and variation. 784 pp. Graficki Zavod Hrvatske, Zagreb, Croatia.

Welch, H.J. (1979). Manual of dwarf conifers. 493 pp. Theophrastus, Little Compton, RI.

Welch, H.J. (1991). The conifer manual. 436 pp. Kluwer Academic Press, Boston.

Welch, H.J. & Haddow, G. (1993). The World Checklist of Conifers. 427 pp. Landsman's Bookshop Ltd., Herefordshire, England on behalf of the World Conifer Data Pool.

PLANT INTRODUCTIONS

10

Stearn, W.T. (1999). Early introduction of plants from Japan into European gardens. In: S. Andrews, A.C. Leslie and C. Alexander (Editors). Taxonomy of Cultivated Plants: Third International Symposium, pp. 337–340. Royal Botanic Gardens, Kew.

EARLY INTRODUCTION OF PLANTS FROM JAPAN INTO EUROPEAN GARDENS

WILLIAM T. STEARN

17 High Park Road, Kew, Richmond, Surrey TW9 4BL, UK

Abstract

A concise survey is presented of contributions to the knowledge of Japanese plants made by Engelbert Kaempfer (1651–1716), Carl Peter Thunberg (1743–1828) and Philip Franz von Siebold (1796–1866), physicians stationed on Deshima Island in Nagasaki harbour while in the employ of the Dutch East India Company. During this feudal Tokugawa period, when Japan was a closed country, Thunberg and Siebold also succeeded in introducing many plants to European gardens.

Introduction

The floristic richness of Japan, making this country a source of hardy plants remarkably distinct from those already known in European gardens, has been determined by a diversity of geographical factors over millions of years. These include intermittent land connections with the East Asiatic mainland, whereby its islands received a varied basic assemblage of plants which evolved into many endemic species and a few endemic genera allied to but distinct from the species and genera of China. Also important is the climatic diversity associated with the latitudinal range of 26°N. to 46°N., from the subarctic conditions of Hokkaido to the warm temperate or subtropical conditions of Kyushu. Moreover the mountainous terrain providing maritime to alpine conditions combined with the long isolation of plant populations have further promoted the evolution of endemic taxa. The glaciation which devastated the formerly rich China-type flora of Europe had a minor effect on the floras of Eastern Asia. In consequence of its uninterrupted evolution, the flora of Japan now comprises about 4,300 species belonging to some 1,000 genera. Among endemic monotypic Japanese genera are *Conandron* Siebold & Zucc., *Fatsia* Decne. & Planch., *Glaucidium* Siebold & Zucc., *Pteridophyllum* Siebold & Zucc., *Ranzania* T. Ito and *Sciadopitys* Siebold & Zucc., all known in British gardens. There are also numerous species of *Asarum* L. *sensu lato* and *Hosta* Tratt.

European access to the floral wealth of Japan was long restricted by Japanese political considerations. Just as the English Wars of the Roses ended in 1483 at the battle of Bosworth Field to be followed by the long Tudor peace, so in Japan the battles of Odawara in 1590 and Sekigahawa in 1600 ended a long period of turmoil and civil war, leading to two and a half centuries of peace under the despotic Tokugawa family.

Meanwhile, well-armed spice-questing ships of three aggressive European powers, Portugal, Spain and the Dutch United Provinces, had reached the Indian and Pacific Oceans and were beginning to establish their far-flung sea-borne empires. The Portuguese were the first to reach Japan, sailing into Nagasaki Bay in 1542. The inquisitive Japanese, unlike the intellectually self-sufficient Chinese, welcomed them, and their missionaries made thousands of Christian converts. By 1640 the Tokugawa shogunate, fearful that the Spaniards and Portuguese might arm disaffected defeated clans to overthrow their regime, had persecuted Christians with harsh severity and decided to close Japan to all foreigners except Chinese and Dutch traders, both restricted to the Nagasaki area. Here the Dutch lived virtually as prisoners on a man-made island, Deshima, in Nagasaki harbour. It occupied about 32 acres and was roughly 200 feet wide, with a seaward side of about 700 ft.; a high palisade surrounded it and a guarded bridge connected the island to the mainland. Deshima as known to the bored Dutch no longer exists. Where they saw empty sea, there now stand high-rise buildings.

Engelbert Kaempfer

This small Dutch community on Deshima always included a doctor. Fortunately, three of these medical men were highly intelligent, with an eager curiosity about matters Japanese including plants, and they wrote accounts of them. The first was Engelbert Kaempfer (1651–1716), a German from Lemgo in Westphalia. After studying at the Universities of Cracow and Königsberg, he went to the University of Uppsala in Sweden, then in 1683 joined a Swedish legation travelling to Persia (Iran), by way of Moscow, which reached Isfahan in March 1684. After a stay in Persia he entered the service of the Dutch East India Company and in September 1690 arrived at Deshima as the community's physician. Every year the Dutch, like Japanese vassals of the Tokugawa Shogun, had to make the long and costly journey of homage to Edo (now Tokyo), a welcome relief from the boredom of Deshima. Kaempfer twice made that journey. His curiosity extended to plants, of which he recorded 420 species, excellently illustrating many himself (Muntschick 1983, Stearn 1949 & 1999). He left Deshima in October 1693 and arrived in Amsterdam in October 1698. He then made the mistake of returning to his birth-place and becoming physician to a local potentate, Count Friedrich zur Lippe. He never found time to publish his numerous observations and illustrations, and not until 1712 did he succeed in publishing his *Amoenitatum exoticarum fasciculi V* (Lemgoviae), which illustrated and portrayed the remarkable Japanese species later named *Camellia japonica* L., *Citrus trifoliata* L. (*Poncirus trifoliata* (L.) Raf.), *Ginkgo biloba* L. and *Ophiopogon japonicus* (L.f.) Ker Gawl.

The name *Ginkgo* L. originated as a misprint for *Ginkjo* (*ginkyo*, silver apricot), as pointed out by Moule (1937) and Thommen (1949). Kaempfer introduced no Japanese plants but his work made known their diversity and horticultural potential.

Carl Peter Thunberg

The next important Deshima doctor was Carl Peter Thunberg (1743–1828), a Swedish student of Carl Linnaeus at Uppsala. On his way to Paris to improve his medical studies, he stopped in Holland and here some wealthy gentlemen interested in gardens provided the means for him to visit Japan and obtain plants for them. He spent three years at the Cape of Good Hope, learning Dutch and collecting South African plants, and did not reach Deshima until August 1775. He had to wait until February 1776 before being allowed to visit Nagasaki. The Dutch kept a few oxen, calves, pigs, deer and sheep on their little island to provide fresh meat, for which Japanese servants brought in local herbage. Thunberg went through this to get specimens for his herbarium. Such often incomplete material provided type specimens for Thunberg's later publications on Japanese plants. He also encouraged Japanese interpreters to bring him specimens from the hills around Nagasaki. Including specimens gathered on the journey to Edo, he acquired material of some 900 Japanese species. He had not forgotten the intent of his visit and got together living plants of *Acer* L., *Celastrus* L., *Citrus* L., *Cupressus* L., *Cycas* L., *Prunus* L., *Viburnum* L., etc. to be sent to Amsterdam via Batavia (now Jakarta). In November 1776 he began the long voyage back to Europe, reaching Amsterdam in October 1778. No list exists of the plants introduced by Thunberg, but he had the satisfaction of seeing some of them in cultivation near Haarlem. Back in Sweden in 1779 he began preparation of a *Flora Japonica* (1784) and an account of his travels, *Resa uti Europa* (1788–1794).

Philipp Franz von Siebold

Much more important for the introduction of Japanese plants into European gardens and for information about Japan in general was the third great Deshima doctor, Philipp Franz von Siebold (1796–1866). He belonged to a distinguished Bavarian medical family of Würzburg, his grandfather, father and two uncles being medical men; he himself studied medicine at the University of Würzburg. In 1822 he entered the service of the Dutch East India Company and was sent to Batavia and then in 1823 to Deshima. Siebold's Germanic pronunciation of Dutch puzzled the Japanese interpreters but was explained as 'yama-granda' (mountain Dutch). Thus as a mountaineer from the Alps of Holland speaking with the strange accent of that region, Siebold entered Japan. His knowledge of western medicine he made so freely available to Japanese medical students that he was permitted to see patients and teach on the mainland outside Deshima.

Although everything Japanese interested Siebold, plants especially attracted him and he employed a talented young Japanese artist, Keiga Kawahara, to portray them. He also sent a consignment of living plants to Holland. However, Siebold's curiosity about so many aspects of Japan had been noticed by the Tokugawa regime and eventually led to his downfall. At Edo he saw the great secret map of Japan and he later obtained a copy. He put the forbidden map aboard a Dutch ship returning to Batavia. Unfortunately, a typhoon wrecked this ship on the Japanese coast. A search of its cargo led to the discovery of Siebold's map and he was arrested as a spy liable to execution. Ultimately he was pardoned but expelled from Japan in January 1829, taking with him his immense Japanese collections and many living plants but sadly leaving behind his mistress O-taki-san and their two-year old daughter Ine, the education of whom he entrusted to one of his best students.

The plants Siebold brought back were planted in the Ghent (Gand) Botanic Garden in what is now Belgium but had formerly been the Spanish Netherlands before being united with the Dutch Netherlands. At the time of his return the Belgians rose in revolt and Siebold fled to Holland but had to leave his plants behind. When the Japanese species of *Epimedium* L. flowered at Ghent, they led Morren and Decaisne to publish in 1834 the first monograph of the genus.

Siebold then settled in Holland, at Leiden, writing an important work, *Nippon*, on Japan, planning an account of its natural history and establishing a Japanese garden near Leiden. Thereafter Bürger at Deshima sent Japanese plants to Holland. The opening of Japan to foreigners in 1854 permitted collectors such as Carl Johann Maximowicz (1827–1891), John Gould Veitch (1839–1876) and Robert Fortune (1812–1880) to enter the interior to begin a new era in introducing plants from Japan.

References and Additional Reading

Bowers, J.Z. (1970). Western medical pioneers in feudal Japan. 245 pp. John Hopkins Press, Baltimore & London.

Haberland, D. (1996). Engelbert Kaempfer, 1651–1716, a biography. 158 pp. The British Library, London.

Henket, M. *et al.* (1993). Philipp Franz von Siebold (1796–1866). Ein Bayer als Mittler zwischen Japan und Europa. 182 pp. Haus der Bayerischen Geschichte, München.

Körner, H. (1967). Die Würzburger Siebold, eine Gelerntefamilie des 18. und 19. Jahrhunderts. 662 pp. J.A. Barth Verlag, Leipzig. (Lebensdarstellungen deutscher Naturforscher, herausgeben von der Naturforscher Leopoldina, Nr. 13).

Meier-Lemgo, K. (1960). Engelbert Kaempfer (1651–1716) erforscht das seltsame Asien. 193 pp. Cram, Hamburg.

Moule, A.C. (1937). The name *Ginkgo biloba* and other names of the tree. *T'oung Pao*, 33: 193–219. Leiden. [Summary in *J. Roy. Hort. Soc.* 63(3): 146 (March 1938)].

Muntschick, W. (1983). Engelbert Kaempfer, Flora Japonica (1712). Reprint des Originals mit Kommentar. 315 pp. Franz Steiner Verlag, Wiesbaden.

Nordenstam, B. (ed.). (1993). Carl Peter Thunberg. Linnean, resenär, naturforskare, 1743–1828. 191 pp. Atlantis, Stockholm.

Stearn, W.T. (1947). The name *Lilium japonicum* as used by Houttuyn and Thunberg. *Lily-Year Book* 11: 101–108.

Stearn, W.T. (1949). Kaempfer and the lilies of Japan. *Lily Year Book* 12 (for 1948): 65–70.

Stearn, W.T. (1970). Philipp Franz von Siebold and the lilies of Japan. *Lily-Year Book* 34: 11–20.

Stearn, W.T. (1994a). The career of Philipp Franz von Siebold. In Y. Kimura & V.I. Grubov (eds). Siebold's Florilegium of Japanese Plants 2, pp. 1–7.

Stearn, W.T. (1994b). Carl Peter Thunberg's visit to Japan. In Y. Kimura & V.P. Leonov (eds). C.P. Thunberg's Drawings of Japanese Plants 2, pp. 347–351.

Stearn, W.T. (1994c). Images of Japan [P.F. von Siebold and Keiga Kawahara]. *Garden (London)* 120(1): 24–27.

Stearn, W.T. (1999). Engelbert Kaempfer (1651–1716), pioneer investigator of Japanese plants. *Bot. Mag.* 16: 103–115.

Thommen, E. (1949). Neues zur Schreibung des Namens *Ginkgo*. *Verh. Naturf. Ges. Basel* 40: 77–103.

Wijnands, D.O. (1990). Correct author citation for the species described on material collected by Thunberg in Japan. *Thunbergia* 12: 1–48.

Dehmer, K.J. & Stracke, S. (1999). Molecular analysis of genebank accessions of the *Solanum nigrum* complex. In: S. Andrews, A.C. Leslie and C. Alexander (Editors). Taxonomy of Cultivated Plants: Third International Symposium, pp. 343–345. Royal Botanic Gardens, Kew.

MOLECULAR ANALYSIS OF GENEBANK ACCESSIONS OF THE *SOLANUM NIGRUM* COMPLEX

KLAUS J. DEHMER AND SILKE STRACKE

Genebank/Molecular Markers Research Group, Institute for Plant Genetics and Crop Plant Research/IPK, Corrensstraße 3, D-06466 Gatersleben, Germany

Introduction

The species of the *Solanum nigrum* L. complex, represented by 51 entries in the Gatersleben Genebank, are members of section *Solanum* (Seithe 1962). They represent a taxonomically difficult complex of very variable forms (Venkateswarlu & Rao 1972) closely related to each other, partially originating from each other and in several cases having no clearly defined interspecific morphological boundaries (Edmonds 1972). They are cosmopolitan in their distribution and are commonly referred to as black nightshades.

Although being considered as troublesome weeds in European and North American agriculture (Edmonds & Chweya 1997), some species are medicinal (Rao & Tandon 1974) or fodder plants (Hammer *et al.* 1992), while selected races are also cultivated because of their edible berries (Hammer 1986).

During recent years, species of this complex have become accessible as sources of resistances for potato breeding (*S. tuberosum* L., section *Petota* Dumart.), e.g. against *Phytophthora infestans* (Mont) de Bary (Colon *et al.* 1993). While initial back-cross experiments with hybrids between both sections generated by protoplast fusion (Binding *et al.* 1982) or 'embryo rescue' (Eijlander & Stiekema 1994) failed because of the high degree of sterility of the F1 plants, Horsman *et al.* (1997) reported on interspecific hybrids developed via protoplast fusions that seem to permit back-crosses.

In order to achieve a thorough characterisation of the genetic diversity present in the Gatersleben black nightshade accessions, a limited number of Genebank accessions was examined by RAPDs (3 species/15 entries) and microsatellite/SSR PCR (4 species/18 entries) in a first stage. This was done in respect to an effective Genebank management and to future collecting strategies, and in order to contribute to a molecular-based taxonomy of these species.

Methods

RAPD PCRs were executed as described in Stracke *et al.* (1996); for SSR analyses, 20 µl reaction assays contained 20 ng template DNA, 400 µM dNTPs, 375 nM per primer, 3 mM MgCl$_2$, 0.4 U Ampli*Taq* polymerase (Perkin Elmer, Foster City/CA) and 25 ng/µl BSA. Amplifications were executed after an initial denaturing step (5 min, 95°C) for 35 cycles with 40 sec at 94°C, 40 sec at 50°C and 20 sec at 72°C, followed by a final 10 min polymerization step.

343

Reaction products were separated by electrophoresis in agarose gels (RAPDs) or on an automatic sequencing apparatus (SSR PCR). The analysis of the RAPD data (65 polymorphic bands out of reactions with 10 primers) was performed using the NTSYS-PC software (version 1.80), with the calculated similarities (Nei & Li 1979) being transformed into a UPGMA dendrogram.

Results and Discussion

In the RAPD dendrogram the 15 accessions examined were grouped into two clusters, one of them being further divided into two subgroups. These were constituted by *S. scabrum* Mill. on the one hand and by *S. nigrum* accessions on the other. In the second cluster, all four *S. americanum* Mill. accessions were encountered, which despite morphological similarities could be viewed as genetically distinct from the two other species.

In the microsatellite analyses with potato SSR primer pairs, a successful amplification could be observed in many cases. By applying two primer pairs (Potato U6snRNA primer, Provan *et al.* 1996; Proteinase Inhibitor I primer, Kawchuk *et al.* 1996) it is easily possible to differentiate all three *S. americanum* accessions from the other three species investigated. This is the case in distinguishing the three *S. villosum* Mill. entries from all *S. nigrum* and *S. scabrum* accessions. Indications that a 1 bp difference between *S. nigrum* and *S. scabrum* generated by one of the primer pairs can be used for separating these two species have yet to be verified further. This — like the differentiation of species in the *S. nigrum* complex in general — should be achieved by employing additional characteristical SSR primers, while the intraspecific relationships can best be covered by AFLPs analyses currently being evaluated.

References

Binding, H., Jain, S.M., Finger, J., Mordhorst, G., Nehls, R. & Gressel, J. (1982). Somatic hybridization of an atrazine resistant biotype of *Solanum nigrum* with *Solanum tuberosum*. Part 1: Clonal variation in morphology and in atrazine sensitivity. *Theor. Appl. Genet.* 63: 273–277.

Colon, L.T., Eijlander, R., Budding, D.J., Ijzendoorn, M.T. van, Pieters, M.M.J. & Hoogendoorn, J. (1993). Resistance to potato late blight (*Phytophthora infestans* (Mont.) de Bary) in *Solanum nigrum*, *S. villosum* and their sexual hybrids with *S. tuberosum* and *S. demissum*. *Ephytica* 66: 55–64.

Edmonds, J.M. (1972). A synopsis of the taxonomy of *Solanum* sect. *Solanum* (*Maurella*) in South America. *Kew Bull.* 27(1): 95–114.

Edmonds, J.M. & Chweya, J.A. (1997). Black Nightshades. *Solanum nigrum* L. and related species. Promoting the conservation and use of underutilized and neglected crops. 15. 113 pp. Institute of Plant Genetics and Crop Plant Research, Gatersleben/ Intenational Plant Genetic Resources Institute, Rome, Italy.

Eijlander, R. & Stiekema, W.J. (1994). Biological containment of potato (*Solanum tuberosum*): outcrossing to the related wild species black nightshade (*Solanum nigrum*) and bittersweet (*Solanum dulcamara*). *Sex. Plant Reprod.* 7(1): 29–40.

Hammer, K. (1986) *Solanaceae*. In J. Schultze-Motel (ed.). Rudolf Mansfelds Kulturpflanzenverzeichnis. pp. 1179-1223. Akademieverlag, Berlin.

Hammer, K., Esquivel, M. & Knüpffer, H. (1992). '... y tienen faxones y fabas muy diversos de los nuestros ...' Origin, evolution and diversity of Cuban plant genetic resources. Vol. 2. 370 pp. IPK Gatersleben.

Horsman, K., Bergervoet, J.E.M. & Jacobsen, E. (1997). Somatic hybridization between *Solanum tuberosum* and species of the *Solanum nigrum* complex: selection of vigorously growing and flowering plants. *Euphytica* 96: 345–352.

Kawchuk, L.M., Lynch, D.R., Thomas, J., Penner, B., Sillito, D. & Kulcsar, F. (1996). Characterization of *Solanum tuberosum* simple sequence repeats and application to potato cultivar identification. *Amer. Potato J.* 73: 325–335.

Nei, M. & Li, W.H. (1979). Mathematical model for studying genetic variation in terms of restriction endonucleases. *Proc. Natl. Acad. Sci. U.S.A.* 76: 5269–5273.

Provan, J., Powell, W. & Waugh, R. (1996). Microsatellite analysis of relationships within cultivated potato (*Solanum tuberosum*). *Theor. Appl. Genet.* 92(8): 1078–1084.

Rao, G.R. & Tandon, S.L. (1974). *Solanum nigrum* L. In J. Hutchinson (ed.). Evolutionary studies in world crops. pp. 109–117. Cambridge University Press, London.

Seithe, A. (1962). Die Haararten der Gattung *Solanum* L. und ihre taxonomische Verwertung. *Bot. Jahr. Syst.* 81(3): 261–336.

Stracke, S., Njoroge, G. & Hammer, K. (1996). Genetic diversity in the collection of *Solanum nigrum* L. in the Gatersleben Genebank. In F. Begemann, C. Ehling & R. Falge (eds). Vergleichende Aspekte der Nutzung und Erhaltung pflanzen- und tiergenetischer Ressourcen (Schriften zu Genetischen Ressourcen, 5). pp. 320–324. IGR/ZADI, Bonn.

Venkateswarlu, J. & Rao, M.K. (1972). Breeding systems, crossability relationships and isolating mechanisms in the *Solanum nigrum* complex. *Cytologia* 37(2): 317–326.

Grant, M.L., Miller, D.M. & Culham, A. (1999). A morphological and molecular investigation into the woody cultivars of *Lavatera* L. In: S. Andrews, A.C. Leslie and C. Alexander (Editors). Taxonomy of Cultivated Plants: Third International Symposium, p. 347. Royal Botanic Gardens, Kew.

A MORPHOLOGICAL AND MOLECULAR INVESTIGATION INTO THE WOODY CULTIVARS OF *LAVATERA* L.

M.L. GRANT[1], D.M. MILLER[1] AND A. CULHAM[2]

[1]RHS Garden, Wisley, Woking, Surrey GU23 6QB, UK
[2]Centre for Plant Diversity and Systematics, School of Plant Sciences, The University of Reading, Whiteknights, Reading, Berkshire RG6 6AS, UK

In recent years, *Lavatera* L. has become a widely grown and popular garden plant. The commonly grown species and cultivars produce seed with abandon and many seedlings and sports have been raised, named and introduced into cultivation. The species may be recognised quite easily using traditional taxonomy but the affinities of most of the cultivars are uncertain. It was decided to investigate this problem using morphological and molecular techniques. With the latter, we were particularly interested in assessing the suitability of randomly amplified polymorphic DNA (RAPD) analysis for differentiating cultivars of ornamental plants.

Morphological studies undertaken so far have focused on a number of characters used for differentiating the species. The cultivars can best be recognised by flower colour and shape.

RAPD analysis has revealed at least 10 primers that will provide reproducible polymorphic amplification fragments. Two in particular reveal species-specific bands that can be used to assess hybrid status. Others have been found which will discriminate between five of the most closely related cultivars.

Preliminary analyses of the morphological and molecular data appears to be showing that the majority of cultivars are intermediate between *L. olbia* L. and *L. thuringiaca* L., suggesting hybrid status.

Kurashige, Y., Mine, M., Eto, J., Kobayashi, N., Handa, T., Takayanagi, K. & Yukawa, T. (1999). Sectional relationships in the genus *Rhododendron* (*Ericaceae*) based on *matK* sequences. In: S. Andrews, A.C. Leslie and C. Alexander (Editors). Taxonomy of Cultivated Plants: Third International Symposium, p. 349. Royal Botanic Gardens, Kew.

SECTIONAL RELATIONSHIPS IN THE GENUS *RHODODENDRON* (*ERICACEAE*) BASED ON *matK* SEQUENCES

Y. KURASHIGE[1], M. MINE[2], J. ETO[3], N. KOBAYASHI[4], T. HANDA[3], K. TAKAYANAGI[3] AND T. YUKAWA[5]

[1]**Laboratory of Botany, Akagi Nature Park, 892 Yuhikami, Minami-Akagisan, Akagi-mura, Seta-gun, Gunma 379-1113, Japan**
[2]**Iwate Biotechnology Research Center, 22-174-4 Narita, Kitakami, Iwate 024-0003, Japan**
[3]**Institute of Agriculture and Forestry, University of Tsukuba, 1-1-1 Tennodai, Tsukuba, Ibaraki 305-0006, Japan**
[4]**Tatebayashi Azalea Research Station, 3258 Hanayama-cho, Tatebayashi, Gunma 374-0005, Japan**
[5]**Tsukuba Botanical Garden, National Science Museum, 4-1-1 Amakubo, Tsukuba, Ibaraki 305-0005, Japan**

Owing to a large number of species and conflicting ideas of classification proposed by many taxonomists, the genus *Rhododendron* L. poses systematic problems at several infrageneric levels. Sequence data derived from *matK* and *trnK* introns were used to examine relationships among all 8 subgenera and 12 sections of *Rhododendron* as well as *Elliottia* Mühl. ex Elliott, *Ledum* L., *Loiseleuria* Desv., *Menziesia* Sm. and *Phyllodoce* Salisb.

The major results from this study are as follows:–

(1) *Menziesia* and *Ledum* are nested within *Rhododendron*

(2) *Rhododendron* is paraphyletic

(3) *Rhododendron camtschaticum* Pall. forms a basal lineage of tribe *Rhodoreae* D.Don

(4) Subgenus *Pentanthera* (G.Don) Pojark. is polyphyletic

(5) Subgenus *Pentanthera* sections *Pentanthera* and *Rhodora* (L.) G.Don show a sister group relationship

(6) Subgenus *Pentanthera* section *Pentanthera* is monophyletic

(7) Subgenus *Pentanthera* section *Sciadorhodion* Rehder & E.H. Wilson is polyphyletic

(8) Subgenus *Rhododendron* is monophyletic, however, sections *Rhododendron*, *Pogonanthum* G.Don and *Vireya* (Blume) H.F. Copel. are para/polyphyletic

(9) Subgenus *Hymenanthes* (Blume) K. Koch is monophyletic

(10) Subgenus *Azaleastrum* Planch. is polyphyletic

(11) Subgenus *Azaleastrum* section *Choniastrum* Franch. is monophyletic and shows a sister group relationship to subgenus *Mumeazalea* (Sleumer) M.N. Philipson

(12) Subgenus *Azaleastrum* section *Azaleastrum* (Planch.) Maxim. is monophyletic

(13) Subgenus *Tsutsusi* (Sweet) Pojark. is monophyletic, however, *Rhododendron tashiroi* Maxim. makes both sections *Tsutsusi* Sweet and *Brachycalyx* Sweet para/polyphyletic

(14) *R. tsusiophyllum* Sugim. is a member of subgenus *Tsutsusi* section *Tsutsusi*.

Yukawa, T., Koga, S. & Handa, T. (1999). DNA uncovers paraphyly of *Dendrobium* (*Orchidaceae*). In: S. Andrews, A.C. Leslie and C. Alexander (Editors). Taxonomy of Cultivated Plants: Third International Symposium, pp. 351–354. Royal Botanic Gardens, Kew.

DNA UNCOVERS PARAPHYLY OF *DENDROBIUM* (*ORCHIDACEAE*)

TOMOHISA YUKAWA[1], SATOSHI KOGA[2] AND TAKASHI HANDA[2]

[1]Tsukuba Botanical Garden, National Science Museum, Tsukuba 305-0005, Japan
[2]Institute of Agriculture and Forestry, University of Tsukuba, Tsukuba 305-0006, Japan

Introduction

Owing to its horticultural importance and great number of species, the phylogeny of *Dendrobium* Sw. and its allied genera has been discussed extensively since the monumental work of Lindley (1830–1840). Our present knowledge, however, is very limited mainly because recurrent convergence of morphological characters has resulted in artificial systems of classification (e.g. Kraenzlin 1910, Schlechter 1912). Also, cytological (Hashimoto 1987), micromorphological (Yukawa *et al.* 1991, 1992; Yukawa unpublished data), and anatomical (Morris *et al.* 1996, Yukawa & Uehara 1996) characters contain few phylogenetic signals.

We have used macromolecular markers such as restriction sites of endonucleases and DNA sequences of the ribulose-bisphosphate carboxylase gene (*rbcL*) to clarify infra- and intergeneric relationships of subtribe *Dendrobiinae* Lindl. (*Cadetia* Gaudich., *Dendrobium*, *Diplocaulobium* (Rchb.f.) Kraenzl., *Epigeneium* Gagnep., *Flickingeria* A.D. Hawkes and *Pseuderia* Schltr.) as well as the sister group position of the *Dendrobiinae* (Yukawa *et al.* 1993, 1996; Yukawa & Uehara 1996). Comparison of DNA sequences of *matK*, the maturase-encoding gene located in an intron of the chloroplast gene *trnK*, the former of which has evolved approximately three times faster than the chloroplast gene *rbcL* (Johnson & Soltis 1994), has proven to be a powerful new tool for phylogenetic reconstruction within angiosperm families and genera (Johnson & Soltis 1994; Steele & Vilgalys 1994; Soltis *et al.* 1996). In this study, we compared *matK* sequences to establish more concrete relationships within subtribe *Dendrobiinae*.

Material and Methods

Tribe *Malaxideae* Lindl. was chosen as an outgroup based on results of our global analyses of sister group candidates of tribe *Dendrobieae* Endl. that comprises subtribes *Dendrobiinae* and *Bulbophyllinae* Schltr. (Yukawa *et al.* unpublished data). *Pseuderia* was not included in this study because our previous studies have shown its placement in tribe *Podochileae* Benth. & Hook. (Yukawa *et al.* 1993, 1996). Methods for analyses were shown in Yukawa *et al.* (1993, 1996). Sequences were determined with PCR-amplified the *matK* gene from a total DNA extract by use of the primers shown in Fig. 1 and Table 1.

TABLE 1. Location and base composition of amplification and sequencing primers used in this study.

Primer	5' sequence 3'	Designed by
OMAT1F	CCGTT(A/C)T(C/G)ACCATATTGC	Yukawa
matK-462F	AATACCCTA(T/C)CCC(A/G)T(C/T)CATC	Chase
matK-841F	CTTTCATACATTATGTTCGA	Chase
OMAT2R	CAAATTATGATATTCGTGA/GA	Yukawa
OMAT3R	ATCGGTCCAGATCGGTTT	Yukawa
OMAT4R	CGTGCTTGCA(A/G)TTTTCATT	Yukawa

Results and Discussion

A single most parsimonious tree of the *matK* phylogeny (Fig. 2) provides the following insights: (1) subtribe *Dendrobiinae* comprises three major clades: Clade 1 (*Cadetia, Diplocaulobium, Flickingeria, Dendrobium,* sections *Dendrocoryne* Lindl., *Eleutheroglossum* Schltr., *Grastidium* (Blume) J.J. Sm., *Latouria* (Blume) Miq., *Phalaenanthe* Schltr., *Rhizobium* Lindl., and *Spatulata* Lindl.); Clade 2 (*Dendrobium* sections *Callista* (Lour.) Schltr., *Crumenata* Pfitzer, *Dendrobium, Formosae* (Benth. & Hook.f.) Hook.f., *Oxyglossum* Schltr., and *Strongyle* Lindl.); Clade 3 *Epigeneium.* (2) *Epigeneium* diverged early from the lineage including Clades 1 and 2. (3) Relative to *Cadetia, Diplocaulobium,* and *Flickingeria, Dendrobium* is shown to be para-/polyphyletic.

We have two alternatives to resolve the para-/polyphyly of *Dendrobium.* (A) To sink *Cadetia, Diplocaulobium,* and *Flickingeria* within *Dendrobium,* the resultant *Dendrobium sensu lato* recovers its monophyly. (B) If *Dendrobium* is subdivided into several genera according to our results, all the genera including *Dendrobium sensu stricto* will be monophyletic. For example, if one wants to conserve *Cadetia, Diplocaulobium,* and *Flickingeria* at generic rank, sections of *Dendrobium* in Clade 1 should be applied to separate generic names. On the other hand, Clade 2 only comprises *Dendrobium* sections; this clade, moreover, includes the type species of the genus, viz. *D. moniliforme* (L.) Sw. Consequently, Clade 2 can be defined as *Dendrobium sensu stricto.*

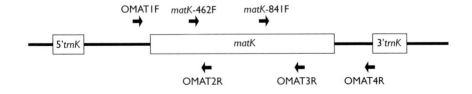

FIG. 1. Relative position of the PCR amplification and sequencing primers used for *matK.* Arrows indicate the direction of strand synthesis. Boxed areas represent coding regions.

**Section in
Dendrobium**

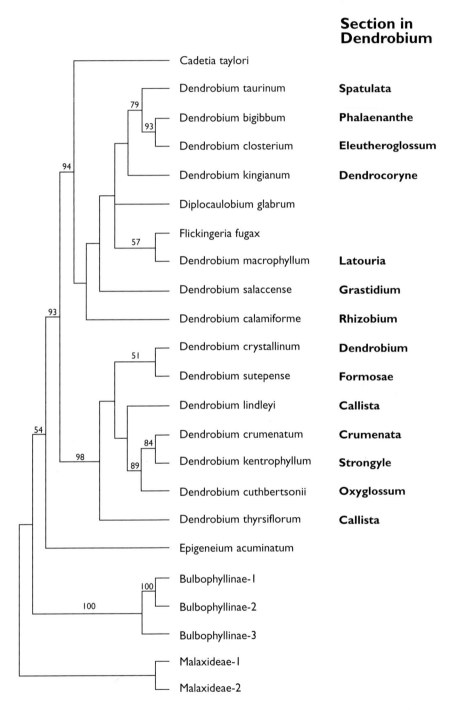

FIG. 2. A single most parsimonious Fitch tree based upon *matK* sequences: length=395, consistency index (excluding uninformative characters)=0.686, retention index=0.804. Numbers above internodes indicate bootstrap values more than 50 percent from 200 replicates.

The (B) option certainly causes unnecessary confusion on the widely accepted usage of plant names. For example, "*Phalaenopsis* type" dendrobiums (*Dendrobium* sections *Phalaenanthe* and *Spatulata*), one of the most important horticultural crops in tropical countries, should be transferred to a separate, new genus. We hence support the (A) option that treats *Dendrobium sensu lato*.

Acknowlegements

We are indebted to Mark W. Chase for providing unpublished information on *matK* primers, Kazuhiro Suzuki for skilful cultivation of plant material for this study, Koichi Kita and Osamu Miikeda for technical assistance. Financial support of this work was provided in part by Fujiwara Natural History Foundation and a Grant-in-Aid (08740673) for Encouragement of Young Scientists from the Ministry of Education, Science, Sports and Culture, Japan to T. Yukawa.

References

Hashimoto, K. (1987). Karyomorphological studies of some 80 taxa of *Dendrobium*, Orchidaceae. *Bull Hiroshima Bot. Gard.* 9: 1–186.

Johnson, L.A. & Soltis, D.E. (1994). *matK* DNA sequences and phylogenetic reconstruction in Saxifragaceae s. str. *Syst. Bot.* 19(1): 143–156.

Kraenzlin, F. (1910). *Orchidaceae-Monandrae-Dendrobiineae* Pars 1. Genera n. 275–277. In A. Engler, (ed.). Pflanzenreich, Heft 45, IV. 50. II. pp. 1–382. W. Engelmann, Leipzig.

Lindley, J. (1830–1840). The genera and species of orchidaceous plants. 553 pp. Ridgeways, London.

Morris, M.W., Stern, W.L. & Judd, W.S. (1996). Vegetative anatomy and systematics of subtribe Dendrobiinae (*Orchidaceae*). *Bot. J. Linn. Soc.* 120(2): 89-144.

Schlechter, R. (1912). 71. *Dendrobium* Sw. In Die Orchidaceen von Deutsch-Neu-Guinea. *Repert. Spec. Nov. Regni Veg. Beih.* 1: 440–643.

Soltis, D.E., Kuzoff, R.E., Conti, E., Gornall, R. & Ferguson, K. (1996). *matK* and *rbcL* gene sequence data indicate that *Saxifraga* (*Saxifrgaceae*) is polyphyletic. *Amer. J. Bot.* 83: 371–382.

Steele, K.P. & Vilgalys, R. (1994). Phylogenetic analyses of *Polemoniaceae* using nucleotide sequences of the plastid gene *matK*. *Syst. Bot.* 19(1): 126–142.

Yukawa, T., Ando, T., Karasawa, K. & Hashimoto, K. (1991). Leaf surface morphology in selected *Dendrobium* species. In Proceedings of the 13th World Orchid Conference 1990. pp. 250–258. 1990 World Orchid Conference Trust, Auckland.

Yukawa, T., Ando, T., Karasawa, K. & Hashimoto, K. (1992). Existence of two stomatal shapes in the genus *Dendrobium* (*Orchidaceae*) and its systematic significance. *Amer. J. Bot.* 79: 946–952.

Yukawa, T., Kurita, S., Nishida, M. & Hasebe, M. (1993). Phylogenetic implications of chloroplast DNA restriction site variation in subtribe Dendrobiinae (*Orchidaceae*). *Lindleyana* 8(4): 211–221.

Yukawa, T., Ohba, H., Cameron, K.M. & Chase, M.W. (1996). Chloroplast DNA phylogeny of subtribe Dendrobiinae (*Orchidaceae*): insights from a combined analysis based on *rbcL* sequences and restriction site variation. *J. Plant Res.* 109(1094): 169–176.

Yukawa, T. & Uehara, K. (1996). Vegetative diversification and radiation in subtribe Dendrobiinae (*Orchidaceae*): evidence from chloroplast DNA phylogeny and anatomical characters. *Pl. Syst. Evol.* 201(1–4): 1–14.

Miikeda, O., Koga, S., Handa, T. & Yukawa, T. (1999). Subgeneric relationships in *Clematis* (*Ranunculaceae*) by DNA sequences. In: S. Andrews, A.C. Leslie and C. Alexander (Editors). Taxonomy of Cultivated Plants: Third International Symposium, pp. 355–358. Royal Botanic Gardens, Kew.

SUBGENERIC RELATIONSHIPS IN *CLEMATIS* (*RANUNCULACEAE*) BY DNA SEQUENCES

O. MIIKEDA[1], S. KOGA[2], T. HANDA[2] AND T. YUKAWA[3]

[1]Tokyo Metropolitan Kitazono High School, 4-14-1, Itabashi, Itabashi-ku, Tokyo 173-0004, Japan.
[2]Institute of Agriculture and Forestry, University of Tsukuba, 1-1-1, Tennodai, Tsukuba-shi, Ibaraki 305-0006, Japan
[3]Tsukuba Botanical Garden, National Science Museum, 4-1-1, Amakubo, Tsukuba-shi, Ibaraki 305-0005, Japan

Introduction

The genus *Clematis* L. (*Ranunculaceae* Juss.) contains approximately 300 species (Tamura 1987). This cosmopolitan genus shows enormous diversification in morphological characters: e.g. shrub or climber, erect or spreading sepals, presence or absence of petals, pilose or glabrous stamens, and alternate or opposite phyllotaxy in seedlings. This morphological diversification has resulted in several different systems of classification of *Clematis*, e.g. de Candolle (1824), Prantl (1887), Tamura (1967), Keener & Dennis (1982), Johnson (1997). These systems are mainly based on the characteristics of the sepals (erect or spreading) and the inflorescences (dichasium or solitary). Tamura (1987) reinvestigated these characters and recognised 4 subgenera (*Campanella* Tamura, *Viorna* (Rchb.) Tamura, *Clematis* and *Flammula* (DC.) Peterm.). In his system, he attached importance to features of phyllotaxy in seedlings (alternate or opposite) and stamens (pilose or glabrous). We, however, have not found conclusive views on which characters reflect the evolutionary history of the genus.

Recently, macromolecular data have been applied to clarify the phylogenetic position of *Clematis* in *Ranunculaceae*, e.g. Johansson & Jensen (1993), Hoot (1995), Jensen *et al.* (1995), Kosuge *et al.* (1995). These results show that the *Clematis–Clematopsis* complex is the sister group to the *Anemone* complex (*Anemone* L., *Hepatica* Mill., *Pulsatilla* Mill. and *Knowltonia* Salisb.), but the infrageneric relationships of *Clematis* have not been investigated yet.

We re-examined the subgeneric relationship in *Clematis* proposed by Tamura (1987) in comparison with aligned sequences of the following regions in chloroplast DNA; *matK* (a maturase-encoding gene) and *trnK* (UUU) introns, (2) *trnL* (UAA) intron and intergenic spacer between *trnL* (UAA) and *trnF* (GAA), (3) intergenic spacer between *atpB* (ATPase beta subunit) and *rbcL* (RuBisCO large subunit). The utility of these regions has been shown by, for example, infrageneric analyses of *Saxifraga* L. (Soltis *et al.* 1996) and *Aconitum* L. (Kita *et al.* 1995) or an intraspecific analysis of *Pedicularis chamissonis* Steven (Fujii *et al.* 1997).

Experimental Procedure

Eight species covering all the subgenera *sensu* Tamura (1987) of *Clematis* were used for the analysis. *Anemone flaccida* F. Schmidt was selected as an outgroup. Total cellular DNA was isolated from fresh leaves by modified CTAB buffer method (Kobayashi *et al.* 1998) or a DNA miniprep method (Lassner *et al.* 1989). The three regions in chloroplast DNA mentioned above were amplified via the polymerase chain reaction (PCR) to obtain sufficient quantities of DNA for sequencing. Each amplification product was purified by a column (CENTRI-SEP COLUMNS, Princeton Separations, Inc.) or a simple column using Sephadex G-50 (Pharmasia) (Nakayama & Nishikata 1995). The product was then used as a template for dye terminator cycle sequencing. After purification, single-stranded DNAs with dye terminators were sequenced by ABI PRISM TM 377 (Perkin Elmer Inc.). Both strands of DNA were sequenced with a nearly complete overlap.

The sequence data were aligned with Sequencing Navigator (Perkin Elmer Inc.). Phyllogenetic analyses were performed using PAUP (Phyllogenetic Analysis Using Parsimony) version 3.1 (Swofford 1993) using the heuristic search option with 100 random additions, and MULPARS in effect (with collapse of zero-length branches). PAUP was also used to perform the bootstrap analysis with 1000 replications.

Results and Discussions

Sequences for *matK* and the *trnK* (UUU) introns (2542 bp (base pairs)), the *trnL* (UAA) intron and the intergenic spacer between *trnL* (UAA) and *trnF* (GAA) (981 bp), and the intergenic spacer between *atpB* and *rbcL* (896 bp) were used for phylogenetic analyses. The heuristic search found a single most parsimonious tree of 231 steps with a consistency index (C.I.) of 0.961 and a retention index (R.I.) of 0.640.

The tree (Fig. 1) shows the following relationships: 1) nested position of subgenus *Viorna* within subgenus *Flammula*, 2) paraphyly of subgenus *Flammula*, 3) nested position of subgenus *Clematis* within subgenus *Campanella*, 4) paraphyly of subgenus *Campanella*. As shown in Fig. 1, distribution of the key characters of the subgenera suggests that the opposite phyllotaxy in seedlings evolved once in the genus and that erect sepals and pilose stamens appeared several times.

Two patterns of phyllotaxy are found in *Clematis* (Tamura 1970): 1) only an opposite phyllotaxy during the whole life history, 2) alternate in seedlings and opposite in shoots. Species in subgenus *Campanella* section *Campanella* such as *C. lasiandra* Maxim. often show alternate phyllotaxy not only in seedlings but also in the basal part of the shoots (Tamura 1980). Similar to the condition of this section, *Anemone* shows alternate phyllotaxy in seedlings as well as in radical leaves (homologous with the basal part of shoots). Our results show that subgenus *Campanella* section *Campanella* comes out as the basal clade and retains a plesiomorphic condition in phyllotaxy.

One thousand bootstrap replications did not result in high values for most clades. This is because few synapomorphic substitutions in the three regions support each clade: 8 in *matK* and the *trnK* introns, 1 in the *trnL–trnF* region, 4 in the spacer between *atpB* and *rbcL*. Synapomorphic insertions/deletions were 0, 2, and 3 in *matK* and the *trnK* introns, the *trnL–trnF* region, and the spacer between *atpB* and *rbcL*, respectively.

However, a sister-group relationship of section *Tubulosae* (Decne.) Kitag. (subgenus *Campanella*) and section *Clematis* (subgenus *Clematis*), is strongly supported by a 98% bootstrap value. The two sections are placed in separate subgenera in Tamura (1987) because of the difference in stamen characters (pilose or glabrous). They are, however, united with such characters as the alternate phyllotaxy in seedlings, the cymose inflorescence on the current season's stems, and the ternate and coarsely serrate leaves. Furthermore, presence of intersectional hybrids from the wild (*C.* × *takedana*

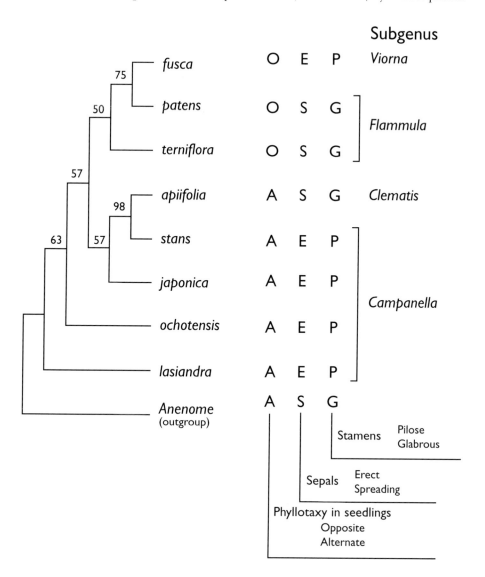

FIG. 1. A single most parsimonious tree resulting from phylogenetic analysis of 3-region sequences in chloroplast DNA for *Clematis*: length = 231, consistency index = 0.961, retention index = 0.640. Bootstrap values from 1000 replications are indicated above the branches. Subgenera are based on the classification of Tamura (1987).

Makino) or from cultivation, e.g. *C. tubulosa* × *C.* 'Jouiniana' (Johnson 1997), also supports a close relationship between the two sections.

Although we can estimate the outline of the evolutionary route of *Clematis*, inclusion of representation from all the sections and the accumulation of more phylogenetic signals by use of DNA regions of more rapid substition rates are necessary to propose a comprehensive view of the phylogeny of the genus.

References

Candolle, A.P. de (1824). *Clematis*. In A.P. de Candolle. Prodromus systematis naturalis regni vegetabilis. Vol. 1. 2–10 pp. Treuttel & Würtz. Paris.

Fujii, N., Ueda, K., Watano, Y. & Shimizu, T. (1997). Intraspecific sequence variation of chloroplast DNA in *Pedicularis chamissonis* Steven (*Schrophulariaceae*) and geographic structuring of the Japanese "alpine" plants. *J. Plant Res.* 110(1098): 195–207.

Hoot, S.B. (1995). Phylogenetic relationships in *Anemone* (*Ranunculaceae*) based on DNA restriction site variation and morphology. *Pl. Syst. Evol., Suppl.* 9: 195–200.

Jensen, U., Hoot, S.B., Johansson, J.T. & Kosuge, K. (1995). Systematics and phylogeny of the *Ranunculaceae* — a revised family concept on the basis of molecular data. *Pl. Syst. Evol., Suppl.* 9: 273–280.

Johansson, J.T. & Jensen, R.K. (1993). Chloroplast DNA variation and phylogeny of the *Ranunculaceae. Pl. Syst. Evol.* 187: 29–49.

Johnson, M. (1997). Släktet Klematis. 881 pp. Magnus Johnsons Plantskola AB. Södertälje.

Keener, C.S. & Dennis, W.M. (1982). The subgeneric classification of *Clematis* (*Ranunculaceae*) in temperate North America north of Mexico. *Taxon* 31(1): 37–44.

Kita, Y., Ueda, K. & Kadota, Y. (1995). Molecular phylogeny and evolution of the Asian *Aconitum* subgenus *Aconitum* (*Ranunculaceae*). *J. Plant Res.* 108(1092): 429–442.

Kobayashi, N., Horikoshi, T., Katsuyama, H., Handa T. & Takayanagi, K. (1998). A simple and efficient DNA extraction method for plants, especially woody plants. *Pl. Tissue Cul. Biotech.* 4(2): 76–80.

Kosuge, K., Sawada, K., Denda, T. & Watanabe, K. (1995). Phylogenetic relationships of some genera in the *Ranunculaceae* based on alcohol dehydrogenase genes. *Pl. Syst. Evol., Suppl.* 9: 263–271.

Lassner, M.W., Peterson, P. & Yoder, J.I. (1989). Simultaneous amplification of multiple DNA fragments by polymerase chain reaction in the analysis of transgenic plants and their progeny. *Pl. Molec. Biol. Reporter* 7: 116–128.

Nakayama, H. & Nishikata, T. (1995). Illustrated Reference in Bio-experiments II — Bases of Gene Analysis, pp. 132–135. Shujunsha Co., Ltd., Tokyo. (In Japanese).

Prantl, K. (1887). Beitrage zur morphologie und systematik der Ranunculaceen. *Bot. Jahrb. Syst.* 9: 225–273.

Soltis, D.E., Kuzoff, R.K., Conti, E., Gornall, R. & Ferguson, K. (1996). *matK* and *rbcL* gene sequence data indicate that *Saxifraga* (*Saxifragaceae*) is polyphyletic. *Am. J. Bot.* 83(3): 371–382.

Swofford, D.L. (1993). PAUP: Phylogenetic analysis using parsimony, version 3.1. Computer program distributed by Illinois Natural History Survey., Champaign, Illinois.

Tamura, M. (1967). Morphology, ecology and phylogeny of the *Ranunculaceae* VII. *Sci. Rep. Coll. Gen. Educ. Osaka Univ.* 16(2): 21–43.

Tamur, M. (1970). *Archiclematis*, a precursory genus of *Clematis. Acta Phytotax. Geobot.* 24(4–6): 146–151. (In Japanese).

Tamura, M. (1980). Change of phyllotaxis in *Clematis lasiandra* Maxim. *J. Jap. Bot.* 55(9): 257–265. (In Japanese).

Tamura, M. (1987). A classification of genus *Clematis. Acta Phytotax. Geobot.* 38: 33–44. (In Japanese).

Russell, J., Booth, A., Fuller, J., Provan, J., Ellis, R., Marshall, D. & Powell, W. (1999). Exploiting barley microsatellites at the Scottish Crop Research Institute (SCRI). In: S. Andrews, A.C. Leslie and C. Alexander (Editors). Taxonomy of Cultivated Plants: Third International Symposium, p. 359. Royal Botanic Gardens, Kew.

EXPLOITING BARLEY MICROSATELLITES AT THE SCOTTISH CROP RESEARCH INSTITUTE (SCRI)

JOANNE RUSSELL, ALLAN BOOTH, JOHN FULLER, JIM PROVAN, ROGER ELLIS, DAVID MARSHALL AND WAYNE POWELL

Scottish Crop Research Institute, Invergowrie, Dundee DD2 5DA, Scotland, UK

Microsatellite-based assays have been developed for both the nuclear and chloroplast genomes in various plant species. Here the potency of this technology with reference to barley (*Hordeum vulgare* L.) is shown. Four main areas of research are summarised.

1. The deployment of previously mapped microsatellites to create a genotypic database for spring barley cultivars. More than 100 spring barley cultivars representing leading European germplasm generated over the past century has been genotyped with 28 microsatellites. In addition to uniquely fingerprinting (profiling) this material, we have generated a two-dimensional, graphical display of the data that provides a retrospective analysis of allelic selection by breeders during cultivar development.

2. The genetic base of spring barley has been quantified to examine changes in levels and patterns of genetic variability over time. More than 77% of the allelic variability present in the cultivated gene-pool can be traced to 15 cultivars which represent significant landraces or their immediate derivatives.

3. Cytoplasmic analysis of *H. vulgare* and its wild progenitor *H. spontaneum* K. Koch using seven polymorphic chloroplast microsatellites has revealed a decrease in cytoplasmic diversity between *H. spontaneum* and *H. vulgare* as well as between *H. vulgare* landraces and cultivars. This is characteristic of domestication processes in many crop species. We have also observed geographical partitioning of chloroplast microsatellite genotypes within *H. vulgare* landraces from Syria and Jordan.

4. Bio-informatics provides the computational and data-handling infrastructure that provides a graphical user interface for laboratory and field scientists. This is shown with reference to graphical genotyping tools developed for our barley genomics programmes.

Treu, R., Holmes, D., Smith, B., Astley, D. & Trueman, L. (1999). RAPD analysis of genetic diversity within *Allium ampeloprasum* with special reference to var. *babingtonii*. In: S. Andrews, A.C. Leslie and C. Alexander (Editors). Taxonomy of Cultivated Plants: Third International Symposium, p. 361. Royal Botanic Gardens, Kew.

RAPD ANALYSIS OF GENETIC DIVERSITY WITHIN *ALLIUM AMPELOPRASUM* WITH SPECIAL REFERENCE TO VAR. *BABINGTONII*

R. Treu[1], D. Holmes[1], B. Smith[2], D. Astley[2] AND L. Trueman[2]

**[1]School of Environmental Sciences and Land Management,
University College Worcester, Worcester WR2 6AJ, UK
[2]Plant Genetics and Plant Breeding Department, Horticulture Research International,
Wellesbourne, Warwick CV35 9EF, UK**

Classified within *Allium ampeloprasum* L. are a number of horticulturally important cultivars including leek, kurrat and great headed garlic. *Allium ampeloprasum* var. *babingtonii* (Borrer) Syme is a little studied endemic variety that, like other wild populations, may hold significant potential for genetic resources. Random amplified polymorphic DNA (RAPD) analysis was applied to a) 21 Cornish populations of var. *babingtonii* to assess genetic variation within the variety and b) to 5 cultivars and 8 wild accessions of *A. ampeloprasum* to assess genetic variation within the species. Southern analysis was used to ascertain the degree of sequence homology between a number of common bands. No genetic variation was found between populations of var. *babingtonii* or between great headed garlic, *A. ampeloprasum* (ex. Flat Holm Island, Bristol Channel) and *A. ampeloprasum* (ex South Stack, near Holyhead, Wales). RAPD band polymorphism between the remaining accessions ranged from 2 to 53%; further analysis of this data allowed inference of phylogenetic relationships. The lack of variation within var. *babingtonii* would suggest obligate asexuality and is thus of limited value to conventional plant breeding. The RAPD data indicates that var. *babingtonii* is well delimited within *A. ampeloprasum* which, together with its distinct morphological characters, indicates the plant may be more accurately classified as *A. ampeloprasum* subsp. *babingtonii* ined. (Treu 1999).

Reference

Treu, R. (1999). The ecology of *Allium ampeloprasum* var. *babingtonii*: biosystematic, cytological and molecular perspectives. Unpublished PhD Thesis. University College Worcester.

Groendijk-Wilders, N., Kardolus, J.P., Zevenbergen, M.J. & Berg, R.G. van den. (1999). AFLP, a new molecular marker technique applied to potato taxonomy. In: S. Andrews, A.C. Leslie and C. Alexander (Editors). Taxonomy of Cultivated Plants: Third International Symposium, pp. 363–365. Royal Botanic Gardens, Kew.

AFLP, A NEW MOLECULAR MARKER TECHNIQUE APPLIED TO POTATO TAXONOMY

NYNKE GROENDIJK-WILDERS, JOUKE P. KARDOLUS, MARTIN J. ZEVENBERGEN AND RONALD G. VAN DEN BERG

Plant Taxonomy, Wageningen Agricultural University, Gen. Foulkesweg 37, 6703 BL Wageningen, The Netherlands

Introduction

The cultivated potato, *Solanum tuberosum* L., is one of about 200 described tuber-bearing *Solanum* L. species (*Solanum* section *Petota* Dumort.), all of which, are from the Americas and mostly from the Andes-region. Section *Petota* is subdivided into 21 series by Hawkes (1990). The taxa are often hard to distinguish morphologically and classification of this group is therefore often troublesome.

Some series, e.g. *Acaulia* Juz., are morphologically relatively well-defined by a unique combination of character state(s). The circumscription of the other series, for example, of series *Tuberosa* (Rydb.) Hawkes, and also specific identification within the series *Tuberosa*, is much more difficult. Molecular techniques, by providing many qualitative characters, could lead to a better understanding of the taxonomy and evolutionary history of this group.

The AFLP Technique

The molecular marker technique AFLP™ (Amplified Fragment Length Polymorphism) (Vos *et al.* 1995) is based on the selective PCR (Polymerase Chain Reaction) amplification of DNA restriction fragments under stringent PCR conditions, making use of restriction site specific primers.

We (Kardolus *et al.* 1998) studied a large set of wild and cultivated potato species with the AFLP technique:

- To determine the applicability of this novel marker technique for plant biosystematics.
- To clarify the taxonomy of *Solanum* section *Petota*.

In total 53 *Solanum* species were analysed, most of them represented by 3 genebank accessions. As outgroup species we chose 5 non-tuber-bearing *Solanum* species.

AFLP Data-analysis

Fragments with a size of 150 to 550 nucleotides were scored. An AFLP fingerprint generated ± 35 scorable fragments per genotype. In total three different primer combinations were examined, resulting in a data-matrix of 997 AFLPs. We analysed this matrix both phenetically (with NTSYS-PC & TREECON) and cladistically (PAUP 3.1.1).

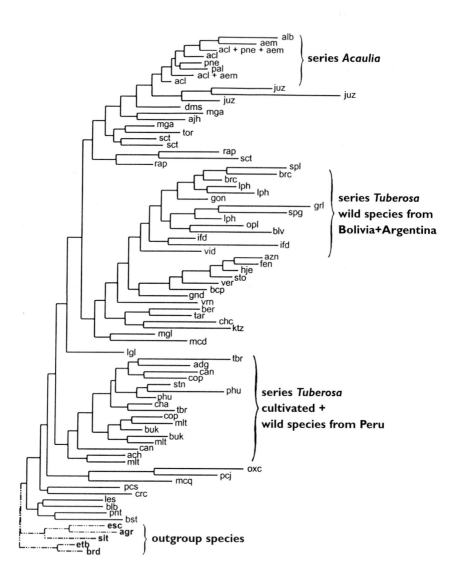

Fig. 1. Phylogram based on the PAUP analysis. Species abbreviations following Hawkes (1990).

New Insights in Potato Taxonomy

- The AFLP technique is valuable for taxonomic studies, at least for related species groups as in the genus *Solanum*.
- The combination of NeighborJoining and Parsimony analysis resulted in informative and comparable phylogenies. The NJ-analysis seemed more appropriate for studying hybridisation and the parsimony method for designating monophyletic groups.
- The species most directly related to cultivated potato (series *Tuberosa pro parte*), although morphologically quite similar, are divided into two branches (**not** sister-groups):
    1. The cultivated potatoes and diploid wild species from Peru and N.W. Bolivia
    2. Wild species from Bolivia and Argentina
- The series classification needs serious reconsideration. Phylogenetic relations are now often poorly reflected.

References

Hawkes, J.G. (1990). The Potato. 259 pp. Belhaven Press, London.

Kardolus, J.P., Van Eck, H.J. & Berg, R.G. van den (1998). The potential of AFLPs in biosystematics: a first application in *Solanum* taxonomy (*Solanaceae*). *Pl. Syst. Evol.* 210(1–2): 87–103.

Vos, P., Hogers, R., Bleeker, M., Reijans, M., Lee, T. van de, Hornes, M., Frijters, A., Pot, J., Peleman, J., Kuiper, M. & Zabeau, M. (1995). AFLP: a new technique for DNA fingerprinting. *Nucl. Acids Res.* 23: 4407-4414.

Lea, V.J., Leigh, F.J., Lee, D., Cooke, R.J., Cooper, W. & Reeves, J.C. (1999). Molecular methods for hybrid cultivar identification in sunflower. In: S. Andrews, A.C. Leslie and C. Alexander (Editors). Taxonomy of Cultivated Plants: Third International Symposium, p. 367. Royal Botanic Gardens, Kew.

MOLECULAR METHODS FOR HYBRID CULTIVAR IDENTIFICATION IN SUNFLOWER

V.J. LEA, F.J. LEIGH, D. LEE, R.J. COOKE, W. COOPER AND J.C. REEVES

National Institute of Agricultural Botany, Huntingdon Road, Cambridge CB3 OLE, UK

Hybrid crops offer potential gains via heterosis (hybrid vigour), resulting in increased reliability, quality and yields. To benefit, there is a need to assess the hybrid (genetic) purity of seed lots, prior to sowing. In addition, new cultivars must be shown to be Distinct, Uniform and Stable (DUS) for protection via Plant Breeders' Rights (PBR) schemes. Currently, this involves the comparison of mainly morphological characteristics which may be multigenic, affected by environmental interactions and requires a full growing season for replicated testing. Morphological differences may not be sufficient for discrimination in some cultivars. Current DUS testing in certain crops may be augmented by biochemical and molecular techniques (i.e. isoenzyme analysis, RFLP, RAPD, AFLP and microsatellites) [Lee *et al.* 1996a]. Microsatellites and AFLPs offer the greatest potential; many loci may be analysed simultaneously, they are suitable for automation and cultivars may be characterised by reproducible differences in known bands or loci. It has been shown that considerable levels of variation exist at some microsatellite loci among hybrid cultivars of sunflower. By testing sets of eighty individuals of several hybrid cultivars, we have shown that there are *intra-* and *inter-*varietal variations. The proportion of 'offtypes' or 'outcrossed' plants differs between cultivars. Parental lines will be examined to determine the cause of this variation (Lee *et al.* 1996b).

References

Lee, D., Reeves, J.C. & Cooke, R.J. (1996a). DNA profiling and plant variety registration 2. Restriction fragment length polymorphisms in varieties of oilseed rape. *Pl. Var. Seeds* 9: 181–190.

Lee, D. Reeves, J.C. & Cooke, R.J. (1996b). DNA profiling for varietal identification in crop plants. 1996 BCPC Symposium proceedings No. 65: Diagnostics in Crop Production. pp. 235–240. British Crop Protection Council.

Berg, R.G. van den, Zevenbergen, M.J., Kardolus, J.P. & Groendijk-Wilders, N. (1999). The origin of *Solanum juzepczukii*. In: S. Andrews, A.C. Leslie and C. Alexander (Editors). Taxonomy of Cultivated Plants: Third International Symposium, pp. 369–370. Royal Botanic Gardens, Kew.

THE ORIGIN OF *SOLANUM JUZEPCZUKII*

RONALD G. VAN DEN BERG, MARTIN J. ZEVENBERGEN, JOUKE P. KARDOLUS AND NYNKE GROENDIJK-WILDERS

Plant Taxonomy, Wageningen Agricultural University, Gen. Foulkesweg 37, 6703 BL Wageningen, The Netherlands

Introduction

The triploid cultivated potato *Solanum juzepczukii* Bukasov has been assumed to be a hybrid between the allotetraploid wild potato species *S. acaule* Bitter of series *Acaulia* Juz. and the diploid (2n=24) cultivated *S. stenotomum* Juz. & Bukasov of series *Tuberosa* (Rydb.) Hawkes (Hawkes 1962). This hypothesis is supported by a strong morphological resemblance, biogeographical data and by a re-synthesis experiment (Hawkes 1962). Until now, no cytogenetic or molecular studies confirmed this assumption. We re-examined this hypothesis in a combined study using both the DNA fingerprinting technique, AFLP and the fluorescent genome painting technique GISH.

AFLP Results

AFLP has proven to be a promising technique in biosystematic studies (Kardolus *et al.* 1998). In this study *S. juzepczukii* has been demonstrated to be closely related to the polyploid series *Acaulia* and *S. demissum* Lindl. of series *Demissa* Bukasov from Mexico. All this material shared many apomorphies, that is AFLP fragments not present in the rest of the material studied. However, *S. juzepczukii* was more closely related to *S. tuberosum* L. and other wild diploid potatoes from Peru than to *S. stenotomum*.

GISH

Genomic *in situ* hybridisation (GISH) is a cytogenetic tool frequently used in biosystematic studies which focus on hybrid speciation. Fluorescently labelled DNA of the supposed parents of a hybrid (taxon) is hybridised on chromosome spread preparations of that hybrid. When there is sufficient DNA-differentiation between the parents, the parental genomes can be made visible in the hybrid.

GISH was performed with both *S. acaule* and *S. stenotomum* as a probe (not simultaneously) on *S. juzepczukii* chromosome preparations. Sonicated herring sperm DNA was used as a block. Controls were also performed (Table 1).

TABLE 1. *In situ* hybridisation experiments.

Spread preparation	Probe		Preliminary results
S. juzepczukii (2n=36)	*S. acaule*		± 24 chr. differentially labelled
	S. stenotomum		≥ 12 chr. differentially labelled
	S. juzepczukii		36 chr. labelled
S. acaule (2n=48)	*S. juzepczukii*	Controls	48 chr. labelled
	S. stenotomum		48 chr. weakly labelled
S. stenotomum (2n=24)	*S. acaule*		24 chr. weakly labelled
	S juzepczukii		24 chr. labelled

With *S. acaule* as a probe on spread preparations of *S. juzepczukii*, there was a clear differentiation and approx. 24 chromosomes were labelled. The differentiation was best in late pro-metaphase/early metaphase. With *S. stenotomum* as a probe, there was also differentiation, but more than the expected 12 chromosomes were labelled in most cells, probably due to the overlap in DNA sequences between *S. acaule* and *S. stenotomum*.

Conclusions

Our AFLP- and GISH-results are consistent with the hypothesis that *S. juzepczukii* has originated from hybridisation between *S. acaule* of series *Acaulia* and a diploid species of series *Tuberosa*, but not necessarily *S. stenotomum*.

References

Hawkes, J.G. (1962). The origin of *Solanum juzepczukii* and *S. curtilobum*. *Z. Pflanzenzüch.* 47: 1–14.

Kardolus, J.P., Eck, H.J. van & Berg, R.G. van den (1998). The potential of AFLPs in biosystematics: a first application in *Solanum* taxonomy. *Pl. Syst. Evol.* 210(1–2): 87–103.

DATABASES AND REGISTERS

B

Sadie, J. (1999). The International *Protea* Register — a report. In: S. Andrews, A.C. Leslie and C. a International Symposium, pp. 373–374. Royal Botanic Gardens, Kew.

THE INTERNATIONAL *PROTEA* REGISTER — A REPORT

JOAN SADIE

National Department of Agriculture, Directorate Genetic Resources, Private Bag X5044, Stellenbosch 7599, South Africa

The International *Protea* Register (IPR) is only applicable to the 8 South African genera of the *Proteaceae* Juss., namely *Aulax* Bergins, *Leucadendron* R.Br., *Leucospermum* R.Br., *Mimetes* Salisb., *Orothamnus* Pappe ex Hook., *Paranomus* Salisb., *Protea* L. and *Serruria* Salisb. When referring to this group in general, they are called proteas.

Due to the increasing international interest in the South African proteas and the simultaneous development of breeding programmes in six parts of the world, the Horticultural Research Institute in South Africa initiated the negotiation for the appointment of an International Registration Authority (IRA) for proteas in 1978. Authority was gained in 1980. It was proposed that the Division Variety Control of the Department of Agriculture be appointed as the IRA for proteas, because they were already responsible for statutory registration in South Africa.

The first applications for registration were already received in 1974, resulting from the breeding programme in South Africa. Plant Breeders' Rights were also granted. Dr G.J. Brits of the Horticultural Research Institute published a checklist containing 142 cultivar names in 1988 (Brits 1988). This list however, also included the five cultivars already registered in 1974. By 1986 nineteen cultivars were registered and published in the *South African Plant Variety Journal* under the heading Additional Information (Anon. 1986). In the period 1987 to 1991 there was no progress with the IPR, due to the absence of a person to develop and maintain the register.

In 1992 a new attempt was made to promote *Protea* cultivar registration internationally and to produce a publication containing the IPR and a checklist. Resulting from this effort, a preliminary first edition of the IPR and checklist was published (Anon. 1993) and distributed at the 7th Conference of the International *Protea* Association (IPA) held in Harare, Zimbabwe in 1993. Since then the IPR and checklist are published annually and distributed to more than 100 producers, originators and interested persons in fifteen countries.

Good response was received following the call for co-operation to register cultivar names and to provide information regarding cultivar names already published in journals, etc. Cultivars from Australia, Hawaii and Zimbabwe have been registered since then. Currently the IPR contains 86 cultivar names while the checklist has expanded to 292 cultivar names (Sadie 1998).

Another aspect of the IPR that was started, is the collection of herbarium specimens of the registered cultivars. The specimens are accommodated by the National Herbarium, Pretoria (PRE). Approximately 40 specimens have already been collected; all of them South African cultivars. A problem to be encountered is to find contact persons in the other countries who will be responsible for the collection and preparation of specimens and to send them to the Registrar of the IPR.

The way forward is to continue with international promotion and asking for co-operation, therefore striving to have as many as possible cultivar names registered. Regional co-operators must be found to help with the promotion and to collect herbarium specimens in their countries. A homepage is also planned to make the IPR as well as the application forms, more accessible. This should encourage registration of cultivar names, as the application forms can be easily obtained.

References

Anon. (1993). International *Protea* Register. Preliminary 1st edition. Directorate Plant & Quality Control, Stellenbosch.

Anon. (1986). Additional Information *S. African Pl. Var. J.* 30: 10.

Brits, G.J. (1988). Sample list of validly published *Protea* cultivar names. *Protea News*, 7: 6–10.

Sadie, J. (1998). The International *Protea* Register. Ed. 5. 58 pp. Directorate of Plant and Quality Control, Stellenbosch.

Jones†, A.W. & other Heather Society members (1999). Stages in the development of the International Register for heathers. In: S. Andrews, A.C. Leslie and C. Alexander (Editors). Taxonomy of Cultivated Plants: Third International Symposium, pp. 375–378. Royal Botanic Gardens, Kew.

STAGES IN THE DEVELOPMENT OF THE INTERNATIONAL REGISTER FOR HEATHERS

A.W. Jones† and other Heather Society members

Denbeigh, All Saints Road, Ipswich, Suffolk IP6 8PJ, UK

The Heather Society was founded by a number of enthusiasts in February 1963 at a meeting organised at Vincent Square with the object of fostering interest in the growing of hardy heaths and heathers. In 1970, it was appointed as the International Registration Authority for cultivars of *Andromeda* L., *Bruckenthalia* Rchb., *Calluna* Salisb, *Daboecia* D. Don and *Erica* L. (including Cape heaths) during the Botanical Congress held in Seattle in August 1969. This was formally approved at the International Horticultural Congress at Tel Aviv in March 1970. At that time, a reviewer (Knight 1971) exclaimed at the 'wealth' of cultivars — around 500 — the result of the preceding fifty years.

Prior to that appointment, David McClintock, who was one of the Society's founder members, had already begun to extract information on all heather species and cultivars from published references and nursery catalogues. For some thirty years he added these details to a card index and was inevitably appointed Registrar in 1970. Yet owing to the reluctance of growers to submit new introductions for registration, only 126 cultivars were formally registered by the time he retired in 1994. However, a succession of articles resulted from his keen interest in the nomenclature of the *Ericaceae* Juss., as well as his *A guide to the naming of plants* (1969) that utilised heather names as examples. The genera of *Ericaceae* undertaken for registration reflect both his practical & whimsical nature for there is no botanical logic for this grouping.

Soon after its appointment as an IRA, the Society's Technical sub-committee briefly vetted new heather cultivar introductions (Anon. 1980). In the 1960s & 1970s members also assisted in the Heather Trials carried out at RHS, Wisley. However, concerned over the proliferation of new names, the Society decided to begin its own trials from 1970 in ground made available at Harlow Carr (Vickers 1973 & 1976, Julian 1990).

Apart from nomenclatorial confusion, a common problem has been that of establishing the origin of the many cultivars, particularly the date of their introduction. Inevitably, hybrids between the species of *Erica* have caused other problems. The Society's two journals publish annual lists of Registrations, New Acquisitions and New Names. These regular listings have assisted in establishing the time of introduction. Other members have contributed articles that have clarified particular names (Lead, 1970, Jones (1979, 1980, 1984, 1985, 1986, 1987, 1988, 1990, 1991), Parris (1976, 1977, 1978, 1980) & Turpin (1983a, 1983b, 1984, 1985, 1988).

In his bibliographies, McClintock (1970, 1993) commented that works on heathers were "remarkably few". Although early references (Bentham 1839, Regel 1843) refer to forms and varieties, only modern books provide lists of cultivars, which are generally incomplete (Underhill 1971). Authors outside the UK have made valuable contributions by recording cultivars introduced in Europe e.g. Denkewitz (1987), van de Laar (1976) and in the U.S.A. Metheny (1991).

After consultation with the taxonomists at Wisley in April 1981, on the nature of information needed in its register, the Society requested Albert Julian (Chairman of a Technical Committee) to undertake the listing of all known heather cultivars as a separate exercise. Thus the first draft Register, or Check List was eventually achieved in 1988. This had then to be checked against the more complete data contained in the Registrar's card-index. Ultimately, in 1988 it was decided to incorporate both sets of data into a computer database, using Microsoft's Access.

The task of in-putting some 3,300 records was accomplished during the early 1990s by David & Anne Small. This information was utilised in the comprehensive *Handy Guide to Heathers* published on behalf of the Heather Society (1992). It has recently been revised and contains additional details on all heather species (1998). Eventually in 1997, a new Registrar — A.W. Jones was able to issue a three volume, draft document *Checklist of Heathers from Europe and adjacent areas* for internal distribution. Members specialising in cultivars of particular genera had the task of correcting the entries in the lists:

1) An alphabetical list of all cultivar names.
2) Separate lists of the epithets used in each genus or species.

On becoming Registrar in 1994, A.W. Jones revised the Society's Registration Application form to conform with the database. On amending the advisory notes he provided explanatory diagrams and emphasized the necessity for providing a description of the character/s that were thought to distinguish a new cultivar from others. To encourage registration he carried out explanatory workshops in the U.K. and U.S.A. during 1996 & 1997.

The various sections of the up-dated database are in the process of correction before part publication of each genus, or for each species of *Erica*. Growers in Europe e.g. Blum and Kramer, have provided details of cultivars recently introduced in Germany and Holland. As a preliminary exercise to establish a format, the genus *Bruckenthalia* was selected as the initial subject for publication, since very few taxa are involved. This was completed during 1998. The rather more complex task of dealing with the genus *Andromeda* is now underway and expected to be finished by 1999.

Using the search facility of the database it has been possible to discover the earliest dates for the introduction of cultivars belonging to each species and confirm fashionable patterns such as the preponderance of golden foliage cultivars during the 1960s. It is apparent that the more obvious descriptive epithets of colour e.g. 'Alba', 'Coccinea', 'Minor', 'Minima', 'Praecox' and 'Rubra' had been used for most of the species placed in this denomination class long before Article 17.2 was formulated (Trehane *et al.* 1995). Turpin (1985) pointed out that nurserymen have often been rather free in the interpretation of 'identical characteristics' in their general use of an established name for any new clone that they encountered, citing examples of *E. vagans* 'Mrs D.F. Maxwell', *C. vulgaris* 'Alportii' and *C. vulgaris* 'Flore Pleno'.

References

Anon. (1980). Secondary Reference Collections. *Year Book Heather Soc.* 2(9): 70–72.
Bentham, G. (1839). *Macnabia, Calluna, Pentapera, Erica, Bruckenthalia.* In A.P. de Candolle Prodromus Systematis Naturalis Regni Vegetabilis. 7(2): 612–694.
Denkewitz, L. (1987). *Heidegärten.* 356 pp. Verlag Eugen Ulmer, Stuttgart.
Jones, A.W. (1979). The classification of hardy winter flowering heaths with notes on *Erica × darleyensis. Year Book Heather Soc.* 2(8): 38–46.

Jones, A.W. (1980). One, two or three cultivars? *Year Book Heather Soc.* 2(9): 59–64.

Jones, A.W. (1984). Heather cultivars from Cornwall. *Year Book Heather Soc.* 3(2): 50–56.

Jones, A.W. (1985). Some notes on *Erica cinerea* var. *rendlei* and six recent finds. *Year Book Heather Soc.* 3(3): 59–66.

Jones, A.W. (1986). First records of the heathers of the British Isles. *Year Book Heather Soc.* 3(4): 34–39.

Jones, A.W. (1987). Notes on *Erica manipuliflora, E. vagans* and their hybrids. *Year Book Heather Soc.* 3(5): 51–57.

Jones, A.W. (1988). The yellow-foliage cultivars of *Erica carnea. Year Book Heather Soc.* 3(6): 59–68.

Jones, A.W. (1990). A note on *Erica carnea* 'Unknown Warrior'. *Year Book Heather Soc.* 3(8): 46–47.

Jones, A.W. (1991).Cultivars of the *Erica manipuliflora* Group. *Year Book Heather Soc.* 3(9): 29–32.

Jones, A.W. (1992). English and scientific names for heathers. *Year Book Heather Soc.* 3(10): 43–44.

Jones, A.W. (1997). From the Registrar – On five sports of *C. vulgaris* 'Marleen'. *Bull. Heather Soc.* 5(10): 9–10.

Julian, T.A. (1990). The Heather Trials plot at Harlow Carr – 1971 to 1990. *Year Book Heather Soc.* 3(8): 25–31.

Knight, F.P. (1971). Book notes. Heaths and heathers. *J. Roy. Hort. Soc.* 96(11): 514.

Laar, H.J. van de (1974). Het Heidetuinboek. 160 pp. Zomer und Keuning, Wageningen.ß

Laar, H.J. van de (1976). Heidegärten, Anlage, Pflege, Pflanzenwahl. 141 pp. Parey, Berlin & Hamburg.

Laar, H.J. van de (1978). The Heather Garden. The plants and their cultivation. 160 pp. Collins, London.

Lead, W.L. (1970). Is it correctly named ? *Year Book Heather Soc.* 1(7): 41–43.

McClintock, D. (1969). A guide to the naming of plants with special reference to heathers. 38 pp. Heather Society.

McClintock, D. (1969). *Daboecia azorica* and its hybrids with *D. cantabrica. J. Roy. Hort. Soc.* 94(10): 449–453.

McClintock, D. (1970). A brief bibliography of heathers. *Year Book Heather Soc.* 1(7): 7–14.

McClintock, D. (1993). A Bibliography of Heathers. First Supplement: 1970–1993. *Year Book Heather Soc.* 4(1): 23–29.

Metheny, D. (1991). Hardy Heather Species. 186 pp. Frontier, Seaside, Oregon.

Parris, A. (1976). Preliminary note on a cross between *Erica erigena* and *E. carnea. Year Book Heather Soc.* 2(5): 48–49.

Parris, A. (1977). Further note on a cross between *Erica erigena* and *E. carnea. Year Book Heather Soc.* 2(6): 10.

Parris, A. (1978). Further notes on the induced *Erica erigena* × *E. carnea* hybrids. *Year Book Heather Soc.* 2(7): 43–43.

Parris, A. (1980). Further notes on *E.* × *darleyensis* induced hybrids. *Year Book Heather Soc.* 2(9): 67–70.

Regel, E.A. (1843). *Die Kultur und Aufzählung in der Deutschen und Englischen Gärten befindlichen Eriken, nebst Synonymie und kurzer Charakterisirung und Beschreibung derselben* ...... 189 pp. Zürich.

Small, D. & Small, A. (1992). Handy Guide to Heathers. 116 pp. Denbeigh Heather Nurseries, Ipswich.

Small, D. & Small, A. (1998). Handy Guide to Heathers. Ed. 2. 168 pp. Denbeigh Heather Nurseries, Ipswich.

Trehane, P. *et al.* (1995). International Code of Nomenclature for Cultivated Plants —
1995. (*Regnum Vegetabile* 133). 175 pp. Quarterjack Publishing, Wimborne.

Turpin, P.G. (1983a). The heather species, hybrids and varieties of the Lizard district
of Cornwall. *Cornish Studies* 10(5): 5–17.

Turpin, P.G. (1983b). A correction *E.* × *watsonii* 'Morning Glow'. *Year Book Heather Soc.*
3(1): 41–42.

Turpin, P.G. (1984). Late-flowering callunas. *Year Book Heather Soc.* 3(2): 37–40.

Turpin, P.G. (1985). From the Chairman. *Year Book Heather Soc.* 3(3): 4–6.

Turpin, P.G. (1988). The cultivars of *Erica* × *watsonii*. *Year Book Heather Soc.* 3(6): 57–59.

Underhill, T. (1971). *Heaths and Heathers.* 256 pp. David & Charles, Newton Abbot.

Vickers, G.P. (1973). Harlow Carr Heather Trials. *Year Book Heather Soc.* 2(2): 18–19.

Vickers, G.P. (1976). Harlow Carr Heather Trials. *Year Book Heather Soc.* 2(5): 40–43.

Fryer, R.A., Horrocks, D.H. & Walter, K.S. (1999). Development of a horticultural database for the Western Australian flora. In: S. Andrews, A.C. Leslie and C. Alexander (Editors). Taxonomy of Cultivated Plants: Third International Symposium, pp. 379. Royal Botanic Gardens, Kew.

DEVELOPMENT OF A HORTICULTURAL DATABASE FOR THE WESTERN AUSTRALIAN FLORA

R.A. FRYER[1], D.H. HORROCKS[1] AND K.S. WALTER[2]

[1]Kings Park and Botanic Garden, West Perth 6005, W Australia
[2]Royal Botanic Garden Edinburgh, Inverleith Row, Edinburgh EH3 5LR, Scotland

The Western Australian flora is one of the world's most diverse and has great potential but there is very little accurate horticultural information about it.

The Western Australia Wildflower Society, Kings Park and Botanic Garden and the Western Australian Herbarium, with finance from the Gordon Reid Foundation, combined to develop a database of plant based information collected directly from the growers of the plants.

Collecting and collating horticultural data is very complex with many interrelated factors which makes it unsuited to most database structures. BG-BASE™ was selected because it has multi-value fields to link interrelated factors.

Data dissemination is of prime concern and discussions indicated that we needed good interrogation ability and output of free text descriptions. To provide this the BG-BASE™ module was designed by Dr Kerry Walter with the ability to output in a format which DELTA™ (DEscriptive Language for TAxonomy) could handle. DELTA™ was developed to code taxonomic descriptions, provide easy interrogation on multiple keys and to provide free text from these descriptions. This is the first instance of using DELTA™ for horticultural data. Using this programme will allow rapid production of published data focused on the user group requirements.

Walter, K.S. (1999). *BG-BASE*™, a database for cultivated plants. In: S. Andrews, A.C. Leslie and C. Alexander (Editors). Taxonomy of Cultivated Plants: Third International Symposium, pp. 381. Royal Botanic Gardens, Kew.

BG-BASE™, A DATABASE FOR CULTIVATED PLANTS

K.S. WALTER

Royal Botanic Garden Edinburgh, Inverleith Row, Edinburgh EH3 5LR, Scotland, UK

BG-BASE™ has been designed specifically to handle the information management needs of institutions holding collections of living and preserved plants. Begun in 1985 for the Arnold Arboretum (USA), it has now been installed in 16 countries, making it the most widely used system in botanic gardens world-wide.

Modern collection management practices require far more than simple inventory control (what do we have?, where did we get it?, etc.). *BG-BASE*™ handles a huge range of information, including nomenclature and taxonomy, geography, wild collecting details, bibliography, images, verification, propagation, horticultural treatments, sources of plants, shipments, seed banks and quarantine.

Data are stored in a variable-length, relational database structure allowing users to enter data once and having that information available throughout the system; data in any field can expand to a maximum of 64,000 characters. The relevant rules in the botanical and horticultural codes of nomenclature are built into the system, which can handle cultivar and cultivar-groups, selling names, commercial synonyms, etc. Taxa can be located by any combination of accepted scientific name, synonym or common name.

In addition to standard reports, it produces many kinds of labels, including engraved, embossed, barcode, herbarium, mailing and tractor-fed pot labels.

SYSTEMATICS, CLASSIFICATION
AND NOMENCLATURE

C

Whiteley, A.C. & Grant, M.L. (1999). A rose by any other name would be misleading! In: S. Andrews, A.C. Leslie and C. Alexander (Editors). Taxonomy of Cultivated Plants: Third International Symposium, p. 385. Royal Botanic Gardens, Kew.

A ROSE BY ANY OTHER NAME WOULD BE MISLEADING!

A.C. WHITELEY AND M.L. GRANT

Botany Department, The Royal Horticultural Society's Garden, Wisley, Woking, Surrey GU23 6QB, UK

Garden plants are generally ignored by plant taxonomists but there is a whole industry based upon them and a large proportion of the population, at least in the UK, has a considerable interest in them.

The Royal Horticultural Society considers it very important that garden plants should be correctly identified in all situations. To that end, the Society's botanists have always provided a plant identification service to the membership and tried to ensure that plants in the Society's Gardens at Wisley, Rosemoor and Hyde Hall are correctly named.

The Society also conducts trials of garden plants which gather together as many species and cultivars of particular genera as possible. Trials provide an ideal opportunity to check taxonomy and nomenclature, leading to many corrections and publications of the findings in the widely-read journal, *The Garden*.

With the advent of a computerised plant records database at Wisley, information about the living and herbarium collections is gradually being integrated. As areas of the Garden are added to the database, the botanists are kept busy verifying the identity of the plants recorded. In conjunction with this, voucher specimens of all plants are being added to the herbarium, enabling verification to continue throughout the year.

Gardner, D. (1999). Naming temperate woody plants at the Royal Botanic Gardens, Kew. In: S. Andrews, A.C. Leslie and C. Alexander (Editors). Taxonomy of Cultivated Plants: Third International Symposium, p. 387. Royal Botanic Gardens, Kew.

NAMING TEMPERATE WOODY PLANTS AT THE ROYAL BOTANIC GARDENS, KEW

DAVID GARDNER

The Herbarium, The Royal Botanic Gardens, Kew, Richmond, Surrey TW9 3AB, UK

The Royal Botanic Gardens, Kew has always had close associations with cultivated plants since it was founded in 1759. The establishment of the Cultivated Section between the early 1960s and 1976 and the creation of the post of Gardens Verifier in 1977 made this association more formal. This appointment allowed Kew's traditionally tropical taxonomic emphasis to encompass the Living Collections Department's (LCD) temperate woody plants, a group which had been somewhat neglected over the years. The Gardens Verifier was able to name large groups of plants in the Living Collections to bring Kew's database and LCD records up to date.

The post and role of Gardens Verifier was expanded in 1987 by the appointment of the Horticultural Taxonomist. The purpose of this position and subsequently the Horticultural Taxonomy Unit, as it became known after the creation of a second post in 1994, was not only to carry on the work of verifying collections in the Gardens but to carry out research into cultivated groups, to establish horticultural links around the world and to encourage links between the Herbarium and LCD. The Horticultural Taxonomist, in conjunction with specialists from other institutions and countries, takes an active part in publicising cultivated plant taxonomy and nomenclature as well helping to make decisions regarding these two disciplines.

Maxwell, H.S. (1999). Identification and verification of cultivated plants at the Royal Botanic Garden Edinburgh. In: S. Andrews, A.C. Leslie and C. Alexander (Editors). Taxonomy of Cultivated Plants: Third International Symposium, pp. 389–392. Royal Botanic Gardens, Kew.

IDENTIFICATION AND VERIFICATION OF CULTIVATED PLANTS AT THE ROYAL BOTANIC GARDEN EDINBURGH

H.S. MAXWELL

Royal Botanic Garden Edinburgh, 20A Inverleith Row, Edinburgh EH3 5LR, Scotland, UK

Introduction

In common with other botanic gardens it is necessary for the Royal Botanic Garden Edinburgh (RBGE) to be able to name its rich and diverse living collections. An accurately named reference collection is essential in order that plants can be propagated and cultivated in the appropriate way and so that research programmes can use them fully. Furthermore, it is the responsibility of botanic gardens to display plants which are accompanied by the correct name for the visiting public. RBGE also provides an important service for the general public in the identification of plants, horticultural pests and diseases and general horticultural advice.

The systematic process involved in dealing with these enquiries is presented in a flow diagram Fig. 1; this highlights the important inter-departmental collaboration which is necessary for this to take place.

Public Identifications

On average RBGE receives *c.* 520 enquiries a year. However, it is estimated that this is only 25–30% of the actual number as many are dealt with by telephone or are sent directly by the general public to the relevant departments. These may include, for example, specimens sent to the Flora of the Lothians project as possible new records for the County Flora. Some specimens prove to be of scientific importance and these may be photographed and pressed for incorporation into the slide library and the herbarium. Unusual enquiries include the entrails of rabbit which were assumed to be the remains of a flatworm. This enquiry, wrapped in 'clingfilm', was unfortunately left unopened over the weekend!

Garden Identifications

The Living Collections held by RBGE have a 328 year history and today they comprise 42,628 accessions of 21,484 taxa from 168 areas of the world. This represents one of the largest collections of living plants in the world, amounting to about 6% of the world's flora. When non-verified plants flower or fruit for the first time, the horticultural staff generate a verification form from the Garden's database — *BG-BASE*™, see Fig. 2. This accompanies the plant and is sent to the relevant member of staff for determination. After the plant has been identified the completed form is passed to the Records Office where the new information is used to update the database. Such information may

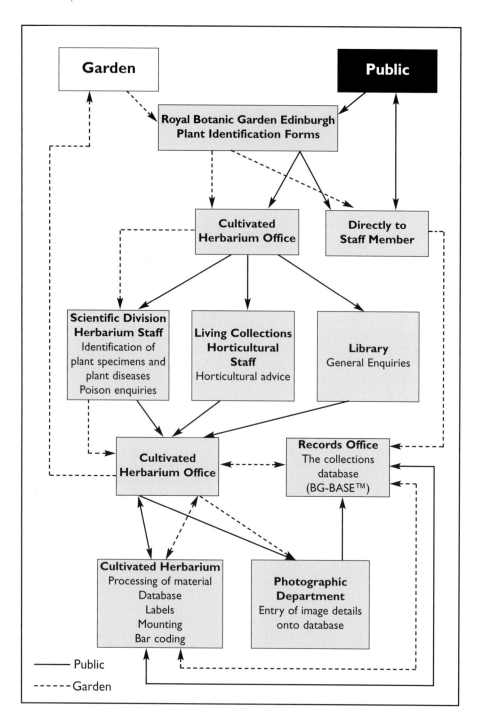

Fɪɢ. 1. Flow diagram of enquiries process at Royal Botanic Garden Edinburgh.

RBG, Edinburgh   **LIVING MATERIAL FOR VERIFICATION**

Accession No 19980481   **Ident No** 24642   **Date** 25 JUN 1998
     Qual A
Current name SCROPHULARIACEAE   **Make specimen** part is available
39480   Mimulus
          Range Chile, Argentina
Initiated by Colin Belton

Taken to   Martin Gardner

Reason sent/suggested name Verification

Location   E44   **All locations** A: PD0

Wild origin details
   Chile: Región VII [Maule]: Prov. de Talca: Valle de Maule. 1750 m. 35 deg
   56'43"S, 70 deg 30'29"W. Margin of stream side - west facing. Fleshy herb
   to 45 cm. Flowers bright yellow, spotted with reddish dots on lower petals.
   Universidad de Chile, RBG Edinburgh Expedition to Southern Chile 77. 5 Feb 1998

Collector Universidad de Chile, RBG Edinburgh   **Collector No** 77
         Expedition to Southern Chile
Donor   **Donor No**

Verified as (please use CAPITAL letters)
                            Subgenus   _____
  Family   _____   Section   _____
                           Subsection _____
  Genus   _____   Series   _____
                           Subseries _____
  Species   _____   Author   _____

  Subspecies _____   Author   _____

  Variety   _____   Author   ____ _____

  Forma/cultivar _____
  (circle one)
  Synonym/parents of hybrid _____

Wild distribution _____

Verified by _____   **Date** _____   **Level** ____

Previous verifications   **Date**   **Level**

Previous specimens or photographs

Specimens or photographs made now   C   H   HF   HR   HS   HV   HW   F   L
                            P   PB   PC   PS   S   SP   W   X
Sent to/awaiting _____   Date _____

More material required (specify) _____

Garden label(s) needed _____   **Processed** _____

References/notes   _____

FIG. 2. Verification form used at Royal Botanic Garden Edinburgh.

include: new name or verified name, complete distribution of the taxon, library references, etc. If necessary a voucher herbarium specimen is taken and the plant is photographed; this information, including the image produced, is also stored on _BG-BASE_™. Currently there are 15,516 verification records which represents 24.7% of the living collections. This figure is increasing by an average of 4.5% each year.

The Cultivated Herbarium

Having a comprehensive, well-curated cultivated herbarium is essential as a reference library for use in public and garden plant determinations. RBGE's herbarium has *c.* 3 million specimens, with an estimated 0.5 million of these being cultivated specimens. Many of the earliest cultivated specimens date from around 1850, but specimens dating back much earlier are also to be found in the herbarium, for example, a cultivated specimen of *Phytolacca icosandra* L. thought to have been grown at Haddington Place, near Leith Walk, Edinburgh — the third location of the Botanic Garden, dates back to 1764.

The final stage in the garden identification process is the generating of a garden label with the relevant details, see Fig. 3.

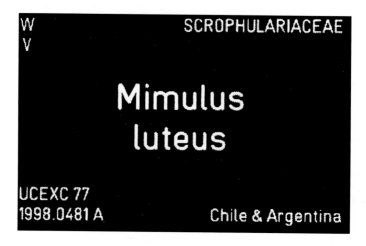

Fig. 3. Garden label from Royal Botanic Garden Edinburgh.

W – Plant of known wild origin

V – Verified

UCEXC 77 – Collectors code and number (Universidad de Chile, RBG Edinburgh expedition to southern Chile)

1998.0481A – Accession number with a qualifier. (A qualifier is used so that unique genotypes under a single accession can be tracked)

Chile & Argentina – Total range of taxon

McLewin, W. (1999). Time to change the type. In: S. Andrews, A.C. Leslie and C. Alexander (Editors). Taxonomy of Cultivated Plants: Third International Symposium, p. 393. Royal Botanic Gardens, Kew.

TIME TO CHANGE THE TYPE

W. McLewin

Phedar Research & Experimental Nursery, Bunkers Hill, Romiley, Stockport, Cheshire SK6 3DS, UK

There is a simple change of nomenclatural convention that could bring enormous benefit in the way plants in commercial horticulture are named. It is to stop using 'type' in its technical sense of the original specimen of a taxon deposited in a herbarium (and with it the term 'type locality'), so that 'type' could be used with its commonly understood meaning.

At present there is no accepted way to say of a plant 'This looks like XX and we believe that for garden purposes it will be effectively the same as XX (or at least similar) but we do not have impeccable provenance for it and we are not completely sure we have ever seen the true XX anyway'. Consequently any such plant gets labelled XX when frequently it is not XX and is actually, for example, a (hybrid) seedling from cultivated seed if XX is a species, or a vaguely similar plant propagated because XX is a desirable cultivar. If the plant could be labelled 'XX type' then the uncertainty about its identity would be clear and inaccuracies would not be ever further propagated (well, less often). This convention would encourage honesty whereas the current situation, where a plant without a name will not sell, encourages dishonesty and inaccuracy.

Unfortunately, for the convention of 'XX type' as a style of plant name to be adopted and recommended, taxonomists will have to choose something to use in its place. Finding another word is no problem. Finding the collective will to do something so down to earth and widely beneficial to everyone except themselves is quite another matter. But if even more incentive is needed then this change would probably mean that the misleading and ambiguous term 'Group' could probably be dispensed with as well.

Cribb, P., Chase, M. and Thomas, S. (1999). Orchid hybrid generic names — time for a change?
In: S. Andrews, A.C. Leslie and C. Alexander (Editors). Taxonomy of Cultivated Plants: Third
International Symposium, pp. 395–396. Royal Botanic Gardens, Kew.

ORCHID HYBRID GENERIC NAMES — TIME FOR A CHANGE?

PHILLIP CRIBB, MARK CHASE AND SARAH THOMAS

Royal Botanic Gardens, Kew, Richmond, Surrey TW9 3AB, UK

The current generic names of hybrids in the orchids are relatively meaningless, especially in hybrids involving three genera or more. New taxonomic analyses are likely to bring about an increasing number of changes in generic concepts and delimitation over the next few years. A simplified system of naming artificial orchid hybrids at the generic level might solve problems produced by the recognition of these new generic concepts. A tentative system is suggested to stimulate debate amongst the orchid fraternity.

Without access to the published list in each volume of the quinquennially (now triennially) produced *Sander's List of Orchid Hybrids*, the information in generic names of hybrid orchids involving three or more genera is minimal. The names, based upon a person's name, usually that of an orchid personality, plus the suffix *-ara*, as recommended by the International Congress in Brussels, are published by the International Registrar in the lists of new registrations that appear in each issue of the *Orchid Review*. Concerns stem from two sources: firstly, the lack of information in most hybrid generic names and, secondly, the changing perception of orchid genera resulting from modern systematic research.

The generic delimitation of orchids is about to undergo a revolution as modern molecular methods and phylogenetic analysis are applied on a broad scale to the orchids. Many genera that are popular in cultivation, particularly in subtribes like the *Oncidiinae* Benth. and *Laeliinae* Benth., will undoubtedly be redefined or even sunk as the results appear. New genera are also likely to be established taking well-known parents with them. Although horticulturalists will resist the demise of their favourite genera, we suspect that the new classification, because it reflects relationships better than present classifications, will prove to be a useful tool to hybridists, thereby accelerating its popularity. The current basis of hybrid nomenclature will rapidly be undermined and it, therefore, seems appropriate to ask if now is the time to rationalise and simplify hybrid generic names.

Of course, we can continue to use the horticultural equivalent concept, ending up with a greatly inflated list of names that must be used in registration but are not accepted scientifically. We, personally, think that this cannot continue indefinitely; as confusion will be the end result. The new generic concepts that are more firmly based in fact than the present ones will show intergeneric relationships more clearly, and will certainly be useful to growers for their breeding programmes.

What is the solution? We appreciate that change is undesirable, particularly as scientists will continue to propose new genera as the suite of techniques, particularly molecular, at their disposal increases. We would like here to submit a different approach from that currently used.

All orchid growers recognise a small number of discrete breeding groups in the *Orchidaceae* Juss. e.g. *Odontoglossum* Kunth and its allies, *Cattleya* Lindl. and its allies,

Cymbidium Sw., *Paphiopedilum* Pfitzer, *Phalaenopsis* Blume, *Vanda* Jones ex R.Br. and allies. Would it not be sensible to consider now a system of generic group nomenclature that recognises the realities of orchid hybridisation and at the same time simplifies the nomenclature and is informative to growers and orchid hobbyists?

We would like to suggest that a name is given to each breeding group involving two or more genera that includes the horticulturally most important parent genus together with an ending that signifies it is a hybrid genus. Thus all intergeneric hybrids involving *Odontoglossum* alliance intergenerics are *Odontoglossum* Group hybrids; all involving the *Cattleya* alliance are *Cattleya* Group hybrids; and so on. Of course, it should be possible to replace the 'Group' term by a suffix attached to the generic name: '-ara' is already used for orchids, might it not be used for the Groups as well? If not, then the Greek language will surely have other suffixes available that might prove suitable.

There would be problems with such a system. For example, the same grex epithets have been used in closely related genera, and new epithets would be required for later homonyms. However, we do not see this as an insurmountable problem, especially as the Orchid Hybrid Register is now computerised.

The main aims of nomenclature are to produce a universal, simple and informative system. For man-made intergeneric hybrid orchids, we would suggest that simplification of generic names along the line suggested would be widely welcomed by growers who are by and large confused by *Holttumara, Vasquezara, Stewartara* and the like and who already abbreviate *Sophrolaeliocattleya* and *Brassolaeliocattleya* to *Slc* and *Blc* respectively. We are not suggesting in any way that the above refers to natural hybrids which do not involve more than two genera anyway. If acceptable to orchid growers such a system would also need approval from the International Commission for the Nomenclature of Cultivated Plants who are responsible for the *International Code of Nomenclature for Cultivated Plants*. We are sure that other solutions are possible and would be pleased to hear from anyone who might contribute to this debate.

Miller, D.M. & Grayer, S.R. (1999). Standard portfolios in the herbarium of the Royal Horticultural Society. In: S. Andrews, A.C. Leslie and C. Alexander (eds). Taxonomy of Cultivated Plants: Third International Symposium, pp. 397–399. Royal Botanic Gardens, Kew.

STANDARD PORTFOLIOS IN THE HERBARIUM OF THE ROYAL HORTICULTURAL SOCIETY

DIANA M. MILLER AND SUSAN R. GRAYER

The Herbarium of the Royal Horticultural Society's Garden, Wisley, Surrey GU23 6QB, UK

What is a Standard?

The Standard for a cultivar is the designated herbarium specimen, or illustration if more appropriate, showing the characters which distinguish that plant from others in cultivation. This is in accordance with Principle 3 of the *International Code of Nomenclature for Cultivated Plants 1995*. (Trehane *et al.* 1995) which states:

> "The selection, preservation and publication of designations of Standards is important in stabilizing the application of cultivar epithets. Particular cultivar epithets are attached to Standards to make clear the precise application of the name and to help avoid duplication of such epithets. Although not a requirement for the establishment of a cultivar epithet, the designation of such Standards is strongly to be encouraged."

Characters not readily discernible from herbarium specimens such as colour, habit, texture and scent are usually necessary for the identification of cultivars and so detailed descriptions are essential. The value of a Standard is increased further if it is augmented with information such as origin or parentage, references to original publications and any other material which confirms the identity of that plant. It is also useful if the whereabouts of the living plant from which the Standard was prepared, is noted together with dates. Reports of further detailed research and even molecular records, seeds or pollen samples could all be added at a later date.

All this information constitutes the Standard portfolio for the plant. It is not practical for all these items to be stored in herbarium folders but with an efficient database, there is no problem in recording the data so the whereabouts of these items may be recorded and traced with ease.

Cultivar Identification

The identification of cultivated plants is notoriously difficult. Nursery catalogues contain thousands of names, often duplicated in successive years for obviously different plants. A plant name may be changed to a more commercially viable alternative and unfortunately this is not only a problem with the very old cultivars. The majority of catalogue descriptions are so brief that it is usually impossible to be certain of the plants' identity.

New species have a well-established type system to anchor their names but cultivated plants do not have this luxury. To name a new cultivar, the catalogue description may be as simple as "flowers red" or "plant taller than 'Patty's Pride'", which is not very helpful. There are thousands of shades of red and perhaps 'Patty's Pride' no longer exists.

397

These problems emphasise the importance of the routine preparation of Standard specimens and the publication of clear descriptions of new cultivated plants. Even more important is the accessibility of the Standards and the sharing of this knowledge. A survey of British nurseries and National Collection holders has indicated good support for the routine deposition of specimens of new cultivars in the Society's herbarium.

Records of Cultivars in RHS Archives

With its collection of herbarium specimens, paintings, photographs and transparencies of cultivated plants, the Society's herbarium is in a unique position to identify cultivars.

The records of the Royal Horticultural Society for nearly 200 years, have incorporated a great deal of important information on the characteristics of cultivated plants. Over the years, numerous new plants have been displayed or submitted for award at Shows held by the Society.

In the early part of this century, awarded plants were painted and the herbarium houses about 4,000 of these paintings, many of which may be regarded as the Standard for the cultivar. All orchids receiving awards are still painted and at the Society's London headquarters in Vincent Square there are over 6,600 paintings of orchids, the majority of which can also be treated as Standards. The herbarium photographic collection provides more recent illustrations of plants with awards and of course there are the pressed specimens in the herbarium itself.

From the early days of the Society, descriptions have been written of awarded plants and published in the *RHS Journal* or *Proceedings*. The information to be gleaned from discussions recorded in committee minutes and correspondence is a source of further useful snippets of historical importance, although none of these sources of knowledge have previously been systematically searched and recorded.

Herbarium Standards Project

An assessment of all the specimen, photographic and painting collections in the RHS herbarium is underway. Where applicable, these are being designated as Standards. In 1999 the first list of Standards was published (Miller & Grayer 1999).

Initially decisions had to be made about how Standards would be recognised retrospectively. Each specimen and illustration is being examined. The information in the *RHS Proceedings*, in entry forms of plants exhibited before the RHS Floral Committees, international registers and monographs are being studied and committee secretaries and registrars are being consulted. In cases of doubt, when it is possible, the exhibitors or raisers are being contacted for further information. Any specimens sent to RHS international registrars and added to the herbarium are automatically included.

Once it is determined that the specimen can validly be designated a Standard, the specimen is filed in a labelled green-edged folder, to distinguish it from a botanical type specimen. If an illustration constitutes the Standard, it is marked appropriately.

Copies of all relevant information and articles are being made to file with the specimens as well as references to illustrations or even living plants.

All this material and information will constitute the Standard portfolio of a cultivar which may be augmented if more information becomes available.

Records are being made on the RHS Horticulture Database making all these details readily accessible.

Disseminating the Information

It is planned that a printed list will be available of all designated Standards maintained in the RHS herbarium. Additionally, this list will be accessible through the RHS website, eventually including thumb nail illustrations, either of a living representative of the cultivar or of the herbarium specimen so that anyone in the world will be able to discover what Standards are housed at Wisley and any associated material and records.

Other herbaria around the world will also certainly contain a number of potential Standards. If no-one knows about them other than the staff in the herbaria concerned, and the information is inaccessible to outsiders who may need it, the time taken to curate the material is wasted. The records need to be made available to all.

In the future, it is hoped that all information on Standards can be centralised. The database utilised by RHS (using *BG-BASE*™) is able to perform this function and international co-operation could achieve this long term aim.

It is proposed therefore to create a single point of access through the Society's web pages for a distributed database of Standards throughout the world.

References

Miller, D.M. & Grayer, S.R. (1999). Designated Standards held in the Royal Horticultural Society's Herbarium, Wisley, 24 pp. RHS Wisley, Woking.

Trehane, R.P. *et al.* (eds) (1995). International Code of Nomenclature for Cultivated Plants — 1995. 175 pp. Quarterjack Publishing, Wimborne.

Tebbitt, M. (1999). A revised classification of selected Asian begonias based on evidence from morphology and molecules. In: S. Andrews, A.C. Leslie and C. Alexander (Editors). Taxonomy of Cultivated Plants: Third International Symposium, pp. 401–402. Royal Botanic Gardens, Kew.

A REVISED CLASSIFICATION OF SELECTED ASIAN BEGONIAS BASED ON EVIDENCE FROM MORPHOLOGY AND MOLECULES

MARK TEBBITT

Brooklyn Botanic Garden, 1000 Washington Avenue, Brooklyn, New York 11225-1099, USA

Introduction

A large number of *Begonia* L. species, although described in sufficient detail, cannot be satisfactorily assigned to any of the existing sections, because they either exhibit few of the morphological characters traditionally utilized in *Begonia* classification or they possess unusual combinations of characters. The lack of a sound taxonomic framework for this large, pantropical genus, means that it is often difficult to identify newly cultivated species. Molecular data from the *trn*C-*trn*D intergenic spacer region, along with new anatomical data, are shown to be ideally suited to determining the phylogenetic affinities, and hence correct taxonomic placement of two problematic species, *Begonia amphioxus* Sands and *B. balansana* Gagnep. A revised sectional classification of these two species is discussed.

Classification

Begonia amphioxus was originally classified in section *Platycentrum* (Klotzsch) A.DC. (Sands 1990) on the basis of its four male tepals, columnar torus, bilocular ovaries with two styles and variable numbers of female tepals ranging from three to five. Sands, however, acknowledges that the species also possesses a number of morphological features that are not characteristic of this section. These features include an erect, rather than creeping, rhizomatous habit and fruits that occasionally possess three rather than two wings. In the original description of *B. balansana*, Gagnepain (1919) suggests that this species may be closely affiliated to *B. silletensis* (A.DC.) C.B. Clarke and *B. tessaricarpa* C.B. Clarke based on shared similarities in ovary anatomy and anther and style morphology. The latter two taxa are currently placed in section *Sphenanthera* (Hassk.) Benth. & Hook. f. The only other reference to the sectional placement of *B. balansana* appears to be Barkley and Golding's (1974) compendium of published names in *Begonia*. These authors list the taxon as being unassigned to a section. In an attempt to determine the correct sectional placement of *B. amphioxus* and *B. balansana*, an in-depth analysis of living and herbarium material of these taxa and other representatives of the genus was carried out and the resultant data was analysed within a cladistic framework.

Morphological, anatomical and molecular data were obtained for the cladistic analysis from 27 species of *Begonia* representing 17 sections and the two problematic species. Cladograms were rooted with *Datisca cannibina* L. (*Datiscaceae* R.Br.), a species closely related to the *Begoniaceae* C. Agard (Swensen 1996). Taxa were chosen to represent a wide range of the geographical and morphological variation found within the genus, as well as taxa thought to be closely related to the two problematic species. The study incorporated 30 multi-state morphological/anatomical characters and 23 restriction site characters from the *trnC-trnD* intergenic spacer region. Heuristic parsimony analyses were performed in PAUP (version 3.1.1.; Swofford 1993) set for TBR branch swapping. All characters were unordered and un-weighted.

Analysis of the data set found 597 equally parsimonious trees with a length of 215. These had a CI of 0.409 and a RI of 0.569. In the strict consensus tree *Begonia amphioxus* appears in a clade otherwise composed of five members of section *Petermannia* (Klotzsch) A.DC. This clade is only distantly related to members of section *Platycentrum*, with which *B. amphioxus* was previously classified. The clade is well supported by both restriction site characters and more traditional characters of the fruit and anthers. The members of this clade are characterized also by the possession of an unusual type of anther anatomy not previously reported from the *Begoniaceae* C. Agardh. The habit and ovary wing characters that Sands (1990) states are uncharacteristic of the section *Platycentrum*, in which he places *B. amphioxus*, are characteristic of the members of section *Petermannia* with which this species is found to be affiliated here. It is therefore recommended that *B. amphioxus* be moved to the section *Petermannia*.

In the cladistic analysis *Begonia balansana* is affiliated with species either of section *Sphenanthera* or the closely related section *Platycentrum*, but cannot be reliably placed within either section. This is partly a result of *B. balansana* sharing several morphological and molecular characters in common with both sections and partly due to the relatively low number of characters used in the analysis. Before *B. balansana* is formally placed within a section it will be imperative to resolve its position relative to these sections. Sequence data from the nuclear ribosomal internal transcribed spacer (ITS) regions are currently being generated for this purpose.

Conclusions

The existing classification of the horticulturally important genus *Begonia* can clearly benefit from rigorous analyses of both traditional and molecular data. It appears that molecular data in particular will prove a promising source of information, especially in the investigation of those species that have undergone rapid morphological evolution and hence exhibit few characters in common with other *Begonia* taxa.

References

Barkley, F.A. & Golding, J. (1974). The species of the *Begoniaceae*. Ed. 2. 142 pp. Northeastern University, Boston.

Gagnepain, M.F. (1919). Nouveaux *Begonia* d'Asie; quelques synonymes. *Bull. Mus. Natl. Hist. Nat.* 25(3): 194–201.

Sands, M.J.S. (1990). Six new begonias from Sabah. *Kew Mag.* 7(2): 57–81.

Swensen, S.M. (1996). The evolution of actinorhizal symbioses: evidence for multiple origins of the symbiotic association. *Amer. J. Bot.* 83(11): 1503–1512.

Swofford, D.L. (1993). PAUP Phylogenetic Analysis Using Parsimony, Version 3.1.1. Computer program distributed by the Illinois Natural History Survey, Champaign, Illinois.

Upson, T.M. (1999). The horticultural taxonomy of the genus *Rosmarinus* (*Lamiaceae*). In: S. Andrews, A.C. Leslie and C. Alexander (Editors). Taxonomy of Cultivated Plants: Third International Symposium, p. 403. Royal Botanic Gardens, Kew.

THE HORTICULTURAL TAXONOMY OF THE GENUS *ROSMARINUS* (*LAMIACEAE*)

T.M. UPSON

Cambridge University Botanic Garden, Cory Lodge, Bateman Street, Cambridge CB2 1JF, UK

Of the three species of *Rosmarinus* L., *R. officinalis* L., *R. eriocalix* Jordan & Fourr. and *R. tomentosus* Hub.-Mor. & Maire, the first has long been cultivated as a herb and many cultivars selected. These include flower colour variants, prostrate and upright variants and fast-growing selections used in the fresh herb trade. This species and its infraspecific taxa in cultivation exemplify some of the problems faced by the horticultural taxonomist. Examples include cultivar names which in practice are now applied to a number of different clones, particularly problematic when the original clone can no longer be identified, e.g. 'Miss Jessopp's Upright', a name applied to almost any rosemary with an upright habit although the original clone to date has not been traced. The misapplication of names to cultivated taxa, e.g. *R.* × *lavandulaceus* Noë ex Debeaux, a wild hybrid (*R. officinalis* × *R. eriocalix*) applied to prostrate variants in cultivation which are correctly *R. officinalis* var. *prostratus* Pasquale. The problem of correctly assigning cultivars to infraspecific taxa is highlighted, e.g. the low-growing cultivars with dome-shaped habits, many of which are questionably and unsatisfactorily assigned to var. *prostratus*. They probably represent low-growing variants of *R. officinalis* var. *officinalis* and not the true var. *prostratus*.

Shaw, J.M.H. (1999). Variation in *Podophyllum hexandrum* and its nomenclatural consequences. In: S. Andrews, A.C. Leslie and C. Alexander (Editors). Taxonomy of Cultivated Plants: Third International Symposium, pp. 405–408. Royal Botanic Gardens, Kew.

VARIATION IN *PODOPHYLLUM HEXANDRUM* AND ITS NOMENCLATURAL CONSEQUENCES.

JULIAN M.H. SHAW

4 Albert Street, Stapleford, Nottingham NG9 8DB, UK

Introduction

Podophyllum hexandrum Royle (*Berberidaceae* Juss.), an important source of anticancer therapy, is a perennial herb distributed along the Himalayas from E. Afghanistan to NW. China. Many collections have been made over the past century and several clones and inbred lines are cultivated as ornamental perennials. It is polymorphic with considerable variation of leaf shape and dissection, pigmentation and flower colour, and consequently, many infraspecific taxa have been described. It has also been confused with *P. aurantiocaule* Hand.-Mazz. — apparently its nearest relative.

Discussion

Foliar dissection and to a lesser degree the position of the flower have been used to attempt infraspecific division, however, these characters show no clear geographical pattern as has been suggested (Chatterjee 1953, Selivanova-Gorodkova 1969). The use of foliar dissection tends to be subjective as it is very difficult to quantify. Therefore, anther length was analysed geographically since it is easy to measure and shows pronounced clinal variation in the related *P. aurantiocaule* — subsp. *furfuraceum* (S.Y. Bao) J.M.H. Shaw *comb. nov.* alliance comprising section *Paradysosma* J.M.H. Shaw *sect. nov.* (Shaw 1999). Also crosses made between different inbred lines of *P. hexandrum* have demonstrated that anther length is genetically controlled. For example, the cross between *Kingdon-Ward* 20 and the inbred line distributed as 'Major' produces progeny with very short anthers, similar to those found on some wild collections.

Anthers from all suitable herbarium sheets at Edinburgh, Kew and The Natural History Museum were measured and the data obtained were used to produce a three dimensional graph, in which anther length (vertical axis) was displayed over the geographical range of the species. From this, it was immediately apparent that the geographical distribution of anther length does not follow any obvious pattern. There are some areas of extreme variation marked by the close proximity of peaks and troughs on the graph. These occur from Himachal Pradesh to western Nepal and from Sikkim to western Bhutan, and may indicate interaction between dissimilar lines of *P. hexandrum* caused by the presence of specific pollinators or possibly segregation within *P. hexandrum*, perhaps indicating a hybrid origin or at least introgression.

Anther length ranges from 2.5–9.5 mm, with most anthers falling between 4–7 mm. There may be a bimodal distribution of anther length since 44% fall within the range 4–5 mm and 34% within 5.8–7 mm. Only 2% fall between these two ranges. In many features individuals of *P. hexandrum* seem to fluctuate between two extremes; short and long anthers, white to red flowers, and leaves with three entire lobes to highly dissected leaves provide three obvious examples. This may indicate that the present *P. hexandrum* has arisen from gene exchange between two other species, one possessing entire-lobed leaves, white flowers and short anthers and another with red flowers, dissected foliage and long anthers. It is tempting to speculate what these may be — perhaps *P. aurantiocaule* and *P. delavayi* Franch.?

Pollination and breeding system provide another group of characters which provide important insights to understand the variation within the genus *Podophyllum* L. Apparently, *P. hexandrum* is unique in the genus with the extent of its self-compatibility and self-pollination. A remarkable mechanism for ensuring self-pollination has recently been described by Xu *et al.* (1997). As the flower opens the gynoecium is upright and surrounded by a ring of six regularly spaced stamens. When the flower is fully open the gynoecium tilts from the base until the stigma makes contact with an adjacent anther, after which the gynoecium returns to a normal upright position in the centre of the flower and immediately enlarges. This tilting process is complete within 4 to 6 hours and results in almost 100% fruit set. Consequently, out-crossing appears to be rare in this species; when attempted it produces mixed results. Crosses between very different lines can result in no seed production or plants with aborted anthers. This behaviour favors the formation within the species of a large number of inbreeding lines which appear different from one another, while the individual plants within each line are fairly uniform. Some cultivar names in use apply to such lines, examples include 'Major' and 'Chinese'. This also helps to explain the maintainence of many different morphs in the wild. Most of the genus consists of self-incompatible, out-breeding species, and in this circumstance, self-compatiblity appears to be a derived feature, which again may suggest a hybrid origin for *P. hexandrum*. The only other species known to have self-compatiblity is *P. peltatum* L., which is variable in this regard. Individual populations, which are often clones, vary from self-incompatible to partly self-compatible (Whistler & Snow 1992). It has been suggested that *P. peltatum* may have been introgressed in the past (Martin 1958), which may have some bearing on this. Unfortunately the breeding behaviour of *P. aurantiocaule* remains unknown, but the flower exhibits some unusual features, such as the petals becoming reflexed with age, like a *Cyclamen* L. This suggests out-breeding behaviour. Another surprise in the genus has been the discovery of sapromyophily amongst the dark red flowered self-incompatible species of *Podophyllum* section *Dysosma* (Woodson) J.M.H. Shaw *stat. nov.* (Shaw 1999), in which the flower emits an odour of decaying proteins to attract carrion flies. This use of indiscriminate pollinators like flies, does much to account for the widespread occurrence of hybrids within the genus. Flies have been observed visiting the flowers of *P. hexandrum* and may therefore be the vector for the introgression by section *Dysosma*.

Synonymy of *Podophyllum hexandrum* Royle

Podophyllum emodi Wall. *nom. nud.*(1829)
Podophyllum emodi Wall. var. *royleana* Wall. *nom. nud.*(1829)
Podophyllum acutifolium Royle *in sched.* (1834)
Podophyllum acutum Royle *in sched.* (1834)
Podophyllum emodi Wall. var. *chinense* Sprague (1920)
Podophyllum sinense Hort. *nom. nud., in sched.* (1921)
Podophyllum leichtlinii Langlet *nom. nud.* (1928)
Podophyllum hexandrum Royle var. *chinense* (Sprague) Stearn *in ms.* (1933)
Podophyllum emodi Wall. var. *hexandrum* (Royle) Chatterjee & Mukerjee *nom. illegit.* (1953)

Podophyllum emodi Wall. var. *axillaris* Chatterjee & Mukerjee (1953)
Podophyllum emodi Wall. var. *bhootanensis* Chatterjee & Mukerjee (1953)
Podophyllum emodi Wall. var. *jaeschkei* Chatterjee & Mukerjee (1953)
Podophyllum indica Chopra *nom. nud.* (1958)
Podophyllum hexandrum Royle var. *emodi* (Wall. ex Honigsb.) Seliv.- Gor. (1969)
Dysosma emodi (Wall. ex Honigsb.) Hiroë *nom. illegit.* (1973)
Podophyllum hexandrum Royle var. *axillare* (Chatterjee & Mukerjee) Browicz (1973)
Podophyllum hexandrum Royle var. *bhootanense* (Chatterjee & Mukerjee) Browicz (1973)
Podophyllum hexandrum Royle var. *jaeschkei* (Chatterjee & Mukerjee) Browicz (1973)
Sinopodophyllum emodi (Wall. ex Honigsb.) Ying *nom. illegit., pro parte excl. P. sikkimensis*
    Chatterjee & Mukerjee, *P. delavayi* Franch. (1979).
Podophyllum hexandrum Royle var. *majus* Hort. *nom. nud.* (1982)
Podophyllum versipelle sensu auctt. non Hance (1982)
Sinopodophyllum hexandrum (Royle) Ying (1985)
Podophyllum emodi 'Major' (1992)
Podophyllum pentaphyllum Deno *nom. nud.* (1993)
Podophyllum heterophyllum sensu Gehenio *non* Raf. (1996)
Podophyllum hexandrum Royle subsp. *substerilis* Hort. *nom. nud.* (1996)
Podophyllum hendersonii Hort. *nom. nud.* (1997)

Conclusion

As a result of these studies it was concluded that no infraspecific taxa could be distinguished within *P. hexandrum*, which appears to be a relictual hybrid swarm, whereas anther length provided a useful character to separate out the two subspecies of *P. aurantiocaule*. It is recommended that cultivar names be used for outstanding clones or inbred lines of *P. hexandrum* in cultivation. A taxonomic revision of the genus *Podophyllum* has been accepted for publication in *The New Plantsman* (Shaw in press).

Differences between *P. hexandrum* and *P. aurantiocaule*

Hara (1971) and Ying (1979) both treat *P. sikkimensis* Chatterjee & Mukerjee (a synonym of *P. aurantiocaule*) as synonyms of *P. hexandrum*. Ying even described a new species, *Dysosma tsayuensis* Ying which is another synonym of *P. aurantiocaule*. While some populations of *P. aurantiocaule* produce solitary flowers with floral presentation similar to *P. hexandrum*, there are specific characters that provide a ready means of distinguishing the two species.

P. hexandrum	*P. aurantiocaule*
Leaves palmate, 3–5 lobed, with one sinus almost penetrating to the petiole apex	Leaves peltate with 5–7 lobes, sinuses equal, penetrating $\frac{1}{2}$ to $\frac{2}{3}$ of radius
Leaf lobes deeply subdivided	Leaf lobes not divided
Leaf upper surface glabrous	Hairs present on leaf upper surface
Petals spreading when fully open	Petals reflexed with age
Pistil green	Pistil bright pink
Anthers 2.5–9.5 mm, away from style	Anthers 5–9 mm, pressed to style
Pollen in tetrads	Pollen in monads

References

Chatterjee, R. (1953). Studies in Indian *Berberidaceae* from botanical, chemical and pharmaceutical aspects. *Rec. Bot. Surv. India* 16(2): i–iv, 1–86.

Hara, H. (1971). Flora of the Eastern Himalaya, second report. *Univ. Mus. Univ. Tokyo Bull.* 2: 1–393.

Martin, F.W. (1958). Variation and morphology of *Podophyllum peltatum.* Ph.D thesis, Washington University. *Dissertation Abstracts* 19: 424–425.

Selivanova-Gorodkova, E.A. (1969). On two Himalayan species of *Podophyllum* L. *Bot Zhurn. (Moscow & Leningrad)* 54(10): 1604–1605.

Shaw, J.M.H. (1999). New taxa, combinations and taxonomic notes on *Podophyllum* L. *New Plantsman* 6(3): 158–165.

Whistler, S.L. & Snow, A.A. (1992). Potential for the loss of self-incompatibility in pollen-limited populations of Mayapple (*Podophyllum peltatum*). *Amer. J. Bot.* 79: 1273–1278.

Xu, Z., Ma, S., Hu, C., Yang, C. & Hu, Z. (1997). The floral biology and its evolutionary significance of *Sinopodophyllum hexandrum* (Royle) Ying (*Berberidaceae*). *J. Wuhan Bot. Res.* 15(3): 223–227.

Ying, T.S. (1979). On *Dysosma* Woodson and *Sinopodophyllum* Ying, *gen. nov.* of the *Berberidaceae. Acta Phytotax. Sin.* 17(1): 15–23.

Shaw, J.M.H. (1999). Specific identities and relationships in the *Iochroma gesnerioides/fuchsioides* complex. In: S. Andrews, A.C. Leslie and C. Alexander (Editors). Taxonomy of Cultivated Plants: Third International Symposium, pp. 409–412. Royal Botanic Gardens, Kew.

SPECIFIC IDENTITIES AND RELATIONSHIPS IN THE *IOCHROMA GESNERIOIDES/FUCHSIOIDES* COMPLEX

JULIAN M.H. SHAW

4 Albert Street, Stapleford, Nottingham NG9 8DB, UK

Introduction

Iochroma Benth. (*Solanaceae* Juss.), a genus of 15–20 species from the Andes of north-western South America, consists of shrubs or small trees, mostly from cloud forest zones with tubular flowers in shades of red, orange, yellow, blue to violet or white. They have been in cultivation since 1844, when John Lindley described *I. cyaneum* (Lindl.) M.L. Green from cultivated plants. Today there are many cultivated forms available and their naming is often confused. In preparation for the European Garden Flora account, the identity of the red flowered *Iochroma* species and their hybrids has been investigated (Shaw 1998b).

History

Humboldt and Bonpland described a red flowered *Iochroma*, as *Lycium fuchsioides* Kunth in their *Plantae equinoctiales* 1: 147 t. 42 (1807). It was a small shrub with glabrous, obovate to bluntly ovate leaves, a calyx 9–15 mm long, and a few flowers. Similar plants were cultivated at Kew and illustrated in the *Bot. Mag.* t. 4149 (1849). Since then the name *I. fuchsioides* (Kunth) Miers has been applied to many wild and cultivated plants. Later in the *Nova Genera et Species Plantarum* 3(1): 52–55 (1818), a further five *Iochroma* were described as *Lycium* L. species. Of these, two had red flowers — *I. gesnerioides* (Kunth) Miers and *I. umbrosa* (Kunth) Miers. Both have been confused with *I. fuchsioides* although they have larger lanceolate, acute to acuminate leaves with a dense felt of branched hairs, a small calyx 3–5 mm long, and many flowered inflorescences.

In 1845 Miers described another collection from Colombia as *Chaenesthes lanceolata* Miers, which later became *Iochroma lanceolatum* (Miers) Miers, stating that the "corolla seems of a crimson colour". Unfortunately, this plant has been regarded a synonym of *I. cyaneum*. The confusion arose when Hooker misapplied the name *C. lanceolata* to a plant of *I. cyaneum* with blue-purple flowers, that had been grown from seed collected in Colombia at the type locality of *C. lanceolata* in the *Bot. Mag.* t. 4338, (1847). Since the foliar characters and corolla shape were very similar to *C. lanceolata*, Hooker assumed that Miers had misinterpreted the corolla colour from his dried herbarium specimen. This prompted Lawrence & Tucker (1955) to reduce *I. lanceolatum* to synonymy under *I. cyaneum*, whereas the type specimen at Kew, which they had not seen, clearly represents *I. gesnerioides*. Miers also described *I. longipes* Miers with unusually long pedicels to 8 cm. It appears to represent a shade form of *I. fuchsioides*.

A glabrescent plant of uncertain origin with red flowers was described from cultivation in 1857 as *I. coccineum* Scheidw. It is still widely available today and similar plants are known from the wild in Colombia. It seems best treated as a cultivar of *I. gesnerioides*, on account of its small calyx and large lanceolate leaves. Experience with cultivated plants has revealed that the colour of flowers in *I. gesnerioides* depends on environmental factors such as temperature and light intensity, which is also known to occur in the sympatric *Brugmansia sanguinea* Ruíz & Pav. (Shaw 1998a).

Three more names were coined for red iochromas during the early 20th century, but they have been ignored by later workers. They are *I. sodiroi* Dammer (1905), *I. solanifolia* Dammer (1905) and *I. puniceum* Werderm. (1935), all based on collections from Ecuador.

Ethnobotanical work and much collecting since the 1950's in Colombia and Ecuador have provided many collections of red flowered iochromas. These have been mostly determined as *I. fuchsioides*. Their variability prompted Schultes (1977) to reduce *I. umbrosa* to synonymy under *I. fuchsioides*. However, his collections are referable to *I. gesnerioides* under which species *I. umbrosa* is now treated as a synonym.

Finally, following exploration in the Andes of northern Peru, Gonzales Leiva (1995) has recently described a small shrub with glabrous leaves and few flowers as *I. edule* G. Leiva, since the berries are collected and eaten by the local peasants. E.K. Balls collected similar plants in which the fruit turn yellow when ripe and are eaten by peasants who call them 'Pepinillo'. This collection is a good match with the type of *I. fuchsioides*.

While Leiva compares *I. edule* with *I. parvifolium* (Roem. & Schult.) D'Arcy, it has much more in common with *I. fuchsioides*, which was not known from the wild in Peru at the time when the catalogue of the Peruvian flora was prepared (Bracho & Zarucchi 1993). It is possible that references to *I. fuchsioides* from Peru are based on plants since described as *I. edule*, rather than on supposedly cultivated plants introduced from Ecuador. It seems doubtful that the two can be maintained at specific level.

Key to Cultivated *Iochroma*

1a. Calyx strongly inflated at flowering, ellipsoidal, longer than 2 cm · · **I. calycinum**
 b. Calyx cup-shaped, tubular or urceolate, 3–15 mm long · · · · · · · · · · · · · · · · 2
2a. Leaves densely glandular hairy, sticky; corolla 4.5–7 cm with lobes · · **I. grandiflorum**
 b. Leaves hairy or hairless, not sticky; corolla 2–4.5 cm, with small teeth · · · · · · · 3
3a. Corolla with 10 prominent, narrowly triangular teeth of almost equal
 size · **I. cyaneum** × **I. australe**
 b. Corolla with 5 broadly triangular teeth, sometimes alternating with 5
 minute teeth · 4
4a. Corolla funnel-to bell-shaped; leaves narrowly lanceolate, hairless · · · **I. australe**
 b. Corolla tubular; leaves ovate, elliptic or lanceolate, acute to acuminate,
 felted beneath · 5
5a. Corolla yellow · **I. gesnerioides** var. **flavum**
 b. Corolla red to orange, white to pink, blue- purple to violet · · · · · · · · · · · · 6
6a. Corolla white to pink, or blue- purple to violet · · · · · · · · · **I. cyaneum** cultivars
 b. Corolla red to orange · 7
7a. Calyx 9–15 mm; flowers 1–12; leaves obovate to elliptic, hairless · · **I. fuchsioides**
 b. Calyx 3–5 mm; flowers *c.* 30–120; leaves lanceolate, felted or sparsely
 hairy beneath · 8
8a. Leaves densely felted on underside · · · · · · · · · · **I. gesnerioides** (typical forms)
 b. Leaves almost hairless · · · · · · · · · · · · · · · · · · · **I. gesnerioides** 'Coccineum'

Iochroma cyaneum and its Relationship to *I. gesnerioides*

The episode related above in which Hooker mistakenly applied the name *I. lanceolatum*, a synonym of *I. gesnerioides*, to a cultivated plant of *I. cyaneum* underscores the similarity between *I. cyaneum* and *I. gesnerioides*. Generally *I. gesnerioides* can be distinguished by more lanceolate leaves, and a smaller calyx, but some forms can only be separated by corolla colour.

Several observations suggest that *I. cyaneum* represents a group of hybrids to which *I. gesnerioides* is a likely contributor.

- It is well known that seed obtained by self-pollination of *I. cyaneum* plants gives rise to heterogenous seedlings with a wide range of flower colours including some with red hues (Vermeulen 1997). This is known to be the origin of several cultivars, including the pale pink 'Woodcote White'.
- Similarly, seedlings derived from the cross *I. cyaneum* × *I. australe* are also very variable, indicating heterogenicity of *I. cyaneum*, since seedlings from self-pollinated *I. australe* Griseb. are comparatively uniform.
- Cytological examination of *I. cyaneum* has revealed heteromorphic chromosome pairs, chromosomes without homologues, incomplete pairing, presence of univalents, and chromosome breakage, all indicative of a hybrid origin for *I. cyaneum* (Madhavadian 1967).
- Madhavadian (1967) also reported a high percentage of pollen sterility, another trait of hybrids.
- The known distribution of *I. cyaneum* in the wild appears to be very localised. Most records, even from South America, refer to cultivated plants.
- Other hybrids have been reported from the wild that involve *I. gesnerioides* as a putative parent. For example, the suspected hybrid with *I. cornifolium* (Kunth) Miers reported by Jorgensen and Ulloa (1994), from Ecuador as *I. fuchsioides sensu auctt.* × *I. cornifolium*. The hybrid *I. grandiflorum* Benth. × *I. gesnerioides* is now known from cultivation and is similar to some wild collections, such as *Plowman 1864*.

Differences between *I. fuchsioides* and *I. gesnerioides*

	I. fuchsioides	*I. gesnerioides*
Synonyms	*Lycium fuchsioides* Kunth (1807)	*Lycium gesnerioides* Kunth (1818)
	Chaenesthes fuchsioides (Kunth) Miers (1845)	*Lycium umbrosum* Kunth (1818)
	?*Iochroma longipes* Miers (1848)	*Valteta gesnerioides* (Kunth) Raf. (1838)
	?*Iochroma sodiroi* Dammer (1905)	*Chaenesthes gesnerioides* (Kunth) Miers (1845)
	Iochroma edule G. Leiva (1995)	*Chaenesthes lanceolata* Miers (1845)
		Chaenesthes umbrosa (Kunth) Miers (1845)
		Iochroma umbrosa (Kunth) Miers (1848)
		Iochroma lanceolatum (Miers) Miers (1848)
		Dunalia gesnerioides (Kunth) Dunal (1852)
		Dunalia umbrosa (Kunth) Dunal (1852)
		Iochroma coccineum Scheidw. (1857)
		Iochroma solanifolia Dammer (1905)
		Acnistus gesnerioides (Kunth) Hunz. (1960)

cont.

	I. fuchsioides	*I. gesnerioides*
Distribution	Colombia – N. Peru	Colombia – Ecuador
Height	0.7–2.5 m	1.5–8 m
Leaves	3.5–7 × 1.5–3 cm	12–18 × 5–7.5 cm
apex	obtuse to acute	acute to acuminate
surface	glabrous	underside tomentose
		upperside glabrescent
Indumentum	simple hairs	forked hairs
Number of flowers	1–12	up to *c.* 120
Calyx	7–15 mm	3–5 mm
Corolla	2–3 cm, glabrous	2.5–4 cm, pubescent exterior

Conclusion

There are two main taxa in the *I. fuchsioides/I. gesnerioides* complex, and also some intermediate specimens which may represent hybrids. Hybridisation appears to be common in the genus suggesting that hummingbirds are indescriminate pollinators. Many of the usually numerous undetermined specimens in larger herbaria, probably represent such hybrids. *Iochroma cyaneum* is probably of hybrid origin with *I. gesnerioides* a likely parent. Further field studies and biosystematic work are needed to elucidate the problem.

References

Bracho, L. & Zarucchi, J.L. (1993). Catalogue of the flowering plants and gymnosperms of Peru. *Monogr. Syst. Bot. Missouri Bot. Gard.* 45, 1286 pp. Missouri Botanical Garden, Missouri.

Jorgensen, P.M. & Ulloa, C.U. (1994). Seed plants of the High Andes of Ecuador – a checklist. *AAU Reports 34.* 443 pp. Aarhus University Press, Aarhus.

Lawrence, G.H.M. & Tucker, J.M. (1955). *Iochroma tubulosum* and *I. lanceolatum* are *I. cyaneum. Baileya* 3(2): 65–67.

Leiva, G. (1995). Una neuva espicies de *Iochroma* del norte Peru. *Arnaldoa* 3(1): 41–44.

Madhavadian, P. (1967). The cytology of *Iochroma tubulosa* Benth. *Caryologia* 20: 309–315.

Schultes, R.E. (1977). A new hallucinogen from Andean Colombia: *Iochroma fuchsioides. J. Psychedelic Drugs* 9(1): 45–49.

Shaw, J.M.H. (1998a). Variation in *Brugmansia sanguinea. New Plantsman* 5(1): 48–60.

Shaw, J.M.H. (1998b). *Iochroma* – a review. *New Plantsman* 5(3): 154–192.

Vermeulen, N. (1997). Encyclopaedia of Container Plants. 280 pp. Rebo productions, Lisse.

Hoffman, M.H.A. (1999). Cultivar classification of *Philadelphus* L. In: S. Andrews, A.C. Leslie and C. Alexander (Editors). Taxonomy of Cultivated Plants: Third International Symposium, pp. 413–414. Royal Botanic Gardens, Kew.

CULTIVAR CLASSIFICATION OF *PHILADELPHUS* L.

M.H.A. HOFFMANN

Research Station for Nursery Stock, PO Box 118, 2770 AC Boskoop, The Netherlands

Introduction

The cultivars of the genus *Philadelphus* L. have previously been classified under five interspecific hybrids, as proposed by A. Rehder some seventy years ago (Rehder 1927). This system is no longer satisfactory and instead a more stable classification of four cultivar-groups is proposed here (Hoffman 1994, 1996). This recent classification still shows the influence of Rehder's system, but differs quite a lot in its principles. It is based on practical application (the culton concept), while the old classification is based on genetic relationships (the taxon concept). The definitions of the cultivar-groups are simple and clear; cultivars can be classified more easily and even non-specialists can understand the system.

Proposed Cultivar Groups

Purpureo-maculatus Group (basionym: *P. × purpureo-maculatus* (Lemoine) Rehder)
DESCRIPTION: Plants 0.3–2 m high. Leaves on the non-flowering shoots 2–8 cm long. Flowers single, with a purple-red spot at the base of each petal.
STANDARD CULTIVAR: *P.* 'Purpureo-maculatus'
CULTIVARS: 'Beauclerk', 'Belle Etoile', 'Bicolore', 'Burkwoodii', 'Galathée', 'Nuage Rose', 'Oeil de Pourpre', 'Purpureo-maculatus' and 'Sybille'.

Lemoinei Group (basionym: *P. × lemoinei* (Lemoine) Rehder)
DESCRIPTION: Plants 0.2–2 m high. Leaves on the flowering shoots always and on the non-flowering shoots usually less than 5 cm long. Flowers single or semi-double, completely white or cream.
STANDARD CULTIVAR: *P.* 'Lemoinei'
CULTIVARS: 'Avalanche', 'Dame Blanche', 'Erectus', 'Fimbriatus', 'Frosty Morn', 'Lemoinei', 'Manteau d'Hermine', 'Mont Blanc', 'Pavillon Blanc', 'Silberregen', 'Snowdwarf', 'Snowgoose' and 'Velléda'.

Virginalis Group (basionym: *P. × virginalis* Rehder)
DESCRIPTION: Plants 1–4 m high. Leaves on the non-flowering shoots predominantly longer than 5 cm. Flowers predominantly double or semi-double, completely white or cream.
STANDARD CULTIVAR: *P.* 'Virginal'
CULTIVARS: 'Albâtre', 'Arctica', 'Audrey', 'Bannière', 'Boule d'Argent', 'Bouquet Blanc', 'Buckley's Quill', 'Enchantement', 'Girandole', 'Glacier', 'Komsomoletz', 'Minnesota Snowflake', 'Pekphil', 'Pyramidal', 'Rusalka', 'Schneesturm', 'Snowbelle', 'Virginal', 'Yellow Hill' and 'Zhemczug'.

Burfordensis Group

DESCRIPTION: Plants 1–4 m high. Leaves on the non-flowering shoots predominantly longer than 5 cm. Flowers predominantly single, completely white or cream.

STANDARD CULTIVAR: *P.* 'Burfordensis'

CULTIVARS: 'Academic Komarov', 'Apollo', 'Atlas', 'Bialy Sopel', 'Burfordensis', 'Conquête', 'Falconeri', 'Favorite', 'Hidden Blush', 'Innocence', 'Kalina', 'Karolinka', 'Kasia', 'Limestone', 'Marjorie', 'Norma', 'Rosace', 'Slavinii', 'Switezianka' and 'Voie Lactée'.

References

Hoffman, M.H.A. (1996). Cultivar classification of *Philadelphus* L. (*Hydrangeaceae*). *Acta Bot. Neerl.* 45(2): 199–209.

Hoffmann, M.H.A. (1994). *Philadelphus*, sortiments en gebruikswaardeonderzoek. *Dendroflora* 31: 44–71. (Dutch article with English summary).

Rehder, A. (1927). *Philadelphus* L. In Manual of cultivated trees and shrubs. pp. 270–281. Macmillan Company, New York.

Snoeijer, W. & Verpoorte, R. (1999). Classification of *Catharanthus* cultivars. In: S. Andrews, A.C. Leslie and C. Alexander (Editors). Taxonomy of Cultivated Plants: Third International Symposium, p. 415. Royal Botanic Gardens, Kew.

CLASSIFICATION OF *CATHARANTHUS* CULTIVARS

W. SNOEIJER AND R. VERPOORTE

International Registration Authority for *Catharanthus*, Division of Pharmacognosy, Leiden/Amsterdam Center for Drug Research, PO Box 9502, 2300 RA Leiden, The Netherlands

The genus *Catharanthus* G.Don (*Apocynaceae* Juss.) comprises eight species and about forty cultivars. *Catharanthus roseus* (L.) G.Don, also known as *Vinca rosea* L. and Madagascar periwinkle, is widely cultivated in tropical and subtropical zones. In temperate zones, the cultivars are grown as pot plants, either indoors or as annuals outdoors. It is also an important medicinal plant.

Recently a classification of the cultivars was published by Snoeijer (1998). This classification is based on flower colour. For commercial purposes, seed companies classify their cultivars in series on the basis of growing conditions and habit. Different flower colours are usually offered in such series.

Reference

Snoeijer, W. (1998). International Register of *Catharanthus* cultivars 1998. 52 pp. Leiden University, Leiden.

Rutherford, A. (1999). Algerian ivy. In: S. Andrews, A.C. Leslie and C. Alexander (Editors). Taxonomy of Cultivated Plants: Third International Symposium, p. 417. Royal Botanic Gardens, Kew.

ALGERIAN IVY

A. RUTHERFORD

19 South King Street, Helensburgh G84 7DU, Scotland, UK

Hedera algeriensis Hibberd or Algerian ivy is frequently mis-named as *H. canariensis* Willd. Some recent authors compromise with calling it Canary or Algerian ivy, giving this taxon false Canary Islands and Algerian origins.

Over 20 years' research has produced some solutions. By growing ivies from as many countries as possible, reading floras, journals and garden works and making extensive searches, it became clear that several names were misapplied and authors had copied older unchecked information.

Hedera canariensis from the Canary Islands is endemic, well-described and very distinct. In North Africa there are two endemic species, *H. maroccana* McAllister from Morocco and *H. algeriensis* from Algeria, the latter being introduced to France in the 1840s.

Victor Rantonnet, an acclimatisation gardener on the Côte d'Azur, is credited with introducing this splendid plant into cultivation. From the mid-1860s however, the name *H. algeriensis* began to disappear. By 1912, F. Tobler, in his monograph did not believe it existed.

This species has a close relationship with Moroccan ivy, while true *H. canariensis* is nearest to *H. colchica* (K. Koch) K. Koch, a native to the Caucasus and Turkey. The mis-naming of *H. algeriensis* is due to copying muddled labels 130 years ago and by Tobler trying to unite the Canary Island and Moroccan ivies (Tobler 1912). He saw no living and no sterile branches of the latter (the part used for identification). The McAllister *Hedera* classification (Rutherford *et al.* 1993) makes much better botanical and geographical sense.

References

Rutherford, A., McAllister, H.A. & Mill, R.R. (1993). New ivies from the Mediterranean area and Macaronesia. *Plantsman* 15(2): 115–128.

Tobler, F. (1912). Die Gattung *Hedera*, Studien über Gestalt und Leben des Efeus, seine Arten und Geschichte. 151 pp. Gustav Fischer, Jena.

Louneva, N.N. (1999). Systematics of cultivated plants and problems of breeding. In: S. Andrews, A.C. Leslie and C. Alexander (Editors). Taxonomy of Cultivated Plants: Third International Symposium, pp. 419–421. Royal Botanic Gardens, Kew.

SYSTEMATICS OF CULTIVATED PLANTS AND PROBLEMS OF BREEDING

N.N. LOUNEVA

N.I. Vavilov Institute of Plant Industry, 42 Bolshaya Morskaya St., St. Petersburg 190000, Russia

Available classifications on cultivated plants show a merger of two approaches; a botanical one, based on essential systematic plant characters, and a practical one, based on the analysis of commercial properties.

For example, practical classification systems have been based on drought resistance (Krivchenko *et al.* 1987), productivity (Shashko 1967) and technological values (Kravtsova *et al.* 1974). There have been also attempts to subdivide the existing genetic diversity according to the levels of photosynthetic efficiency, that is by plant productivity (Krivchenko *et al.* 1987). The possibility of classifying plant genetic diversity on the basis of a systematised energy-oriented approach, i.e. proceeding from the content of energy in a unit of the commercially valuable part of the plant (Krivchenko *et al.* 1988). Such approaches to the classification of plants aim to ensure certain results in breeding and plant production practice. Despite their artificial nature, such classification concepts are quite convenient in application and are recognised by plant scientists who use these classifications as guidelines for their research (Omarov 1978).

At the same time, only critical revision of the specific and intraspecific composition of the species and a proper understanding of the process of form development and hybridisation can provide a clearer insight into the potential of any genus for its fullest use in breeding. Numerous collecting missions undertaken by N.I. Vavilov All-Russian Research Institute of Plant Industry (VIR) pursued the goal of studying diversity among wild relatives of cultivated plants and their polymorphism from the viewpoint of morphological characters, times of flowering and ripening, taste, etc. Analysis of these and similar characters in populations provided grounds for identification of plant forms recommended for utilisation in breeding.

> "Clear concept of original materials for breeding can be obtained only after the analysis of populations. It is with this method only that selection turns into a deliberate procedure and makes it possible to test a diverse variety of ways in individual and group selection."
>
> (E.N. Sinskaya 1958)

N.I. Vavilov regarded the Linnean species as a complex morpho-physiological system linked in its origins to particular environments and areas of distribution. E.N. Sinskaya's conclusion that "the species is never uniform within the scope of its distribution and represents by itself a system of ecotypes" (Sinskaya 1948), helped to develop a new appraisal of intraspecific variability and to reject the trend of subdividing species into smaller ones. The detailed study of species acquired greater importance from the viewpoint of searching for valuable materials to be used in breeding.

> "Utilisation of such representatives of varieties, subvarieties and forms which better correspond to breeding goals provides more favourable prerequisites for making hybrids with prescribed traits."
>
> (V.L. Vitkovsky *et al.* 1991)

Varietal diversity in crops were differentiated using ecological principles. It helped to identify groups of varieties which had been formed by natural means under the influence of climatic factors during the spread of a crop (Fursa & Stepanova 1991).

Plant breeders are interested from a practical point of view in intraspecific genetic variability of plants. This depends on numerous factors. The 'mechanismal classification' proposed by V.A. Dragavtsev is based on the description of multilevel regulation of intraspecific genetic variability, e.g. intracellular, organismal, organism-environmental, populational and community levels (Dragavtsev 1997). At the population level several systems have been reported which regulate intraspecific diversity. These are Darwinian selection, proembryonal selection, genetic drift, gene migration (flows), inbreeding and stabilisation of gene frequencies. It is difficult to identify 'the main level' of regulation which would be 'responsible for the given diversity, and it is also difficult to organise on this basis an efficient technology of a certain species in certain environments' (Dragavtsev 1997). The last remark demonstrates an attempt to revive the principles of practical convenience in the problems of classification and systematics of cultivated plants. It is also confirmed by the new classification offered in the final part of this publication (Dragavtsev 1997), that is a 'breeding-oriented' classification of intraspecific genetic variability, which may help 'a plant breeder to attain genetic improvement of a species in the process of breeding'. Actually, the author offers the principles to be used as the foundation of a classification of intraspecific genetic variability. V.A. Dragavtsev names seven physio-genetic systems: the system of attraction, microdistributions, adaptivity, polygenous immunity, payback for feed, tolerance to crowdedness of phytocoenosis, and genetic variability of the length of ontogenetic phases. The author is quite right to assume that the use of these systems in plant classification makes it possible to select plant samples for placement in the core collection, and to organise efficient breeding technologies. However, there are two aspects which escaped the author's attention. First of all, plant samples should be included in this selection, and these developed in the territory of the area of the species concerned within definite populations and ecotypes and should be attributable to particular ecological factors. If the species is recognised as a system, one cannot deny that it should have a certain structure, organisational level, subsystems and a regular correlation between the system and subsystems. The hierarchical principle is one of the basic regularities in general biology (Dyuldin 1974). All steps of the hierarchy differ from each other in pecularities of their structure. According to Sinskaya (1979):

> "Organisms (individuals, specimens) are members of a population where they are united in structural groups corresponding to this level, thus turning into a integral part of the entity and acquiring specific nature of interrelations both between themselves and with the entity."

When a plant sample is distinguished, a plant breeder should predict, if possible, all its properties, including those related to its future existence within an agrocoenosis, where it will be a structural unit of a subsystem at a higher level. These properties can be predicted if the structure of the species, its phylogeny and relationships with other taxa have been successfully studied. Secondly, any sample selected for further work should bear a name which must show its place in the structure of the species. Even from the practical point of view "it is much more convenient to attribute them to certain taxa than make extensive descriptions" (Borkovskaya *et al.* 1979).

Finally it should be stressed that only the combination of comprehensive knowledge in general biology with the proper understanding of agricultural needs can make it possible to address the problem of intraspecific systematics of cultivated plants, and to develop convenient classifications which reflect the natural structure of a species.

References

Borkovskaya, V.A. & Filatenko, A.A. (1979). About intraspecific taxa of cultivated plants. *Byull. Vsesoyuzn. Inst. Rastaniev.* 91: 60–68.

Dragavtsev, V.A. (1997). Problems of classification of intraspecific plant genetic variability. 8 pp. N.I. Vavilov All-Russian Research Institute of Plant Industry, St. Petersburg.

Dyuldin, A.A. (1974). Non-uniformity of statistical structure in taxonomic hierarchic systems. In Systems approach in plant biology, pp. 39–47. Naukova Dumka, Kiev.

Fursa, T.B. & Stepanova, V.M. (1991). Eco-geographic differentiation of table watermelon. *Byull. Vsesoyuzn. Inst. Rastaniev.* 216: 54–59.

Kravtsova, B.E., Nikitskaya, K.I. & Podgornyi, E.A. (1974). Classification of bread wheat according to technological value. *Works of the All-Union Research Institute of Cereals and Processed Cereal Products* 79: 1–9.

Krivchenko, V.I., Burenin, V.I. & Korneyev, V.A. (1987). N.I. Vavilov's classification of agricultural plants according to drought resistance in view of modern breeding tasks. *Sbom. Nauchn. Trudov. Prikl. Bot. Benet. Genet. Selekts.* 100: 215–222.

Krivchenko, V.I., Burenin, V.I., Korinets, V.V., Shalygina, O.M. & Grushin, A.A. (1988). Systematised energy-oriented approach to methodoligical principles of plant classification. *Nauchno-Tekhn. Byull. Vsesoyuzn Ordene Lenina Ordena Druzhby Narodov Nauchno-Issl. Inst. Rasteniev N.I. Vavilova* 186: 3–6.

Omarov, D.S. (1978). New awnless broad-glume botanical forms of barley. *Byull. Vsesoyuzn Inst. Rastaniev.* 81: 39–41.

Shashko, D.I. (1966). Agro-climatic zoning in the USSR. *Vestn. Selskokhoz Nauki.* 4: 66–68.

Sinskaya, E.N. (1948). Dynamics of the species. 494 pp. Sekhozgiz, Moscow/Leningrad.

Sinskaya, E.N. (1958). Doctrine of populations and its significance for plant science. *Vestn. Selskokhoz Nauki.* 1: 52–61.

Sinskaya, E.N. (1979). Species and its structural parts at different levels of the organic world. *Byull. Vsesoyuzn. Inst. Rastaniev.* 91: 7–23.

Vitkovsky, V.L., Yushev, A.A. & Denisov, V.P. (1991). Modern concepts of systematic situation with stone-fruit plants. *Byull. Vsesoyuzn. Inst. Rastaniev* 216: 43–49.

Louneva, N.N. (1999). Infraspecific systematics of cultivated plants and a case study of *Prunus* L. (*Rosaceae*). In: S. Andrews, A.C. Leslie and C. Alexander (Editors). Taxonomy of Cultivated Plants: Third International Symposium, pp. 423–425. Royal Botanic Gardens, Kew.

INTRASPECIFIC SYSTEMATICS OF CULTIVATED PLANTS AND A CASE STUDY OF *PRUNUS* L. (*ROSACEAE*)

N.N. LOUNEVA

N.I. Vavilov Research Institute of Plant Industry, 42 Bolshaya Morskaya St., St Petersburg 190000, Russia

Introduction

It is impossible to comprehend the genetic diversity of breeding resources without addressing the research experience in the systematics of both cultivated and wild plants. Thus, it is necessary to formulate what is recognised here as the wild relatives of cultivated plants (WRCP). The most generalised approach suggests that WRCP should encompass all species of a genus, where at least one species was introduced into cultivation either through domestication, or by being included in breeding programmes (Maxted *et al.* 1997). A more detailed approach demands differentiation of the species within a genus into gene pools. The primary pool is represented by wild samples of the same species which were introduced into cultivation. Closely related species of the same genus are in the secondary pool, while those of a more remote relationship constitute the tertiary pool (Gadgil *et al.* 1996). This information is to be taken as input data when selecting priorities in breeding resources. It is obvious that the identification of pools of different relationships would require a profound knowledge of the systematics of the studied genera.

When analysing WRCP genetic diversity it is necessary to take into account that, although the samples of the populations within a wild species, which represents a primary pool, are attributed to the same species as the domesticated samples, in the course of time they will become genetically more and more different from the latter. This happens, because in natural populations selection is preconditioned by a complex of environmental factors in the natural habitat and in the cultivation process by the type of agriculture, cultivation technologies, farming techniques and tendencies in local plant breeding (Shvytov *et al.* 1995). Therefore, it was proposed to identify the following main levels of gene pools, i.e. varietal, specific and supraspecific, within the genetic diversity of breeding (Konarev 1995).

Case Study

In its turn the varietal pool may be systematically subdivided into two groups. The first group includes such cultivated forms which came into existence through the introduction of wild plants into cultivation (see Table 1). Many wild fruit plants were introduced into cultivation in this manner. For example, the process of selecting numerous different wild fruit samples, which are extremely valuable for commercial utilisation, has been and is still going on in many areas within the Caucasus. Targeted selection of certain taxa has resulted, for instance in the wide distribution of local cultivars of sour myrobalan plum all over Georgia. These cultivars are actually the domesticated forms of *Prunus cerasifera* Ehrh. subsp. *georgica* Erem. & Louneva occurring in local forests (Louneva 1991). Such cultivars should be regarded as having the same specific (and in this case also subspecific) attribution as the initial species of their wild relative.

TABLE 1. Specific pool level.

Taxa included in the pool	Cultivars based on the taxa	Botanical taxa and their cultivars
P. cerasifera Ehrh. subsp. *caspica* Louneva	'Gejga Arasi'	*P. cerasifera* subsp. *caspica* 'Gejga Arasi'
P. cerasifera Ehrh. subsp. *divaricata* (Ledeb.) Schneid.	'Shuntukskaya'	*P. cerasifera* subsp. *divaricata* 'Shuntukskaya'
P. cerasifera Ehrh. subsp. *georgica* Erem. & Louneva	'Rioni'	*P. cerasifera* subsp. *georgica* 'Rioni'
P. cerasifera Ehrh. subsp. *georgica* Erem. & Louneva var. *pissardii* Louneva	'Citeli Drosha'	*P. cerasifera* subsp. *georgica* var. *pissardii* 'Citeli Drosha'
P. cerasifera Ehrh. subsp. *iranica* Louneva & Erem.	'Gejga'	*P. cerasifera* subsp. *iranica* 'Gejga'
P. cerasifera Ehrh. subsp. *pontica* Louneva	'Pionerka'	*P. cerasifera* subsp. *pontica* 'Pionerka'

The other group of taxa developed on the basis of hybridisation with other species (and maybe even with other genera) will be genotypically different from the initial species of WRCP, since this group contains elements of other species (see Table 2). Such taxa should not be regarded as belonging to the initial species of WRCP. Instead, they must be recognised as interspecific hybrids. An example of these is the Nakhichevan myrobalan plum which represents a result of long-term breeding in the primary centre of myrobalan plum cultivation. The Nakhichevan myrobalan plum is an interspecific hybrid between *P. cerasifera* and the Japanese plum *P. salicina* Lindl. Its botanical name is *Prunus* × *nachitschevanica* Louneva (Louneva 1987).

TABLE 2. Supraspecific pool level.

Names of taxa included in the pool	Hybrids within the pool	Cultivars based on the hybrids	Botanical taxa and their cultivars
P. salicina Lindl.	*P. salicina* × *P. cerasifera*	'Aziza'	*P. salicina* × *P. cerasifera* 'Aziza'
P. angustifolia Marshall	*P. cerasifera* × *P. angustifolia*	'Marianna'	*P. cerasifera* × *P. angustifolia* 'Marianna'
?	*P.* × *nachitschevanica* Louneva	'Aresh'	*P.* × *nachitschevanica* 'Aresh'
?	*P.* × *taurica* Louneva	'Tauritsheskaya'	*P.* × *taurica* 'Tauritsheskaya'
?	*P.* × *foveata* Louneva	Local form	*P.* × *foveata* (local form)

References

Gadgil, M., Singh, S.N., Nagendra, H. & Chandran, M.D.S. (1996). *In-situ* conservation of wild relatives of cultivated plants: guiding principles and a case study. 22 pp. Centre for Ecological Sciences, Indian Institute of Science & Jawaharlal Nehru Centre for Advanced Scientific Research, Bangalore.

Konarev, V.G. (1995). The species as a biological system in evolution and breeding. 180 pp. N.I. Vavilov All-Russian Research Institute of Plant Industry, St. Petersburg.

Louneva, N.N. (1987). Naknichevan form of myrobalan plum and its taxonomic location. *Sbom. Nauchn. Trudov. Prikl. Bot. Genet. Selekts.* 112: 57–62.

Louneva, N.N. (1991). Wild-growing myrobalan plum in West Georgia. *Sbom. Nauchn. Trudov. Prikl. Bot. Genet. Selekts.* 139: 72–78.

Maxted, N., Ford-Lloyd, B.V. & Hawkes, J.G. (eds). (1997). Plant genetic conservation: the *in-situ* approach. 446 pp. Chapman & Hall, London.

Shvytov, I.A., Dzyubenko, N.I., Kurlovich, B.S. & Burmistrov, L.A. (1995). Methodological approaches to the development of criteria in evaluation of genetic diversity during plant introduction. 76 pp. N.I. Vavilov All-Russian Research Institute of Plant Industry, St. Petersburg.

Smekalova, T.N. (1999). Major aspects concerning the systematics of *Lathyrus* L. subgenus *Cicercula* (Medik.) Czefr. In: S. Andrews, A.C. Leslie and C. Alexander (Editors). Taxonomy of Cultivated Plants: Third International Symposium, pp. 427–428. Royal Botanic Gardens, Kew.

MAJOR ASPECTS CONCERNING THE SYSTEMATICS OF *LATHYRUS* L. SUBGENUS *CICERCULA* (MEDIK.) CZEFR.

T.N. SMEKALOVA

Herbarium, 42 Bolshaya Morskaya Str., St Petersburg 190000, Russia

There are several viewpoints concerning the systematics of the genus *Lathyrus* L. particularly the position, rank, size and structure of subgenus *Cicercula* (Medik.) Czefr. This entity was recognised by Medicus as long ago as 1787 as a distinct genus *Cicercula* Medik. It included the species morphologically similar to *L. cicera* L.

We attempted to critically analyse this group of species to define more precisely its rank and size as well as the degree of differentiation between species. The results of our morphological and anatomical research together with data from available publications, confirmed that researchers were correct to treat this group of species at the rank of subgenus (Table 1). Subgenus *Cicercula* consists of 21 species, circumscribed on the basis of the following criteria:

- Annual life cycle.
- Inhibition of growth of the main stem at early phases of plant development.
- Peculiarities in the structure of the spermoderm, e.g. thickness of major tissues, position of the light line, degree of swelling in cell coating of the hypodermis and epidermis (which is more so than in other subgenera), the shape of tissue cells and the shape of the cavity in epidermal cells.
- Common features of the racemes, e.g. relatively small number of flowers in the axillary raceme.

With these in mind, the most important traits were found to be the anatomical characters of vegetative and especially generative organs as they are more conservative and less changeable in the process of evolution.

In the identification of supraspecific taxa within the subgenus, the following characters were most important:

- Life cycle, short in ephemeroids and long tending towards a biennial cycle.
- Nature of shoot development, compact or spreading.
- Features of the stem conducting system, presence/absence of stem cavity, number and position of conducting bundles in the stem, wings and ridges.
- Nature of nervation in leaflets of which there are four main types.
- Number of flowers in the raceme.
- Colour of flowers and seeds.
- Features of the pod structure.
- The nature of the seed cover surface, e.g. large-tubercular, with/without microtubercles and small-tubercular types.

Some of the species studied displayed a rather wide spectrum of variability in their basic morphological and biological characters, which influenced the composition of their intraspecific systems (Table 2).

TABLE 1. The systematics of *Lathyrus* L. subgen. *Cicercula* (Medick.) Czefr.

I. Sect. *Cicercula*	II. Sect. *Hirsuta* Czefr.

1. Subsect. *Cicera*
   1) Ser. *Cicera*
      L. cicera L.
      L. pseudocicera Pamp.
   2) Ser. *Annua* Czefr.
      L. annuus L.
      L. gorgoni Parl.
      L. hierosolymitanus Boiss.
   3) Ser. *Colchyci* Czefr.
      L. colchycus Lipsky
      L. cassius Boiss.
 	
2. Subsect. *Sativa* Czefr.
      L. sativus L.
      L. stenophyllus Boiss.
      L. marmoratus Boiss.
      L. blepharicarpus Boiss.
      L. amphycarpus L.
	3. Subsect. *Hirta* Smekal.
	      L. hirsutus L.
	      L. chloranthus Boiss.
	      L. odoratus L.
	      L. lycicus Boiss.
	      L. phaselitanus Hub.-Mor.
	      L. trachicarpus (Boiss.) Boiss.
	
	4. Subsect. *Tingitana* Smekal.
	      L. tingitanus L.
	
	III. Sect. *Tetrafoliolata* Smekal.
	
	      L. bijugus Boiss.

TABLE 2. Intraspecific structure of some of the species from *Lathyrus* L. subgen. *Cicercula* (Medick.) Czefr.

L. cicera L.
1. *L. cicera* subsp. *cicera*
   1) var. *cicera*
   2) var. *negevensis* Plitm.
2. *L. cicera* subsp. *asiaticus* Smekal.
   L. tingitanus L.
   1) var. *tingitanus*
   2) var. *albus* Smekal.
   3) var. *roseus* Smekal.
   L. sativus L.
1. *L. sativus* subsp. *sativus*
   A. convar. *sativus*
   1) var. *sativus*
   2) var. *violascens* Smekal.
   B. convar. *cyaneus* Smekal.
   3) var. *cyaneus* Hov. & Khan ex
       Smekal.
   4) var. *parviflorus* Smekal.
   5) var. *azureus* Korsh. ex Smekal.
   6) var. *pulchrus* Smekal.
	C. convar. *coeruleus* Smekal.
	   7) var. *coeruleus* Smekal.
	   8) var. *pisiformis* Smekal.
	2. *L. sativus* subsp. *albus* Smekal.
	   D. convar. *albus* Smekal.
	   9) var. *albus* L.
	   10) var. *platyspermus* Smekal.
	   11) var. *orbiculatus* Smekal.
	   E. convar. *coloratus* Ser.
	   12) var. *coloratus* Ser.
	   13) var. *variegatus* Smekal.
	   14) var. *comitans* Smekal.
	   15) var. *depressus* Smekal.

Schifino-Wittmann, M.T. & Weber, L.H. (1999). The *Vicia sativa* aggregate in southern Brazil. In: S. Andrews, A.C. Leslie and C. Alexander (Editors). Taxonomy of Cultivated Plants: Third International Symposium, p. 429. Royal Botanic Gardens, Kew.

THE *VICIA SATIVA* AGGREGATE IN SOUTHERN BRAZIL

M.T. Schifino-Wittmann and L.H. Weber

Departamento de Plantas Forrageiras e Agrometeorologia, Faculdade de Agronomia, Universidade Federal do Rio Grande do Sul, Caixa Postal 776, 91501-970 Porto Alegre, RS, Brazil

Vicia sativa L. *sensu strictu* was introduced into Rio Grande do Sul, southern Brazil by Italian settlers and it is cultivated as forage in backyards and larger areas. The naturalised and highly polymorphic *V. angustifolia* L. is a widespread ruderal. Both taxa grow sympatrically and morphologically intermediate types are often found. The taxonomical situation of the aggregate is controversial (Maxted 1995, Potokina 1997) and the group is considered to be in active evolution (Hanelt & Mettin 1989). Thirty seven accessions of *V. sativa*, *V. angustifolia*, intermediate types, and *V. cordata* Wulf. ex Hoppe were analysed from different regions of Rio Grande do Sul, regarding karyotypes, corolla and legume colour, seed colour pattern and legume constrictions. Karyotypes allowed a clear distinction between the taxa: *V. sativa* (2n=12) had a metacentric marker chromosome, *V. angustifolia* and intermediate types (2n=12) only had acrocentric chromosomes, while *V. cordata* presented 2n=10. Qualitative characters grouped the taxa by Jaccard coefficients into three groups: *V. sativa* at 0.5 similarity level, *V. angustifolia* and intermediate types at 0.6 and *V. cordata* joined *V. sativa* at 0.12. These results agree with previous isoenzyme analysis (Schifino-Wittmann & Lange 1997) and show that cytological and qualitative characters are diagnostic for the aggregate and should be used instead of quantitative ones.

References

Hanelt, P. & Mettin, D. (1989). Biosystematics of the genus *Vicia* L. (*Leguminosae*). *Annual Rev. Ecol. Syst.* 20: 199–223.

Maxtead, N. (1995). An ecogeographical study of *Vicia* subgenus *Vicia*. Systematic and ecogeographic studies on crop genepools 8. 184 pp. IPGRI, Rome.

Potokina, E. (1997). *Vicia sativa* L. aggregate (*Fabaceae*) in the flora of the former USSR. *Genet. Resources Crop Evol.* 44: 199–209.

Schifino-Wittmann, M.T. & Lange, O. (1997). Isoenzymatic characterisation of native and cultivated forage legume species of Rio Grande do Sul (Southern Brazil). In International Grassland Society. Proceedings of the XVIII International Grassland Congress, Winnipeg and Saskatoon, Canada. 48 pp.

Davey, J.C. (1999). Deploying male-sterility as a grouping character in carrot distinctness, uniformity and stability trials. In: S. Andrews, A.C. Leslie and C. Alexander (Editors). Taxonomy of Cultivated Plants: Third International Symposium, pp. 431–433. Royal Botanic Gardens, Kew.

DEPLOYING MALE-STERILITY AS A GROUPING CHARACTER IN CARROT DISTINCTNESS, UNIFORMITY AND STABILITY TRIALS

J.C. DAVEY

Scottish Agricultural Science Agency, East Craigs, Edinburgh EH12 8NJ, Scotland, UK

Introduction

Since the early 1960s, when cytoplasmic male-sterility (CMS) was introduced for breeding F1 carrot cultivars, hybrids have come to dominate the market for early and late cultivars in Europe and, for special purposes, almost exclusively in the USA (Stein & Nothnagel 1995). In response to this development and the introduction of the character in the Carrot DUS Test Guideline (UPOV 1990), an assessment of male-sterility was undertaken for the inclusion of the character in Distinctness, Uniformity and Stability (DUS) test procedures.

Method

The floral morphology of 229 seedlots, containing 141 hybrids, from the United Kingdom Carrot Reference Collection was examined. Stecklings were grown in over-wintered trials during 1992 and flowering was observed the following spring. 20 plants were selected randomly for observation from plots containing three 6 metre rows with a target density of 40 roots per metre. The floral structure of flowers within the primary umbel was noted (1: open-pollinated, 2: CMS brown-anther or 3: CMS petaloid). If the petaloid form was present, further observation of the floral structure was made according to Eisa & Wallace (1969). Inflorescence colour was also scored visually (1: White, 2: Cream or 3: Green).

Results and Discussion

Two CMS forms, brown-anther and petaloid were clearly identified. Male-sterile flowers were found in 99% of cultivars listed as hybrids in the Common Catalogue of Varieties of Vegetable Plant Species. Since 54% of the hybrids were petaloid, this supports the view that petaloidy is a more widely used CMS system than the brown-anther form (Peterson & Simon 1986).

Both forms were found in all the major distinctness groups. They were found in similar proportions amongst the medium and late maturing oblong groups, which contain the horticultural types 'Nantes' and 'Berlicum'. Deploying male-sterility will aid cultivar differentiation in these large cultivar groups. Amongst the other main groups, there were more petaloids in the early and late obtriangular groups (containing 'Chantenay' and 'Autumn King') and more brown-anther forms in the early oblong group (predominately 'Amsterdam Forcing' types).

431

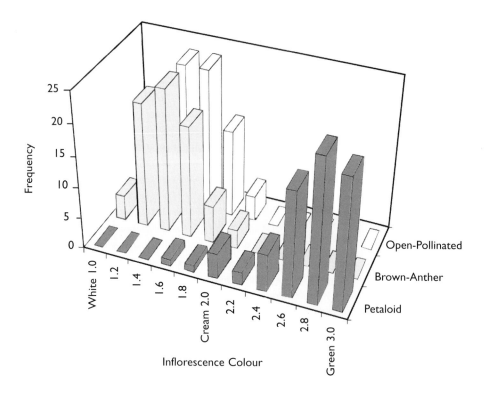

FIG. 1. Inflorescence colour of cultivars screened from the UK Carrot Reference Collection.

Petaloid hybrids exhibited a wide range of morphological expression. Entire and segmented petals replacing stamens, and or transformations of either the anther or the filament producing spoon, filament and strap structures, were found in differing frequencies. Although 38% of the petaloid hybrids differed in these frequencies (p=0.010); it was considered impractical to classify cultivars by this variation.

Open-pollinated cultivars and brown-anther hybrids were predominately white and cream (off-white) in colour (Fig. 1). Petaloid hybrids also differed in inflorescence colour. The majority were pale green or green. A small number of cultivars had an off-white/cream colour. Others have also reported white petaloid lines (Park 1995). There is breeding interest in this form of white petaloidy for improving seed set, as bees have a foraging preference for white petaloid lines (Erickson *et al.* 1979). As petal colour is apparently under genetic control of the nucleus, these associations may not be related to cytoplasmic effects (Morelock *et al.* 1996).

Other forms of male-sterility also exist. The 'gum' type, characterized by a total reduction of the anthers and petals, originating from a cross between the wild carrot *Daucus carota* L. var. *gummifer* Hook.f. and the cultivated carrot *D. carota* var. *sativus* Hoffm., has also been reported (Nothnagel 1992). This system has not been widely introduced into commercial material.

This study has shown male-sterility can be used to subdivide existing groups used in DUS tests, thereby reducing the number of cultivars required to be compared for establishing distinctness. Further differentiation of petaloid hybrids can be made using inflorescence colour. Inflorescence colour would merit inclusion in the next revision of the Carrot UPOV Guideline. This work illustrates the value of these collections as a source of germplasm and emphasises their recognition as national assets (Green 1997).

References

Eisa, H.M. & Wallace, D.H. (1969). Morphological and anatomical aspects of petaloidy in carrot, *D. carota* L. *J. Amer. Soc. Hort. Sci.* 94: 545–548.

Erickson, E.H., Peterson, C.E. & Werner, P. (1979). Honeybee foraging and resultant seed set among male-fertile and cytoplasmic male-sterile carrot inbreds and hybrid seed parents. *J. Amer. Soc. Hort. Sci.* 104: 635–638.

Green, F.N. (1997). The value of the United Kingdom statutory seed collections from a genetic resource perspective. *Pl. Var. Seeds* 10: 195–204.

Morelock, T.E., Simon, P.W. & Peterson, C.E. (1996). Wisconsin Wild: another petaloid male-sterile cytoplasm for carrot. *HortScience* 31(5): 887–888.

Nothnagel, T. (1992). Results in the development of alloplasmic carrots (*D. carota sativus* Hoffm.). *Pl. Breed.* 109: 67–74.

Park, Y. (1995). Selection of petaloid type male-sterile lines with white petal colour in carrots. *J. Korean Soc. Hort. Sci.* 36(1): 10–20.

Peterson, C.E. & Simon, P.W. (1986). Carrot Breeding. In M.J. Basset (ed.). Breeding Vegetable Crops 9. pp. 322–356. Avi, Connecticut.

Stein, M. & Nothnagel, T. (1995). Some remarks on carrot breeding (*Daucus carota sativus* Hoffm.). *Pl. Breed.* 114: 1-11.

UPOV (1990). Carrot, (*Daucus carota* L.). 'Guidelines for the conduct of tests for Distinctness, Homogeneity and Stability'. International Union for the Protection of New Varieties of Plants. TG/49/6. Genéve.

Davey, J.C. & Nevison, I. (1999). Selection of close controls in cabbage distinctness, uniformity and stability (DUS) trials using similarity coefficients. In: S. Andrews, A.C. Leslie and C. Alexander (Editors). Taxonomy of Cultivated Plants: Third International Symposium, pp. 435–438. Royal Botanic Gardens, Kew.

SELECTION OF CLOSE CONTROLS IN CABBAGE DISTINCTNESS, UNIFORMITY AND STABILITY (DUS) TRIALS USING SIMILARITY COEFFICIENTS

JONATHAN C. DAVEY[1] AND IAN NEVISON[2]

[1]Scottish Agricultural Science Agency, East Craigs, Edinburgh EH12 8NJ, Scotland, UK
[2]Biomathematics and Statistics Scotland, James Clerk Maxwell Buildings, Kings Buildings, Edinburgh EH9 3JZ, Scotland, UK

Introduction

Coefficients of similarity derived from descriptive data are used as a tool for matching candidates with cultivars held in a statutory reference collection. The selection of 'similar' listed cultivars enables registration authorities to concentrate on controls required to establish distinctness. The initial selection can be verified by substituting field data collected during the first year of test. If necessary, additional cultivars can then be grown alongside those already identified as close controls in the initial selection. This approach is a cost-effective way of managing registration trials where the ratio of candidates to listed cultivars in a cultivar-group is high.

Limiting the number of cultivar comparisons in Distinctness, Uniformity and Stability (DUS) trials is a challenge associated with cross-pollinated crops when cultivar-groups are large, and whose boundaries are often indistinct. Coefficients of similarity can be used as a tool for reduing the number of cultivars required to establish distinctness. The selection of the most similar cultivars overcomes the need to grow more than one trial group if candidates lie close to a group boundary. This approach is illustrated using Savoy Cabbage, *Brassica oleracea* L. convar. *capitata* (L.) Alef. var. *subauda* L. The initial selection, following assessment in the field, can be refined with subsequent sowings.

Method

The method involves compiling a cultivar × character matrix, calculating similarity coefficients using the Euclidean distance method and then using a weighted average of these character similarity matrices to identify the most similar cultivars to the candidates. The steps are as follows:

A. Compiling Data Matrices

Cultivar × character matrices can be taken from a variety of different sources, e.g. breeder's technical questionnaires, official descriptions or data taken from field trials. In this paper, grouping characters were taken from official descriptions of listed cultivars and data of candidate cultivars provided by plant breeders.

B. Calculation of Similarity Coefficients

Similarity coefficients were calculated from the data matrix using the Euclidean distance method after standardising by range as described by Krzanowski et al. (1994). Since some grouping characters are more important than others, an overall similarity matrix was calculated by weighting the average coefficients for all these characters. Pairwise similarities of all cultivars were ranked against the candidates and the first twenty cultivars with the highest coefficients were selected for each candidate.

C. Field Trials

Selected cultivars were compared with the candidates in DUS field trials using an internationally-agreed character set (UPOV 1992). To balance any omissions in selection, marker cultivars were included in registration trials that cover the range of cultivar expression for the crop.

D. Principal Coordinate Analysis

To illustrate this approach, principal coordinate analysis (PCO) was used to represent the similarities between 103 Savoy Cabbage cultivars in as few dimensions (axes) as possible. Plots of the axes illustrate patterns of variation between cultivars. Each axis is often correlated with different subsets of characters.

Results

The first five axes accounted for 79% of the total variation but none of these individually exceeded 24%. Fig. 1 shows that Candidate 'A' is closely associated with 15 of the 20 selected cultivars on these two axes, the first of which is strongly correlated with maturity and leaf colour. Fig. 2 shows a situation where the selected controls for Candidate 'B' cannot be clearly separated from other cultivars on these axes.

When the selected cultivars are compared with the two candidates in the field, using data recorded for the full character set, Candidate 'A' was found to be a unique cultivar whereas Candidate 'B' was most similar to the 1st, 7th, 11th and 15th ranked cultivars.

Discussion

On completion of the first year of trialling, data recorded at the test centre can be substituted for the breeder's data to assess whether the selection of 'similar' cultivars had been appropriate. Additional cultivars that show a high similarity, using this second analysis, can then be grown alongside those cultivars already identified as close controls in the initial selection. If necessary a third year of test may then be required to establish distinctness.

A weakness of this method is that matrices derived from official cultivar descriptions may not be a reliable source of data. As these descriptions will have been produced from different environments, character expression may differ in another environment. An alternative approach is to use data compiled from a single trial site. Such data can be adjusted by the technique of fitted constants (FITCON) as described by Patterson (1997). The technique deals with incomplete data since not all cultivars in a collection are grown every year. The variety means are then converted into descriptive scores. Finally, descriptive data of listed cultivars not previously grown in field trials, and breeder's information on candidates, are added to these scores. This approach has been applied to registration trials using leek.

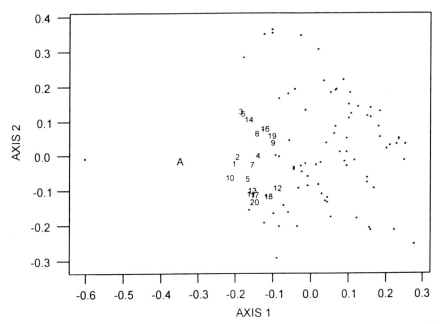

FIG. 1. Principal coordinate scores for Candidate A, 20 most similar cultivars to A and 82 unselected cultivars. Low numbers indicate a closer similarity to the candidate. Axis 1 strongly correlates with plant maturity and leaf size, as does Axis 2 with leaf and head size. Axes 1 and 2 account for 47% of the total variation.

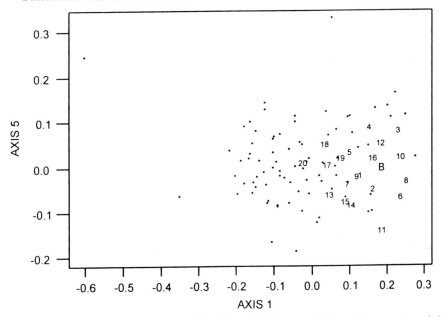

FIG. 2. Principal coordinate scores for Candidate B, 20 most similar cultivars to B and 82 unselected cultivars. Low numbers indicate a closer similarity to the candidate. Axis 1 strongly correlates with plant maturity and leaf size, as does Axis 5 with head shape. Axes 1 and 5 account for 32% of the total variation.

437

Cultivar-matching tools using images are also being developed to assist in the selection of close controls in registration trials (Davey *et al.* 1997). However, until this technology is accepted in registration procedures, similarity coefficients can provide a cost-effective matching tool for identifying control cultivars for registration testing. This is particularly important where the ratio of candidates to the number of listed cultivars in a cultivar-group is high.

References

Davey, J.C., Horgan, G.W. & Talbot, M. (1997). Image Analysis: a tool for assessing plant uniformity and cultivar matching. *J. Appl. Genet.* 38(A): 120–135.

Krzanowski, W.J. & Marriott, F.H.C. (1994). Multivariate Analysis Part 1. Distributions, ordination and inference. 280 pp. Edward Arnold, London.

Patterson, H.D. (1997). Analysis of series of variety trials. In R.A. Kempton & P.N. Fox (eds). Statistical methods for plant variety evaluation. pp. 139–161. Chapman & Hall, London.

UPOV (1992). Cabbage, (*Brassica oleracea* L. convar. *capitata* (L.) Alef.). Guidelines for the conduct of tests for Distinctness, Homogeneity and Stability. International Union for the Protection of New Cultivars of Plants. TG/48/6. Genéve.

Thomson, C.M. (1999). Classification of Brussels sprout cultivars in the UK — *Brassica oleracea* L. convar. *oleracea* var. *gemmifera* DC. In: S. Andrews, A.C. Leslie and C. Alexander (Editors). Taxonomy of Cultivated Plants: Third International Symposium, pp. 439–442. Royal Botanic Gardens, Kew.

CLASSIFICATION OF BRUSSELS SPROUT CULTIVARS IN THE UK — *BRASSICA OLERACEA* L. CONVAR. *OLERACEA* VAR. *GEMMIFERA* DC.

C.M. THOMSON

Scottish Agricultural Science Agency, East Craigs, Edinburgh EH12 8NJ, Scotland, UK

Introduction

Brussels sprout cultivars are classified to enable the selection of controls for candidates entered in Distinctness, Uniformity and Stability (DUS) tests, and for genetic resource purposes. DUS tests are undertaken for the registration of cultivars on a National List and/or for an award of Plant Breeders' Rights (MAFF 1973,1964). The International Union for the Protection of New Varieties of Plants (UPOV) publishes internationally agreed methods and technical guidelines for Plant Breeders' Rights tests (UPOV 1979, 1991); these define the characters used to describe cultivars.

The UK Brussels Sprout Cultivar Collection, which is held at the Scottish Agricultural Science Agency (SASA), has two components:

1. The statutory seed reference collection containing UK and EU registered cultivars, example cultivars and candidates submitted for UK DUS tests (130 accessions).
2. The genetic resource collection containing uncommercialised lines, and cultivars no longer registered (120 accessions).

Brussels sprout cultivars are classified using three main sources of data:

A. Administrative and Passport Data
Breeding method, hybridity, origin, registered status, synonymy and seed collection details are held.

B. Cumulative Cultivar Data
Complete or partial records taken at Cambridge for 24 characters × 194 cultivars × 6 years, and at Edinburgh for 18 characters × 130 cultivars × 4 years are being collated into a UK over-year dataset.

C. Cultivar Descriptions
UK descriptions are compiled from both cumulative means of scored characters and scores derived from measured characters. Non-UK descriptions are translated and descriptive scores are aligned with current UPOV characters, where necessary.

439

Use of Data from Different Sources for Classification

Scored data on 33 cultivars over 14 characters recorded at Edinburgh and Cambridge (UK) and other EU member states (mainly The Netherlands), were statistically analysed. Between-site correlations were computed for each character to test consistency of expression (Table 1). Start of harvest maturity, plant height and leaf blade cupping (UPOV grouping characters) and leaf blade blistering, sprout size, sprout colour, leaf blade size (UPOV asterisked characters) were highly correlated between sites, though sprout colour had a weaker correlation. Thus, SASA can confidently use data from different sources for these characters, as they are least affected by environment. However, two grouping characters, leaf blade colour and leaf blade intensity of colour were variably expressed between sites, as were two asterisked characters, petiole attitude and petiole anthocyanin colour. For these characters it was preferable only to use data recorded at SASA as they were consistent over years at that site.

TABLE 1. Principal characters used in UK DUS Brussels sprout tests.

UPOV Character Number	UPOV Character	Ranked Discriminating Ability at SASA	Ranked Consistency Over Sites
Discontinuously expressed characters			
*4G	Leaf blade: colour	-	11
–	Hybridity	-	1
Continuously expressed characters			
*1G	Plant: height	7	8
2	Plant: tendency to form a head	6	6
*3	Leaf blade: size	8	10
*5G	Leaf blade: intensity of colour	9	5
*7G	Leaf blade: cupping	2	4
*8	Leaf blade: blistering	4	3
*10	Petiole: attitude	10	14
*12	Petiole: anthocyanin coloration	11	13
*13	Sprout: size	3	7
14	Sprout: shape in longitudinal section	12	12
*15	Sprout: colour	5	9
*18G	Start of harvest maturity	1	2

* UPOV asterisked characters
G UPOV grouping characters

Classification of Cultivars

The aim of classification for DUS testing is to split the range of variation into the largest number of small groups based on the most discriminating characters (Kelly 1968). The UPOV Guideline (UPOV 1990) lists five grouping and six asterisked characters, all of which must be recorded for a DUS test (Table 1). Leaf blade colour is the only discontinuous character recommended for grouping. A collection of 107 cultivars was divided into five major groups using leaf blade colour and breeding method (Table 2). Within the three largest groups, cultivar groupings were sought using Furthest Neighbour Cluster Analysis, based on SASA data from five of the most discriminating characters. Data for characters which had poor between-site correlations were omitted.

TABLE 2. Major groups in the UK Brussels Sprout Cultivar Collection

Group	Breeding Method	Leaf Blade Colour	Number of Cultivars
1	F_1 hybrid	Green foliage	66
2	F_1 hybrid	Blue-green foliage	19
3	Open-pollinated	Green foliage	19
4	Open-pollinated	Blue-green foliage	2
5	Open-pollinated	Purple foliage	1

For the two medium-sized groups (2 and 3), it was possible to relate clusters and the expression of specific characters in the field. However, in the largest group (1) the clusters appeared to be a mixture of visually dissimilar cultivars. These cultivars will be grown in the field to see if they can be classified into sub-groups on good visual characters. Refinement of the classification would be sought by cluster analysis on the sub-groups.

Conclusion

The current UK classification of 107 Brussels sprout cultivars is based on over-year data recorded at SASA. The addition of data for consistent characters from other sites will enhance the selection of controls for DUS tests and the compilation of over-year descriptions. The existing classification will be refined with the help of clustering analysis and further examination of cultivars in field trials.

Acknowledgements

Thanks are due to Ian Nevison (Biomathematics & Statistics Scotland), for providing statistical advice, correlation of character data from different sources, evaluation of character discrimination and cluster analyses, and to Niall Green for advice and assistance with this paper.

References

Kelly, A.F. (1968). The work of systematic botany branch. *J. Natl. Inst. Agric. Bot.* 11: 246–252.

MAFF. (1973). The Seeds (National List of Varieties) Regulations 1973. Statutory Instrument 1973/994 London. HMSO.

MAFF. (1964). The Plant Varieties and Seeds Act 1964 CH. 14. London. HMSO.

UPOV. (1979). Revised general introduction to the guidelines for the conduct of tests for Uniformity, Homogeneity and Stability of new varieties of plants. TG/1/2. International Union for the Protection of New Varieties of Plants. Geneva, Switzerland.

UPOV. (1990). Guidelines for the conduct of tests for Distinctness, Homogeneity and Stability: Brussels Sprout: TG/54/6. International Union for the Protection of New Varieties of Plants. Geneva, Switzerland.

UPOV. (1991). International Convention for the Protection of New Varieties of Plants. International Union for the Protection of New Varieties of Plants. Geneva, Switzerland.

Massie, I.H., Astley, D., King, G.J. & Ingrouille M. (1999). Patterns of ecogeographic variation in the Italian landrace cauliflower and broccoli (*Brassica oleracea* L. var. *botrytis* L. and var. *italica* Plenck). In: S. Andrews, A.C. Leslie and C. Alexander (Editors). Taxonomy of Cultivated Plants: Third International Symposium, pp. 443–445. Royal Botanic Gardens, Kew.

PATTERNS OF ECOGEOGRAPHIC VARIATION IN THE ITALIAN LANDRACE CAULIFLOWER AND BROCCOLI (*BRASSICA OLERACEA* L. VAR. *BOTRYTIS* L. AND VAR. *ITALICA* PLENCK)

I.H. MASSIE[1], D. ASTLEY[2], G.J. KING[2] AND M. INGROUILLE[4]

[1]c/o 59 Chiltley Way, Liphook, Hants GU30 7HE, UK
[2]Horticulture Research International, Wellesbourne, Warwick CV35 9EF, UK
[3]Dept of Biology, Birkbeck College, University of London, Malet Street, London WC1E 7HX, UK

Introduction

There is a large diversity between the traditional varieties of cauliflower and broccoli grown throughout Italy, which have been selected for local conditions and preferences. Heterogeneous populations offer a greater potential variation, and possible permutations for recombination between different sets of genes for use in breeding, due to their wide genetic base. The conservation of a diversity of hereditary forms is therefore important for the future development of a crop. Vavilov (1951) stated that under certain conditions the local assortment of hereditary material is represented by heterogeneous populations consisting of many forms differing from one another morphologically and physiologically. However in other cases the local varieties may consist of populations of uniform physiological and morphological characters.

Historical diversification of cauliflower and broccoli in Italy has been due to natural and planned artificial selection within local landraces. Vavilov (1951) recognised generally that selective pressure has given rise to local varieties that are ecologically suited to their local environment. Different provinces of Italy have been associated with specific varietal types of cauliflower and broccoli. The following varietal types are perceived as distinct regionalised forms: Romanesco cauliflower in the Lazio region; Di Jesi, Macerata and Tardivo di Fano varieties in the Marche region; Cavolfiore Violetto di Sicilia (Sicilian Purple cauliflower) in Sicily, except for the Palermo region where a green cauliflower is typical.

In this study morphological diversity is assessed using cluster analysis to identify the existence of geographical cultivar groupings. Each accession was considered as an Operational Taxonomic Unit (OTU). Hierarchical Cluster analysis and principal components analysis were used to identify clustering patterns in the landraces of cauliflower and broccoli studied.

A total of 101 accessions of cauliflower and broccoli landraces were analysed over a two-year trial. Plants were raised in polytunnels. Leaves were harvested from plants grown in polytunnels in randomised plots. The leaves were pressed and photocopied in order to record data using a digitiser. Head characters were scored from plants transferred to randomised field plots. Forty-two accessions were also sampled for DNA which was used in a parallel RAPD analysis.

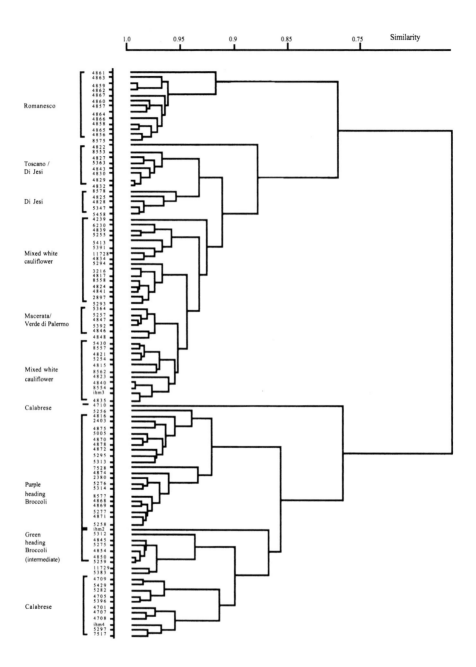

FIG. 1. Average Linkage dendrogram from cluster analysis of cauliflower and broccoli accessions from Italy analysed for 15 leaf and head variables.

Results

In the first year of trials, the 55 accessions which were grown produced four clear cluster groups based on 23 leaf and head variables (Massie *et al.* 1996). In the second year the number of accessions was increased to assess a wider range of landraces and combining the results from each year (reference landraces were grown in each year), the four main clusters were retained as shown in Fig. 1. Using principal components analysis and canonical variate analysis the clusters could also be identified.

In parallel with the morphological study a RAPD analysis was carried out on a subgroup of 42 accessions of cauliflower and broccoli. The results from the RAPD analysis were not conclusive, however, they did support the results of the morphological study with respect to the degree of variation within and between the different groups.

Conclusions

Phenotypic analysis, using multivariate techniques and cluster analysis, confirmed Gray's classification of the Sicilian purple heading broccoli as a broccoli based on comparative ontogeny (Gray 1989). Romanescos were confirmed as being cauliflowers. Traditionally in Italy, Romanescos have been classed as broccoli, and Sicilian purple heading broccoli as cauliflowers. This still persists in the classification in the Italian breeding company catalogues. Classification of the green heading broccoli from south-east Italy (Puglia) suggests they are related to the Sicilian purple heading broccoli and Calabrese, and are not intermediates between the Sicilian purple heading broccoli and the di Macerata cauliflowers. Genotypic analysis supports the phenotypic results, though a more extensive study would be required to generate more markers and possibly specific markers for the morphological variants. The way forward suggested by approaches to other crop germplasm is that AFLP technology or microsatellite markers may be the key to assessing the diversity and genetic relationships within a germplasm collection.

References

Gray, A.R. (1989). Taxonomy and evolution of broccolis and cauliflowers. *Baileya* 23(1): 28–46.

Massie, I.H., Astley, D. & King, G.J. (1996). Patterns of genetic diversity and relationships between regional groups and populations of Italian landrace cauliflower and broccoli (*Brassica oleracea* L. var. *botrytis* L. and var. *italica* Plenck). *Acta Hort.* 407: 45–53.

Vavilov, N.I. (1951). The origin, variation, immunity, and breeding of cultivated plants. Translation by K. S. Chester. *Chron. Bot.* 13(1/6): 1–366.

Lofti, M. & Kashi, A. (1999). The Iranian melon as a new cultivar-group. In: S. Andrews, A.C. Leslie and C. Alexander (Editors). Taxonomy of Cultivated Plants: Third International Symposium, pp. 447–449. Royal Botanic Gardens, Kew.

THE IRANIAN MELON AS A NEW CULTIVAR-GROUP

M. LOTFI AND A. KASHI

Department of Horticulture, Faculty of Agriculture, Tehran University, Karaj 31585/4111, Iran

Introduction

Cucumis melo L. with its many cultivars is one of the world's most important cucurbits. It plays a great role in the nutrition of people especially in the Middle East. Iran has the third largest area under melon cultivation in the world, over 80,000 ha. (FAO 1997). The great diversity within certain groups of *C. melo* and their lack of historical background have led investigators to look for their native habitat within Iran. For thousands of years, certain melon cultivars have been growing there in the desert margins, these are notable for their drought resistance, a relative resistance to salinity and the ability to tolerate high temperatures. As there is limited knowledge regarding the cultivation of these cultivars in other countries, it was thought that they should now be classified.

Recognised Cultivar-groups of Melon

All the taxa studied belong to *C. melo* as they readily cross with each other. Naudin (1859) classified them as ten botanical varieties, e.g. *C. melo* var. *cantaloupensis* Naudin, var. *reticulatus* Naudin, var. *inodorus* Naudin, var. *flexuosus* Naudin, var. *dudaim* Naudin, var. *chito* Naudin, var. *utillisimus* Naudin, var. *saccharinus* Naudin, var. *acidulus* Naudin and var. *agrestis* Naudin. In addition, Filov (1960) divided these varieties into six subspecies based on the areas where they were growing. Other classifications have been presented on the taxonomy of melons, e.g. Pangalo (1951), Jeffrey (1961), Ashizawa & Yamato (1965), Fujishita & Oda (1965) and Pratt (1971), but all have been incomplete. These occurred after the appearance of the *International Code of Nomenclature for Cultivated Plants* (Brickell *et al.* 1980) and in view of the many cultivars, it is possible that mistakes occurred.

Today, Naudin's varieties with only slight modifications are regarded as horticultural groups based on their fruit characteristics and uses (Robinson & Decker-Walters 1997). They are not regarded as botanical varieties based on phylogeny (Munger & Robinson 1991), nor do they conform to the modern rules for cultivated plant nomenclature (Trehane *et al.* 1995). The groups are:

- Cantalopensis Group (cantaloupe and muskmelon)
- Indorus Group (winter melon)
- Flexuosus Group (snake melon)
- Conomon Group (pickling melon)
- Dudaim Group (pomegranate melon or Queen Anne's pocket melon)
- Momordica Group (phoot or snap melon)

The cultivars within the Cantalopensis Group are the most important commercial melons in Europe and America today. However, in Iran they are of lesser importance. Iran is a major centre of melon diversity and domestication and there are five distinct groups of melon recognised:

- The Talebi Group and the Garmac Group. Both of these have characteristics which accord mostly to the Cantalopensis Group. They have globular-shaped fruits with meridian stripes and soft, spongy flesh. Those in the Garmac Group are larger than the Talebi Group, are less sweet, always have orange flesh and do not have stripes.
- The Curve Cucumber Groups which falls within the Flexuosus Group.
- The Dastanboo Group which is similar to the Dudaim Group.
- The main commercial melon in Iran is 'Kharbazeh', which does not fall into any of the aforementioned groups. Although, it appears very near to the Inodorus Group as it has the large fruits, which matures late and a long shelf life, it differs in its netted skin surface and also in the occasional rugby-shaped fruits.

The most popular member of the Indorus Group worldwide is the cultivar 'Honey Dew'. It differs from Iranian melons in its shape and smooth rind surface. There is great diversity within this group in Iran where major types occur. One has an oval shape, with a completely netted skin, while the other is spindle-shaped with longitudinal stripes with netting between these stripes. Other important characteristics of the Inodorus Group are the crisp flesh and that the pedicel does not release the fruit at maturity.

Conclusions

As it is now clear that there is an important group of melons within Iran, which up to now has not been classified, it has been suggested to call this the Iraniansis Group. Within the Iraniansis Group falls 'Kharbozeh'. Further research will be carried out to refine this group and to work further on the Iranian cultivars of *C. melo*.

References

Ashizawa, M. & Yamato, S. (1965). Soviet Melon. *Agric. & Hort.* 40: 1355–1361. (In Japanese).

Brickell, C.D. *et al.* (eds). (1980). International code of nomenclature for cultivated plants – 1980. 32 pp. Bohn, Scheltema & Holkema, Utrecht.

FAO (1997). Statistics Series, No. 142, (1997), vol. 51.

Filov, A.I. (1960). The problem of melon systematics. *Pl. Breed. Abstr.* 31: 5499.

Fujishita, N. & Oda, Y. (1965). Melons from Pakistan, Afghanistan and Iran. Results of the Kyoto University Scientific Expedition to the Karakovum and Hindukushi. Vol. 1. pp. 233–256. *Plant. Breed. Abstr.* CAB International (Oxon. – New York) Japan.

Jeffrey, C. (1961). Notes on *Cucurbitaceae*, including a proposed new classification of the family. *Kew Bull.* 15(3): 337–371.

Munger, H.M. & Robinson, R.W. (1991). Nomenclature of *Cucumis melo* L. *Cucurbit Genetics Cooperative Report* 14: 43–44.

Naudin, C.H. (1859). Essais d'une monographie des espèces et des variétés du genre *Cucumis. Ann. Sci. Nat., Bot. Biol. Vég.* 11: 5–87.

Pangalo, K.I. (1951). Melons as an independent genus *Melo* Adams. *Bot. Zhurn. (Moscow & Leningrad).* 36: 571–580.

Pratt, H.K. (1971). Melons. In A.C. Hulme (ed.). The biochemistry of fruits and their products. Part II. pp. 207–232. Academic Press, London.

Robinson, R.W. & Decker-Walters, D.S. (1997). Cucurbits.? 208 pp. CAB International (Oxon., UK).

Trehane, P. *et al.* (1995). International Code of Nomenclature for Cultivated Plants – 1995. 175 pp. Quarterjack Publishing, Wimborne.

Rivera, D., Alcaraz, F., Inocencio, C., Obón, C. & Carreño, E. (1999). Taxonomic study of cultivated *Capparis* sect. *Capparis* in the western Mediterranean. In: S. Andrews, A.C. Leslie and C. Alexander (Editors). Taxonomy of Cultivated Plants: Third International Symposium, pp. 451–455. Royal Botanic Gardens, Kew.

TAXONOMIC STUDY OF CULTIVATED *CAPPARIS* SECT. *CAPPARIS* IN THE WESTERN MEDITERRANEAN

D. Rivera, F. Alcaraz, C. Inocencio, C. Obón and E. Carreño

Dep. Biología Vegetal, Fac. Biología, Univ. Murcia, 30100 Murcia, Spain

Introduction

The large tropical and subtropical genus *Capparis* L. is represented in the Mediterranean and the Near East countries by a few species and a number of varieties included within section *Capparis*. This section is characterised by the presence of well-developed and persistent leaves, sepals free in bud, with one sepal of the outer pair, deeply saccate. *Capparis spinosa* L. is the type species of section *Capparis* and it is also the type species of the genus.

Here section *Capparis* is circumscribed to include *C. spinosa* and its related species (*C. sicula* Duhamel, *C. orientalis* Duhamel and *C. aegyptia* Lam.) and it extends from the Mediterranean region to East Malesia and Australia. The centre of diversity for this section is situated in the Mediterranean region and the Near East.

Capers (flower buds), young shoots and tender fruits are commonly eaten as pickles, after several weeks of preserving them in salt and vinegar. Most of the material used for commercial purposes is gathered from the wild but there are several areas where caper plants are cultivated.

Caper consumption started very early in antiquity. It has been documented in Mesopotamia as far as in the Mesolithic and Aceramic Neolithic levels (Rivera, Matilla & Obón in press).

Caper cultivation has been associated with the use of vegetative propagation (cuttings), which has contributed to maintain a relatively high diversity of cultivars. In fact it seems that *C. spinosa* is a cultigen, as it is almost sterile and rarely escapes from cultivation.

Taxonomic Assessment of *Capparis spinosa* 'Mallorquina'

In his Flora of Turkey account, Coode (1965) reported *C. spinosa* as almost exclusively cultivated, thus casting doubts about where it can be said to be truly wild. The situation appears similar for the Western Mediterranean.

We have considered the possibility of determining relationships on morphological grounds between the commonest caper cultivar in the Iberian Peninsula and Balearic Islands coloquially known as fina and its wild relatives.

This taxonomic study was based on the comparison of thirteen populations, including both wild and cultivated.

Balearic Islands: MALLORCA: Llubí (9), Cathedral of Palma (8 and 10). IBIZA: Ibiza (12).

451

Iberian Peninsula: TARRAGONA: Aldover (11), MURCIA: Albudeite (2), Llano del Beal (13). LÉRIDA: Lérida (7). GRANADA: Lanjarón (6). JAÉN: Pozo Alcón (3). ALMERÍA: Cabo de Gata (5), Albox (4). ALICANTE: Villajoyosa (1).

Characters available in the field were considered (numbers of stamens, diameter of the stipules at the base, stipule length, leaf base type, leaf apex type, leaf hair covering, width of nectaries, length of nectaries, anther apex type, ovary length and ovary width). These characters were compared using UPGMA analysis (Podani 1991).

The cultivar 'Mallorquina' from Llubí (9) was shown to be closely related to the wild population at Villajoyosa (1). Both appear related to the group *C. sicula* (2, 3, 7, 4, 5, 6, 8), see Fig. 1.

Cultivars	Features
'ROSA'	Prolific flowering. Inflorescence without intercalary vegetative shoots. High quality flower buds. Not fruiting. Upper part of the shoot almost glabrous. Decaying leaves purple tinged.
'FIGUES SEQUES'	Poor flowering. Inflorescence with intercalary vegetative shoots. Poor quality flower buds (large, flat and empty). Sometimes fruiting. Upper part of the shoot almost glabrous.
'DE LAS MURADAS'	Prolific flowering. Inflorescence without intercalary vegetative shoots. Poor quality flower buds. Fruiting abundantly. Upper part of the shoot glabrous.
'REDONA'	Poor flowering. Inflorescence with intercalary vegetative shoots. High quality flower buds. Not fruiting. Upper part of the shoot almost glabrous.
'PELUDA'	Prolific to normal flowering. Inflorescence with intercalary vegetative shoots. High quality flower buds. Fruiting. Upper part of the shoot woolly.
'COLORA'	Poor flowering. Inflorescence without intercalary vegetative shoots. Poor quality flower buds. Not fruiting. Upper part of the shoot almost glabrous.
'MALLORQUINA'	Prolific flowering. Inflorescence without intercalary vegetative shoots. High quality flower buds. Not fruiting. Upper part of the shoot almost glabrous. Decaying leaves yellow.

FIG. 1. Preliminary descriptions of cultivars of *Capparis spinosa* L.

Morphological Features Recognised for Cultivars within *Capparis* sect. *Capparis*

Available descriptions of caper cultivars are relatively poor. Since caper cultivation is becoming rare in the Western Mediterranean region, it was urgent to carry out an ethnobotanical study concerning the main areas of caper cultivation in order to localise the cultivars and to determine their features.

Caper harvesting in this area was a task exclusive to rural women and marginal groups of the population (e.g. the gypsies). The knowledge concerning characteristics of capers is socially and sexually discriminated, as the scarce amount of information available in the literature (Luna & Pérez 1985) was based on interviewing only male farmers and technicians, never the females.

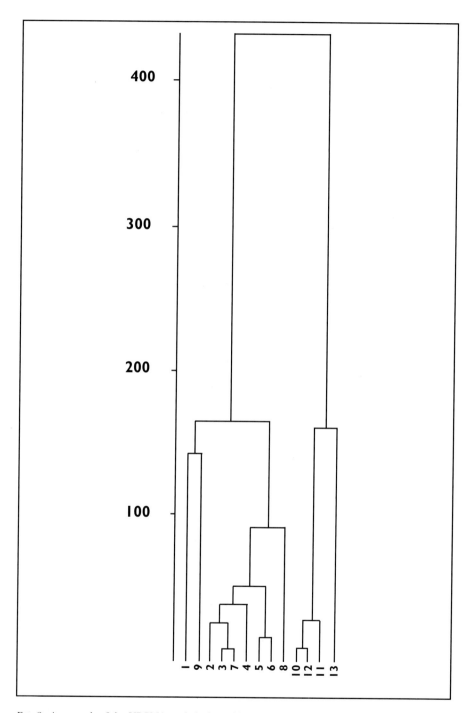

FIG. 2. As a result of the UPGMA analysis the cultivar 'Mallorquina' from Llubí (9) was shown to be closely related to the wild population at Villajoyosa (1). Both appear related to the group *C. sicula* (2, 3, 7, 4, 5, 6, 8).

We have not only collected the morphological data information but also the plant material for propagation and for studying molecular systematics.

Seed Morphology and Recognition of Wild and Cultivated *Capparis*

For the purpose of documenting the origin of caper cultivation in an archaeological context, the study of the taxonomy of the morphological features available in the seeds has been started. The seeds are the most commonly preserved plant parts found in an archaeological context. They are found charred and sometimes lack the outer layer of the testa.

The seeds are obovate in outline, tapering from to the base, more or less wedge shaped in side-view. The prominent curved radicle is a noteworthy feature. Sometimes only the "inner seed" has been preserved and this resembles some chenopodiaceous seeds.

Several features such as the prominence of the radicle were considered promising in terms of recognition of the seeds belonging to one or other taxa. Up to sixteen populations were investigated:

Balearic Islands: MALLORCA: Llubí (1), Cathedral de Palma (4 and 12), Cruce de Sines (6), Sóller (9), Alcudia (11).
Iberian Peninsula: TARRAGONA: Aldover (2), MURCIA: Aguilas (3 and 14), Atamaría (5), Llano del Beal (15). HUESCA: Fraga (7). GRANADA: Lanjarón (8), Guadix (13 and 16). ALMERÍA: Cabo de Gata (10).

The characters considered were: length, width, breadth, maximum diameter of radicle, and length of the free prominent part of the radicle.

An UPGMA analysis was carried out using the SynTax set of programmes (Podani 1991).The results were relatively disappointing since no clear correlation was detected between the clusters obtained and taxonomic categories or division of cultivated and wild populations (Fig. 2). The analysis recognised firstly a group of populations (7, 16, 8, 13, 9?) which were identified as *C. sicula*.

A second separation was obtained between a large group of populations and population 5, which presumably corresponds to *C. aegyptia*. The above large group included mainly seeds of *C. spinosa* cultivars (1, 3, 4, 14), mixed with *C. rupestris* Sibth. & Sm. (2, 6, 9, 11, 12) and only two populations of presumably *C. sicula* (10 and 15).

The most relevant characters for this study were the width and breadth of the seeds. Radicle features were difficult for this purpose because the radicle breaks off easily in seeds preserved in archaeological contexts or even during the cleaning of freshly collected fruits.

Acknowledgements

The authors are indebted to those who helped us in collecting firsthand information on caper cultivars — Sr. J. Rosselló (Llubí), Sra. María Fullana (Campos) and the staff of Agrucapers (Aguilas). We are also indebted to Prof. M. Honrubia for his help in the digital processing of images. This research has been partly financed by the project AGF96 – 1040 from CICYT (I+D).

References

Coode, M. (1965). *Capparis*. In P.H. Davis (ed.). Flora of Turkey Vol. 1, pp. 496–498. Edinburgh University Press, Edinburgh.

Luna, F. & Perez, M. (1985). La tapenera o alcaparra. Cultivo y aprovechamiento. 127 pp. Ministerio de Agricultura, Pesca y Alimentación, Madrid.

Podani, J. (1991). Syn-Tax IV. In E. Feoli & L. Orioci (eds). Computer Assisted Vegetation Analysis, pp. 437–452. Kluwer Academic Publishers, Den Haag.

Rivera, D., Matilla, G. & Obón, C. (In press). Palaeoethnobotany of Mesopotamia: *Euphrates* and *Tigris* Region. B.A.R. Oxford.

Campbell, G.D. (1999). The development of new uniformity standards for Turnip Rape in UK distinctness, uniformity and stability tests. In: S. Andrews, A.C. Leslie and C. Alexander (Editors). Taxonomy of Cultivated Plants: Third International Symposium, pp. 457–459. Royal Botanic Gardens, Kew.

THE DEVELOPMENT OF NEW UNIFORMITY STANDARDS FOR TURNIP RAPE IN UK DISTINCTNESS, UNIFORMITY AND STABILITY TESTS

G.D. CAMPBELL

Scottish Agricultural Science Agency, East Craigs, Edinburgh, EH12 8NJ, Scotland, UK

Introduction

Turnip Rape (*Brassica rapa* L. var. *silvestris* (Lam.) Briggs) is a valuable oilseed crop for north European and Canadian regions and has been an expanding commercial crop in the United Kingdom (UK). Breeders have focused on improving oil quality and yield, specialist chemical components, disease and lodging resistance and crop yield, rather than morphological characteristics.

Most cultivars destined for commercial use in the European Union (EU) have been entered for Distinctness, Uniformity and Stability (DUS) testing in the UK. Candidate numbers for testing in the UK have therefore increased greatly since 1991, with a consequent increase in uniformity problems. The Scottish Agricultural Science Agency (SASA) is the UK DUS test centre for Turnip Rape, and is a centre of expertise for this crop in the EU.

Distinctness, Uniformity and Stability

A. Distinctness: A requirement for cultivar registration in the UK is that all cultivars must be Distinct. To facilitate Distinctness testing, the Turnip Rape collection is classified into groups using five discontinuous characters — erucic acid content, leaf type, time of flowering from an autumn sown trial, ploidy and flower colour.

In the UK, cultivars with a similar breeding background are grouped together and compared only with each other. Groups can therefore be sub-divided into 3 types: Hybrid, Synthetic and Conventional. Cultivars within these sub-groups are distinguished using up to 24 continuous characters (UPOV 1988). The expression of characters which are essential for Distinctness must be uniform (UPOV 1979).

B. Uniformity: Separate standards are applied to the Hybrid, Synthetic and Conventional cross-pollinating sub-groups to reflect their breeding background.

C. Stability: Stability, as such, is not tested, but is assumed if cultivars are Uniform.

Uniformity Problems in Discontinuously-expressed Characters

As breeders had little experience of DUS requirements in Turnip Rape they did not apply any selection pressure to achieve Uniformity in important classification characters. There was, therefore, a high failure rate for candidates in early registration tests.

A. Leaf Type: This is considered to be a simply-inherited single gene character, from which off-types can be easily rogued. Despite this, some early candidates had serious problems with strap-leaved off-types in lobe-leaved cultivars. As a result of discussions between SASA, breeders and test authorities in the EU and Canada, this problem is now recognised and is being addressed.

B. Seed Colour: Seed colour was defined as a discontinuously-expressed character (UPOV 1988). Since then, cultivars with a proportion of yellow seed have been introduced commercially, principally due to increased oil yield. If UPOV standards were applied rigidly, these cultivars would fail the DUS test for lack of Uniformity.

The UK has applied a new character, 'Yellow Seededness', in order to accommodate the large number of partially-yellow-seeded candidates. On a sample of approximately 500 seeds, the percentage of seed with yellow colouration present on the testa is recorded. Registered cultivars are not compared with candidate cultivars in the DUS test if their yellow seededness value falls outwith a 20% tolerance of the candidate percentage. This allows the registration of cultivars with various percentages of yellow seed.

C. Flower Colour: Although some cultivars have failed Uniformity owing to flower colour off-types, this problem is uncommon.

Development and Application of Uniformity Standards

In 1992, the Uniformity standard of a 1% tolerance level at 5% probability was applied in the UK for assessing discontinuous off-types in Conventional Turnip Rape. This practical standard was recognised by breeders and applicants but did not conform to internationally-agreed standards. In order to achieve conformity, the UPOV standard was introduced, replacing the existing UK standard.

The criterion now applied is the mean (statistically adjusted) percent 'off-types' for comparable known cultivars plus a Least Significant Difference (one tailed 5%). The Least Significant Difference (LSD) is derived from an analysis of variance of the tests (trials) X cultivars, and is calculated according to the following formula:

$$\text{LSD } (5\%) = t \sqrt{\frac{\text{Tests} \times \text{Cultivar sMS}}{n_t} \left(1 + \frac{1}{n_c}\right)}$$

where: n_c is the number of comparable known cultivars listed on the UK National List or EU Common Catalogue

$\quad$ n_t is the number of tests

$\quad$ t is the tabulated value of t at 5% level with $(n_t-1)(n_c-1)$ degrees of freedom

In the absence of internationally-agreed Uniformity standards for Hybrids, Synthetics and their parents, a temporary fixed standard of 5% plus a sampling tolerance has been introduced for discontinuous off-types.

Conclusions

There are no internationally-agreed Uniformity standards for testing Hybrid and Synthetic cultivars and their Parent lines in Turnip Rape. Temporary standards have been introduced in the UK until internationally-agreed standards have been set.

Mass selection of early generations has resulted in poor discontinuous and continuous uniformity. Early recognition and removal of discontinuous off-types is essential, particularly in parent or breeding lines, if breeders are to achieve statutory uniformity standards, and register their cultivars.

A fixed standard may well be more practical than the UPOV standard in Turnip Rape where numbers of cultivars are small and their commercial life is short. Fixed standards are predictable and may be applied to any site or trial without the need to grow comparable cultivars.

A revision of the UPOV guideline for Turnip and Turnip Rape is necessary to define clear classification groups to which Hybrids, Synthetics and parental lines can be assigned, and to which Uniformity standards can be applied.

References

UPOV (1979). Revised general introduction to the guidelines for the conduct of tests for Distinctness, Homogeneity and Stability of new varieties of plants. TG/1/2. International Union for the Protection of New Varieties of Plants. Geneva, Switzerland.

UPOV (1988). Guidelines for the conduct of tests for Distinctness, Uniformity and Stability: TG/37/7. Turnip and Turnip Rape. International Union for the Protection of New Varieties of Plants. Geneva, Switzerland.

Abugalieva, S.I., Skokbaev, S.A., Abugalieva, A.I., Turuspekov, Ye. K., Dracheva, L.M. & Savin, V.N. (1999). Classification of wheat and barley cultivars in Kazakstan. In: S. Andrews, A.C. Leslie and C. Alexander (Editors). Taxonomy of Cultivated Plants: Third International Symposium, pp. 461–463. Royal Botanic Gardens, Kew.

CLASSIFICATION OF WHEAT AND BARLEY CULTIVARS IN KAZAKSTAN

S.I. ABUGALIEVA[1], S.A. SKOKBAEV[2], A.I. ABUGALIEVA[3], YE. K. TURUSPEKOV[1], L.M. DRACHEVA[2] AND V.N. SAVIN[3]

[1]Institute of Plant Physiology, Genetics and Bioengineering, 45 Timiryazev st., 480090 Almaty, Kazakstan
[2]Kazakh State Cultivar Trial Committee, 56 Shemyakin st., 480018 Almaty, Kazakstan
[3]Institute of Agriculture, 37 Yerlepesov st., Almalybak, 483133 Almaty region, Kazakstan

In order to set up a database of the cultivated *Triticum aestivum* L. and *Hordeum vulgare* L. from Kazakstan, it is necessary to study and describe its available germplasm. Our goal was the study and description of the cultivar resources of wheat and barley from different regions of Kazakstan using 22 morphological characters according to the UPOV system. Also studied were: protein markers (Peruansky *et al.* 1986), 22 isozyme coding loci of barley (Turuspekov *et al.* 1993) and 11 isozyme coding loci of wheat (Abugalieva *et al.* 1991). Each cultivar was tested on yield and grain quality for 3 years minimum in several ecological localities. Having retrospective data of all the original and tested cultivars identified on 40 morphological traits according to the standards accepted by the State Trial Committee since 1981, we tried to characterise them by using the unified UPOV system. Since 1981, 25 barley cultivars bred in Kazakstan were identified in comparison with 7 standard cultivars. The constancy of such characters as heading (time of appearance of ears), anthocyanin colour of awn tops, filminess (hulled v. naked), and number of rows per spike were recorded. The most differences between the cultivars were found on the flag leaf features.

Unfortunately, due to the lack of some data describing spike, spikelets and flag leaf (## 3, 10, 14, 19–25, UPOV), the testing of all available cultivars were conducted only on 22 out of 33 morphological characters. Almost all the wheat and barley cultivars are morphologically uniform, stable and differ each from another, excluding the identical 'Asyl', 'Azyk', 'Nutans 89' and 'Pastbyshny 1'.

Protein and molecular markers have confirmed a complex genetic structure of cultivar breeding by hybridisation. This complex of characters was not tested for cultivar identification and classification before this.

The majority of cereal cultivars were found to be polymorphic on prolamines separated in polyacrilamid gel (Peruansky *et al.* 1986). The level of polymorphism in both wheat and barley cultivars was different (Abugalieva *et al.* 1996a). From 1 to 4 hordein biotypes and from 1 to 11 gliadin biotypes as well as 1–5 glutenin (on HMW – High Molecular Weight) were observed. The frequency of occurrence of a typical barley biotype ranged from 42% for 'Nutans 86' to 100% for 'Medicum 85', 'Nutans 88', 'Nutans 89' and 'Solontsovy'. Cultivars were variable on their composition of prolamin components. Only 4 barley and 5 wheat cultivars were uniform, the rest of those were found to be heterogeneous. The stability of the component composition varied depending on the environmental conditions. A classification of barley and wheat cultivars on prolamine spectrum was made.

461

Since 1986 more than 70 barley and 40 wheat commercial and tested cultivars grown in Kazakstan were analysed on their isozyme variation (Turuspekov *et al.* 1993, Abugalieva *et al.* 1991, Abugalieva *et al.* 1996a, b, c). 14 enzyme coding loci among the 22 studies were found to be polymorphic in mature seed of barley.

Allozyme analysis showed 31,78% of total variation was intracultivar and 68,21% between barley cultivars. A study of the genetic variation of barley cultivars which were grown in Kazakstan between 1992–1996 allowed us to reveal distribution of alleles of 14 enzyme coding loci in different state trial farms around the country. Geographic distribution of the occurrence of allele frequencies of Prx 2 and Pgd 2 loci were observed by using ANOVA (Analysis of Variance). Obviously, these genetic factors themselves or their linkage with other genes belong to gene associations controlling adaptivity of organisms in definite grown environment. On the other hand, this correlation might be considered as peroxidase and 6-phosphogluconate dehydrogenase themselves taking part in the complex processes of adaptation mechanisms to changes of environmental conditions.

The genetic diversity of 10 commercial barley cultivars was assessed using RAPD UBC primers and PCR. The variation level of DNA was about five times more than that in isozymes. The total quantity of loci on 11 RAPDs was 15,9 with average 10,41 loci per primer.

On the basis of genetic markers and system approach applying for interspecific variation of *Triticum aestivum* and *Hordeum vulgare*, we have received the following results:

- Catalogues of barley and wheat cultivars were created (Abugalieva *et al.* 1996a, c).
- A classification of cultivated barley and wheat:– on technological, development type, agroecotype, morphotype, criptoelement, protein biotypes, near isogenic lines, and dihaploids on the base of variation of ecological-geographical group, local population, cultivar population, biotypes, etc. (Kozhemyakin *et al.* 1994).
- Cultivars of winter bread wheat 'Akterekskaya', 'Anyuta' and 'Kastekskaya' were bred and characterised as uniform, stable and different on morphological, isozyme and protein markers.

References

Abugalieva, S.I., Biyashev, R.M. & Kozhemyakin, Ye.V. (1991). The polymorphism of wheat enzyme proteins. In F. Polimbetova (ed.). Increasing of resistance and productivity of crop plants. pp. 186–191. Science, Alma-Ata. (In Russian).

Abugalieva, A., Turuspekov, Ye. & Abugalieva, S. (1996a). Genetic markers of wheat and barley and their use in breeding. *Novosty Nauki Kazakstana*. N1: 33–37. (In Russian).

Abugalieva, A., Abugalieva, S. & Kozhemyakin, Ye.V. (1996b). System approach in genetic markers in bread wheat breeding and seeds production. 2. Genetic structure of variety. *Vestn. Selskokhoz. Nauki Kazakstana*. N11: 14–24. (In Russian).

Abugalieva, A.I., Turuspekov, Ye.K., Morunova, G.M., Skokbaev, S., Dracheva, L.M., Abugalieva, S. & Savin, V.N. (1996c). Cultivar gene fund of Kazakstan barley *Hordeum vulgare* L. (methods of investigations, identification and catalogue of cultivars). 134 pp. Kazzhol, Alma-Ata. (In Russian).

Kozhemyakin, Ye.V., Abugalieva, A.I., Savin, V.N., Nokolaev, N.N. & Abugalieva, S.I. (1994). The genetic markers and system method in breeding and seed production of bread wheat. 217 pp. Kazzhol, Alma-Ata. (In Russian).

Peruansky, Yu.V., Abugalieva, A.I., Bulatova, K. & Nekhorosheva, L.M. (1986). Gliadin-glutenin biotypes of wheat. *Dokl. VASKhNIL.* N6.

Turuspekov, Ye.K., Biyashev, R. & Abugalieva, S. (1993). Genetical diversity and geographical distribution of isozyme coded genes of barley *Hordeum vulgare* L. *Izv. Akad. Nauk Kazakhstana* N3: 25–29.

Braithwaite, A. & Gregory, P.A. (1999). Determination of odoriferous compounds in plants by direct thermal desorption (DTD) and GC-MS, Part 1: *Artemisia*. In: S. Andrews, A.C. Leslie and C. Alexander (Editors). Taxonomy of Cultivated Plants: Third International Symposium, p. 465. Royal Botanic Gardens, Kew.

DETERMINATION OF ODORIFEROUS COMPOUNDS IN PLANTS BY DIRECT THERMAL DESORPTION (DTD) AND GC-MS, PART 1: *ARTEMISIA*

A. BRAITHWAITE AND P.A. GREGORY

Department of Chemistry and Physics, Nottingham Trent University, Nottingham NG11 8NS, UK. In association with Dr J. Twibell, NCCPG *Artemisia* **Collection Holder, 31 Smith Street, Elsworth, Cambridge CB3 8HY**

Introduction

Many plants produce a range of volatile oils which give the various species their characteristic odours. Each oil comprises many volatile organic compounds (VOCs); the compounds present and the amount of each produces the overall subtle odour of the plant and contributes to the taste of infused drinks. The specific VOCs present and their ratio can vary between species within a particular genus and therefore determination of the individual VOCs present and the amount of each can form the basis of a scientific fingerprinting system for the characterisation of individual plants, determination of parentage and to study the influence of climatic variations on volatile oil composition.

Methods

Various methods have been employed to analyse the VOCs. These include purge and trap, where the volatiles are purged from samples heated to 35°C and collected on a sorbent, and placing a sample in a nylon bag of known volume, volatiles are collected by drawing the enclosed air through a tube containing the sorbent. The volatiles are then analysed by thermal desorption into a GC (gas chromatography) system.

The rapid improved analytical procedure, DTD-GC-MS, avoids the intermediate transfer of the volatiles into a sorbent and the artefacts that can occur from poor recoveries and degradation of both the co-polymer sorbent (e.g. Tenax, poly-2,6-diphenyl-p-phenylene oxide) and the VOCs on the sorbent surface during the thermal desorption process at temperatures of 250–300°C. Pre-weighed samples of the plant were placed directly into a thermal desorption tube, this was then desorbed at 50°C in an automated thermal desorption instrument connected to a GC-MS (gas chromatography–mass spectrometry) system. The separated VOCs were identified and their ratios determined. Comparative chromatograms and data were included for fresh and frozen samples and for purge and trap analyses.

Various data reduction techniques were described which facilitate an easy, rapid comparison of the samples and enable a database of information to be established for fingerprinting plants and species. The procedure is applicable to fingerprinting in many genera and species.

Clennett, C. (1999). A taxonomic review of *Cyclamen* subgenus *Gyrophoebe*. In: S. Andrews, A.C. Leslie and C. Alexander (Editors). Taxonomy of Cultivated Plants: Third International Symposium, p. 467. Royal Botanic Gardens, Kew.

A TAXONOMIC REVIEW OF *CYCLAMEN* SUBGENUS *GYROPHOEBE*

CHRIS CLENNETT

Royal Botanic Gardens Kew, Wakehurst Place, Ardingly, West Sussex RH17 6TN, UK

Research undertaken over two years into the taxa comprising *Cyclamen* L. subgenus *Gyrophoebe* O. Schwarz used morphological, anatomical, palynological and phytochemical techniques of examination on material in cultivation from known source (Clennett 1997). Results in each area of study suggest that the limits of the group have been incorrectly set, and that *Cyclamen coum* Mill. should no longer incorporate the isolated taxon from north Iran, with the latter being re-established at specific rank as *C. elegans* Boiss. & Buhse.

Morphological studies show most of the taxa in the subgenus to be closely allied, forming two clearly defined groups but with three taxa remaining to varying degrees isolated. Palynological examinations show an increase in the number of colpi from three in all taxa of other subgenera to four for most taxa in subgenus *Gyrophoebe*, and a further increase to five colpi for a proportion of individuals within two of the study taxa. Phytochemical studies of flavonols and anthocyanins in the leaves showed many similarities between taxa, but some unexpected differences were also revealed.

Phenetic and cladistic analyses of study results strongly suggest *C. cyprium* Kotschy should be removed from the subgenus, and cladistic analysis further implies *C. libanoticum* Hildebr. should also be removed and with *C. cyprium* form a new grouping, basal to subgenus *Cyclamen*.

A more fully expanded version of this work will be submitted to *Kew Bulletin* in due course.

Reference

Clennett, C. (1997). A taxonomic review of *Cyclamen* L. subgenus *Gyrophoebe* O. Schwarz. Unpublished MSc. Thesis. University of Reading.

Glen, H.F. (1999). The state of horticultural taxonomy in South Africa. In: S. Andrews, A.C. Leslie and C. Alexander (Editors). Taxonomy of Cultivated Plants: Third International Symposium, pp. 469–470. Royal Botanic Gardens, Kew.

THE STATE OF HORTICULTURAL TAXONOMY IN SOUTH AFRICA

H.F. Glen

National Herbarium, Private Bag X101, Pretoria 0001, South Africa

How does one know what plants are grown for the purposes of amenity horticulture, forestry and agriculture in southern Africa? The two best sources of information are surely herbarium specimens and nursery catalogues, with economic reports adding some detail for cash agriculture, but not necessarily subsistence farming. Subsistence crops, which will be significant in terms of disease resistance, adaptation to local conditions and possibly other desirable traits, are known only from herbarium specimens. In this last respect the National Herbarium in South Africa is fortunate to have a tradition, albeit not an extensive one, of studying indigenous uses of plants. There are eight herbaria in South Africa with significant collections of cultivated material, totalling some 40,000 specimens. These are, in decreasing order of size:–

PRE	National Herbarium, Pretoria	21,5000 cultivated specimens
JBG	Johannesburg Botanic Garden Herbarium	6,000 cultivated specimens
GRA	Schonland Botanical Laboratories, Grahamstown	2,500 cultivated specimens
NH	Natal Herbarium, Durban	2,500 cultivated specimens
J	C.E. Moss Herbarium, Johannesburg	2,000 cultivated specimens
NBG	Compton Herbarium, Cape Town	2,000 cultivated specimens
BLFU	University of the Orange Free State, Bloemfontein	1,500 cultivated specimens
PRU	Schweickerdt Herbarium, Pretoria	1,000 cultivated specimens

The remaining cultivated specimens are held in small collections in most South African herbaria. Details of about 32,000 of these are held in a database at the National Herbarium, Pretoria (PRE). Additional facilities of benefit to horticultural taxonomy are the *International Protea Register* (Sadie 1998) at Stellenbosch and the collection of agricultural cultivars at Roodeplaat, near Pretoria. Nursery catalogues are generally hard to come by and necessarily of uncertain accuracy as regards the names used. However there is a small collection of these at PRE.

Information from the specimen database is being supplemented by names from some nursery catalogues to compile a checklist of taxa believed to be cultivated in southern Africa (as far north as Angola). Publication of the first edition of this checklist is scheduled for April 2000. The initial print run will probably be small, so as to allow for the possibility of rapid revisions and extensions. The first edition will include references to literature useful in verifying the names used in the list, and keys helpful in identifying taxa grown in southern Africa, both to genus and to family. Common names are included where known, and these are indexed as well as the genera. It is intended to add places of origin of cultivated taxa to either the first or second edition of this list. In the long term, it is hoped that this list will evolve into a garden flora by successive approximations, each containing more details than the preceding. Reasons for wanting a guide to cultivated plants that is different to *The European Garden Flora* are discussed by Glen (1995a, 1997).

This checklist is seen as the first step towards a southern African garden flora. A second step is the production of precursor treatments of certain groups of trees of horticultural and later silvicultural interest. Although only two of these precursors have so far been published (Glen 1995b, 1996), data have been gathered in collaboration with ENVIRONMENTEK for several large groups, notably *Eucalyptus* L'Hér. and *Pinus* L., the two largest genera of cultivated trees in southern Africa with 320 and 80 species respectively, and the palms (about 300 species). In total, therefore, data on about 40% of the 2,000-odd tree species grown in southern Africa are available in computerised (DELTA) format.

Long-term aims which are seen as desirable and which can be met using these facilities include:

- A 'complete' southern African garden flora, both on paper and as interactive computer files.
- A map, derived from known localities of successful cultivation of plants with different tolerances, showing horticultural zones in the region.
- Identification and information services available to forestry, agriculture, horticulture and the public, stationed where they are most needed.

The second of these is made desirable by the existing maps of this kind in our region. Here we have nothing between the oversimplified thumbnail maps printed on seed packets and that of Poynton (1971), in which so many variables are plotted explicitly on the same map that the result is almost incomprehensible. The others are probably self-evident in the present context.

References

Glen, H.F. (1995a). Light in the darkness: towards a guide to exotic trees grown in southern Africa. *Trees South Africa* 45: 44–50.

Glen, H.F. (1995b). Notes towards a southern African garden flora 1: Introduction, *Nyssaceae, Cornaceae* and *Bixaceae. Trees South Africa* 45: 51–60.

Glen, H.F. (1996). Notes towards a southern African garden flora 2: *Aceraceae. Trees South Africa* 46: 46–61.

Glen, H.F. (1997). The Paper Chase. *SABONET News* 2: 114–118.

Poynton, R.J. (1971). A silvicultural map of southern Africa. *S. African J. Sc.* 67: 58–60.

Sadie, J. (1998). The International *Protea* Register. Ed. 5. 58 pp. Directorate of Plant and Quality Control, Stellenbosch.

Frodin, D.G. (1999). Classification of *Araliaceae*: progress and prospects. In: S. Andrews, A.C. Leslie and C. Alexander (Editors). Taxonomy of Cultivated Plants: Third International Symposium, pp. 471–475. Royal Botanic Gardens, Kew.

CLASSIFICATION OF *ARALIACEAE*: PROGRESS AND PROSPECTS

DAVID G. FRODIN

Herbarium, Royal Botanic Gardens, Kew, Richmond, Surrey TW9 3AE, UK

Introduction

The *Araliaceae* Juss. have acquired a perception in botanical and especially horticultural circles as 'difficult'. On the one hand, there have been continuing problems in establishing – along with *Apiaceae* Lindl. (*Umbelliferae* Juss.) as traditionally circumscribed – its position and affinities among angiosperms as well as its relationship with its sister family; on the other – an issue of more immediate concern to horticulture – generic limits have often been perceived as being in flux. In this presentation I would like briefly to examine these questions and indicate what, if any, progress has been made in recent decades.

Position in the *Magnoliophyta*

Araliales Reveal (or *Apiales* Nakai) historically have mostly been included within *Cornales* Dumort. or in more or less close association with them (Rodriguez 1971, Takhtajan 1980, 1997, Thorne 1992), largely on the basis of floral morphology. This led to their inclusion in or near subclass *Rosidae* Takht. (Takhtajan 1980, Cronquist 1981), although Cronquist (1981, 1988) argued for an association not with *Cornales* but with *Sapindales* Dumort. An alternative view, postulating a close relationship with *Pittosporaceae* R.Br. and – more generally – an association with asterids, was suggested by Dahlgren (1980) on the basis of anatomical, embryological and chemical evidence. These viewpoints were partly reconciled by Takhtajan (1997) wherein *Araliales* (and *Pittosporales* Lindl.) were included in *Cornidae* Frohne & Jensen ex Reveal, a relatively new subclass interposed between *Rosidae* and *Asteridae* Takht. Plunkett *et al.* (1996, 1997) rejected any rosid or cornalean links – although all their aralialean outgroups are members of Takhtajan's *Cornidae*. The Angiosperm Phylogeny Group (1998) placed the order (as *Apiales* Nakai, expanded to include as independent families all of Plunkett, Soltis & Soltis' outgroups) in their 'Euasterids II'; here it is one of three in an unresolved clade also including *Dipsacales* Dumort. and *Asterales* Lindl. The *Cornales*, by contrast, are an outgroup of their main asterid clade, and not close to *Araliales*. Altogether, this represents a distinct advance upon earlier concepts.

Family Limits

Less well resolved is the matter of *Araliaceae* vs. *Apiaceae* (*Umbelliferae*). The two represent a classical 'family pair' (Judd *et al.* 1994): the *Araliaceae* basically tropical, the *Apiaceae* broadly temperate and desertic. Although but weakly supported morphologically and with several demonstratable 'intermediates' (notably *Myodocarpus* Brongn. & Gris (*Araliaceae*); cf. Baumann 1946), over more than 200 years this duality has developed a strong ontology. Proposals to unite the families (Baillon 1878, 1879; Thorne 1973; Judd *et al.* 1994) have generally been resisted; equally, division of *Araliaceae* (Seemann 1868) and separation of *Hydrocotyloideae* Link at family rank from *Apiaceae* (Hylander 1945, Thorne 1983) have not found favour. The latter in particular was increasingly shown to be unnatural (Tseng 1967; Shoup & Tseng 1977; Froebe 1979). Plunkett *et al.* (1996, 1997), with cladograms based upon *rbcL* and *matK* chloroplast gene sequences, argue for recognition of two families with a derived *Apiaceae* and the demonstratably paraphyletic *Hydrocotyloideae* distributed among them (with *Hydrocotyle* L., *Centella* L. and *Spananthe* Jacq. being included in different monophyletic clades within their *Araliaceae*). They admit, however, that support from *rbcL* evidence for a monophyletic (as opposed to a paraphyletic) *Araliaceae* is weak, with low decay and bootstrap values. Internal support also was generally low. Moreover, some traditionally araliad but problematic genera were not included in their analysis, notably *Astrotricha* DC., *Harmsiopanax* Warb., *Oplopanax* (Torr. & A. Gray) Miq. and *Stilbocarpa* (Hook. f.) Decne. & Planch. The potential significance of *Astrotricha* was, however, highlighted by Mitchell & Wagstaff (1997) who showed that for their analyses – based on ITS regions of *r*DNA – it was sister to other *Araliaceae* examined. Of the other genera named, *Harmsiopanax* appears to be morphologically quite isolated, *Stilbocarpa* exhibits several 'umbelliferoid' features, and *Oplopanax* seems to have been overlooked.

Given the weak basis for two families as presented by Plunkett *et al.* (1996, 1997) as well as the distinctiveness of some of the genera not covered by them, there would seem to be two possible directions within existing hierarchical and rank conventions: 1) recognition of a single family following Baillon (1879), Thorne (1973), or Judd *et al.* (1999), or 2) limitation of *Araliaceae* to *Schefflereae* Harms and (part of) *Aralieae* Rchb. *sensu* Harms (Harms 1894–97) with removal of the remaining genera to distinct families, most of the *Hydrocotyloideae* being distributed among these or forming separate clades. Before seriously advancing any new scheme, however, all genera of the present *Araliaceae* and *Hydrocotyloideae* must be included in molecular as well as additional classical studies. Careful consideration should also be given to the suggestion of Plunkett *et al.* (1996) that bicarpelly is plesiomorphic in *Araliales*, given that this is contrary to received wisdom (cf. Baumann 1946).

Generic Limits

The final question is that which has so many people in a muddle: generic limits. A small range of useful floral attributes and limited knowledge of living plants slowed development of useful generic concepts until the mid-nineteenth century. A craze for conservatory plants which then developed in Europe and elsewhere resulted in the introduction of many araliads to cultivation and their availability for study. The majority of our present genera were effectively established through the work of Decaisne & Planchon (1854), Koch (1859), Seemann (1864–68), and Bentham (1867), with a further – and lasting – consolidation by Harms (1894–97) who accepted 51 genera in three tribes. Viguier (1906), utilising an additional range of anatomical attributes, accepted 60 (with some new) in ten tribes; his results were reviewed by Harms (1908). A trickle of further new or segregated genera has followed, with several proposed or revived by Hutchinson (1967).

Hutchinson's reclassification, radical in its recognition of 84 genera in seven tribes, may be seen as an opening to the modern era. Research in several directions was stimulated, with some further light being shed on character trends (cf. Eyde & Tseng 1971; Rodríguez 1971; Tseng & Shoup 1978; Oskolski 1994, 1996). A more dynamic view of characters generally came to be espoused; this facilitated reduction of several segregates to fewer but more broadly conceived genera, notably in *Schefflera* J.R. Forst. & G. Forst. (Frodin 1975), *Polyscias* J.R. Forst. & G. Forst. (Philipson 1979) and *Aralia* L. (Wen 1993). Generic concepts, however, remained predicated on classical notions of overall similarity as well as some sense of pragmatism. Some value continued to be given to certain characters, such as the presence or absence of articulations under the flower. This latter, for example, has helped to deter a union of *Polyscias* and *Gastonia* Comm. ex Lam., while conservatism or insufficient knowledge seems to have stayed a union of *Tetraplasandra* A. Gray with *Gastonia* or *Sciadodendron* Griseb. with an expanded *Aralia* although few real differences are involved. At present I recognise some 50 or so genera; there are presently 1412 published species (Govaerts & Frodin in prep.). Strict application of phylogenetic criteria (and revised family limits) may in future change the first figure.

The greatest diversity of the family is found in monsoonal Asia (including Malesia). Much work has been done in recent years to clarify genera and their relationships in this region. The limits of *Eleutherococcus* Maxim. are, for example, now well understood, although that name has had to take priority over the better-known *Acanthopanax* (Decne. & Planch.) Miq. A close relationship has been observed among *Brassaiopsis* Decne. & Planch., *Grushvitzkya* Skvortsova & Aver. and *Trevesia* Vis. (Jebb 1998), between *Macropanax* Miq. (and *Cromapanax* Grierson) and the Asiatic species referred by Philipson (1965) to *Pseudopanax* K. Koch (Frodin *pers. obs.*), and between *Gamblea* C.B. Clarke and *Evodiopanax* Nakai (Frodin *pers. obs.*). *Aralidium* Miq. and *Diplopanax* Hand.-Mazz. were definitively excluded from the family respectively by Philipson *et al.* (1980) and Eyde & Xiang (1990). In my opinion, the best approach to what has been a complex situation has been to work from the species level upwards, without preconceptions. Problems do remain; examples include a satisfactory understanding of *Brassaiopsis* (H.-J. Esser pers. comm.) or whether or not *Boerlagiodendron* Harms is best revived (Frodin 1998); also needed is a better idea of generic relationships (which may be wanting among the modern flora or, as with *Lardizabalaceae* Decne., found in unexpected quarters). Additional 'good' characters are also desired, it is their apparent lack which has made the resolution of limits and relationships difficult, not only in Asia but elsewhere (e.g. in *Oreopanax* Decne. & Planch.).

It is still too early for there to be much help from gene sequences; phylogenies published so far show little congruence. Some evidence for relationships within and between *Araliaceae* and *Apiaceae* has been secured from chloroplast *rbcL* (Plunkett *et al.* 1996) and *matK* (Plunkett *et al.* 1997) genes; but, as already indicated, resolution was often equivocal and the results in places superficial. Sequences from ITS regions of *rDNA* (Mitchell & Wagstaff 1997, Wen *et al.* 1998, J. Wen *pers. comm.*) appear to yield better results. The work of Mitchell and Wagstaff has, for example, helped significantly to clarify *Pseudopanax*; they showed, for example, that one of its groups – a species of which was once segregated by Seemann as *Raukaua* Seem. – is in fact closer to *Cheirodendron* Nutt. ex Seem. than to *Pseudopanax* proper.

Conclusions

In conclusion, I would like to emphasise that at this stage of our knowledge of *Araliaceae* establishment of a definitive system is premature, not only within its traditional limits but also with respect to *Apiaceae*. While the study of gene sequences should be pursued (with emphasis on araliad and hydrocotyloid genera not so far examined), as much, if not more, attention should also be paid to the careful study of visual characters and probable trends – especially in relation to evolutionary ecology. Gene sequences will simply add another range of characters whose relationship to morphological changes at present remains unclear. At the same time, there should be less overt emphasis on attempts to link together the modern genera and more on re-examining their concepts from species level upwards, without preconceptions.

References

Angiosperm Phylogeny Group. (1998). An ordinal classification for the families of flowering plants. *Ann. Missouri Bot. Gard.* 85(4): 531–553.

Baumann, M.G. (1946). *Myodocarpus* und die Phylogenie der Umbelliferen-Frucht. Umbellifloren – Studies 1. *Ber. Schweiz. Bot. Ges.* 56: 13–112.

Baillon, H. (1878). Recherches nouvelles sur les Araliées et sur la famille des Ombellifères en général. *Adansonia* 12: 125–178.

Baillon, H. (1879). Ombellifères. Histoire des plantes, 7, pp. 84–256. Paris.

Bentham, G. (1867). *Araliaceae*. In G. Bentham & J.D. Hooker. Genera plantarum, 1(3), pp. 931–947. London.

Cronquist, A. (1981). An integrated system of classification of flowering plants. 1262 pp. Columbia University Press, New York.

Cronquist, A. (1988). The evolution and classification of flowering plants. (Ed. 2). 555 pp. The New York Botanical Garden, New York.

Dahlgren, R.M.T. (1980). A revised system of classification of the angiosperms. *Bot. J. Linn. Soc.* 80(2): 91–124.

Decaisne, J. & Planchon, J.-E. (1854). Esquisse d'une monographie des Araliacées. *Rev. Hort.* ser. 4, 3: 104–109.

Eyde, R.H. & Tseng, C C. (1971). What is the primitive floral structure of *Araliaceae? J. Arnold Arbor.* 52(2): 205–239.

Eyde, R.H. & Xiang, Q.-Y. (1990). Fossil mastixioid (*Cornaceae*) alive in eastern Asia. *Amer. J. Bot.* 77(5): 689–692.

Frodin, D.G. (1975). Studies in *Schefflera* (*Araliaceae*): the *Cephaloschefflera* complex. *J. Arnold Arbor.* 56(4): 427–448.

Frodin, D.G. (1998). Notes on *Osmoxylon* (*Araliaceae*), II. *Fl. Males. Bull.* 12(4): 153–156.

Froebe, H.A. (1979). Die Infloreszenzen der Hydrocotyloideen (*Apiaceae*). *Trop. Subtrop. Pflanzenwelt.* 29: 501–679.

Harms, H. (1894–97). *Araliaceae*. In A. Engler & K. Prantl (eds). Die natürlichen Pflanzenfamilien, 3(8): pp. 1–62. Engelmann, Leipzig.

Harms, H. (1908). Verzeichnis der neuen Namen und Beschreibung der neuen Gattungen aus: René Viguier, Recherches anatomiques sur la classification des Araliacées. *Repert. Spec. Nov. Regni Veg.* 6: 45–48.

Hutchinson, J. (1967). *Araliaceae*. The genera of flowering plants, 2: 52–81, 622–624. Oxford University Press, London.

Hylander, N. (1945). Nomenklatorische und systematische Studien über nordische Gefässpflanzen. *Uppsala Univ. Årsskr.* 7: 5–337.

Jebb, M.H.P. (1998). A revision of the genus *Trevesia* (*Araliaceae*). *Glasra* 3: 85–113.

Judd, W.S., Campbell, C.S., Kellogg, E.A. & Stevens, P.F. (1999). Plant systematics: a phylogenetic approach. 464 pp. Sunderland, Massachusetts.

Judd, W.S., Sanders, R.W. & Donoghue, M.J. (1994). Angiosperm family pairs: preliminary phylogenetic analyses. *Harvard Pap. Bot.* 5: 1–51.

Koch, K. (1859). Die Araliaceen im Allgemeinen und Aufzählung der in den Gärten kultivirten Arten. *Wochenschr. Gärtnerei Pflanzenk.* 2: 354–356, 363–367, 370–372.

Mitchell, A.D. & Wagstaff, S.J. (1997). Phylogenetic reationships of *Pseudopanax* species (*Araliaceae*) inferred from parsimony analysis of rDNA sequence data and morphology. *Pl. Syst. Evol.* 208: 131–138.

Oskolski, A.A. (1994). Anatomija drevesiny aralievykh. 105 pp. St. Petersburg. (Proc. Komarov Bot. Inst. 10).

Oskolski, A.A. (1996). Survey of the wood anatomy of the *Araliaceae*. In L.A. Donaldson *et al.* Recent advances in wood anatomy. pp. 99–119. Rotorua, New Zealand.

Philipson, W.R. (1965). The New Zealand genera of the *Araliaceae*. *New Zealand J. Bot.* 3(4): 333–341.

Philipson, W.R. (1979). *Araliaceae,* I. In C.G.G.J. van Steenis (ed.). Flora Malesiana 9(1): pp. 1–105, illus. Alphen aan den Rijn, Netherlands.

Philipson, W.R., Stone, B.C., Butterfield, B.G., Tseng, C.C., Jensen, S.R., Nielsen, B.J., Bate-Smith, E.C., & Fairbrothers, D.E. (1980). The systematic position of *Aralidium* Miq.: a multidisciplinary study. *Taxon* 29(4): 391–416.

Plunkett, G.M., Soltis, D.E. & Soltis, P.S. (1996). Higher level relationships of *Apiales* (*Apiaceae* and *Araliaceae*) based on phylogenetic analysis of *rbcL* sequences. *Amer. J. Bot.* 83(4): 499–515.

Plunkett, G.M., Soltis, D.E. & Soltis, P.S. (1997). Clarification of the relationship between *Apiaceae* and *Araliaceae* based on *matK* and *rbcL* sequence data. *Amer. J. Bot.* 84(4): 565–580.

Rodriguez, R. (1971). The relationships of the Umbellales. In V.H. Heywood (ed.). The biology and chemistry of the *Umbelliferae,* pp. 63–91. London. (*Bot. J. Linn. Soc.* 64, Suppl. 1.).

Seemann, B. (1868). Revision of the natural order *Hederaceae.* 107 pp. London.

Shoup, J.R. & Tseng, C.C. (1977). Pollen of *Klotzschia* (*Umbelliferae*): a possible link to *Araliaceae. Amer. J. Bot.* 64(4): 461–463.

Takhtajan, A. (1980). Outline of the classification of flowering plants (*Magnoliophyta*). *Bot. Rev.* 46(3): 225–359.

Takhtajan, A. (1997). Diversity and classification of flowering plants. 643 pp. Columbia University Press, New York.

Thorne, R.F. (1973). Inclusion of the *Apiaceae* (*Umbelliferae*) in the *Araliaceae. Notes Roy. Bot. Gard. Edinburgh* 32(2): 161–165.

Thorne, R.F. (1983). Proposed new realignments in the angiosperms. *Nordic J. Bot.* 3(1): 85–117.

Thorne, R.F. (1992). An updated phylogenetic classification of the flowering plants. *Aliso* 13(2): 365–389.

Tseng, C.C. (1967). Anatomical studies of flower and fruit in the *Hydrocotyloideae* (*Umbelliferae*). *Univ. Calif. Publ. Bot.* 42: 1–79.

Tseng, C.C. & Shoup, J.R. (1978). Pollen morphology of *Schefflera* (*Araliaceae*). *Amer. J. Bot.* 65(4): 384–394.

Viguier, R. (1906). Recherches anatomiques sur la classification des Araliacées. 210 pp. Paris.

Wen, J. (1993). Generic delimitation of *Aralia* (*Araliaceae*). *Brittonia* 45(1): 47–55.

Wen, J., Shi, S., Jansen, R.K. & Zimmer, E.A. (1998). Phylogeny and biogeography of *Aralia* sect. *Aralia* (*Araliaceae*). *Amer. J. Bot.* 85(6) : 866–875.

Phang, C.I. (1999). A short note on cultivated plant nomenclature in South-East Asia. In: S. Andrews, A.C. Leslie and C. Alexander (Editors). Taxonomy of Cultivated Plants: Third International Symposium, pp. 477–478. Royal Botanic Gardens, Kew.

A SHORT NOTE ON CULTIVATED PLANT NOMENCLATURE IN SOUTH-EAST ASIA

C.I. PHANG

66 Jln. Dato Dollah (aff. Jln. Mastika), 41100 Klang, Selangor, Malaysia

Introduction

Cultivated plant nomenclature is quite neglected and little understood in South-East Asian countries. This is because most plant taxonomists in the region are gainfully employed to attend to the needs of wild plants and timber trees. The scientists associated with cultivated plants are mainly agronomists, geneticists, plant physiologists and other professionals, who are more interested in the plant's performance rather than its name. To them, plant nomenclature ends with providing the scientific name at species or varietal level, probably by referring to an accepted list of plant names. They normally have no interest whatsoever in the proper nomenclature of the plants they work on. A plant list of scientific names is just a necessary aid for the 'identification' of these plants.

In research, however, the plant breeder may be more interested in the physical and genetic characters of his plants; however, he may not have the necessary taxonomic expertise to fully appreciate his findings.

Cultivated plants are often managed by agronomists, geneticists, engineers, plant physiologists and economists, who, as mentioned earlier, have little understanding of the importance of plant nomenclature. This failure to appreciate the expertise of taxonomists in agriculture is to be lamented as there is such a wide range of application for taxonomic know-how especially in the verification of cultivars and clones (or the culton).

Current Practice

In many South-East Asian countries, most cultivated plants or crops, are identified by their scientific names at the species level only. Cultivated plants of great importance to the economy of a country (like paddy, rubber, oil-palm, fruits, some vegetables, groundnuts, tuber crops, etc.), usually have selected cultivars, varieties or clones which are given special 'fancy' names according to their agronomic performance or after their breeder or some-one special like the Prime Minister or Minister or even the farmer from which the original germplasm was obtained. More often, numbers are given to each selection made for ease of identification. The current turnover of new cultivars is not so rapid as in developed countries, to merit the formulation of a special nomenclature for these cultivated plants, except, perhaps, for orchids, where private breeders in Malaysia and Thailand, have developed many fascinating new hybrids.

Most South-East Asian countries do not have a massive breeding programme as in developed countries, since they do not breed cultivars for trade as their developed counterparts do. They have limited human and financial resources which have to be utilised for more important 'bread and butter' activities. Trained skilled experts are limited in numbers and as mentioned earlier, their expertise is centred on more important activities and crops.

Except for certain economically important crops, most plant improvement programmes are concerned with mass selection of cultivated plants. Improved cultivars of short-term crops like vegetables and flowers are imported in seed form. This is sometimes more practical considering the large number of good cultivars available in the trade and also the ease of importation of the relatively light seeds. There is generally little need to indulge in an expensive local breeding programme which is already being undertaken by other more experienced countries.

The Future

Though this situation seems to be optimal under present circumstances, it is too early to predict what the future may bring. With current economic and global changes around the world, one may expect some changes in trends. Foreign companies currently involved in the seed trade may choose to expand their plant breeding activities in South-East Asia — a region little exploited as yet but with a great potential for a hungry market! With a more intense breeding programme and the production of more new varieties and cultivars, it is anticipated that greater interest could be generated for the nomenclature of cultivated plants.

CONSERVATION AND COLLECTIONS

D

Ferguson, A.R. & McNeilage, M.A. (1999). Germplasm for *Actinidia* breeding. In: S. Andrews, A.C. Leslie and C. Alexander (Editors). Taxonomy of Cultivated Plants: Third International Symposium, p. 481. Royal Botanic Gardens, Kew.

GERMPLASM FOR *ACTINIDIA* BREEDING

A.R. FERGUSON AND M.A. MCNEILAGE

HortResearch, Private Bag 92 169, Auckland, New Zealand

HortResearch has the single most comprehensive *ex situ* collection of *Actinidia* Lindl. germplasm in the world with nearly 300 separate accessions of budwood or seed, including representatives of 21 species. There are almost 100 accessions of *A. chinensis* Planch., about 60 accessions of *A. deliciosa* (A. Chev.) Liang & A.R. Ferguson and 10–15 accessions of both *A. arguta* (Siebold & Zucc.) Miq. and *A. polygama* (Siebold & Zucc.) Maxim. In addition there is a collection of nearly 50 cultivars of *A. chinensis* and *A. deliciosa.*

The collection is the basis for much of the research on kiwi fruit and other *Actinidia* species carried out in New Zealand and provides the raw material for breeding. The HortResearch breeding program has almost 40,000 *Actinidia* seedlings occupying 25 ha of land with 20 staff employed. The aim is to provide a number of options for future commercial development.

Liberato, M.C., Vasconcelos, T. & Caixinhas, M.L. (1999). The gymnosperm germplasm collections in some botanic gardens and parks in Lisbon. In: S. Andrews, A.C. Leslie and C. Alexander (Editors). Taxonomy of Cultivated Plants: Third International Symposium, pp. 483–488. Royal Botanic Gardens, Kew.

THE GYMNOSPERM GERMPLASM COLLECTIONS IN SOME BOTANIC GARDENS AND PARKS IN LISBON

M.C. Liberato[1], T. Vasconcelos[2] & M.L. Caixinhas[2]

[1]Jardim-Museu Agrícola Tropical, Instituto de Investigação Científica Tropical
1400 Largo dos Jerónimos, 209 Lisboa, Portugal
[2]Departamento de Protecção das Plantas e de Fitoecologia, Instituto Superior de Agronomia, Tapada da Ajuda, 1349–018 Lisboa Codex, Portugal

Introduction

The botanic gardens and parks in Lisbon are privileged places, by virtue of their particular ecology and for the conservation of living plants from several parts of the world. The germplasm collections of these green areas are a very important heritage to our *ex situ* biodiversity conservation.

The aim of the current research projects in our Institutions is the taxonomic study of these plant collections that are growing in Lisbon. However this paper is only concerned with the gymnosperms at the Jardim Botânico da Ajuda, Jardim-Museu Agrícola Tropical, Tapada da Ajuda and Estufa Fria de Lisboa.

The gymnosperms are a very interesting group of plants in which the ovules are freely exposed and are not enclosed in the ovary. They appeared in the Upper Devonion (Melchior & Werdermann 1954, Stewart & Rothwell 1993) and have been abundant during successive geological periods. They were the dominant plants during the Mesozoic, but are now more restricted and many species have become extinct (Lawrence 1965).

Today most gymnosperms are under environmental pressure around the world. These populations are threatened by two processes: habitat destruction and removal by collectors, this includes logging and medicinal e.g. *Torreya taxifolia* Arn. The International Union for Conservation of Nature and Natural Resources (IUCN) recognizes that some taxa are threatened. Studies concerning threatened conifers were carried out by Farjon *et al.* (1993).

The botanic gardens and parks in Lisbon are enriched by several taxa of gymnosperms, some of which are considered endangered, or vulnerable, rare or indeterminate categories as used by Walter & Gillett (1998).

Material and Methods

The taxonomic study of gymnosperms has been carried out in the following botanic gardens and parks in Lisbon: Jardim Botânico da Ajuda (JBA), Jardim-Museu Agrícola Tropical (JMAT), Tapada da Ajuda (TA) and Estufa Fria de Lisboa (EFL).

The Jardim Botânico da Ajuda was built in 1768 by the order of King D. José I and it is considered the oldest botanic garden in Portugal and the 15th oldest in Europe. A restoration programme was launched in 1994 by the Instituto Superior de Agronomia

(Technical University of Lisbon). In the upper terrace 1100 beds were built to lodge the botanical collections; these species were arranged according to 8 geographic areas while respecting the existing old trees in the garden.

The Jardim-Museu Agricola Tropical was created in 1906 as Jardim-Colonial on another site in Lisbon. In 1914 it was transferred to the present site in Belém and now includes a botanical park of about 50,000 m². Since 1919 it has promoted the introduction of economic and exotic plants, the majority occurring from tropical and subtropical regions. Nowadays it has about 400 perennial species. At the moment it is a Research Centre of the Instituto de Investigação Científica Tropical, and the study, development and the maintenance of these plant collections are among the objectives of this Centre (Instituto de Investigação Científica Tropical 1983).

The Tapada da Ajuda has served as a hunting-ground for kings and aristocrats since the 18th century. Today the botanical area consists of some 80 hectares and about 400 taxa of trees and shrubs. The wild flora of Lisbon is well represented in the Botanic Reserve of D. António Xavier Pereira Coutinho (Coutinho 1956).

The Estufa Fria de Lisboa was opened to the public in 1933 and nowadays it consists of more than one hectare protected by a fence and includes about 250 taxa. It is considered to be a living museum of vegetation. Inside, in a calm and sheltered atmosphere, plants of several origins grow and often produce fruit.

All the living species of gymnosperms were identified using the morphological external characters according to Liberty Hyde Bailey Hortorium (1976), Walters *et al.* (1986), Huxley *et al.* (1992), Tutin *et al.* (1993), Mabberley (1997) and compared with verified herbarium specimens. The author's names follow Brummitt & Powell (1992). A study concerning the propagation of these species has also been carried out.

Results

The IUCN categories, the wild origin and uses of the gymnosperm taxa that are growing in some parks and gardens in Lisbon are presented in Table 1.

It was verified that plants of *Abies alba* Mill., *Afrocarpus mannii* (Hook. f.) C.M.Page, *Agathis robusta* (C. Moore ex F.Muell.) F.Muell., *Araucaria bidwillii* Hook., *Cedrus deodara* (Roxb.) G.Don, *Chamaecyparis lawsoniana* (A.Murray bis) Parl, *Cryptomeria japonica* (Thunb. ex L.f.) D.Don, *Cupressus arizonica* Greene, *Cupressus lusitanica* Mill., *Cupressus macnabiana* A.Murray bis, *Cupressus macrocarpa* Hartw. ex Gordon, *Cupressus sempervirens* L. var. *horizontalis* (Mill.) Loudon, *Cupressus sempervirens* var. *sempervirens*, *Ginkgo biloba* L., *Juniperus virginiana* L., *Pinus canariensis* C.Sm., *Pinus halepensis* Mill. ssp. *halepensis*, *Pinus nigra* J.F.Arnold, *Pinus pinaster* Aiton, *Pinus pinea* L., *Pinus sylvestris* L., *Platycladus orientalis* (L.) Franco, *Sequoia sempervirens* (D.Don) Endl., *Taxus baccata* L., *Tetraclinis articulata* (Vahl) Mast. and *Thuja occidentalis* L. have viable seeds.

Some specimens of other species are easily vegetatively propagated such as *Cycas revoluta* Thunb., *Encephalartos* spp. and *Ephedra* spp.

Other species living in the above mentioned localities are dioecious, with only the male or the female specimen present, e.g. *Podocarpus* spp., *Encephalartos* spp. and *Cephalotaxus* spp. For this reason they cannot reproduce themselves. Lastly, several specimens are only young plants.

TABLE 1. Gymnosperm germplasm collections in Lisbon. IUCN Red List Categories: E – Endangered, I – Indeterminate, R – Rare, V – Vulnerable

TAXA	IUCN Categories	ORIGIN	USE	DISTRIBUTION in some green areas in Lisbon
CYCADOPSIDA				
CYCADALES				
Cycadaceae				
Cycas revoluta Thumb.		E Asia	food, ornamental	JBA, JMAT, TA, EFL
Zamiaceae				
Ceratozamia mexicana Brongn.	I	Mexico	ornamental	TA, EFL
Dioon edule Lindl.	R	Mexico	ornamental	JMAT
Encephalartos altensteinii Lehm.	V	South Africa	ornamental	JMAT
Encephalartos horridus (Jacq.) Lehm.	V	South Africa	ornamental	JMAT, EFL
Encephalartos lebomboensis Verd.	R	SE Africa	ornamental	JMAT
Encephalartos lehmannii Lehm.	R	South Africa	ornamental	JMAT
Encephalartos paucidentatus Stapf & Burtt Davy	V	South Africa	ornamental	JMAT
Encephalartos villosus Lem.	V	South Africa, Mozambique	ornamental	JMAT
Lepidozamia peroffskyana Regel		E Australia	ornamental	JBA
GINKGOALES				
Ginkgoaceae				
Ginkgo biloba L.		China	food, medicinal, soap, ornamental, timber	JBA, JMAT, TA
GNETOPSIDA				
GNETALES				
Ephedraceae				
Ephedra americana Humb. & Bonpl.		Andes (Ecuador to Patagonia)		JBA
Ephedra distachya L.		S Europe to Siberia	medicinal	JBA
Ephedra fragilis Desf.		Mediterranean	medicinal	JBA
PINOPSIDA				
CONIFERALES				
Araucariaceae				
Agathis robusta (C. Moore ex F. Muell.) F. Muell.		Australia	resin timber	JMAT
Araucaria angustifolia (Bertol.) Kuntze	R	Brazil	food, ornamental, timber	JMAT, TA, JBA

TABLE 1 continued

TAXA	IUCN Categories	ORIGIN	USE	DISTRIBUTION in some green areas in Lisbon
Araucaria bidwillii Hook.		NE Australia	food, ornamental, timber	JMAT, TA, JBA
Araucaria columnaris (J.R.Forst.) Hook.		New Caledonia, New Hebrides (Polynesia), Pine Island	ornamental	JMAT
Araucaria cunninghamii Aiton ex D. Don		New Guinea, Australia (Queensland)	ornamental, timber	JBA, JMAT, TA
Araucaria heterophylla (Salisb.) Franco	V	New Zealand (Norfolk Island)	ornamental	JBA, JMAT, TA
Cephalotaxaceae				
Cephalotaxus harringtonii (Knight ex J. Forbes) K. Koch var. *drupacea* (Siebold & Zucc.) Koidz.		Japan, C China	oil, ornamental	JMAT
Cupressaceae				
Callitris macleayana (F. Muell.) F. Muell.		Australia	essential oil, resin, tannin, timber	JBA
Callitris rhomboidea R. Br. ex Rich. & A. Rich.		SE Australia	essential oil, resin, tannin, timber	JBA
Calocedrus decurrens (Torr.) Florin		USA	ornamental	JBA
Chamaecyparis lawsoniana (A. Murray bis) Parl.	R	USA	ornamental	JBA, JMAT
Chamaecyparis obtusa 'Nana'		Cultivated	ornamental	EFL
Cupressus arizonica Greene	V/R	USA	timber	JMAT, TA
Cupressus lusitanica Mill.		Mexico, Guatemala	ornamental, timber	JBA, JMAT, TA
Cupressus macnabiana A. Murray bis	R	USA	ornamental	JMAT
Cupressus macrocarpa Hartw. ex Gordon		USA	ornamental, timber	JBA, JMAT, TA
Cupressus sempervirens L. var. *horizontalis* (Mill.) Loudon		S Europe, Mediterranean region, SW Asia	essential oil, ornamental, timber	JBA, JMAT, TA
Cupressus sempervirens L. var. *sempervirens*		SW Asia	essential oil, ornamental, timber	JBA, JMAT, TA
Juniperus cedrus Webb & Berthel.	E	Canary Islands	essential oil, ornamental	JBA
Juniperus chinensis L.		Himalaya, China, Japan	essential oil, ornamental	JBA, JMAT, TA
Juniperus turbinata Guss. ssp. *turbinata*		W Mediterranean region	essential oil, ornamental	JBA
Juniperus virginiana L.		N America	essential oil	TA
Platycladus orientalis (L.) Franco	R	N China	ornamental	JBA, JMAT, TA
Tetraclinis articulata (Vahl) Mast.	R	NW Africa	resin, timber	JBA, TA
Thuja occidentalis L.		N America	ornamental	JBA, JMAT
Pinaceae				
Abies alba Mill.		C & S Europe	medicinal, ornamental, timber	TA
Cedrus atlantica (Endl.) Manetti ex Carrière		Atlas Mountains	ornamental, timber	JMAT, TA
Cedrus deodara (Roxb.) G. Don		Himalaya	ornamental, timber	JMAT, TA

TABLE 1 continued

Gymnosperm germplasm collections in some botanic gardens and parks in Lisbon

TAXA	IUCN Categories	ORIGIN	USE	DISTRIBUTION in some green areas in Lisbon
Larix decidua Mill.		Europe (Alps, W Carpathians)	medicinal, timber	JBA
Picea abies (L.) H. Karst.		N Europe to N Greece & Bulgaria		JMAT, TA
Picea smithiana (Wall.) Boiss.		Himalaya	timber	JBA, JMAT
Pinus banksiana Lamb.		N America	timber	JBA
Pinus canariensis C. Sm.		Canary Islands	timber	JBA, TA
Pinus halepensis Mill. ssp. *halepensis*		Mediterranean region	ornamental	TA
Pinus mugo Turra		Europe		JBA
Pinus nigra J.F. Arnold		S Europe, SW Asia	ornamental, timber	JMAT
Pinus pinaster Aiton		Europe	ornamental, timber	JBA
Pinus pinea L.		Mediterranean region	food, ornamental	JMAT, TA
Pinus ponderosa C. Lawson & Lawson		N America	timber	JBA
Pinus radiata D. Don		USA	timber	TA
Pinus sylvestris L.		Europe	medicinal	TA
Pinus teocote Schltdl. & Cham.		Mexico	medicinal	JMAT
Pinus wallichiana A.B. Jacks.		Himalaya	timber	JBA, TA
Podocarpaceae				
Afrocarpus mannii (Hook. f.) C.N. Page		S. Tome Islands (endemic)	ornamental	JBA, JMAT
Nageia nagi (Thunb.) Kuntze		China, Taiwan, Japan	ornamental	JBA
Podocarpus laurencii Hook. f.		SW Australia	ornamental	JMAT
Podocarpus macrophyllus (Thunb.) Sweet		China, Japan	ornamental	JMAT, JBA
Podocarpus cunninghamii Colenso		New Zealand (endemic)	ornamental	JMAT
Taxodiaceae				
Cryptomeria japonica (Thunb. ex L. f.) D. Don		C & S Japan	ornamental, timber	JBA, JMAT, TA
Metasequoia glyptostroboides Hu & W.C. Cheng		China	ornamental	JBA
Sequoia sempervirens (D. Don) Endl.		N America	ornamental, timber	JMAT
Taxodium distichum (L.) Rich.		USA	timber	JBA, TA
TAXOPSIDA				
TAXALES				
Taxaceae				
Taxus baccata L.		Europe, N Africa	medicinal	JBA, JMAT, TA, EFL

Conclusions

The following conclusions were reached:
- Within the areas studied, taxa of gymnosperms were found from all over the world
- Sixty-six taxa were identified that provided a wide range of uses to mankind. These included food, timber, essential oils, resin, tannin, soap and ornamental plants
- Among the taxa studied were found: 1 *Endangered*, 1 *Indeterminate*, 9 *Rare* and 6 *Vulnerable*.
- Endemic to a single country were *Afrocarpus mannii* and *Podocarpus cunninghamii*.
- The role of the institutes concerned is considered relevant with regard to the conservation of *ex situ* species, particularly those *endangered, rare* or/and *vulnerable* in their original region. These botanic gardens and parks in Lisbon are extremely important for research and instructive programmes.

References

Brummitt, R.K. & Powell, C.E. (1992). Authors of Plant Names. 732 pp. Royal Botanic Gardens, Kew.

Coutinho, M.P. (1956). A Tapada da Ajuda. Da Tapada Real a Parque Botânico. *Agros.* 19: 137–144.

Farjon, A., Page, C.N. & Schellevis, N. (1993). A preliminary world list of threatened conifer taxa. *Biodiversity Conservation* 2: 304–326.

Huxley, A., (ed.). (1992). The New Royal Horticultural Society Dictionary of Gardening. 4 vols. The Macmillan Press, London.

Instituto de Investigação Científica Tropical. (ed.). (1983). O Jardim-Museu Agrícola Tropical. Da Comissão de Cartografia. Tropical. 100 anos de História. pp. 181–193. Instituto de Investição Científica Tropical, Lisboa.

Lawrence, G. (1965). Taxonomy of Vascular Plants. (Ed. 10) 823 pp. The Macmillan Company, New York.

Liberty Hyde Bailey Hortorium. (eds). (1976). Hortus Third. A concise dictionary of plants cultivated in the United States and Canada. 1290 pp. Macmillan, New York.

Mabberley, D.J. (1997). The Plant-Book. (Ed. 2) 858 pp. Cambridge University Press, Cambridge.

Melchior, H. & Werdermann, E. (1954). A. Engler's Syllabus der Pfanzenfamilien (Ed. 12). Vol. I. 367 pp. Gebrüder Borntraeger, Berlin.

Stewart, W.N. & Rothwell, G.W. (1993). Paleobotany and evolution of plants. (Ed. 2) 521 pp. Cambridge University Press, Cambridge.

Tutin, T.G., Burges, N.A., Chater, A.O., Edmondson, J.R., Heywood, V.H., Moore, D.M., Valentine, D.H., Walters S.M. & Webb, D.A. (eds). (1993). Flora Europaea. (Ed. 2) Vol. I. 574 pp. Cambridge University Press, Cambridge.

Walter, K.S. & Gillett, H.J. (eds). (1998). 1997 IUCN Red List of Threatened Plants. 862 pp. IUCN Gland, Switzerland & Cambridge, UK.

Walters, S.M., Brady, A., Brickell, C.D., Cullen, J., Green, P.S., Lewis, J., Matthews, V.A., Webb, D.A., Yeo, P.F. & Alexander, J.C.M. (eds). (1986). The European Garden Flora. Vol. I. 430 pp. Cambridge University Press, Cambridge.

McKendrick, M.E. (1999). The National Collection of Japanese anemones. In: S. Andrews, A.C. Leslie and C. Alexander (Editors). Taxonomy of Cultivated Plants: Third International Symposium, p. 489. Royal Botanic Gardens, Kew.

THE NATIONAL COLLECTION OF JAPANESE ANEMONES

M.E. McKendrick

4 Knighton Road, Otford, Sevenoaks, Kent TN14 5LF, UK
(formerly of Hadlow College, Hadlow, Tonbridge, Kent TN11 OAL, UK

The National Collection of Japanese anemones, under the auspices of the National Council for the Conservation of Plants and Gardens, is held jointly, by Hadlow College where the plants are incorporated into the ornamental grounds, and by Miss McKendrick, who holds a check collection in pots

The collection is based on plants derived from the cross *Anemone hupehensis* (Lemoine) Lemoine var. *japonica* (Thunb.) Bowles & Stearn × *Anemone vitifolia* Buch.-Ham. ex DC. It also includes *A.* × *hybrida* 'Elegans', with other hybrid cultivars, as well as *A. vitifolia*, *A. tomentosa* (Maxim.) P'ei, *A. hupehensis* and their cultivars.

The aim of the collection is to maintain stocks of all available taxa which can be used for propagation and comparison, while a constant search is maintained for new cultivars and old, forgotten forms. It is pleasing to record that two cultivars, believed lost, have been recovered and several more rescued as they were disappearing from the nurserymen's lists.

The detailed records that need to be kept prove an excellent opportunity for monitoring pest and disease resistance, reliability and ease of propagation, while the fact that the collection can be made open to the public allows an assessment of popularity. There are currently 60 taxa in the collection, and 27 of these correspond to the earliest descriptions and are assumed to be correctly named.

Pattison, G.A. & Cook, L.B. (1999). The National Plant Collections scheme: an important tool in plant conservation. In: S. Andrews, A.C. Leslie and C. Alexander (Editors). Taxonomy of Cultivated Plants: Third International Symposium, p. 491. Royal Botanic Gardens, Kew.

THE NATIONAL PLANT COLLECTIONS SCHEME: AN IMPORTANT TOOL IN PLANT CONSERVATION

G.A. PATTISON AND L.B. COOK

National Council for the Conservation of Plants & Gardens, Stable Courtyard, Wisley, Woking, Surrey, GU23 6QP, UK

Britain has always played a vital role in the furtherance of science, not purely through its institutes, but also through the enthusiasm and dedication of the amateur. This is clearly portrayed by the work of the National Council for the Conservation of Plants & Gardens (NCCPG).

Set up some twenty years ago, its mission is to 'conserve, document, promote and make available Britain's great biodiversity of garden plants for the benefit of horticulture, education and science'. It has sought to achieve this by channelling its efforts through a country-wide volunteer base and through its National Collections Scheme.

The latter has successfully focused the enthusiasm of its participants into well-documented and researched plant collections, each vetted to comply with high standards of horticulture. Mainly genus-based, these Collections are not only expected to hold as wide a representation of the genus or plant group as is feasible, but to include descriptive and photographic plant records with specific acquisition, historical, reference and cultivation detail. The final complement of herbarium specimens of each plant, with botanical descriptions and colour coding, enhances these important research bases.

Today the NCCPG holds in excess of 600 registered Collections dispersed over 420 sites and spanning over 50,000 taxa within 321 genera. The 120 duplicate Collections within this figure further safeguard the genetic base against the causal effects of man and nature.

It is heartening that the escalating impetus and development of the Scheme has encouraged the evolvement of similar initiatives overseas.

Rodríguez-Acosta, M. (1999). Introducing Mexican oaks to cultivation. In: S. Andrews, A.C. Leslie and C. Alexander (Editors). Taxonomy of Cultivated Plants: Third International Symposium, pp. 493–496. Royal Botanic Gardens, Kew.

INTRODUCING MEXICAN OAKS TO CULTIVATION

MARICELA RODRÍGUEZ-ACOSTA

Herbario y Jardín Botánico,Benemerita Universidad Autónoma de Puebla, Ed. 76. C.U. 14 Sur. esq. Av. San Claudio, Puebla, México

Introduction

It is well known that the genus *Quercus* L. reaches its greatest diversity in Mexico. Nearly 200 species are distributed throughout its territory, occupying all types of habitat. Although oaks are widely distributed and diverse in the wild, they are only rarely seen in cultivation and it is unusual to see oaks in Mexican botanical gardens, unless they are part of the natural flora, and even more unusual to see them used in public parks or streets.

The main reason that oaks are infrequently cultivated is that they have a reputation for slow growth. This, however, is only partially true, as some grow much faster than others. Also the complexity of *Quercus* taxonomy makes it difficult to recognise the species and many hybrids.

Puebla University and Louise Wardle de Camacho Botanical Garden are working together to establish an Oak National Collection and during the last three years have cultivated about 100 species. Some of these show good ornamental features, like fast growth, hardiness and attractive foliage. A particularly interesting group of hybrids are those between *Q. acutifolia* Née and *Q. mexicana* Benth., some which have leaves turning a beautiful purple during winter. We are only just starting to work on selecting, propagating and naming good forms that have potential as ornamental plants in Mexico.

Oak Distribution in Mexico

In Mexico *Quercus* species are called "encinos" and they occur in almost all the states excluding only Yucatán and Quintana Roo. Most species are concentrated in the high mountains but they also extend to arid and tropical areas. They occur mainly in oak-pine forest where they are a major component in terms of biomass, and range from low rhizomatous shrubs on dry slopes and high mountain peaks to massive forest trees with buttress roots in wet lowland forests.

As a result of herbarium work, the occurrence and distribution of 194 oak species have been recorded. The richest states in species diversity are Chihuahua, Oaxaca, Nuevo León, Jalisco, Durango, Veracruz, Hidalgo, San Luis Potosí, Tamaulipas, Puebla, Guanajuato, Michoacán, estado de Mexico and Chiapas, all of which have 40 or more species recorded (Table 1). Chihuahua in the north and Oaxaca in the south are by far the most important, with a total of 100 species, 16 of which are found in both states.

TABLE 1. Mexican states with the richest number of oak species.

STATE	NUMBER OF SPECIES
Chihuahua	62
Oaxaca	62
Nuevo León	55
Jalisco	55
Durango	53
Veracruz	50
Hidalgo	48
San Luis Potosí	46
Coahuila	43
Tamaulipas	43
Puebla	42
Guanajuato	41
Michoacán	40
Edo. de Mexico	40
Chiapas	40

Oak Cultivation

In spite of their abundance and diversity in Mexico, oaks are not used as ornamental trees. Indeed, field workers consider oaks as a pest in the pine-oak forests, so they eliminate them and give priority to pines.

However oaks are of exceptional value for high quality timber and provide numerous products including firewood, charcoal, cork, tannins, dyes, food for humans and livestock, ornamental shade trees, and habitats for wildlife and human recreation.

In 1995 the Mexican Association of Botanical Gardens invited its garden members to participate in the establishment of a national collection of plants, and thereby contribute to the conservation effort taken by the association and the Mexican government. The gardens mentioned above decided to establish a *Quercus* collection with the aim of growing as wide a range of oak species as possible and in this way to increase their knowledge of the genus in Mexico.

So far 100 species have been grown, but only 85 have succeeded in cultivation. It has been shown that oaks do not grow as slowly as is often believed and that white oaks often grow faster than red oaks under the same conditions.

Some of the fast growing species are: *Q. deserticola* Trel., *Q. germana* Schltdl. & Cham., *Q. lancifolia* Benth., *Q. magnoliifolia* Née, *Q. polymorpha* Schltdl. & Cham., *Q. rhysophylla* Weath. and *Q. rugosa* Née.

Oak Hybridisation

Another reason why oaks are not grown in Mexico is the complex taxonomy of the genus and the occurrence of many hybrids. Species concept greatly affect estimates of diversity for any group. Oaks are often cited as somewhat unusual because of the great number of reported natural hybrids in the genus. Such hybrids often occur between species that are morphologically and phyllogenetically distant (Nixon 1993).

In Puebla, two red oaks *Q. acutifolia* and *Q. mexicana* are common and where they grow together often produce hybrids. Some of the offspring between these species give plants which have leaves turning a beautiful purple during winter. Work on selecting, propagating and naming good forms that have potential as ornamental plants in Mexico is only just starting. We are working closely with Mr Allen Coombes of the Sir Harold Hillier Gardens and Arboretum in the UK and hope there will be more interesting discoveries in the future.

Oak Conservation

Most of the oak forest in Mexico is located in places suitable for human settlement, and the impact of their activity is very strong. The forest is being destroyed very quickly and is mostly cut for charcoal and fuel . As a result of this, not only are habitats being lost but many species of oak are now threatened. The recent 1997 IUCN Red List of Threatened Plants (Walter & Gillett 1998) includes the following 44 Mexican oak species and one hybrid (Table 2). In it 12 species and one hybrid are included in a indeterminate category and 19 are considered as rare species. 7 of them are considered vulnerable and 5 more are endangered.

TABLE 2. Threatened *Quercus* in Mexico from Walter & Gillett (1998).

Indeterminate

Q. × basaseachicensis C.H. Mull.
Q. cedrosensis C.H. Mull.
Q. chuhuichupensis C.H. Mull.
Q. convallata Trel.
Q. excelsa Liebm.
Q. galeanensis C.H. Mull.
Q. gravesii Sudw.
Q. liebmannii Oerst. ex Trel.
Q. miquihuanensis Nixon & C.H. Mull.
Q. praeco Trel.
Q. subspathulata Trel.
Q. toumeyi Sarg.
Q. undata Trel.

Q. martinezii C.H. Mull.
Q. peninsularis Trel.
Q. planipocula Trel.
Q. prainiana Trel.
Q. rhysophylla Weath.
Q. skinneri Benth.
Q. skutchii Trel.
Q. tomentella Engelm.
Q. uxoris McVaugh

Vulnerable

Q. crispipilis Trel.
Q. depressipes Trel.
Q. dumosa Nutt.
Q. fulva Liebm.
Q. hinckleyi C.H. Mull.
Q. hintoniorum Nixon & C.H. Mull.
Q. macdougallii Martínez

Rare

Q. benthamii A. DC.
Q. brandegeei Goldman
Q. crispifolia Trel.
Q. deliquescens C.H. Mull.
Q. depressa Bonpl.
Q. devia Goldman
Q. engelmannii Greene
Q. germana Schltdl. & Cham.
Q. hypoxantha Trel.
Q. invaginata Trel.

Endangered

Q. hintonii E.F. Warb.
Q. sebifera Trel.
Q. vicentensis Trel.
Q. xalapensis Bonpl.
Q. zempoaltepecana Trel.

One of the aims of our work is to increase public knowledge and awareness of oaks through courses and workshops. One course in Puebla involved participants from 20 different institutions, while tree planting courses for people from the regions have been very successful in increasing awareness of the importance of oak forest conservation.

References

Nixon, K. (1993). The genus *Quercus* in Mexico. In T.P. Ramamoorthy, R. Bye & J.E. Fa (eds). Biological Diversity of Mexico, pp. 447–458. Oxford University Press.

Walter, K.S. & Gillett, H.J. (eds). (1998). *Quercus*. The 1997 IUCN Red List of Threatened Plants, pp. 278–280. The IUCN Species Survival Commission.

Zupancic, A. & Baricevic, D. (1999). Genepool collection of medicinal and aromatic plants in Slovenia. In: S. Andrews, A.C. Leslie and C. Alexander (Editors). Taxonomy of Cultivated Plants: Third International Symposium, pp. 497–500. Royal Botanic Gardens, Kew.

GENEPOOL COLLECTION OF MEDICINAL AND AROMATIC PLANTS IN SLOVENIA

A. ZUPANCIC[1] AND D. BARICEVIC[2]

[1]M-KZK Kmetijstvo Kranj, Mlakarjeva 70, 4208 Sencur, Slovenija
[2]Biotehnical Faculty, Agronomy Department, Jamnikarjeva 101, 1000 Ljubljana, Slovenija

Introduction

Slovenia is a small Central European country, and despite its size, it is very rich in plant diversity. Among the species, which are regarded as endangered, are many medicinal and aromatic plants (also known as MAP). This mainly due to the massive exploitation from nature and the invasion of natural habitats. The expanding use of chemicals in agriculture and the change of climate also count as reasons for loosing the biodiversity of MAP. To preserve the germplasm of some of these plants, the gene bank for MAP has been set up in 1994. Within the genebank, 640 accessions of plants from world gene banks and from our natural habitats, especially *Salvia officinalis* L., *Origanum vulgare* L. ssp. *vulgare*, *Hypericum perforatum* L., *Gentiana lutea* L. ssp. *symphyandra* Murb. and *Cynara scolymus* L. are held. The plants are maintained as seeds at 5°C, as plantations *ex situ* and as *in vitro* culture. Different propagation techniques are considered with the objectives of maintaining accessions and of increasing the plants for future field production. Together with *ex situ* conservation, the monitoring of natural populations and *in situ* conservation of authochthonous material is made. A multi-user relational database MEDPLANT has been put forward for collecting and evaluating the data, while the content and biological activity of secondary metabolites of plants, grown in different habitats is investigated.

Phytogeography of Slovenia

The Republic of Slovenia is located between the northern 45° and 46° parallel and between the 13° and 16° eastern meridian. The surface area measures 20,251 km² and it has 1,965,986 inhabitants. In the north and north-western part of Slovenia lie the Alps, in the southwest the Adriatic Sea, in the east the Pannonia Plain, and in the south there are the Dinaric Mountains. The diversity of the Slovene flora is due to the diversity of soil and climate. There are four different types of flora in Slovenia: the Central European Alpine (30% of the territory) in the north, the Mediterranean in the south west (10%), the Pannonian in the east (30%) and the Illiric-Dinaric in the south (30%) (Cerne *et al.* 1996). According to the division made by a group of botanists from the Floristic Section of the Biological Society, seven phytogeographical regions can be found in Slovenia (Zupancic & Zagar 1995).

Of approximately 3,000 plant species known to be authochthonous or well adapted to the Slovenian climate, about 10% are estimated to be endangered (34 are extinct, 77 vulnerable, 192 are rare) (Wraber & Skoberne 1989).

The extremely heterogenous natural conditions in Slovenia account, together with the common domestic policy and small farm size, for the different development of the Slovenian agricultural branches. In Slovenian agriculture more than half of the production gross income is derived from animal husbandry (54.5%) followed by crop production (39.1%), fruit growing (3.1%) and viticulture (3.3%). Presently MAP are cultivated in Slovenia on 20 ha only (Baricevic 1997).

In situ Preservation

In Slovenia there are over 100 authochthonous species with potential medicinal properties, and the preservation of their natural resources is necessary.

The first stage of the programme on the production, processing and quality control foresees the monitoring of natural populations, their characterisation, and *in situ* conservation of evaluated and authochthonous plant material. In obtaining distribution data on wild accessions of MAP in Slovenia, the herbarium at Ljubljana literature data and floristic/vegetation inventory are used.

Conservation of this natural heritage for future generations and attractive landscapes is the first aim of the conservation strategy of MAP, and this is planned to be achieved by:

- Inventory of natural resources and the estimation of their vulnerability in nature.
- Their active conservation (conservation *in situ*).
- Reasonable and sustainable use of natural resources (limited on exploitation of germplasm for *ex situ* genebank).
- Prevention of massive exploitation of natural resources through the successive introduction of cultivation of known genotypes in a suitable environment.

Ex situ Preservation

In order to ensure genepools for future investigations, the genebank (a National Genebank Collection of medicinal and aromatic plants), that contains 640 authochthonous or foreign/introduced medicinal and aromatic plant accessions, was set up in 1994. This was officially recognised in 1995 and is annually supported by the Slovenian government. The *ex situ* gene bank aims at:

- Maintenance of germplasm (as seed, as *in vitro* and as *in vivo* collections), seed propagation (for cultivation purposes).
- Evaluation of morphological and chemotaxonomic characteristics of MAP.
- Evaluation of medicinal and aromatic plant ecotypes for quantitative and qualitative differences in secondary metabolites with regard to growth and development.
- Evaluation of susceptibility of germplasm descendants to environmental stress (drought, low temperature, impoverished soils, etc.) in pot trials under the conditions of controlled environment.

Because of high morphological and chemical variability in medicinal and aromatic plants, *in vitro* culture technique has been applied for practical purposes. Thus, the screening of optimal (considering velocity, morphological uniformity and low cost input) *in vitro* conditions is considered. The genebank accessions are successively introduced in the micropropagation procedure, where sterilisation of plant material, culture media and rooting capacity are observed. After 4 years of experience, the micropropagation of medicinal and aromatic plants has been recognised as an essential tool in obtaining homogenous descendants in foreign pollinator species, in low-rate and long-period germinating species as well as in virus-infected plant material.

Evaluation and Documentation

For efficient analysis of parameters, that influence the quality of MAP, the available unidimensional databases were not sufficient. So, in order to manage the extensive data/information network, derived from studies of each of the above mentioned sectors, our working group has recently developed a relational database, named MEDPLANT (Baricevic *et al.* 1994 & 1996).

The main objective of the MEDPLANT relational database is the screening of high potential genotypes for selection/breeding work and for the production of raw material requested by processing industries. The MEDPLANT system works on relational databases principles i.e. collected data are arranged according to their characters in appropriate databases (systematics, geography, habitat, pedology, phytocoenosis, chemical analyses, varieties/cultivars). Pointers, which connect specific data in these databases allow establishment of mutual relationships between required data in the network.

The advantages of this database, which offers user-friendly managing of received data are:

- Rapid and easy input or output of desired/required data.
- Large database dimension enables fast processing of large data mass.
- MEDPLANT is a multi-user system for network.
- Open system enables addition of new data types without changing present structure.
- MEDPLANT application is opened for the addition of processing modules (models).
- MEDPLANT graphic module enables the presentation of geographical data as maps as well as presentation of pictures and the establishment of their correlations with inscribed data.

Future Plan of Action

In situ inventarisation/monitoring and estimation of population density as well as their quality control with the objective of defining the particular/optimal market use of raw material, are planned to be done by the multi-user relational data base MEDPLANT. Considering the principle of sustainable use, the natural populations will be included in the National Collection of medicinal and aromatic plants, when further activities (multiplication of plant materials and breeding of seed/seedling) needed for future cultivation are foreseen. One of the most urgent actions in the near future is to prepare an official list of descriptors for MAP. These represent the documentation background for future selection work and prebreeding of plant material in the direction of uniformity of chemical characteristics of plant raw material needed by the food/pharmaceutical industries.

References

Cerne, M. & Kraigher, H. (1996). Slovenia: Country report to the FAO International technical conference on plant genetic resources, Leipzig, 1996, FAO. The state of the world's plant genetic resources for food and agriculture, CD-ROM.

Baricevic, D. (1996). Biodiversity preservation on the field of medicinal and aromatic plants — the role in development of new herbal remedies. In I. Maher (ed.). Proceedings of International Symposium Agriculture that preserves biodiversity. pp. 113–119. Slovenian Fund for Nature, Ljubljana.

Baricevic, D. (1997). Plant genetic resources in Slovenia, with special reference to neglected medicinal and aromatic plants. In L. Mont (ed.). Neglected plant genetic resources with landscape and cultural importance for the Mediterranean region. pp. 108–111. Office for Scientific and Technical Cooperation with Mediterranean Countries, Rome. Italy.

Baricevic, D. & Rode, J. (1996). National Program for cultivation, processing and quality control of drug plants and herbal remedies. 12 pp. Proposal. University of Ljubljana, Biotechnical Faculty & IHP Zalec.

Baricevic, D., Cernila, M., Bartol, T., Zupancic, A. & Rode, J. (1996). Managing information data on medicinal and aromatic plants by MEDPLANT relational database. 5 pp. Proceedings of EUCARPIA International Meeting, Hungary, Oct. 17–20, 1996.

Baricevic, D., Cernila, M., Gomboc, S. & Habeler, H. (1994). MEDPLANT — monitoring, characterisation and evaluation software for medicinal and aromatic plants. 12 pp. SAA, Patent Nr. R 186/94, Ljubljana, Slovenia.

Wraber, T. & Skoberne, P. (1989). The Red Data List of Threatened Vascular Plants in Slovenia. — Nature Conservation, 14–15, pp. 9–428. Institute for the Conservation of Natural and Cultural Heritage of Slovenia, Ljubljana.

Zupancic, M. & Zagar, V. (1995). New views about the phytogeographic division of Slovenia I, Razprave. pp. 3–30. Classis 4, Historia Naturalis.

Green, F.N. (1999). The United Kingdom Swede Cultivar Collection (*Brassica napus* L. var. *napobrassica* (L.) Rchb.). In: S. Andrews, A.C. Leslie and C. Alexander (Editors). Taxonomy of Cultivated Plants: Third International Symposium, pp. 501–503. Royal Botanic Gardens, Kew.

THE UNITED KINGDOM SWEDE CULTIVAR COLLECTION (*BRASSICA NAPUS* L. VAR. *NAPOBRASSICA* (L.) RCHB.)

F.N. GREEN

Scottish Agricultural Science Agency, East Craigs, Edinburgh EH12 8NJ, Scotland, UK

Introduction

The UK Swede Cultivar Collection is stored at the Scottish Agricultural Science Agency (SASA). It was formed in the early 1970s to serve as a definitive reference collection for UK statutory Distinctness, Uniformity and Stability (DUS) tests. The creation of this collection enabled the development of DUS tests, the definition of characters and the classification of cultivars (Green & Winfield 1984).

The UK Swede Cultivar Collection has two components:

1. The UK DUS collection containing definitive seed of registered cultivars, example cultivars and candidates submitted for DUS testing. Seed is safely duplicated at SASA.

2. A genetic resource collection containing commercially obsolete cultivars and uncommercialised breeding lines. Seed is safely duplicated at the UK Vegetable Genebank, Horticulture Research International, Wellesbourne in Warwickshire.

Cultivars in Commerce

Since 1974, the number of cultivars in commerce has fallen from approximately 140 to 25 today. This decline has resulted from a rationalisation of cultivars by breeding and marketing companies within the UK and Europe. There are now 182 taxa (cultivars and lines) in the seed collection, of which 25 are currently-registered cultivars within the European Community. 8 of these cultivars have been maintained by more than one company and have been regenerated as 35 unique taxa. 42 cultivars have been tested in UK DUS tests since 1973.

Origin of Cultivars

Most cultivars (75%) originate from the UK, though some of the older cultivars may have been imported and renamed as UK cultivars. The remaining cultivars originate from Europe (16%) and New Zealand, Canada and the United States of America (9%).

Use of Cultivars

Traditionally, Swede has been used as a fodder crop and this is reflected in the large percentage (87%) of fodder cultivars in the collection. In the late 1970s, some cultivars were bred for processing, mainly for the production of canned soups. In the 1980's, attention was focused on developing the crop as a supermarket vegetable. Initially the most suitable fodder cultivars were selected, but these dual-purpose cultivars are now being replaced by cultivars with more uniform root shape and size, more attractive root skin colour, low skin cork and better disease resistance. Dual-purpose and commercial vegetable cultivars now account for 45% of the UK registered cultivars. Only 4% of cultivars are used as garden vegetables.

Taxonomy and Description

5 discontinuous characters are used to classify the collection into 9 major groups, within which cultivars are distinguished by comparing 17 measured and scored continuous variables (Green & Campbell 1996). Official descriptions, based on a 2 year DUS test, have been compiled for 40% of the collection. These are defined according to guidelines published by the International Union for the Protection of New Varieties of Plants (UPOV 1984).

The collation of over-year data was carried out to aid the selection of similar cultivars for DUS tests (Winfield & Green 1985), but also to provide a cumulative source of data for generating long-term 'Dynamic' descriptions for genetic resource purposes (Green 1994).'Dynamic' descriptions, which change if the range of variation is extended, are derived from 25 years of statistically-adjusted data (FITCON) (Patterson 1978, Finney 1980); these descriptions are relative, enabling cultivars which have never been grown together, to be compared (Green & Campbell 1996). 'Dynamic' descriptions have been compiled for 60% of the collection.

Regeneration and Conservation for Genetic Resource Purposes

In 1976, SASA sent 80 of the oldest cultivars to the Scottish Crop Research Institute, Dundee for regeneration. About 90% were successfully regenerated and are now stored at the Vegetable Genebank at Horticulture Research International at Wellesbourne, with duplicates at SASA.

Definition of a Cultivar 'Core Collection'

Multivariate analyses of over-year descriptive data have been used to define a 'core collection'. Further recording is planned to check and revise descriptions of the oldest cultivars, thus improving data quality.

Acknowledgements

Ian Nevison, Biomathematics and Statistics Scotland, for statistical advice and cluster analyses of the data, and George Campbell for help with nomination of the 'core collection'.

References

Finney, D.J. (1980). The estimation of parameters by least squares from unbalanced experiments. *J. Agric. Sci.* (*Cambridge*) 95: 181–189.

Green, F.N. (1994). Cultivar registration: an under-utilised source of information and variation for plant genetic resources. In F. Balfourier & M.R. Perretant (eds). Evaluation and Exploitation of Genetic Resources Pre-Breeding. Proceedings of the Genetic Resources Section Meeting of EUCARPIA. 15–18 March 1994. pp. 261–263. Clermont Ferrand, France.

Green, F.N. & Campbell, G.D. (1996). 'Dynamic' statistically-adjusted descriptions: a better tool for comparing cultivars, using Swede (*Brassica napus* L. var. *napobrassica* (L.) Rchb.) as an example. *Acta Hort.* 407: 137–145.

Green, F.N. & Winfield, P.J. (1984). The Development of Distinctness, Uniformity and Stability tests for Turnip, Turnip Rape and Swede in the United Kingdom. In W.H. Macfarlane Smith, T. Hodgkin & A.B. Wills (eds). Procedures of Better Brassicas '84 Conference. St. Andrews, September 1984. pp. 96–107. Scottish Crop Research Institute, Dundee.

Patterson, H.D. (1978). Routine least squares estimation of variety means in incomplete tables. *J. Natl. Inst. Agric. Bot.* 14: 401–412.

UPOV. (1984). Guidelines for the conduct of tests for Distinctness, Homogeneity and Stability: Swede. TG/89/3. International Union for the Protection of New Varieties of Plants. Geneva, Switzerland.

Winfield, P.J. & Green, F.N. (1985). The development of a cultivar database for Swede. *Cruciferae Newslett. EUCARPIA* 10: 146.

Bertrand, H. & Lambert, C. (1999). The *Hydrangea* reference collection at Angers: some examples of management. In: S. Andrews, A.C. Leslie and C. Alexander (Editors). Taxonomy of Cultivated Plants: Third International Symposium, pp. 505–507. Royal Botanic Gardens, Kew.

THE *HYDRANGEA* REFERENCE COLLECTION AT ANGERS: SOME EXAMPLES OF MANAGEMENT

H. BERTRAND AND C. LAMBERT

I.N.H. 2 rue Le Nôtre, 49045 Angers cedex 01, France

Introduction

The UPOV *Hydrangea* L. collection has been situated in Angers since 1975. Today the collection holds about 450 taxa. According to the classification of McClintock (1957), 13 species are present and *H. macrophylla* (Thunb. ex Murray) Ser. together with ssp. *macrophylla* and ssp. *serrata* (Thunb. ex Murray) Makino number about 350 cultivars. The main objectives of the research are variability analysis and management of the collection, as well as cultivar identification and detection of synonyms.

Material and Methods

Morphological studies with macroscopic characters were carried out, especially on *H. macrophylla* ssp. *macrophylla* (Bertrand 1992). They concentrated on:

- Plant height.
- Size, shape and ornamentation of the leaves.
- Size, shape and colour of the sterile flowers (and fertile flowers for the lacecaps).

Molecular markers based on enzymes were used to improve our knowledge of the taxa (Lambert *et al.* 1998). Five bands on polyacrylamide gel (ACP, EST, SOD, AAT, LAP) and three on starch (PGI, MNR, PGD) showed polymorphism. This method highlighted enough variability to distinguish the genotypes.

For some cultivars, chromosome numbers were counted on root tips with a microscope, while ploidy level was evaluated on more cultivars with flow cytometry (Demilly *et al.* 1998).

Examples of Synonymy

1) 'Möwe' and 'Geoffrey Chadbund'

'Geoffrey Chadbund' is a well known cultivar in European nurseries, while 'Möwe' is not so often cultivated.

These two lacecap cultivars are of medium height, 0.7–1 m. The ray flower calyx comprises 4 purple-pink (RHS Red-Purple Group 60B–60C), overlapping sepals which are entire or sometimes toothed. The fertile flowers are composed of 5 small sepals, 5 pink petals, 10–12 stamens with white filaments, yellow anthers, and 2–3 carpels. The ovate, acuminate leaves are not glossy. The red colour at the base of the petiole and on the nodes is very characteristic.

As far as morphological characters are concerned, these two cultivars are identical.

For the six isoenzymes which are polymorphic, 'Möwe' and 'Geoffrey Chadbund' show identical patterns (ACP, EST, SOD, AAT, LAP and PGI). In addition, both cultivars are triploid 3x = 54, and this is a rare cytological character.

According to Clarke (1988), 'Geoffrey Chadbund' was found in England by G. Chadbund "growing in a pot in a florist's shop" and was exhibited by Messrs. L.R. Russell Ltd. in 1974 and given an Award of Merit (Anon. 1974).

'Möwe' belongs to the Teller Series of 26 lacecaps bred by F. Meier in Wädenswill, Switzerland. It was introduced in 1964. Thus we have concluded that there are two names for the same genotype, and so the valid one is 'Möwe'.

2) 'Sea Foam', 'Azisai', 'Variegata' and 'Sir Joseph Banks'

'Sea Foam', according to Haworth-Booth (1984) is a mutation from 'Sir Joseph Banks', which is an old mophead hydrangea brought back to England in 1789 by Sir Joseph Banks. The cultivar 'Sea Foam' which is present in European gardens differs from 'Sir Joseph Banks' by its leaves, which are thinner, more elliptic and less glossy.

'Azisai' is an old cultivar introduced into Europe and described by P.F. von Siebold. Our specimen comes from Hiroshima Botanical Garden. 'Variegata' was mentioned in 1849 by van Houtte and in 1861 by de Talou and our specimen looks like their coloured illustrations.

The two lacecaps 'Sea Foam' and 'Azisai' are robust plants, over 1 m in height. The sterile ray flowers are made up of 4 entire, nearly rhombic, pale pink sepals which are partly overlapping. The 5 petals of the fertile flowers are pink, with long, lilac pink filaments and grey anthers. There are 3 carpels. The leaves are elliptic-ovate and acuminate at the apex.

As far as morphological characters are concerned, these two lacecap cultivars are very similar. The only difference between 'Variegata' and 'Azisai' is the cream variegation on the leaves, as 'Variegata' is a variegated mutation of 'Azisai'.

'Sir Joseph Banks' is a pale pink mophead hydrangea with bright green, rather elliptic leaves.

The allozymes studied (ACP, EST, PGI) show similar profiles for 'Sea Foam', 'Azisai' and 'Variegata', but 'Sir Joseph Banks' differs from them by 5 alleles. These four cultivars are diploid; 2x = 36 as are the majority of the analysed *Hydrangea*.

Isoenzyme studies confirm that 'Sea Foam' and 'Azisai' are of the same genotype. There are two names for the same plant, the valid one is 'Azisai'. 'Sea Foam' cannot be a mutation of 'Sir Joseph Banks' because of the difference of the 5 alleles. But, have we got the correct 'Sea Foam' in our European collections?

Conclusion

We have shown that methods such as morphological characterisation, molecular markers based on enzymes and ploidy level evaluation are available to reduce the enormous list of *Hydrangea* cultivar names.

Acknowledgments

Our thanks to D. Daury, E. Mortreau and the students under instruction for helping with the chromosome counting and enzymatic extractions, and to M. Laffaire and D. Relion for taking care of the *Hydrangea* plants in the collection.

References

Anon. (1974). Awards to Plants, Floral Committee B, July 9, 1974. *Proc. Roy. Hort. Soc. London* 119(2): 59.

Bertrand, H. (1992). Identification of *Hydrangea macrophylla* Ser. cultivars. *Acta Hort.* 320: 209–212.

Clarke, D.L. (ed.). (1988). *Hydrangea.* In W.J. Bean. Trees and Shrubs Hardy in the British Isles, Supplement. pp. 272–274. John Murray, London.

Demilly, D., Lambert, C. & Bertrand, H. (1998). Diversity of nuclear DNA contents of *Hydrangea. EUCARPIA* 19th International Symposium, Angers, France, 27–30 July 1998. Poster.

Haworth-Booth, M. (1984). The Hydrangeas. (Ed. 5) 217 pp. Constable, London.

Houtte, L. van (1849). *Hydrangea japonica foliis variegatis. Fl. Serres Jard. Eur.*, VII: 139, 1 tab.

Lambert, C., Bertrand, H., Lallemand, J. & Bourgoin, M. (1998). Characterization of a collection of *Hydrangea macrophylla* using isoenzyme analysis. *EUCARPIA* 19th International Symposium, Angers, France, 27–30 July 1998. Poster.

Lawson-Hall, T. & Rothera, B. (1996). Hydrangeas. A gardeners' guide. 160 pp. Batsford Ltd., London.

Meier, F. (1990). Tellerhortensien-Züchtungen, Flugschrift 120, 24 pp. Eidgenössische Forschungsanstalt für Obst-Wein- und Gartenbau, Wädenswil.

Siebold, P.F. von & Zuccarini, J.S. (1841). Icones et descriptiones *Hydrangearum* in Japonica hucusque detectarum, Flora Japonica, sect I: 101–122, tab. 51–66. Lugduni Batavorum.

Talou de, A. (1861). Les *Hydrangea. Horticultur Franç.* série 2, 3: 107–111, 1 tab.

PUBLICATIONS

E

Sier, A.R.J. & Miller, D.M. (1999). The Collector Codes Index Project. In: S. Andrews, A.C. Leslie and C. Alexander (Editors). Taxonomy of Cultivated Plants: Third International Symposium, pp. 511–513. Royal Botanic Gardens, Kew.

THE COLLECTOR CODES INDEX PROJECT

ANDREW R.J. SIER AND DIANA M. MILLER

The Royal Horticultural Society's Garden, Wisley, Woking, Surrey GU23 6QB, UK

What is a Collector Code?

A collector code can be any code (usually non-numeric) that is used to refer to an individual collector, a group of collectors, or a specific expedition. A code can include an abbreviation of a name e.g. WIL (E.H. Wilson), B&L (C.D. Brickell & A.C. Leslie, Yunnan, China 1987) and ACE (Alpine Garden Society Expedition to China, 1994).

In order to record the provenance of plant material (in the form of living plants, seeds, herbarium material, etc.) collected from around the world, it is important to know the following details:

- Where the material was found (preferably grid reference and altitude)?
- When it was collected?
- Who collected it?

In addition descriptions of, for example, the locality, plant community, vernacular names, local uses of plant, etc. are helpful. By citing both collector code and the number given by the collector to the material in question, a cross-reference is easily made to this information. Collector codes are widely used wherever it is important to identify the provenance of plant material. Examples include:

- Records for living plants (accessions) in a garden collection.
- Nursery stock records.
- Herbarium records.
- Labels on slides, drawings or paintings of the material.
- Field notes related to the plant or material.
- Identifying a particular garden-worthy selection from the wild.

The code will appear on labels attached to items or plants and in the records for a collection (paper- or computer-based). Increasing use is made of computer databases to track holdings of plant material and related items. This is particularly true of large botanic gardens, but nurserymen and private individuals are also using computers to manage their information. There is also a growing number of reference works available on CD-ROM and the Internet.

There is a Problem!

A **unique** code for each collector, group of collectors or expedition makes sense. When used in combination with unique numbers, they provide an easy reference to all the provenance information and additional notes about the plant material collected. But collector codes are not unique. There is no international standard for assigning codes, which means, in some cases more than one collector per code, or more than one code per collector. Examples include:

- F which has been used for collections made by George Forrest and Reginald Farrer (both collected extensively in China and the Himalayas).
- C. & A. Chadwell and A. McKelvie are cited variously using these codes C&Mc, CC&McK, CHMC and CMC.

This creates particular difficulties if a computer database is used to store the codes. Because codes are not unique, many institutions adopt their own set of codes which **are** unique. But this only serves to propagate more synonymous codes per collector. The result is potential confusion, particularly when material or information is exchanged between institutions.

Why not Adopt a Standard Set of Codes?

This may seem the obvious solution. But there are many problems with a standard set of codes, for example:

- Who would administer it? (It involves time, money and commitment).
- Would everyone adopt it? (It may mean recoding a large existing dataset).
- How would it work in practice? (Collectors would have to register their trips in advance or upon return — would people bother if, say, they collected some plants whilst on vacation?).
- Which code format do you choose? (Some commonly used codes date back over a hundred years and are seen in a multitude of books, journals and herbarium collections; should these be replaced by new codes?).

An international standard is, in reality, difficult to devise and operate. It is probably not the best solution, at least in the short-term. A more practical solution is to publish a cross-reference of known, existing codes. It would list collectors by code, and cross-refer between codes. Brief information could be listed for each collector, group of collectors or expedition. The index would be updated regularly to incorporate additions and corrections.

The Collector Codes Index Project

The idea of a collector codes index stemmed from a discussion between ourselves and Alan Hardy, a rhododendron collector deeply involved with the RHS. Diana Miller has encountered a number of problems with non-unique codes, taking substantial time and effort to resolve. In 1996 Andrew Sier raised the subject at a meeting of the Taxonomic Databases Working Group (TDWG*) in Toronto. Stephen Jury (University of Reading), offered to present the idea at this meeting. This resulted in the formation of the Expedition Acronyms Subgroup, convened by Andrew Sier.

* TDWG is an international group, set up to explore ideas on standardisation and collaboration between major taxonomic database projects. It is affiliated with the International Union of Biological Sciences (IUBS). The TDWG web site is at: http://www.tdwg.org/.

Since then we have written to colleagues in UK botanic gardens, herbaria and horticultural societies, seeking support. The response was very positive: many of those questioned agreed that a cross-reference of collectors codes would be a useful tool. There was support for both a paper and electronic version. We therefore plan to publish the index in two formats — a booklet, and as a searchable database on the Internet.

Phase 2 of the project has just begun; we have written to all those people who offered data for inclusion in the index. Work should begin later this year to build up the database of collectors codes. We will use *BG-BASE*™† software to maintain these data.

Our aim is to publish a first edition of the index in early 1999. This will inevitably be incomplete, but we hope it will encourage further contributions. With time we will build up a more extensive cross-reference. The Internet database can be updated on a regular basis. The paper version of the index will be revised whenever a sufficient number of changes merit a new edition. The Internet database could be hosted by one or more sites. The RHS plans to provide access from its own web pages, and the data would probably be maintained at the RHS server. Access would be free.

Scope

The primary purpose of the index will be to enable the user to identify a collector (or group or expedition) from a code. As many codes as possible will be included, so that the collector can be identified from any one of them.

The index will not necessarily provide detailed information about each entry, though if we are supplied with information we will try to include it. A comprehensive database of collector information already exists, maintained by the Harvard Herbaria. The primary source of data in this index is the International Association for Plant Taxonomy's (IAPT) *Index Herbariorum series, Index Herbariorum, Part II, Collectors A–Z*, available in book form. The Harvard Herbaria Collectors Database is available on the web at http://www.herbaria.harvard.edu/Data/Collectors/collectors.html. We are looking at the possibility of providing a link between the two databases, but feel that at this stage they should remain separate.

Further Information

For further information on the Collector Codes Index project, or to offer data for inclusion, please contact; Dr Andrew Sier, Senior Plant Database Administrator, Royal Horticultural Society, Wisley Garden, Woking, Surrey GU23 6QB, UK. Tel: +44(0) 1483 224234, e-mail: andrews@rhs.org.uk, RHS web site: http://www.rhs.org.uk.

Acknowledgements

We are grateful to the Royal Botanic Garden Edinburgh (RBGE) for supplying much of the collector information on our database. The RHS and RBGE have agreed to co-operate over the allocation of collector codes in their respective databases. Thus, we have adopted the use of RBGE codes, and inform RBGE whenever we need to designate a new code. We are also grateful to Kerry Walter and Robert Cubey for *BG-BASE*™ support, and for their (almost certain!) assistance with aspects of this project.

† *BG-BASE*™ Collections Management and Conservation Software. The †*BG-BASE*™ web site is at http://www.tdwg.org/.

SYMPOSIUM DELEGATES
Key to acronyms at end of delegates

Abugalieva, S.I., Institute of Plant Physiology, Genetics & Bioengineering, 45 Timiryavez Street, Almaty 480090, KAZAKSTAN.

Alanko, P., PO Box 7, FIN-00014, Helsinki University, FINLAND.

Alexander, C., Royal Botanic Garden, Inverleith Row, Edinburgh EH3 5LR, Scotland, UK.

Amerson, B., (American *Hemerocallis* Society), 13339 Castleton Circle, Dallas, Texas 75234-5111, USA.

Andrews, S., Royal Botanic Gardens, Kew, Richmond, Surrey TW9 3AE, England, UK.

Astley, D., Plant Genetics and Plant Breeding Department, HRI, Wellesbourne, Warwick CV35 9EF, England, UK.

Austin, J.E., Plant Variety Rights Office and Seeds Division, MAFF, White House Lane, Huntingdon Road, Cambridge CB3 OLF, England, UK.

Bachmann, K., IPK, Corrensstraße 3, D-06466 Gatersleben, GERMANY.

Bartels, M.J., Fleuroselect, Parallel Boulevard 214d, NL 2202 HT Noordwijk, THE NETHERLANDS.

Batdorf, L.R., US National Arboretum, 3501 New York Avenue NE, Washington DC 20002, USA.

Baum, B.R., ECORC, AAFC, Research Branch, Neatby Building, Central Experimental Farm, Ottawa, Ontario K1A OC6, CANADA.

Becker, B., ZADI/IGR, PO Box 20 14 15, 53144 Bonn, GERMANY.

Bennett, M.A., (University of Aberdeen), 49 Harris Drive, Aberdeen AB24 2TF, Scotland, UK.

Berg, R.G. van den, Department of Plant Taxonomy, Agricultural University, Gen. Foulkesweg 37, 6703 BL Wageningen, THE NETHERLANDS.

Bertrand, H., INH, 2 rue le Nôtre, 49045 Angers-cedex, FRANCE.

Borys, J., COBORU, 63-022 Słupia Wielka, POLAND.

Braimbridge, E., Langley Boxwood Nursery, Rake, Liss, Hants GU33 7JL, England, UK.

Brandenburg, W.A., CPRO, PO Box 16, 6700AA Wageningen, THE NETHERLANDS.

Brickell, C.D., (RHS, ICNCP & International Society for Horticultural Science), The Camber, The Street, Nutbourne, Pulborough, West Sussex RH20 2HE, England, UK.

Bulwer, B., SASA, East Craigs, Craigs Road, Edinburgh EH12 8NJ, Scotland, UK.

Burrows, J., Pro-Veg Seeds Ltd., 6 Shingay Lane, Sawston, Cambridge CB2 4SS, England, UK.

Butler, S., Field House, Church Street, Wing, Rutland LE15 8RS, England, UK.

Campbell, G.D., SASA, East Craigs, Craigs Road, Edinburgh EH12 8NJ, Scotland, UK.

Cann, D., Halbury, Golden Joy, Crediton, Devon, England, UK.

Chalmers, A., (Kirkley Hall College), 23 St Oswin's Avenue, Cullercoats, North Shields, Tyne & Weir, England, UK.

Chaudry, S.A., Department of Botany, Forman Christian College, Lahore 54600, PAKISTAN.

Chisholm, D., Floranova Ltd., Norwich Road, Foxley, Dereham, Norfolk NR20 4SS, England, UK.

Cirtautas, V.G., Arboretum Dubravae, Girionys, Kaunas reg., 4312 LITHUANIA.

Cleevely, R.J., (Heather Society), High Croft, Gunswell Lane, South Molton, Devon EX36 4DH, England, UK.

Clennett, C., Wakehurst Place (Royal Botanic Gardens, Kew), Ardingly, West Sussex RH17 6TN, England, UK.

Cook, A., Longwood Gardens, PO Box 501, Kennett Square, Pennsylvania 19348-0501, USA.

Cook, L.B., NCCPG, The RHS Garden, Wisley, Woking, Surrey GU23 6QB, England, UK.

Coombes, A., The Sir Harold Hillier Gardens & Arboretum, Jermyns Lane, Ampfield, nr. Romsey, Hampshire SO51 OQA, England, UK.

Cooper, W., 51 Longdell Hills, Contessey, Norwich NR5 OPB, England, UK.

Culham, A., Centre for Plant Diversity and Systematics, Department of Botany, The University of Reading, Whiteknights, Reading RG6 6AS, England, UK.

Darling, D., SASA, East Craigs, Craigs Road, Edinburgh EH12 8NJ, Scotland, UK.

Davey, J.C., SASA, East Craigs, Craigs Road, Edinburgh EH12 8NJ, Scotland, UK.

Davidson, C., AAFC, Morden Research Centre, Unit 100-101, Route 100, Morden, Manitoba, CANADA.

Davies, B.J., (National Collection of *Nymphaea*), PO Box 62438, 8046 Paphos, CYPRUS.

Dawson, I., Australian National Botanic Gardens, GPO Box 1777, Canberra, ACT 2614, AUSTRALIA.

Dehmer, K.J., IPK, Corrensstraße 3, D-06466 Gatersleben, GERMANY.

Donald, D., The National Trust for Scotland, 5 Charlotte Square, Edinburgh EH2 4DU, Scotland, UK.

Dourado, A., The RHS Garden, Wisley, Woking, Surrey GU23 6QB, England, UK.

Ettekoven, K. van, NAKG, Sotaweg 22, Postbus 27, 2370 AA Roelofarendsveen, THE NETHERLANDS.

Fantz, P.R., Department of Horticultural Science, Box 7609, North Carolina State University, Raleigh, North Carolina 27695-7609, USA.

Ferguson, A.R., HortResearch, Mt Albert Research Centre, Private Bag 92 169, Auckland, NEW ZEALAND.

Foster, M., Whitehouse Farm, Ivy Hatch, Sevenoaks, Kent TN15 ONN, England, UK.

Gardner, D., Royal Botanic Gardens, Kew, Richmond, Surrey TW9 3AB, England, UK.

Gardner, M., Royal Botanic Garden, Inverleith Row, Edinburgh EH3 5LR, Scotland, UK.

Gioia, S., Christie, Parker & Hale, LLP, 5 Park Plaza, Irvine, California 92614, USA.

Gioia, V.G., Christie, Parker & Hale, LLP, 5 Park Plaza, Irvine, California 92614, USA.

Glen, H.F., National Botanical Institute, Private Bag X101, Pretoria 0001, SOUTH AFRICA.

Glendoick Gardens Ltd., Glencarse, Perth, Tayside PH2 7NS, Scotland, UK.

Grant, M.L., The RHS Garden, Wisley, Woking, Surrey GU23 6QB, England, UK.

Green, F.N., SASA, East Craigs, Craigs Road, Edinburgh EH12 8NJ, Scotland, UK.

Greengrass, B., UPOV, 34 chemin des Colombettes, CH-1211 Geneva 20, SWITZERLAND.

Gregory, P.A., (Maple Society), Trefyclawdd, 3 Park Close, Tetbury, Gloucestershire GL8 8HS, England, UK.

Groendijk-Wilders, N., Department of Plant Taxonomy, Agricultural University, Gen. Foulkesweg 37, 6703 BL Wageningen, THE NETHERLANDS.

Handa, T., Institute of Agriculture & Forestry, University of Tsukuba, 1-1-1 Tennodai, Tsukuba, Ibaraki 305-0006, JAPAN.

Haston, E., Plant Science Laboratories, The University, Whiteknights, PO Box 221, Reading RG6 2AS, England, UK.

Hazekamp, T., IPGRI, Via delle Sette Chiese 142, 00145 Rome, ITALY.

Heitz, A., UPOV, 34 chemin des Colombettes, CH-1211 Geneva 20, SWITZERLAND.

Hender, A.B., Floranova Ltd., Norwich Road, Foxley, Dereham, Norfolk NR20 4SS, England, UK.

Hennipman, E., (Botanical Garden, Ghent University), Zweerslaan 34, 3723HP Bilthoven, THE NETHERLANDS.

Hetterscheid, W.L.A., VKC, Linnaeuslaan 2a, 1431 JV Aalsmeer, THE NETHERLANDS.

Hewitt, J., (British *Iris* Society), Haygarth, Cleeton St Mary, Cleobury Mortimer, Kidderminster DY14 OQU, England, UK.

Hoffman, M.H.A., Research Station for Nursery Stock, PO Box 118, 2770 AC Boskoop, THE NETHERLANDS.

Hulden, M., Nordic Gene Bank, Box 41, 23053 Alnarp, SWEDEN.

Hunt, D.B., (Orchid Registrar, RHS), PO Box 1072, Frome, Somerset BA11 5NY, England, UK.

Hunt, P.F., (Orchid Registrar, RHS), PO Box 1072, Frome, Somerset BA11 5NY, England, UK.

Huntley, D.N., (Myerscough College), 9 Jepps Avenue, Barton, Preston, Lancashire, England, UK.

Huttunen, M., Botania, PO Box 111, FIN-80101 Joensuu, FINLAND.

Jakobsdottir, D., Reykjavik Botanic Garden, Laugardal, 104 Reykjavik, ICELAND.

Kington, S., The RHS, 80 Vincent Square, London SW1P 2PE, England, UK.

Knees, S., The RHS, at Royal Botanic Garden, Inverleith Row, Edinburgh EH3 5LR, Scotland, UK.

Knüpffer, H., IPK, Corrensstraße 3, D-06466 Gatersleben, GERMANY.

Kobayashi, N., Tatebayashi Azalea Research Station, 3258 Hanayama-cho, Tatebayashi, Gunma 374-0005, JAPAN.

Le Duc, A., Department of Horticulture, Kansas State University, 2021 Throckmorton, Manhattan, Kansas 66506, USA.

Lea, V.J., NIAB, Huntingdon Road, Cambridge CB3 OLE, England, UK.

Lean, A., National Fruit Collections, Brogdale Road, Faversham, Kent ME13 8XZ, England, UK.

Lemmers, W., Fresia Straat 10, 2161 XM Lisse, THE NETHERLANDS.

Leslie, A.C., The RHS Garden, Wisley, Woking, Surrey GU23 6QB, England, UK.

Lewis, P.E., (National Collection of *Campanula*), Padlock Croft, Padlock Road, West Wratting, Cambridge CB1 5LS, England, UK.

Liberato, M.C., Jardim-Museu Agricola Tropical, Calçada do Galvão, 1400 Lisboa, PORTUGAL.

Lord, W.A., (RHS Plant Finder), Trafalgar House, High Street, Tewkesbury, Gloucestershire GL20 5BJ, England, UK.

Louneva, N.N., VIR, 42 Bolshaya Morskaya St, St Petersburg 190000, RUSSIA.

Malecot, V., Arboretum de Chevreloup, 1 rue de Chevreloup, 78150 Le Chesnay, FRANCE.

Manners, M.M., Department of Citrus & Environmental Horticulture, Florida Southern College, 111 Lake Hollingsworth Drive, Lakeland, Florida 33801-5698, USA.

Matthews, V.A., (RHS *Clematis* Registrar), Denver Botanic Gardens, 909 York Street, Denver, Colorado 80206-3799, USA.

Maxwell, H.S., Royal Botanic Garden, Inverleith Row, Edinburgh EH3 5LR, Scotland, UK.

McAllister, H.A., Applied Ecology Research Group, School of Biological Sciences, The University, Liverpool L69 3BX, England, UK.

McClintock, E., (Herbarium, University of California, Berkeley), 1551 9th Avenue, San Francisco, California 94122, USA.

McGregor, M., (Saxifrage Society), 16 Mill Street, Hutton, nr. Driffield, East Yorkshire YO25 9PU, England, UK.

McKendrick, M.E., 4 Knighton Road, Otford, Sevenoaks, Kent TN14 5LF, England, UK.

McLewin, W., Phedar Research & Experimental Nursery, Bunkers Hill, Romiley, Stockport, Cheshire SK6 3DS, England, UK.

McNeill, J., The Rowans, Drem, East Lothian EH39 5BN, Scotland, UK.

Miikeda, O., Tokyo Metropolitan Kitazono High School, 4-14-1, Itabashi, Itabashi-ku, Tokyo 173-0004, JAPAN.

Miller, D.M., The RHS Garden, Wisley, Woking, Surrey, GU23 6QB, England, UK.

Möller, M., Royal Botanic Garden, Inverleith Row, Edinburgh EH3 5LR, Scotland, UK.

Morgan, C.S., Bedgebury National Pinetum, Goudhurst, Cranbrook, Kent TN17 2SL, England, UK.

Moseley, J., The National Botanic Garden of Wales, Middleton Hall, Llanarthne, Carmarthenshire SA32 8BA, Wales, UK.

Munson, R., The Holden Arboretum, 9500 Sperry Road, Kirtland, Ohio 44094-5172, USA.

Nurse, M.C., Flat B, 34 Marylebone High Street, London W1M 3PF, England, UK.

Oakeley, H.F., (National Collection of *Lycaste* and *Anguloa*), 77 Copers Cope Road, Beckenham, Kent BR3 1NR, England, UK.

Ochsmann, J., IPK, AG Experimentelle Taxonomie, Corrensstraße 3, D-06466 Gatersleben, GERMANY.

Olwell, P., Endangered Species Co-ordinator, National Park Service, 1849 C Street, NW, MSMIB 3223, Washington, DC 20240, USA.

Page, R.G., (National Collection of *Cistus*), Thornfield, Layton Avenue, Rawdon, Leeds LS19 6QQ, England, UK.

Paris, H.S., Agricultural Research Organization, PO Box 1021, Ramat Yishay, 30-095, ISRAEL.

Parzies, H., SAC, West Mains Road, Edinburgh EH9 3JG, Scotland, UK.

Pattison, G.A., NCCPG, The RHS Garden, Wisley, Woking, Surrey GU23 6QP, England, UK.

Pearson, K.M., SASA, East Craigs, Craigs Road, Edinburgh EH12 8NJ, Scotland, UK.

Perrin, M.E.B., The Cottage, Churchtown, Ruan Minor, Helston, Cornwall TR12 7JL, England, UK.

Pickersgill, B., Department of Agricultural Botany, The University, Whiteknights, PO Box 221, Reading, Berkshire RG6 6AS, England, UK.

Plovanich-Jones, A.E., Department of Botany & W.J. Beal Botanical Garden, Michigan State University, East Lansing, Michigan 48824-1312, USA.

Price, G., (Pan American Seed Research), 1S.861 Green Road, Elburn, Illinois 60199, USA.

Rivera, D., Departamento de Biología Vegetal, Facultad de Biología, Universidad de Murcia, 30100 Murcia, SPAIN.

Rodrígues-Acosta, M., Herbario de la BUAP, Edificio 76, Conjuncta de Ciencias Cd Universitaria, Av San Claudio y 14 sur col San Manuel, CP 72590 Puebla, MEXICO.

Rutherford, A., (*Hedera* Project), 19 South King Street, Helensburgh, Dunbartonshire GL84 7DU, Scotland, UK.

Saavedra, G., SAC, West Mains Road, Edinburgh EH9 3JG, Scotland, UK.

Sadie, J., National Department of Agriculture, Directorate Genetic Resources, Private Bag X5044, Stellenbosch 7599, SOUTH AFRICA.

Sandved, M., (Agricultural University of Norway), Solberglivei 10, 0671 Oslo, NORWAY.

Schifino-Wittmann, M.T., Depto Plantas Forrageiras e Agrometeorologica, Faculdada de Agronomia, Universidade Federal do Rio Grande do Sol, Caixa Postal 776, 91501-970 Porto Alegre, RS, BRAZIL.

Scoble, G., (*Hebe* Society), Maes Glas, Penrhyd, Amlwch, Angelsea LL68 9TF, Wales, UK.

Scott, E.M.R., Ornamental Plants Section, NIAB, Huntingdon Road, Cambridge CB3 0LE, England, UK.

Shaw, J.M.H., 4 Albert Street, Stapleford, Nottingham, NG9 8DB England, UK.

Sier, A.R.J., The RHS Garden, Wisley, Woking, Surrey GU23 6QB, England, UK.

Simpson, M., The National Trust, 33 Sheep Street, Cirencester, Gloucestershire GL7 1RQ, England, UK.

Simpson, N., The RHS Garden, Wisley, Woking, Surrey GU23 6QB, England, UK.

Singh, B., Division of Floriculture & Landscaping, IARI, New Delhi-110012, INDIA.

Sinnott, M.W., Royal Botanic Gardens, Kew, Richmond, Surrey TW9 3AB, England, UK.

Smekalova, T., VIR, 42 Bolshaya Morskaya St, St Petersburg 190000, RUSSIA.

Snoeijer, W., Div. of Pharmacognosy, Leiden/Amsterdam Center for Drug Research, PO Box 9502, 2300 RA Leiden, THE NETHERLANDS.

Spencer, R.D., Royal Botanic Gardens, Birdwood Avenue, South Yarra, Melbourne, Victoria 3141, AUSTRALIA.

Spoor, W., SAC, West Mains Road, Edinburgh EH9 3JG, Scotland, UK.

Staples, G., Bishop Museum: Botany, 1525 Bernice Street, Honolulu, Hawai'i 96817, USA.

Stearn, W.T., 17 High Park Road, Kew, Richmond, Surrey TW9 4BL, England, UK.

Stirton, C.H., The National Botanic Garden of Wales, Middleton Hall, Llanarthne, Carmarthenshire SA32 8HW, Wales, UK.

Stork, A.L., Conservatoire et Jardin Botaniques, Case Postale 60, CH-1292 Chambésy/GE, SWITZERLAND.

Strachan, J.M., US Department of Agriculture: Plant Variety Protection Office, NAL Building, Room 500, 10301 Baltimore Avenue, Beltsville, MD 20705, USA.

Stuart-Rogers, C.M., Department of Biochemistry, University of Dundee, Nethergate, Dundee DD1 4HN, Scotland, UK.

Stützel, T., Verband Botanischer Gärten, Botanischer Gärten RUB Bochum, Ruhr-Universität Bochum, 44780 D-Bochum, GERMANY.

Tebbitt, M., Brooklyn Botanic Garden, 1000 Washington Avenue, Brooklyn, New York 11225-1099, USA.

Thomson, C.M., SASA, East Craigs, Craigs Road, Edinburgh EH12 8NJ, Scotland, UK.

Thornton-Wood, S.P., The RHS Garden, Wisley, Woking, Surrey GU23 6QB, England, UK.

Thorvaldsdottir, E.G., Icelandic Institute of Natural History, PO Box 5320, IS-125 Reykjavik, ICELAND.

Tramposch, A., WIPO, 34 chemin des Colombettes, 1211 Geneva 20, SWITZERLAND.

Tredici, P. del, The Arnold Arboretum of Harvard University, 125 Arborway, Jamaica Plain, Massachusetts 0213-3500, USA.

Trehane, P., (ICNCP), 13 Westborough, Wimborne, Dorset BH21 1LT, England, UK.

Treu, R., School of Environmental Sciences and Land Management, University College Worcester, Worcester WR2 6AJ, England, UK.

Twibell, J.D., (National Collection of *Artemisia*), 31 Smith Street, Elsworth, Cambridge CB3 8HY, England, UK.

Vasconcelos, T., Instituto Superior Agronomia, Departamento de Protecçâo das Plantas e de Fitoecologia, Tapada da Ajuda, 1399 Lisboa, PORTUGAL.

Victor, D.X., (British *Pelargonium* & *Geranium* Society), The Old Stables, Church Lane, Hockliffe, Leighton Buzzard, Bedfordshire LU7 9NL, England, UK.

Walter, K.S., (*BG-BASE*™), The Royal Botanic Garden, Inverleith Row, Edinburgh EH3 5LR, Scotland, UK.

Waltrip, C., (All-America Selections), 10392 Boulder Ct, Ventura, CA 93004, USA.

Waltrip, J., (All-America Selections), 10392 Boulder Ct, Ventura, CA 93004, USA.

Waters, T., 8 Woodberry Crescent, Muswell Hill, London N10 1PH, England, UK.

Whiteley, A.C., The RHS Garden, Wisley, Woking, Surrey GU23 6QB, England, UK.

Whiteman, J., (Waterperry Gardens), 63 Elthorne Avenue, Hanwell, London W7 2JZ, England, UK.

Wiersema, J.H., US Department of Agriculture: Agric Research Service, Systematic Botany & Mycology Laboratory, Building 011A, Room 304, BARC-West, Beltsville, Maryland 20705-2350, USA.

Wilford, R., Royal Botanic Gardens, Kew, Richmond, Surrey TW9 3AE, England, UK.

Woods, P., Gowanbrae Cottage, 42 Perth Street, Blairgowrie PH10 7PL, Scotland, UK.

Yukawa, T., Tsukuba Botanical Garden, National Science Museum, 4-1-1 Amakubo, Tsukuba, Ibaraki 305-0005, JAPAN.

Zhang, D., Landscape Horticulture, University of Maine, Orono, ME 04469-5722, USA.

Zhou, Z., Kunming Institute of Botany, Heilongtan, Kunming, Yunnan 650204, CHINA.

Zupancic, A., M-KZK Kmetijstvo Kranj, LFVB, Mlakarjeva 70, 4280 Sencur, SLOVENIJA.

AAFC	Agriculture and Agri-Food Canada
BUAP	Benemérita Universidad Autónoma de Puebla
COBURU	Research Centre for Cultivar Testing
CPRO	Centre for Plant Breeding and Reproduction Research
ECORC	Eastern Cereal & Oilseed Research Center
HRI	Horticulture Research International
IARI	Indian Agricultural Research Institute
ICNCP	International Commission for the Nomenclature of Cultivated Plants
INH	Institute National d'Horticulture
IPGRI	International Plant Genetic Resources Institute
IPK	Institut für Pflanzengenetik und Kulturpflanzenforschung
MAFF	Ministry of Agriculture, Fisheries & Food
NAKG	Nederlandse Algemene Keuringsdienst Groentezaden
NCCPG	National Council for the Conservation of Plants & Gardens
NIAB	National Institute of Agricultural Botany
RHS	Royal Horticultural Society
SAC	Scottish Agricultural College
SASA	Scottish Agricultural Science Agency
UPOV	International Union for the Protection of New Varieties of Plants
VIR	NI Vavilov Research Institute of Plant Industry
VKC	Vaste Keurings Commissie
WIPO	World Intellectual Property Organisation
ZADI/IGR	Zentralstelle für Agrardokumentation und Information/Informations Zentrum für Genetisch Ressourcen

INDEX